Studies in Big Data

Volume 77

Series Editor

Janusz Kacprzyk, Polish Academy of Sciences, Warsaw, Poland

The series "Studies in Big Data" (SBD) publishes new developments and advances in the various areas of Big Data- quickly and with a high quality. The intent is to cover the theory, research, development, and applications of Big Data, as embedded in the fields of engineering, computer science, physics, economics and life sciences. The books of the series refer to the analysis and understanding of large, complex, and/or distributed data sets generated from recent digital sources coming from sensors or other physical instruments as well as simulations, crowd sourcing, social networks or other internet transactions, such as emails or video click streams and other. The series contains monographs, lecture notes and edited volumes in Big Data spanning the areas of computational intelligence including neural networks, evolutionary computation, soft computing, fuzzy systems, as well as artificial intelligence, data mining, modern statistics and Operations research, as well as self-organizing systems. Of particular value to both the contributors and the readership are the short publication timeframe and the world-wide distribution, which enable both wide and rapid dissemination of research output.

** Indexing: The books of this series are submitted to ISI Web of Science, DBLP, Ulrichs, MathSciNet, Current Mathematical Publications, Mathematical Reviews, Zentralblatt Math: MetaPress and Springerlink.

More information about this series at http://www.springer.com/series/11970

Aboul Ella Hassanien · Ashraf Darwish
Editors

Machine Learning and Big Data Analytics Paradigms: Analysis, Applications and Challenges

 Springer

Editors
Aboul Ella Hassanien
Faculty of Computer and Artificial
Intelligence
Cairo University
Giza, Egypt

Ashraf Darwish
Faculty of Science
Helwan University
Helwan, Egypt

ISSN 2197-6503 ISSN 2197-6511 (electronic)
Studies in Big Data
ISBN 978-3-030-59340-7 ISBN 978-3-030-59338-4 (eBook)
https://doi.org/10.1007/978-3-030-59338-4

This Springer imprint is published by the registered company Springer Nature Switzerland AG
The registered company address is: Gewerbestrasse 11, 6330 Cham, Switzerland

This work is dedicated to my wife Nazaha El Saman, and to the Scientific Research Group in Egypt members

Preface

Artificial intelligence and machine techniques have shown an increasing impact and presence on a wide variety of applications and research areas. In recent years, the machine learning approaches have been used in solving domain-specific problems in various fields of applications. The success in artificial intelligence and machine learning combined with the advent of deep learning and big data has sparked a renewed interest in data science in many technical fields. Recently, deep learning is widely applied because of the improvements in computing hardware and the discovery of fast optimization algorithms. This technique can learn many representations at various abstraction levels for text, sound and image data to analyze large data and obtain relevant solutions. This book is intended to present the state of the art in research on machine learning and big data analytics. We accepted 30 submissions. The accepted papers covered the following four themes (parts):

Artificial Intelligence and Data Mining Applications
Machine Learning and Applications
Deep Learning Technology for Big Data Analytics
Modeling, Simulation, Security with Big Data

Each submission is reviewed by the editorial board. Evaluation criteria include correctness, originality, technical strength, significance, quality of presentation, and interest and relevance to the book scope. Chapters of this book provide a collection of high-quality research works that address broad challenges in both theoretical and application aspects of machine learning and big data and its applications. We acknowledge all that contributed to the staging of this edited book (authors, committees and organizers). We deeply appreciate their involvement and support that was crucial for the success of the *Machine Learning and Big Data Analytics Paradigms: Analysis, Applications and Challenges* edited book.

Giza, Egypt   Aboul Ella Hassanien
                                                  Scientific Research Group in Egypt (SRGE)

Helwan, Egypt   Ashraf Darwish
                                                  Scientific Research Group in Egypt (SRGE)

Contents

Artificial Intelligence and Data Mining Applications

Rough Sets and Rule Induction from Indiscernibility Relations Based on Possible World Semantics in Incomplete Information Systems with Continuous Domains

Michinori Nakata, Hiroshi Sakai, and Keitarou Hara

Abstract Rough sets and rule induction in an incomplete and continuous information table are investigated under possible world semantics. We show an approach using possible indiscernibility relations, whereas the traditional approaches use possible tables. This is because the number of possible indiscernibility relations is finite, although we have the infinite number of possible tables in an incomplete and continuous information table. First, lower and upper approximations are derived directly using the indiscernibility relation on a set of attributes in a complete and continuous information table. Second, how these approximations are derived are described applying possible world semantics to an incomplete and continuous information table. Lots of possible indiscernibility relations are obtained. The actual indiscernibility relation is one of possible ones. The family of possible indiscernibility relations is a lattice for inclusion with the minimum and the maximum indiscernibility relations. Under the minimum and the maximum indiscernibility relations, we obtain four kinds of approximations: certain lower, certain upper, possible lower, and possible upper approximations. Therefore, there is no computational complexity for the number of values with incomplete information. The approximations in possible world semantics are the same as ones in our extended approach directly using indiscernibility relations. We obtain four kinds of single rules: certain and consistent, certain and inconsistent, possible and consistent, and possible and inconsistent ones from certain lower, certain upper, possible lower, and possible upper approximations, respectively. Individual objects in an approximation support single rules. Serial sin-

M. Nakata (✉)
Faculty of Management and Information Science,
Josai International University, 1 Gumyo, Togane, Chiba 283-8555, Japan
e-mail: nakatam@ieee.org

H. Sakai
Department of Mathematics and Computer Aided Sciences, Faculty of Engineering,
Kyushu Institute of Technology, Tobata, Kitakyushu 804-8550, Japan
e-mail: sakai@mns.kyutech.ac.jp

K. Hara
Department of Informatics, Tokyo University of Information Sciences,
4-1 Onaridai, Wakaba-ku, Chiba 265-8501, Japan
e-mail: hara@rsch.tuis.ac.jp

© The Author(s), under exclusive license to Springer Nature Switzerland AG 2021
A. E. Hassanien et al. (eds.), *Machine Learning and Big Data Analytics Paradigms: Analysis, Applications and Challenges*, Studies in Big Data 77,
https://doi.org/10.1007/978-3-030-59338-4_1

gle rules from the approximation are brought into one combined rule. The combined rule has greater applicability than single rules that individual objects support.

Keywords Neighborhood rough sets · Rule induction · Possible world semantics · Incomplete information · Indiscernibility relations · Continuous values

1 Introduction

The Information generated in the real world includes various types of data. When we deal with character string data, the data is broadly classified into discrete data and continuous data.

Rough sets, constructed by Pawlak [18], are used as an effective method for feature selection, pattern recognition, data mining and so on. The framework consists of lower and upper approximations. This is traditionally applied to complete information tables with nominal attributes. Fruitful results are reported in various fields. However, when we are faced with real-world objects, it is often necessary to handle attributes that take a continuous value. Furthermore, objects with incomplete information ubiquitously exist in the real world. Without processing incomplete and continuous information, the information generated in the real world cannot be fully utilized. Therefore, extended versions of the rough sets have been proposed to handle incomplete information in continuous domains.

An approach handling incomplete information, which is often adopted [7, 20–22], is to use the way that Kryszkiewicz applied to nominal attributes [8]. This approach gives in advance the indistinguishability of objects that have incomplete information with other objects. However, it is natural that there are two possibilities for incomplete information objects. One possibility is that an object with incomplete information may have the same value as another object. That is, the two objects may be indiscernible. The other possibility is that the object may have a different value from another object. That is, they may be discernible. Giving in advance the indiscernibility corresponds to neglecting one of the two possibilities. Therefore, the approach leads to loss of information and creates poor results [11, 19].

Another approach is to directly use indiscernibility relations extended to handle incomplete information [14]. Yet another approach is to use possible classes obtained from the indiscernibility relation on a set of attributes [15]. These two approaches have no computational complexity for the number of values with incomplete information. We need to give some justification to these extended approaches. It is known in discrete data tables that an approach using possible class has some justification from the viewpoint of possible world semantics [12]. We focus on an approach directly using indiscernibility relations.[1] To give it some justification, we need to develop an approach that is based on possible world semantics. The previous approaches are developed under possible tables derived from an incomplete and continuous informa-

---

[1] See reference [16] for an approach using possible classes.

tion table. Unfortunately, an infinite number of possible tables can be generated from an incomplete and continuous information table. Possible world semantics cannot be applied to an infinite number of possible tables.

The starting point for a rough set is the indiscernibility relation on a set of attributes. When an information table contains values with incomplete information, we obtain lots of possible indiscernibility relations in place of the indiscernibility relation. The number is finite, even if the number of possible tables is infinite, because the number of objects is finite. We note this finiteness and develop an approach based on the possible indiscernibility relations, not the possible tables.

The paper is constructed as follows. Section 2 describes an approach directly using indiscernibility relations in a complete and continuous information table. Section 3 develops an approach applying possible world semantics to an incomplete and continuous information table. Section 4 describes rule induction in a complete and continuous information table. Section 5 address rule induction in an incomplete and continuous information table. Section 6 mentions the conclusions.

2 Rough Sets by Directly Using Indiscernibility Relations in Complete and Continuous Information Systems

A continuous data set is represented as a two-dimensional table, called a continuous information table. In the continuous information table, each row and each column represent an object and an attribute, respectively. A mathematical model of an information table with complete and continuous information is called a complete and continuous information system. The complete and continuous information system is a triplet expressed by $(U, AT, \{D(a) \mid a \in AT\})$. U is a non-empty finite set of objects, which is called the universe. AT is a non-empty finite set of attributes such that $a : U \rightarrow D(a)$ for every $a \in AT$ where $D(a)$ is the continuous domain of attribute a.

We have two approaches for handling continuous values. One approach is to discretize a continuous domain into disjunctive intervals in which objects are considered as indiscernible [4]. How to discretize has a heavy influence over results. The other approach is to use neighborhood [10]. The indiscernibility of two objects is derived from the distance of the values that characterize them. A threshold is given, which is the indiscernibility criterion. When the distance between two objects is less than or equal to the threshold, they are considered as indiscernible. As the threshold changes, the results change gradually. Therefore, we take the neighborhood-based approach.

Binary relation $R_A{}^2$ that represents the indiscernibility between objects on set $A \subseteq AT$ of attributes is called the indiscernibility relation on A:

$$R_A = \{(o, o') \in U \times U \mid |A(o) - A(o')| \leq \delta_A\}, \tag{1}$$

$^2 R_A$ is formally $R_A^{\delta_A}$. δ_A is omitted unless confusion.

where $A(o)$ is the value sequence for A of object o and $(|A(o) - A(o')| \leq \delta_A) = (\wedge_{a \in A}|a(o) - a(o')| \leq \delta_a)$ and $\delta_a{}^3$ is a threshold indicating the range of indiscernibility between $a(o)$ and $a(o')$.

Proposition 1 *If* $\delta 1_A \leq \delta 2_A$, *equal to* $\wedge_{a \in A}(\delta 1_a \leq \delta 2_a)$, *then* $R_A^{\delta 1_A} \subseteq R_A^{\delta 2_A}$, *where* $R_A^{\delta 1_A}$ *and* $R_A^{\delta 2_A}$ *are the indiscernibility relations with thresholds* $\delta 1_A$ *and* $\delta 2_A$, *respectively and* $R_A^{\delta 1_A} = \cap_{a \in A} R_a^{\delta 1_a}$ *and* $R_A^{\delta 2_A} = \cap_{a \in A} R_a^{\delta 2_a}$.

From indiscernibility relation R_A, *indiscernible class* $[o]_A$ *for object o is obtained*:

$$[o]_A = \{o' \mid (o, o') \in R_A\}, \tag{2}$$

where $[o]_A = \cap_{a \in A}[o]_a$.

Directly using indiscernibility relation R_A, lower approximation $\underline{apr}_A(\mathcal{O})$ and upper approximation $\overline{apr}_A(\mathcal{O})$ for A of set $\mathcal{O}$ of objects are:

$$\underline{apr}_A(\mathcal{O}) = \{o \mid \forall o' \in U \, (o, o') \notin R_A \vee o' \in \mathcal{O}\}, \tag{3}$$

$$\overline{apr}_A(\mathcal{O}) = \{o \mid \exists o' \in U \, (o, o') \in R_A \wedge o' \in \mathcal{O}\}. \tag{4}$$

Proposition 2 *[14] Let* $\underline{apr}_A^{\delta 1_A}(\mathcal{O})$ *and* $\overline{apr}_A^{\delta 1_A}(\mathcal{O})$ *be lower and upper approximations under threshold* $\delta 1_A$ *and let* $\underline{apr}_A^{\delta 2_A}(\mathcal{O})$ *and* $\overline{apr}_A^{\delta 2_A}(\mathcal{O})$ *be lower and upper approximations under threshold* $\delta 2_A$. *If* $\delta 1_A \leq \delta 2_A$, *then* $\underline{apr}_A^{\delta 1_A}(\mathcal{O}) \supseteq \underline{apr}_A^{\delta 2_A}(\mathcal{O})$ *and* $\overline{apr}_A^{\delta 1_A}(\mathcal{O}) \subseteq \overline{apr}_A^{\delta 2_A}(\mathcal{O})$.

For object o in the lower approximation of $\mathcal{O}$, all objects with which o is indiscernible are included in $\mathcal{O}$; namely, $[o]_A \subseteq \mathcal{O}$. On the other hand, for objects in the upper approximation of $\mathcal{O}$, some objects indiscernible o are in $\mathcal{O}$. That is, $[o]_A \cap \mathcal{O} \neq \emptyset$. Thus, $\underline{apr}_A(\mathcal{O}) \subseteq \overline{apr}_A(\mathcal{O})$.

3 Rough Sets from Possible World Semantics in Incomplete and Continuous Information Systems

An information table with incomplete and continuous information is called an incomplete and continuous information system. In incomplete and continuous information systems, $a : U \rightarrow s_a$ for every $a \in AT$ where s_a is the union of or-sets of values over domain $D(a)$ of attribute a and sets of intervals on $D(a)$. Note that an or-set is a disjunctive set [9]. Single value $v \in a(o)$ is a possible value that may be the actual value of attribute a in object o. The possible value is the actual one if $a(o)$ is single; namely, $|a(o)| = 1$.

---

[3] Subscript a of δ_a is omitted if no confusion.

We have lots of possible indiscernibility relations from an incomplete and continuous information table. The smallest possible indiscernibility relation is the certain one. Certain indiscernibility relation CR_A is:

$$CR_A = \cap_{a \in A} CR_a, \tag{5}$$

$$CR_a = \{(o, o') \in U \times U \mid (o = o') \vee (\forall u \in a(o) \forall v \in a(o') | u - v | \leq \delta_a)\}. \tag{6}$$

In this binary relation, which is unique on A, two objects o and o' of $(o, o') \in CR_A$ are certainly indiscernible on A. Such a pair is called a certain pair. Family $\mathcal{F}(R_A)$ of possible indiscernibility relations is:

$$\mathcal{F}(R_A) = \{e \mid e = CR_A \cup e' \wedge e' \in \mathcal{P}(MPPR_A)\}, \tag{7}$$

where each element is a possible indiscernibility relation and $\mathcal{P}(MPPR_A)$ is the power set of $MPPR_A$ and $MPPR_A$ is:

$$MPPR_A = \{\{(o', o), (o, o')\} \mid (o', o) \in MPR_A\},$$

$$MPR_A = \cap_{a \in A} MPR_a, \tag{8}$$

$$MPR_a = \{(o, o') \in U \times U \mid \exists u \in a(o) \exists v \in a(o') | u - v | \leq \delta_a)\} \backslash CR_a. \tag{9}$$

A pair of objects in MPR_A is called a possible one. $\mathcal{F}(R_A)$ has a lattice structure for set inclusion. CR_A is the minimum possible indiscernibility relation in $\mathcal{F}(R_A)$ on A, which is the minimum element, whereas $CR_A \cup MPR_A$ is the maximum possible indiscernibility relation on A, which is the maximum element. One of possible indiscernibility relations is actual. However, we cannot know it without additional information.

Example 1 Or-set $< 1.25, 1.31 >$ means 1.25 or 1.31. Let threshold δ_{a_1} be 0.05 in T of Fig. 1. The set of certain pairs of indiscernible objects on a_1 is:

$$\{(o_1, o_1), (o_1, o_3), (o_3, o_1), (o_1, o_5), (o_5, o_1), (o_2, o_2), (o_3, o_3), (o_4, o_4), (o_5, o_5)\}.$$

The set of possible pairs of indiscernible objects is:

Fig. 1 Incomplete and continuous information table T

U	a_1	a_2
o_1	0.71	$< 1.25, 1.31 >$
o_2	$[0.74, 0.79]$	$[1.47, 1.53]$
o_3	0.73	1.51
o_4	$[0.85, 0.94]$	1.56
o_5	$< 0.66, 0.68 >$	$< 1.32, 1.39 >$

T

$$\{(o_1, o_2), (o_2, o_1), (o_2, o_3), (o_3, o_2), (o_3, o_5), (o_5, o_3)\}.$$

Applying formulae (5)–(7) to these sets, the family of possible indiscernibility relations and each possible indiscernibility relation pr_i with $i = 1, \ldots, 8$ are:

$$\mathcal{F}(R_{a_1}) = \{pr_1, \cdots, pr_8\},$$
$$pr_1 = \{(o_1, o_1), (o_1, o_3), (o_3, o_1), (o_1, o_5), (o_5, o_1), (o_2, o_2), (o_3, o_3),$$
$$(o_4, o_4), (o_5, o_5)\},$$
$$pr_2 = \{(o_1, o_1), (o_1, o_3), (o_3, o_1), (o_1, o_5), (o_5, o_1), (o_2, o_2), (o_3, o_3),$$
$$(o_4, o_4), (o_5, o_5), (o_1, o_2), (o_2, o_1)\},$$
$$pr_3 = \{(o_1, o_1), (o_1, o_3), (o_3, o_1), (o_1, o_5), (o_5, o_1), (o_2, o_2), (o_3, o_3),$$
$$(o_4, o_4), (o_5, o_5), (o_2, o_3), (o_3, o_2)\},$$
$$pr_4 = \{(o_1, o_1), (o_1, o_3), (o_3, o_1), (o_1, o_5), (o_5, o_1), (o_2, o_2), (o_3, o_3),$$
$$(o_4, o_4), (o_5, o_5), (o_3, o_5), (o_5, o_3)\},$$
$$pr_5 = \{(o_1, o_1), (o_1, o_3), (o_3, o_1), (o_1, o_5), (o_5, o_1), (o_2, o_2), (o_3, o_3),$$
$$(o_4, o_4), (o_5, o_5), (o_1, o_2), (o_2, o_1), (o_2, o_3), (o_3, o_2)\},$$
$$pr_6 = \{(o_1, o_1), (o_1, o_3), (o_3, o_1), (o_1, o_5), (o_5, o_1), (o_2, o_2), (o_3, o_3),$$
$$(o_4, o_4), (o_5, o_5), (o_1, o_2), (o_2, o_1), (o_3, o_5), (o_5, o_3)\},$$
$$pr_7 = \{(o_1, o_1), (o_1, o_3), (o_3, o_1), (o_1, o_5), (o_5, o_1), (o_2, o_2), (o_3, o_3),$$
$$(o_4, o_4), (o_5, o_5), (o_2, o_3), (o_3, o_2), (o_3, o_5), (o_5, o_3)\},$$
$$pr_8 = \{(o_1, o_1), (o_1, o_3), (o_3, o_1), (o_1, o_5), (o_5, o_1), (o_2, o_2), (o_3, o_3),$$
$$(o_4, o_4), (o_5, o_5), (o_1, o_2), (o_2, o_1), (o_2, o_3), (o_3, o_2), (o_3, o_5), (o_5, o_3)\}.$$

The family of these possible indiscernibility relations has the lattice structure for set inclusion like Fig. 2. pr_1 is the minimum element, whereas pr_8 is the maximum element.

We develop an approach based on possible indiscernibility relations in an incomplete and continuous information table. Applying formulae (3) and (4) to a possible indiscernibility relation pr, Lower and upper approximations in pr are:

Fig. 2 Lattice structure

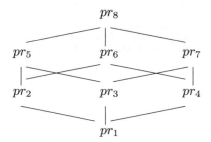

$$\underline{apr}_A(\mathcal{O})^{pr} = \{o \mid \forall o' \in U \ ((o, o') \notin pr \wedge pr \in \mathcal{F}(R_A)) \vee o' \in \mathcal{O}\}, \quad (10)$$

$$\overline{apr}_A(\mathcal{O})^{pr} = \{o \mid \exists o' \in U \ ((o, o') \in pr \wedge pr \in \mathcal{F}(R_A)) \wedge o' \in \mathcal{O}\}. \quad (11)$$

Proposition 3 *If $pr_k \subseteq pr_l$ for possible indiscernibility relations $pr_k, pr_l \in \mathcal{F}(R_A)$, then $\underline{apr}_A(\mathcal{O})^{pr_k} \supseteq \underline{apr}_A(\mathcal{O})^{pr_l}$ and $\overline{apr}_A(\mathcal{O})^{pr_k} \subseteq \overline{apr}_A(\mathcal{O})^{pr_l}$.*

From this proposition the families of lower and upper approximations in possible indiscernibility relations also have the same lattice structure for set inclusion as the family of possible indiscernibility relations.

By aggregating the lower and upper approximations in possible indiscernibility relations, we obtain four kinds of approximations: certain lower approximation $\underline{Capr}_A(\mathcal{O})$, certain upper approximation $\overline{Capr}_A(\mathcal{O})$, possible lower approximation $\underline{Papr}_A(\mathcal{O})$, and possible upper approximation $\overline{Papr}_A(\mathcal{O})$:

$$\underline{Capr}_A(\mathcal{O}) = \{o \mid \forall pr \in \mathcal{F}(R_A)o \in \underline{apr}_A(\mathcal{O})^{pr}\}, \quad (12)$$

$$\overline{Capr}_A(\mathcal{O}) = \{o \mid \forall pr \in \mathcal{F}(R_A)o \in \overline{apr}_A(\mathcal{O})^{pr}\}, \quad (13)$$

$$\underline{Papr}_A(\mathcal{O}) = \{o \mid \exists pr \in \mathcal{F}(R_A)o \in \underline{apr}_A(\mathcal{O})^{pr}\}, \quad (14)$$

$$\overline{Papr}_A(\mathcal{O}) = \{o \mid \exists pr \in \mathcal{F}(R_A)o \in \overline{apr}_A(\mathcal{O})^{pr}\}. \quad (15)$$

Using Proposition 3,

$$\underline{Capr}_A(\mathcal{O}) = \underline{apr}_A(\mathcal{O})^{pr_{\max}}, \quad (16)$$

$$\overline{Capr}_A(\mathcal{O}) = \overline{apr}_A(\mathcal{O})^{pr_{\min}}, \quad (17)$$

$$\underline{Papr}_A(\mathcal{O}) = \underline{apr}_A(\mathcal{O})^{pr_{\min}}, \quad (18)$$

$$\overline{Papr}_A(\mathcal{O}) = \overline{apr}_A(\mathcal{O})^{pr_{\max}}, \quad (19)$$

where $pr_{\min}$ and $pr_{\max}$ are the minimum and the maximum possible indiscernibility relations on A.

Using formulae (16)–(19), we can obtain the four approximations without the computational complexity for the number of possible indiscernibility relations, although the number of possible indiscernibility relations has exponential growth as the number of values with incomplete information linearly increases.

Definability on set A of attributes is defined as follows:

Set $\mathcal{O}$ of objects is certainly definable if and only if $\forall pr \in \mathcal{F}(R_A) \exists S \subseteq U \ \mathcal{O} = \cup_{o \in S}[o]_A^{pr}$.

Set $\mathcal{O}$ of objects is possibly definable if and only if $\exists pr \in \mathcal{F}(R_A) \exists S \subseteq U \ \mathcal{O} = \cup_{o \in S}[o]_A^{pr}$.

These definition is equivalent to:

Set $\mathcal{O}$ of objects is certainly definable if and only if $\forall pr \in \mathcal{F}(R_A) \ \underline{apr}_A(\mathcal{O})^{pr} = \overline{apr}_A(\mathcal{O})^{pr}$.

Set $\mathcal{O}$ of objects is possibly definable if and only if $\exists pr \in \mathcal{F}(R_A) \ \underline{apr}_A(\mathcal{O})^{pr} = \overline{apr}_A(\mathcal{O})^{pr}$.

Example 2 We use the possible indiscernibility relations in Example 1. Let set $\mathcal{O}$ of objects be $\{o_2, o_4\}$. Applying formulae (10) and (11) to $\mathcal{O}$, lower and upper approximations from each possible indiscernibility relation are:

$$\underline{apr}_{a_1}(\mathcal{O})^{pr_1} = \{o_2, o_4\}, \overline{apr}_{a_1}(\mathcal{O})^{pr_1} = \{o_2, o_4\},$$
$$\underline{apr}_{a_1}(\mathcal{O})^{pr_2} = \{o_4\}, \overline{apr}_{a_1}(\mathcal{O})^{pr_2} = \{o_1, o_2, o_4\},$$
$$\underline{apr}_{a_1}(\mathcal{O})^{pr_3} = \{o_4\}, \overline{apr}_{a_1}(\mathcal{O})^{pr_3} = \{o_2, o_3, o_4\},$$
$$\underline{apr}_{a_1}(\mathcal{O})^{pr_4} = \{o_2, o_4\}, \overline{apr}_{a_1}(\mathcal{O})^{pr_4} = \{o_2, o_4\},$$
$$\underline{apr}_{a_1}(\mathcal{O})^{pr_5} = \{o_4\}, \overline{apr}_{a_1}(\mathcal{O})^{pr_5} = \{o_1, o_2, o_3, o_4\},$$
$$\underline{apr}_{a_1}(\mathcal{O})^{pr_6} = \{o_4\}, \overline{apr}_{a_1}(\mathcal{O})^{pr_6} = \{o_1, o_2, o_4\},$$
$$\underline{apr}_{a_1}(\mathcal{O})^{pr_7} = \{o_4\}, \overline{apr}_{a_1}(\mathcal{O})^{pr_7} = \{o_2, o_3, o_4\},$$
$$\underline{apr}_{a_1}(\mathcal{O})^{pr_8} = \{o_4\}, \overline{apr}_{a_1}(\mathcal{O})^{pr_8} = \{o_1, o_2, o_3, o_4\}.$$

By using formulae (16)–(19),

$$\underline{Capr}_{a_1}(\mathcal{O}) = \{o_4\},$$
$$\overline{Capr}_{a_1}(\mathcal{O}) = \{o_2, o_4\},$$
$$\underline{Papr}_{a_1}(\mathcal{O}) = \{o_2, o_4\},$$
$$\overline{Papr}_{a_1}(\mathcal{O}) = \{o_1, o_2, o_3, o_4\}.$$

$\mathcal{O}$ is possibly definable on a_1.

As with the case of nominal attributes [12], the following proposition holds.

Proposition 4 $\underline{Capr}_A(\mathcal{O}) \subseteq \underline{Papr}_A(\mathcal{O}) \subseteq \mathcal{O} \subseteq \overline{Capr}_A(\mathcal{O}) \subseteq \overline{Papr}_A(\mathcal{O})$.

Using the four approximations denoted by formulae (16)–(19), lower approximation $\underline{apr}^{\bullet}_A(\mathcal{O})$ and upper approximation $\overline{apr}^{\bullet}_A(\mathcal{O})$ are expressed in interval sets, as is described in [13][4]:

$$\underline{apr}^{\bullet}_A(\mathcal{O}) = [\underline{Capr}_A(\mathcal{O}), \underline{Papr}_A(\mathcal{O})], \tag{20}$$
$$\overline{apr}^{\bullet}_A(\mathcal{O}) = [\overline{Capr}_A(\mathcal{O}), \overline{Papr}_A(\mathcal{O})]. \tag{21}$$

The two approximations $\underline{apr}^{\bullet}_A(\mathcal{O})$ and $\overline{apr}^{\bullet}_A(\mathcal{O})$ are dependent through the complementarity property $\underline{apr}^{\bullet}_A(\mathcal{O}) = U - \overline{apr}^{\bullet}_A(U - \mathcal{O})$.

Example 3 Applying four approximations in Example 2 to formulae (20) and (21),

---

[4]Hu and Yao also say that approximations are described by using an interval set in information tables with incomplete information [5].

$$\underline{apr}^{\bullet}_{a_1}(\mathcal{O}) = [\{o_4\}, \{o_2, o_4\}],$$
$$\overline{apr}^{\bullet}_{a_1}(\mathcal{O}) = [\{o_2, o_4\}, \{o_1, o_2, o_3, o_4\}].$$

Furthermore, the following proposition is valid from formulae (16)–(19).

Proposition 5

$$\underline{Capr}_A(\mathcal{O}) = \{o \mid \forall o' \in U \ (o, o') \notin (CR_A \cup MPR_A) \vee o' \in \mathcal{O}\},$$
$$\overline{Capr}_A(\mathcal{O}) = \{o \mid \exists o' \in U \ (o, o') \in CR_A \wedge o' \in \mathcal{O}\},$$
$$\underline{Papr}_A(\mathcal{O}) = \{o \mid \forall o' \in U \ (o, o') \notin CR_A \vee o' \in \mathcal{O}\},$$
$$\overline{Papr}_A(\mathcal{O}) = \{o \mid \exists o' \in U \ (o, o') \in (CR_A \cup MPR_A) \wedge o' \in \mathcal{O}\}.$$

Our extended approach directly using indiscernibility relations [14] is justified from this proposition. That is, approximations from the extended approach using two indiscernibility relations are the same as the ones obtained under possible world semantics. A correctness criterion for justification is formulated as

$$q(R_A) = \bigodot q'(\mathcal{F}(R_A)),$$

where q' is the approach for complete and continuous information, which is described in Sect. 2, and q is an extended approach of q', which directly handles with incomplete and continuous information, and $\bigodot$ is an aggregate operator. This is represented in Fig. 3.

This kind of correctness criterion is usually used in the field of databases handling incomplete information [1–3, 6, 17, 23].

When objects in $\mathcal{O}$ are specified by a restriction containing set B of nominal attribute with incomplete information, elements in domain $D(B)(= \cup_{b \in B} D(b))$ are used. For example, $\mathcal{O}$ is specified by restriction $B = X(= \wedge_{b \in B}(b = x_b))$ with $B \in AT$ and $x_b \in D(b)$. Four approximations: certain lower, certain upper, possible lower, and possible upper ones are:

Fig. 3 Correctness criterion of extended method q

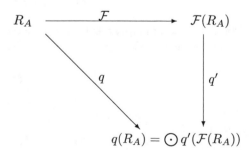

$$Capr_A(\mathcal{O}) = apr_A(CO_{B=X})^{pr_{max}}, \tag{22}$$

$$\overline{Capr}_A(\mathcal{O}) = \overline{apr}_A(CO_{B=X})^{pr_{min}}, \tag{23}$$

$$Papr_A(\mathcal{O}) = apr_A(PO_{B=X})^{pr_{min}}, \tag{24}$$

$$\overline{Papr}_A(\mathcal{O}) = \overline{apr}_A(PO_{B=X})^{pr_{max}}. \tag{25}$$

where

$$CO_{B=X} = \{o \in \mathcal{O} \mid B(o) = X\}, \tag{26}$$

$$PO_{B=X} = \{o \in \mathcal{O} \mid B(o) \cap X \neq \emptyset\}. \tag{27}$$

When $\mathcal{O}$ is specified by a restriction containing set B of numerical attributes with incomplete information, set $\mathcal{O}$ is specified by an interval where precise values of $b \in B$ are used.

$$Capr_A(\mathcal{O}) = apr_A(CO_{\wedge_{b \in B}[b(o_{m_b}), b(o_{n_b})]})^{pr_{max}}, \tag{28}$$

$$\overline{Capr}_A(\mathcal{O}) = \overline{apr}_A(CO_{\wedge_{b \in B}[b(o_{m_b}), b(o_{n_b})]})^{pr_{min}}, \tag{29}$$

$$Papr_A(\mathcal{O}) = apr_A(PO_{\wedge_{b \in B}[b(o_{m_b}), b(o_{n_b})]})^{pr_{min}}, \tag{30}$$

$$\overline{Papr}_A(\mathcal{O}) = \overline{apr}_A(PO_{\wedge_{b \in B}[b(o_{m_b}), b(o_{n_b})]})^{pr_{max}}, \tag{31}$$

where

$$CO_{\wedge_{b \in B}[b(o_{m_b}), b(o_{n_b})]} = \{o \in \mathcal{O} \mid \forall b \in B \ b(o) \subseteq [b(o_{m_b}), b(o_{n_b})]\}, \tag{32}$$

$$PO_{\wedge_{b \in B}[b(o_{m_b}), b(o_{n_b})]} = \{o \in \mathcal{O} \mid \forall b \in B \ b(o) \cap [b(o_{m_b}), b(o_{n_b})] \neq \emptyset\}, \tag{33}$$

where $b(o_{m_b})$ and $b(o_{n_b})$ are precise and $\forall b \in B \ b(o_{m_b}) \leq b(o_{n_b})$.

Example 4 In incomplete information table T of Example 1, let $\mathcal{O}$ be specified by values $a_2(o_3)$ and $a_2(o_4)$. Using formulae (32) and (33),

$$CO_{[a_2(o_3), a_2(o_4)]} = \{o_3, o_4\},$$

$$PO_{[a_2(o_3), a_2(o_4)]} = \{o_2, o_3, o_4\}.$$

Possible indiscernibility relations pr_{min} and pr_{max} on a_1 is pr_1 and pr_8 in Example 1. Using formulae (28)–(31),

$$Capr_{a_1}(\mathcal{O}) = \{o_4\},$$

$$\overline{Capr}_{a_1}(\mathcal{O}) = \{o_3, o_4\},$$

$$Papr_{a_1}(\mathcal{O}) = \{o_2, o_3, o_4\},$$

$$\overline{Papr}_{a_1}(\mathcal{O}) = \{o_1, o_2, o_3, o_4\}.$$

4 Rule Induction in Complete and Continuous Information Systems

Let single rules that are supported by objects be derived from the lower and upper approximations of O specified by restriction $B = X$.

- Object $o \in \underline{apr}_A(O)$ supports rule $A = A(o) \rightarrow B = X$ consistently.
- Object $o \in \overline{apr}_A(O)$ supports rule $A = A(o) \rightarrow B = X$ inconsistently.

The accuracy, which means the degree of consistency, is $|[o]_A \cap O|/|[o]_A|$. This degree is equal to 1 for $o \in \underline{apr}_A(O)$.

In the case where a set of attributes that characterize objects has continuous domains, single rules supported by individual objects in an approximation usually have different antecedent parts. So, we obtain lots of single rules. The disadvantage of the single rule is that it lacks applicability. For example, let two values $a(o)$ and $a(o')$ be 4.53 and 4.65 for objects o and o' in $\underline{apr}_a(O)$. When O is specified by restriction $b = x$, o and o' consistently support single rules $a = 4.53 \rightarrow b = x$ and $a = 4.65 \rightarrow b = x$, respectively. By using these single rules, we can say that an object with value 4.57 of a, which is indiscernible with 4.53 under $\delta_a = 0.05$, supports $a = 4.57 \rightarrow b = x$. However, we cannot at all say anything for a rule consistently supported by an object with value 4.59 discernible with 4.53 and 4.65 under $\delta_a = 0.05$. This shows that the single rule has low applicability.

To improve applicability, we bring serial single rules into one combined rule. Let $o \in U$ be arranged in ascending order of $a(o)$ and be given a serial superscript from 1 to $|U|$. $\underline{apr}_A(O)$ and $\overline{apr}_A(O)$ consist of collections of serially superscripted objects. For instance, $\underline{apr}_A(O) = \{\cdots, o_{i_h}^h, o_{i_{h+1}}^{h+1}, \cdots, o_{i_{k-1}}^{k-1}, o_{i_k}^k, \cdots\}$ ($h \leq k$). The following processing is done to each attribute in A. A single rule that $o^l \in \underline{apr}_A(O)$ has antecedent part $a = a(o^l)$ for attribute a. Then, antecedent parts of serial single rules induced from collection $(o_{i_h}^h, o_{i_{h+1}}^{h+1}, \cdots, o_{i_{k-1}}^{k-1}, o_{i_k}^k)$ can be brought into one combined antecedent part $a = [a(o_{i_h}^h), a(o_{i_k}^k)]$. Finally, a combined rule is expressed in $\wedge_{a \in A}(a = [a(o_{i_h}^h), a(o_{i_k}^k)] \rightarrow B = X)$. The combined rule has accuracy

$$\min_{h \leq j \leq k} |[o_{i_j}^j]_A \cap O|/|[o_{i_j}^j]_A|. \tag{34}$$

Proposition 7 *Let $\underline{r}$ be the set of combined rules obtained from $\underline{apr}_A(O)$ and $\overline{r}$ be the set from $\overline{apr}_A(O)$. If $(A = [l_A, u_A] \rightarrow B = X) \in \underline{r}$, then $\exists l'_A \leq l_A, \exists u'_A \geq u_A$ $(A = [l'_A, u'_A] \rightarrow B = X) \in \overline{r}$, where O is specified by restriction $B = X$ and $(A = [l_A, u_A]) = \wedge_{a \in A}(a = [l_a, u_a])$.*

Proof A single rule obtained from $\underline{apr}_A(O)$ is also derived from $\overline{apr}_A(O)$. This means that the proposition holds.

Example 7 Let continuous information table $T0$ in Fig. 3 be obtained, where U consists of $\{o_1, o_2, \cdots, o_{19}, o_{20}\}$. Tables $T1$, $T2$, and $T3$ in Fig. 4 are created from

T0					T1				T2				T3	
U	a_1	a_2	a_3	a_4	U	a_1	a_4	U	a_2	a_3		U	a_3
o_1	4.12	1.97	4.03	x	o_3	3.34	z	o_5	1.34	3.68		o_5	3.68
o_2	3.95	2.64	4.45	x	o_{18}	3.46	z	o_{20}	1.34	3.68		o_{20}	3.68
o_3	3.34	2.68	4.29	z	o_{12}	3.64	y	o_{14}	1.77	3.89		o_{14}	3.89
o_4	5.79	1.97	4.53	w	o_7	3.82	y	o_7	1.94	3.92		o_7	3.92
o_5	4.43	1.34	3.68	x	o_{17}	3.90	y	o_1	1.97	4.03		o_1	4.03
o_6	4.04	3.51	5.08	y	o_2	3.95	x	o_4	1.97	4.53		o_{15}	4.23
o_7	3.82	1.94	3.92	y	o_{16}	3.96	x	o_8	2.10	4.50		o_3	4.29
o_8	5.37	2.10	4.50	w	o_{11}	3.98	x	o_{15}	2.28	4.23		o_{17}	4.33
o_9	4.23	3.62	5.22	x	o_6	4.04	y	o_{19}	2.43	4.58		o_2	4.45
o_{10}	4.08	2.77	4.58	y	o_{15}	4.06	y	o_{17}	2.50	4.33		o_{16}	4.45
o_{11}	3.98	2.97	4.69	x	o_{10}	4.08	y	o_2	2.64	4.45		o_8	4.50
o_{12}	3.64	3.80	5.17	y	o_1	4.12	x	o_{16}	2.64	4.45		o_{18}	4.52
o_{13}	4.92	2.70	4.78	w	o_{14}	4.13	x	o_3	2.68	4.29		o_4	4.53
o_{14}	4.13	1.77	3.89	x	o_9	4.23	x	o_{13}	2.70	4.78		o_{19}	4.58
o_{15}	4.06	2.28	4.23	y	o_5	4.43	x	o_{10}	2.77	4.58		o_{10}	4.58
o_{16}	3.96	2.64	4.45	x	o_{20}	4.43	x	o_{18}	2.95	4.52		o_{11}	4.69
o_{17}	3.90	2.50	4.33	y	o_{19}	4.87	x	o_{11}	2.97	4.69		o_{13}	4.78
o_{18}	3.46	2.95	4.52	z	o_{13}	4.92	w	o_6	3.51	5.08		o_6	5.08
o_{19}	4.87	2.43	4.58	x	o_8	5.37	w	o_9	3.62	5.22		o_{12}	5.17
o_{20}	4.43	1.34	3.68	x	o_4	5.79	w	o_{12}	3.80	5.17		o_9	5.22

Fig. 4 $T0$ is an incomplete and continuous information table. $T1$, $T2$, and $T3$ are derived from $T0$

$T0$. $T1$ where set $\{a_1, a_4\}$ of attributes is projected from $T0$, $T2$ where $\{a_2, a_3\}$ is projected, and $T3$ where $\{a_3\}$ is projected. In addition, objects included in $T1$, $T2$, and $T3$ are arranged in ascending order of values of attributes a_1, a_2, and a_3, respectively.

Indiscernible classes on a_1 of each object under $\delta_{a_1} = 0.05$ are:

$$[o_1]_{a_1} = \{o_1, o_{10}, o_{14}\}, [o_2]_{a_1} = \{o_2, o_{11}, o_{16}, o_{17}\},$$
$$[o_3]_{a_1} = \{o_3\}, [o_4]_{a_1} = \{o_4\}, [o_5]_{a_1} = \{o_5, o_{20}\},$$
$$[o_6]_{a_1} = \{o_6, o_{10}, o_{15}\}, [o_7]_{a_1} = \{o_7\}, [o_8]_{a_1} = \{o_8\},$$
$$[o_9]_{a_1} = \{o_9\}, [o_{10}]_{a_1} = \{o_1, o_6, o_{10}, o_{14}, o_{15}\},$$
$$[o_{11}]_{a_1} = \{o_2, o_{11}, o_{16}\}, [o_{12}]_{a_1} = \{o_{12}\}, [o_{13}]_{a_1} = \{o_{13}, o_{19}\},$$
$$[o_{14}]_a = \{o_1, o_{10}, o_{14}\}, [o_{15}]_a = \{o_6, o_{10}, o_{15}\},$$
$$[o_{16}]_{a_1} = \{o_2, o_{11}, o_{16}\}, [o_{17}]_{a_1} = \{o_2, o_{17}\}, [o_{18}]_{a_1} = \{o_{18}\},$$
$$[o_{19}]_{a_1} = \{o_{13}, o_{19}\}, [o_{20}]_{a_1} = \{o_5, o_{20}\}.$$

When O is specified by $a_4 = x$, $O = \{o_1, o_2, o_5, o_9, o_{11}, o_{14}, o_{16}, o_{19}, o_{20}\}$. Let O be approximated by objects characterized by attribute a_1 whose values are continuous.

Using formulae (3) and (4), two approximations are:

$$\underline{apr}_{a_1}(O) = \{o_5, o_9, o_{11}, o_{16}, o_{20}\},$$
$$\overline{apr}_{a_1}(O) = \{o_1, o_2, o_5, o_9, o_{10}, o_{11}, o_{13}, o_{14}, o_{16}, o_{17}, o_{19}, o_{20}\}.$$

In continuous information table $T1$, which is created from $T0$, objects are arranged in ascending order of values of attribute a_1 and each object is given a serial superscript from 1 to 20. Using the serial superscript, the two approximations are rewritten:

$$\underline{apr}_{a_1}(O) = \{o_{16}^7, o_{11}^8, o_9^{14}, o_5^{15}, o_{20}^{16}\},$$
$$\overline{apr}_{a_1}(O) = \{o_{17}^5, o_2^6, o_{16}^7, o_{11}^8, o_{10}^{11}, o_1^{12}, o_{14}^{13}, o_9^{14}, o_5^{15}, o_{20}^{16}, o_{19}^{17}, o_{13}^{18}\},$$

The lower approximation creates consistent combined rules:

$$a_1 = [3.96, 3.98] \rightarrow a_4 = x, a_1 = [4.23, 4.43] \rightarrow a_4 = x,$$

from collections $\{o_{16}^7, o_{11}^8\}$ and $\{o_9^{14}, o_5^{15}, o_{20}^{16}\}$, respectively, where $a_1(o_{16}^7) = 3.96$, $a_1(o_{11}^8) = 3.98$, $a_1(o_9^{14}) = 4.23$, and $a_1(o_{20}^{16}) = 4.43$. The upper approximation creates inconsistent combined rules:

$$a_1 = [3.90, 3.98] \rightarrow a_4 = x, a_1 = [4.08, 4.92] \rightarrow a_4 = x,$$

from collections $\{o_{17}^5, o_2^6, o_{16}^7, o_{11}^8\}$ and $\{o_{10}^{11}, o_1^{12}, o_{14}^{13}, o_9^{14}, o_5^{15}, o_{20}^{16}, o_{19}^{17}, o_{13}^{18}\}$, respectively, where $a_1(o_{17}^5) = 3.90$, $a_1(o_{10}^{11}) = 4.08$, and $a_1(o_{13}^{18}) = 4.92$.

Next, let O be specified by a_3 that takes continuous values. In information table $T3$ projected from $T0$ the objects are arranged in ascending order of values of a_3 and each object is given a serial superscript from 1 to 20. Let lower and upper bounds be $a_3(o_{15}^6) = 4.23$ and $a_3(o_8^{11}) = 4.50$, respectively. Then, $O = \{o_{15}^6, o_3^7, o_{17}^8, o_2^9, o_{16}^{10}, o_8^{11}\}$. We approximate O by objects restricted by attribute a_2. Under $\delta_{a_2} = 0.05$, indiscernible classes of objects $o_1, \ldots o_{20}$ are:

$$[o_1]_{a_2} = \{o_1, o_4, o_7, o_8\}, [o_2]_{a_2} = \{o_2, o_3, o_{16}\},$$
$$[o_3]_{a_2} = \{o_2, o_3, o_{13}, o_{16}\}, [o_4]_{a_2} = \{o_1, o_4, o_7, o_8\},$$
$$[o_5]_{a_2} = \{o_5, o_{20}\}, [o_6]_{a_2} = \{o_6\}, [o_7]_{a_2} = \{o_1, o_4, o_7\},$$
$$[o_8]_{a_2} = \{o_8\}, [o_9]_{a_2} = \{o_9\}, [o_{10}]_{a_2} = \{o_{10}\},$$
$$[o_{11}]_{a_2} = \{o_{11}, o_{18}\}, [o_{12}]_{a_2} = \{o_{12}\}, [o_{13}]_{a_2} = \{o_3, o_{13}\},$$
$$[o_{14}]_{a_2} = \{o_{14}\}, [o_{15}]_{a_2} = \{o_{15}\}, [o_{16}]_{a_2} = \{o_2, o_3, o_{16}\},$$
$$[o_{17}]_{a_2} = \{o_{17}\}, [o_{18}]_{a_2} = \{o_{11}, o_{18}\}, [o_{19}]_{a_2} = \{o_{19}\}, [o_{20}]_{a_2} = \{o_5, o_{20}\}.$$

Formulae (3) and (4) derives the following approximations:

$$\underline{apr}_{a_2}(O) = \{o_2, o_8, o_{15}, o_{16}, o_{17}\},$$

$$\overline{apr}_{a_2}(O) = \{o_1, o_2, o_3, o_4, o_8, o_{13}, o_{15}, o_{16}, o_{17}\}.$$

In continuous information table $T2$, objects are arranged in ascending order of values of attribute a_2 and each object is given a serial superscript from 1 to 20. Using objects with superscripts, the two approximations are rewritten:

$$\underline{apr}_{a_2}(O) = \{o_8^7, o_{15}^8, o_{17}^{10}, o_2^{11}, o_{16}^{12}\},$$

$$\overline{apr}_{a_2}(O) = \{o_1^5, o_4^6, o_8^7, o_{15}^8, o_{17}^{10}, o_2^{11}, o_{16}^{12}, o_3^{13}, o_{13}^{14}\},$$

Consistent combined rules from collections $\{o_8^7, o_{15}^8\}$ and $\{o_{17}^{10}, o_2^{11}, o_{16}^{12}\}$ are

$$a_2 = [2.10, 2.28] \rightarrow a_3 = [4.23, 4.50],$$
$$a_2 = [2.50, 2.64] \rightarrow a_3 = [4.23, 4.50],$$

where $a_2(o_8^7) = 2.10$, $a_2(o_{15}^8) = 2.28$, $a_2(o_{17}^{10}) = 2.50$, and $a_2(o_{16}^{12}) = 2.64$. Inconsistent combined rules from collections $(o_1^5, o_4^6, o_8^7, o_{15}^8)$ and $\{o_{17}^{10}, o_2^{11}, o_{16}^{12}, o_3^{13}, o_{13}^{14})$ are

$$a_2 = [1.97, 2.28] \rightarrow a_3 = [4.23, 4.50],$$
$$a_2 = [2.50, 2.70] \rightarrow a_3 = [4.23, 4.50],$$

where $a_2(o_1^5) = 1.97$ and $a_2(o_{13}^{14}) = 2.70$.

Example 7 shows that a combined rule has higher applicability than single rules. For example, by using the consistent combined rule $a_2 = [2.10, 2.28] \rightarrow a_3 = [4.23, 4.50]$, we can say that an object with attribute a_2 value 2.16 supports this rule, because 2.16 is included in $[2.10, 2.28]$. On the other hand, by using single rules $a_2 = 2.10 \rightarrow a_3 = [4.23, 4.50]$ and $a_2 = 2.28 \rightarrow a_3 = [4.23, 4.50]$, we cannot say what rule the object supports, because 2.16 is discernible with both 2.10 and 2.28 under threshold 0.05.

5 Rule Induction in Incomplete and Continuous Information Tables

When O is specified by restriction $B = X$, we can say for rules induced from objects in approximations as follows:

- Object $o \in \underline{Capr}_A(O)$ certainly supports rule $A = A(o) \rightarrow B = X$ consistently.
- Object $o \in \overline{Capr}_A(O)$ certainly supports rule $A = A(o) \rightarrow B = X$ inconsistently.
- Object $o \in \underline{Papr}_A(O)$ possibly supports $A = A(o) \rightarrow B = X$ consistently.
- Object $o \in \overline{Papr}_A(O)$ possibly supports $A = A(o) \rightarrow B = X$ inconsistently.

We create combined rules from these single rules. Let U_a^C be the set of objects with complete and continuous information for attribute a and U_a^I be one with incomplete and continuous information.

$$U_A^C = \cap_{a \in A} U_a^C, \tag{35}$$
$$U_A^I = \cup_{a \in A} U_a^I. \tag{36}$$

A combined rule is represented by:

$$(A = [l_A, u_A] \rightarrow B = X) = (\wedge_{a \in A}(a = [l_a, u_a]) \rightarrow B = X). \tag{37}$$

The following treatment is done for each attribute $a \in A$ $a \in U_a^C$ is arranged in ascending order of $a(o)$ and is given a serial superscript from 1 to $|U_a^C|$. Objects in $(Capr_A(O) \cap U_a^C)$, in $(\overline{Capr}_A(O) \cap U_a^C)$, in $(Papr_A(O) \cap U_a^C)$, and in $(\overline{Papr}_A(O) \cap U_a^C)$ are arranged in ascending order of attribute a values, respectively. And then the objects are expressed by collections of objects with serial superscripts like $\{\cdots, o_{i_h}^h, o_{i_{h+1}}^{h+1}, \cdots, o_{i_{k-1}}^{k-1}, o_{i_k}^k, \cdots\}$ ($h \le k$). From collection $(o_{i_h}^h, o_{i_{i+1}}^{h+1}, \cdots, o_{i_{k-1}}^{k-1}, o_{i_k}^k)$, the antecedent part for a of the combined rule expressed by $A = [l_A, u_A] \rightarrow B = X$ is created. For a certain and consistent combined rule,

$$l_a = \min(a(o_{i_h}^h), \min_Y e) \text{ and } u_a = \max(a(o_{i_k}^k), \max_Y e),$$

$$Y = \begin{cases} e < a(o_{i_{k+1}}^{k+1}), & \text{for } h = 1 \wedge k \ne |U_a^C| \\ a(o_{i_{h-1}}^{h-1}) < e < a(o_{i_{k+1}}^{k+1}), & \text{for } h \ne 1 \wedge k \ne |U_a^C| \\ a(o_{i_{h-1}}^{h-1}) < e, & \text{for } h \ne 1 \wedge k = |U_a^C| \end{cases}$$

$$\text{with } e \in a(o') \wedge o' \in Z, \tag{38}$$

where Z is $(Capr_A(O) \cap U_a^I)$.

In the case of certain and inconsistent, possible and consistent, possible and inconsistent combined rules, Z is $(\overline{Capr}_A(O) \cap U_a^I)$, $(Papr_A(O) \cap U_a^I)$, and $(\overline{Papr}_A(O) \cap U_a^I)$, respectively.

Proposition 8 *Let Cr be the set of combined rules induced from $Capr_A(O)$ and Pr the set from $Papr_A(O)$. When O is specified by restriction $B = X$, if $(A = [l_A, u_A] \rightarrow B = X) \in Cr$, then $\exists l_A' \le l_A, \exists u_A' \ge u_A$ $(A = [l_A', u_A'] \rightarrow B = X) \in Pr$.*

Proof A single rule created from $Capr_A(O)$ is also derived from $Papr_A(O)$ because of $Capr_A(O) \subseteq Papr_A(O)$. This means that the proposition holds.

Proposition 9 *Let $\overline{Cr}$ be the set of combined rules induced from $\overline{Capr}_A(O)$ and $\overline{Pr}$ the set from $\overline{Papr}_A(O)$. When O is specified by restriction $B = X$, if $(A = [l_A, u_A] \rightarrow B = X) \in \overline{Cr}$, then $\exists l_A' \le l_A, \exists u_A' \ge u_A$ $(A = [l_A', u_A'] \rightarrow B = X) \in \overline{Pr}$.*

Proof The proof is similar to one for Proposition 8.

$IT2$

U	a_1	a_2	a_3	a_4
o_1	$< 4.07, 4.12 >$	2.98	$[3.02, 3.17]$	$\{x, y\}$
o_2	3.95	$< 3.64, 3.65 >$	3.44	x
o_3	3.34	$[3.69, 3.72]$	3.28	z
o_4	5.79	$[2.98, 3.12]$	3.52	w
o_5	4.43	2.35	2.67	x
o_6	4.04	4.52	4.07	y
o_7	3.82	2.95	2.91	y
o_8	5.37	3.11	3.49	w
o_9	$< 3.98, 4.23 >$	4.63	4.21	x
o_{10}	4.08	3.78	3.57	y
o_{11}	$[3.97, 3.98]$	3.98	3.68	x
o_{12}	3.64	4.81	4.16	y
o_{13}	4.92	3.71	3.77	w
o_{14}	4.13	2.78	2.88	x
o_{15}	4.06	3.29	3.22	y
o_{16}	3.96	$\{3.35, 3.65\}$	3.44	x
o_{17}	$[3.90, 3.93]$	3.51	$[3.32, 3.40]$	$\{x, y\}$
o_{18}	$[3.46, 3.56]$	3.96	$< 3.49, 3.51 >$	z
o_{19}	$[4.87, 4.93]$	3.44	3.57	$< w, x >$
o_{20}	4.43	2.35	2.67	x

Fig. 5 Information table $IT2$ containing incomplete information

Proposition 10 *Let $C\underline{r}$ be the set of combined rules induced from $C\underline{apr}_A(O)$ and $C\overline{r}$ the set from $C\overline{apr}_A(O)$. When O is specified by restriction $B = X$, if $(\overline{A} = [l_A, u_A] \rightarrow B = X) \in C\underline{r}$, then $\exists l'_A \leq l_A, \exists u'_A \geq u_A$ $(A = [l'_A, u'_A] \rightarrow B = X) \in C\overline{r}$.*

Proof The proof is similar to one for Proposition 8.

Proposition 11 *Let $P\underline{r}$ be the set of combined rules induced from $P\underline{apr}_A(O)$ and $P\overline{r}$ the set from $P\overline{apr}_A(O)$. When O is specified by restriction $B = X$, if $(\overline{A} = [l_A, u_A] \rightarrow B = X) \in P\underline{r}$, then $\exists l'_A \leq l_A, \exists u'_A \geq u_A$ $(A = [l'_A, u'_A] \rightarrow B = X) \in P\overline{r}$.*

Proof The proof is similar to one for Proposition 8.

Example 8 Let O be specified by restriction $a_4 = x$ in $IT2$ of Fig. 5.

$$CO_{a_4=x} = \{o_2, o_5, o_9, o_{11}, o_{14}, o_{16}, o_{20}\},$$
$$PO_{a_4=x} = \{o_1, o_2, o_5, o_9, o_{11}, o_{14}, o_{16}, o_{17}, o_{19}, o_{20}\}.$$

Each $C[o_i]_{a_1}$ with $i = 1, \ldots, 20$ is, respectively,

$$C[o_1]_{a_1} = \{o_1, o_{10}\}, C[o_2]_{a_1} = \{o_2, o_{11}, o_{16}, o_{17}\},$$
$$C[o_3]_{a_1} = \{o_3\}, C[o_4]_{a_1} = \{o_4\}, C[o_5]_{a_1} = \{o_5, o_{20}\},$$
$$C[o_6]_{a_1} = \{o_6, o_{10}, o_{15}\}, C[o_7]_{a_1} = \{o_7\},$$
$$C[o_8]_{a_1} = \{o_8\}, C[o_9]_{a_1} = \{o_9\},$$
$$C[o_{10}]_{a_1} = \{o_1, o_6, o_{10}, o_{14}, o_{15}\},$$
$$C[o_{11}]_{a_1} = \{o_2, o_{11}, o_{16}\}, C[o_{12}]_{a_1} = \{o_{12}\},$$
$$C[o_{13}]_{a_1} = \{o_{13}, o_{19}\}, C[o_{14}]_{a_1} = \{o_{10}, o_{14}\},$$
$$C[o_{15}]_{a_1} = \{o_6, o_{10}, o_{15}\}, C[o_{16}]_{a_1} = \{o_2, o_{11}, o_{16}\},$$
$$C[o_{17}]_{a_1} = \{o_2, o_{17}\}, C[o_{18}]_{a_1} = \{o_{18}\},$$
$$C[o_{19}]_{a_1} = \{o_{13}, o_{19}\}, C[o_{20}]_{a_1} = \{o_5, o_{20}\}$$

Each $P[o_i]_{a_1}$ with $i = 1, \ldots, 20$ is, respectively,

$$P[o_1]_{a_1} = \{o_1, o_6, o_{10}, o_{14}, o_{15}\},$$
$$P[o_2]_{a_1} = \{o_2, o_9, o_{11}, o_{16}, o_{17}\},$$
$$P[o_3]_{a_1} = \{o_3\}, P[o_4]_{a_1} = \{o_4\}, P[o_5]_{a_1} = \{o_5, o_{20}\}$$
$$P[o_6]_{a_1} = \{o_1, o_6, o_{10}, o_{15}\}, P[o_7]_{a_1} = \{o_7\}, P[o_8]_{a_1} = \{o_8\},$$
$$P[o_9]_{a_1} = \{o_2, o_9, o_{11}, o_{16}, o_{17}\},$$
$$P[o_{10}]_{a_1} = \{o_1, o_6, o_{10}, o_{14}, o_{15}\},$$
$$P[o_{11}]_{a_1} = \{o_2, o_9, o_{11}, o_{16}, o_{17}\}, P[o_{12}]_{a_1} = \{o_{12}\},$$
$$P[o_{13}]_{a_1} = \{o_{13}, o_{19}\}, P[o_{14}]_{a_1} = \{o_1, o_{10}, o_{14}\},$$
$$P[o_{15}]_{a_1} = \{o_1, o_6, o_{10}, o_{15}\},$$
$$P[o_{16}]_{a_1} = \{o_2, o_9, o_{11}, o_{16}, o_{17}\},$$
$$P[o_{17}]_{a_1} = \{o_1, o_2, o_9, o_{11}, o_{16}, o_{17}\}, P[o_{18}]_{a_1} = \{o_{18}\},$$
$$P[o_{19}]_{a_1} = \{o_{13}, o_{19}\}, P[o_{20}]_{a_1} = \{o_5, o_{20}\}.$$

Four approximations are:

$$\underline{Capr}_{a_1}(O) = \{o_5, o_{20}\},$$
$$\underline{Papr}_{a_1}(O) = \{o_2, o_5, o_9, o_{11}, o_{16}, o_{17}, o_{20}\},$$
$$\overline{Capr}_{a_1}(O) = \{o_2, o_5, o_9, o_{10}, o_{11}, o_{14}, o_{16}, o_{17}, o_{20}\},$$
$$\overline{Papr}_{a_1}(O) = \{o_1, o_2, o_5, o_6, o_9, o_{10}, o_{11}, o_{13}, o_{14}, o_{15}, o_{16}, o_{17}, o_{19}, o_{20}\}.$$

$$U^C_{a_1} = \{o_2, o_3, o_4, o_5, o_6, o_7, o_8, o_{10}, o_{12}, o_{13}, o_{14}, o_{15}, o_{16}, o_{20}\},$$
$$U^I_{a_1} = \{o_1, o_9, o_{11}, o_{17}, o_{18}, o_{19}\}$$

Objects in $U_{a_1}^C$ are arranged in ascending order of $a_1(o)$ like this:

$$o_3, o_{12}, o_7, o_2, o_{16}, o_6, o_{15}, o_{10}, o_{14}, o_5, o_{20}, o_{13}, o_8, o_4$$

Giving serial superscripts to these objects,

$$o_3^1, o_{12}^2, o_7^3, o_2^4, o_{16}^5, o_6^6, o_{15}^7, o_{10}^8, o_{14}^9, o_5^{10}, o_{20}^{11}, o_{13}^{12}, o_8^{13}, o_4^{14}.$$

And then, the four approximations are rewritten like these:

$$\underline{Capr}_{a_1}(O) = \{o_5^{10}, o_{20}^{11}\},$$
$$\underline{Papr}_{a_1}(O) = \{o_2^4, o_{16}^5, o_5^{10}, o_{20}^{11}, o_9, o_{11}, o_{17}\},$$
$$\overline{Capr}_{a_1}(O) = \{o_2^4, o_{16}^5, o_{10}^8, o_{14}^9, o_5^{10}, o_{20}^{11}, o_9, o_{11}, o_{17}\},$$
$$\overline{Papr}_{a_1}(O) = \{o_2^4, o_{16}^5, o_6^6, o_{15}^7, o_{10}^8, o_{14}^9, o_5^{10}, o_{20}^{11}, o_{13}^{12}, o_1, o_9, o_{11}, o_{17}, o_{19}\}.$$

Objects are separated into two parts: ones with a superscript and ones with only a subscript; namely, ones having complete information and ones having incomplete information for attribute a_1, respectively. That is,

$$\underline{Capr}_{a_1}(O) \cap U_{a_1}^C = \{o_5^{10}, o_{20}^{11}\},$$
$$\underline{Capr}_{a_1}(O) \cap U_{a_1}^I = \emptyset,$$
$$\underline{Papr}_{a_1}(O) \cap U_{a_1}^C = \{o_2^4, o_{16}^5, o_5^{10}, o_{20}^{11}\},$$
$$\underline{Papr}_{a_1}(O) \cap U_{a_1}^I = \{o_9, o_{11}, o_{17}\},$$
$$\overline{Capr}_{a_1}(O) \cap U_{a_1}^C = \{o_2^4, o_{16}^5, o_{10}^8, o_{14}^9, o_5^{10}, o_{20}^{11}\},$$
$$\overline{Capr}_{a_1}(O) \cap U_{a_1}^I = \{o_9, o_{11}, o_{17}\},$$
$$\overline{Papr}_{a_1}(O) \cap U_{a_1}^C = \{o_2^4, o_{16}^5, o_6^6, o_{15}^7, o_{10}^8, o_{14}^9, o_5^{10}, o_{20}^{11}, o_{13}^{12}\},$$
$$\overline{Papr}_{a_1}(O) \cap U_{a_1}^I = \{o_1, o_9, o_{11}, o_{17}, o_{19}\}.$$

From these expressions and formula (38), four kinds of combined rules are created. A certain and consistent rule is:

$$a_1 = 4.43 \rightarrow a_4 = x.$$

Possible and consistent rules are:

$$a_1 = [3.90, 3.98] \rightarrow a_4 = x,$$
$$a_1 = [4.23, 4.43] \rightarrow a_4 = x.$$

Certain and inconsistent rules are:

$$a_1 = [3.90, 3.98] \rightarrow a_4 = x,$$
$$a_1 = [4.08, 4.43] \rightarrow a_4 = x.$$

A possible and inconsistent rule is:

$$a_1 = [3.90, 4.93] \rightarrow a_4 = x.$$

6 Conclusions

We have described rough sets that consist of lower and upper approximations and rule induction from the rough sets in continuous information tables.

First, we have handled complete and continuous information tables. Rough sets are derived directly using the indiscernibility relation on a set of attributes.

Second, we have coped with incomplete and continuous information tables under possible world semantics. We use a possible indiscernibility relation as a possible world. This is because the number of possible indiscernibility relations is finite, although the number of possible tables, which is traditionally used under possible world semantics, is infinite. The family of possible indiscernibility relations has a lattice structure with the minimum and the maximum elements. The families of lower and upper approximations that are derived from each possible indiscernibility relation also have a lattice structure for set inclusion. The approximations are obtained by using the minimum and the maximum possible indiscernibility relations. Therefore, we have no difficulty of computational complexity for the number of attribute values with incomplete information, although the number of possible indiscernibility relations increases exponentially as the number of values with incomplete information grows linearly.

Consequently, we derive four kinds of approximations. These approximations are the same as those obtained from an extended approach directly using indiscernibility relations. Therefore, this justifies the extended approach in our previous work.

From these approximations, we derive four kinds of single rules that are supported by individual objects. These single rules have weak applicability. To improve the applicability, we have brought serial single rules into one combined rule. The combined rule has greater applicability than the single ones that are used to create it.

References

1. Abiteboul, S., Hull, R., Vianu, V.: Foundations of Databases. Addison-Wesley Publishing Company (1995)
2. Bosc, P., Duval, L., Pivert, O.: An initial approach to the evaluation of possibilistic queries addressed to possibilistic databases. Fuzzy Sets and Syst. **140**, 151–166 (2003)

3. Grahne, G.: The problem of incomplete information in relational databases. Lect. Notes Comput. Sci. **554** (1991)
4. Grzymala-Busse, J.W.: Mining numerical data a rough set approach. In: Peters, J.F., Skowron, A. (eds.) Transactions on Rough Sets XI. LNCS, vol. 5946, pp. 1–13. Springer, Heidelberg (2010). https://doi.org/10.1007/978-3-642-11479-3_1
5. Hu, M.J., Yao, Y.Y.: Rough set approximations in an incomplete information table. In: Polkowski, L., Yao, Y., Artiemjew, P., Ciucci, D., Liu, D., Ślęzak, D., Zielosko, B. (eds.) IJCRS 2017. LNCS (LNAI), vol. 10314, pp. 200–215. Springer, Cham (2017). https://doi.org/10.1007/978-3-319-60840-2_14
6. Imielinski, T., Lipski, W.: Incomplete information in relational databases. J. ACM **31**, 761–791 (1984)
7. Jing, S., She, K., Ali, S.: A universal neighborhood rough sets model for knowledge discovering from incomplete heterogeneous data. Expert Syst. **30**(1), 89–96 (2013). https://doi.org/10.1111/j.1468-0394.2012.00633_x
8. Kryszkiewicz, M.: Rules in incomplete information systems. Inf. Sci. **113**, 271–292 (1999)
9. Libkin, L., Wong, L.: Semantic representations and query languages for or-sets. J. Comput. Syst. Sci. **52**, 125–142 (1996)
10. Lin, T.Y.: Neighborhood systems: a qualitative theory for fuzzy and rough sets. In: Wang, P. (ed.) Advances in Machine Intelligence and Soft Computing, vol. IV, pp. 132–155. Duke University (1997)
11. Nakata, M., Sakai, H.: Applying rough sets to information tables containing missing values. In: Proceedings of 39th International Symposium on Multiple-Valued Logic, pp. 286–291. IEEE Press (2009). https://doi.org/10.1109/ISMVL.2009.1
12. Nakata, M., Sakai, H.: Twofold rough approximations under incomplete information. Int. J. Gen. Syst. **42**, 546–571 (2013). https://doi.org/10.1080/17451000.2013.798898
13. Nakata, M., Sakai, H.: Describing rough approximations by indiscernibility relations in information tables with incomplete information. In: Carvalho, J.P., Lesot, M.-J., Kaymak, U., Vieira, S., Bouchon-Meunier, B., Yager, R.R. (eds.) IPMU 2016, Part II. CCIS, vol. 611, pp. 355–366. Springer, Cham (2016). https://doi.org/10.1007/978-3-319-40581-0_29
14. Nakata, M., Sakai, H., Hara, K.: Rules induced from rough sets in information tables with continuous values. In: Medina, J., Ojeda-Aciego, M., Verdegay, J.L., Pelta, D.A., Cabrera, I.P., Bouchon-Meunier, B., Yager, R.R. (eds.) IPMU 2018, Part II. CCIS, vol. 854, pp. 490–502. Springer Cham (2018). https://doi.org/10.1007/978-3-319-91476-3_41
15. Nakata, M., Sakai, H., Hara, K.: Rule induction based on indiscernible classes from rough sets in information tables with continuous Values. In: Nguyen, H.S., et al. (eds.) IJCRS 2018, LNAI 11103, pp. 323-336. Springer (2018). https://doi.org/10.1007/978-3-319-99368-3_25
16. Nakata, M., Sakai, H., Hara, K.: Rough sets based on possibly indiscernible Classes in Incomplete Information Tables with Continuous Values. In: Hassanien, A.B., et al. (eds.) Proceedings of the International Conference on Advanced Intelligent Systems and Informatics 2019, Advances in Intelligent Systems and Computing 1058, pp. 13–23. Springer (2019). https://doi.org/10.1007/978-3-030-31129-2_2
17. Paredaens, J., De Bra, P., Gyssens, M., Van Gucht, D.: The Structure of the Relational Database Model. Springer-Verlag (1989)
18. Pawlak, Z.: Rough Sets: Theoretical Aspects of Reasoning about Data. Kluwer Academic Publishers, Dordrecht (1991). https://doi.org/10.1007/978-94-011-3534-4
19. Stefanowski, J., Tsoukiàs, A.: Incomplete information tables and rough classification. Comput. Intell. **17**, 545–566 (2001)
20. Yang, X., Zhang, M., Dou, H., Yang, Y.: Neighborhood systems-based rough sets in incomplete information system. Inf. Sci. **24**, 858–867 (2011). https://doi.org/10.1016/j.knosys.2011.03.007
21. Zenga, A., Lia, T., Liuc, D., Zhanga, J., Chena, H.: A fuzzy rough set approach for incremental feature selection on hybrid information systems. Fuzzy Sets Syst. **258**, 39–60 (2015). https://doi.org/10.1016/j.fss.2014.08.014

22. Zhao, B., Chen, X., Zeng, Q.: Incomplete hybrid attributes reduction based on neighborhood granulation and approximation. In: 2009 International Conference on Mechatronics and Automation, pp. 2066–2071. IEEE Press (2009)
23. Zimányi, E., Pirotte, A.: Imperfect Information in Relational Databases. In: Motro, A., Smets, P. (eds.) Uncertainty Management in Information Systems: From Needs to Solutions, pp. 35–87. Kluwer Academic Publishers (1997)

Big Data Analytics and Preprocessing

Noha Shehab, Mahmoud Badawy, and Hesham Arafat

Abstract Big data is a trending word in the industry and academia that represents the huge flood of collected data, this data is very complex in its nature. Big data as a term used to describe many concepts related to the data from technological and cultural meaning. In the big data community, big data analytics is used to discover the hidden patterns and values that give an accurate representation of the data. Big data preprocessing is considered an important step in the analysis process. It a key to the success of the analysis process in terms of analysis time, utilized resources percentage, storage, the efficiency of the analyzed data and the output gained information. Preprocessing data involves dealing with concepts like concept drift, data streams that are considered as significant challenges.

Keywords Big data · Big data analytics · Big data preprocessing · Concept drift

1 Introduction

In recent years internet application usage is rapidly increasing in most life aspects as it makes life simpler. This results in overwhelming quantities of data that needed to be processed and shared between people, applications, and companies to make the best benefit of the current technology revolution. Data's overwhelming quantity poses a great challenge regarding the means of managing and efficiently using the data for different applications which led to the "Big data" term [1]. Big data refers to any set of data that, with traditional systems, would require large capabilities in

N. Shehab (✉) · M. Badawy · H. Arafat
Computers Engineering and Control Systems Department, Faculty of Engineering, Mansoura University, Mansoura, Egypt
e-mail: noha.a.shehab@students.mans.edu.eg

M. Badawy
e-mail: engbadawy@gmail.com

H. Arafat
e-mail: h_arafat_ali@mans.edu.eg

© The Author(s), under exclusive license to Springer Nature Switzerland AG 2021   25
A. E. Hassanien et al. (eds.), *Machine Learning and Big Data Analytics Paradigms: Analysis, Applications and Challenges*, Studies in Big Data 77,
https://doi.org/10.1007/978-3-030-59338-4_2

terms of storage space and time to be analyzed [2]. It is the big volume of the data sets in large numbers which may be inconsistence or have incorrect values or both [3].

Big data also produces new opportunities for discovering new values, helps us to deeply understand of the hidden values, and incurs new challenges, e.g., how to effectively organize and manage such datasets [4]. These collected datasets consists of structured, unstructured and semi-structured ones that need to be processed [5], but processing these huge amounts is a trail of madness because of two mainly reasons the processing itself is not an easy task and data may not be important as it may include noise, uncompleted or incorrect messages which consume processing power and time without any advantage, so data preprocessing and reduction have become essential techniques in current knowledge discovery scenarios, dominated by increasingly large datasets [6]. These data cannot be straightforwardly treated by people or manual applications to acquire sorted out knowledge which prompted the improvement of data management and data mining systems. The rest of this chapter is organized as follows; Sect. 2 discusses big data as a concept. Section 3 discusses data analytics related concepts. Section 4 goes deep in describing the big data preprocessing and terms related to data streams and concept drift.

2 Big Data

The fast development rate of information technology, electronic gadgets mushroomed in all fields of life like health care, education, environment service, agriculture, spacing, climate figures, social media, military, web-based life, and industry. Data are collected from everything in the nonstop behavior process as shown in Fig. 1, low-cost

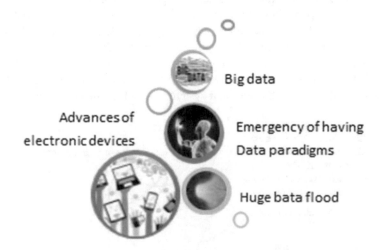

Fig. 1 What caused big data

and widespread sensing and a dramatic drop in data storage costs have significantly expanded the amount of effortlessly extractable information with unprecedented speed and volume. In the past decade, the volume of data has increased sharply, which is currently categorized as "big data" [7].

As indicated by a report from International Business Machine (IBM), Two quintillion and half bytes of data are created every day, and 90% of the data in the world today were produced within the past two years [1]. Big data is not a standalone technology, but it is an aftereffect of the last fifty years or more of the technology revolution. Big data is the amount of data beyond the ability of technology to store, manage and process efficiently [2]. "Big data is a term that describes large volumes of high velocity, complex and variable data that require advanced techniques and technologies to enable the capture, storage, distribution, management, and analysis of the information." [6]. Big data is a process of gathering, management and analysis of data to generate knowledge and reveal hidden patterns [8]. Datasets that could not be captured, managed and processed by general computers within an acceptable scope, Hadoop definition [9]. All Data that be collected around of us is considered a big data source. Table 1 shows some sources of unstructured data that boosted the big data era [9].

It's clearly observed that big data differs from traditional data in volume, data structure, rate and other factors illustrated in Table 2 [9]. Addressing big data is a

Table 1 Unstructured big data sources

Data source	Production
Apple devices	47,000 applications/minute are downloaded
Facebook application	34,722 likes/minute are registers 100 Terabytes/day of data is uploaded
Google products	Over 2 million search queries/minute are executed
Instagram	40 million photos/day are shared by users
Twitter	More than 654 users are using twitter More than 175 million tweets/day are generated
WordPress application	350 blogs/minute is published by bloggers

Table 2 Differences between traditional and big data

Comparison	Traditional data	Big data
Volume	In GBs	TBs, PBs, and ZBs
Data generation rate	Per hour/day	Faster than traditional data
Data structure	Structured	Structured, unstructured, semi-structured
Data source	Centralized/distributed	Fully distributed
Data integration	Easy	Difficult
Storage technology	RDBMS	HDPS + NoSQL
Access	Interactive	Batch/real-time

challenging and time-demanding task that requires a huge computational infrastructure to guarantee effective and successful data processing and analysis [10]. To be addressed, we must know more about its characteristics the best model that could clearly present them is Big Data multi V Model.

Managing these overwhelming quantities of data cannot be efficiently processed without understanding the characteristics of big data which can be summarized in the Vs. model (Variety, Velocity, Volume, Veracity, Value) of data at which scalability begins to bind. Dealing with the multi v model gives the measure of the correctness and accuracy of information retrieved as shown in Fig. 2 [11].

Extracting valuable information from data via big data value chain that can be summarized into four stages as shown in Fig. 3. Data generation, data storage, data acquisition, and data analysis [12]. First data generation describes how data is being generated from their resources. This generated data needs to be stored so the data

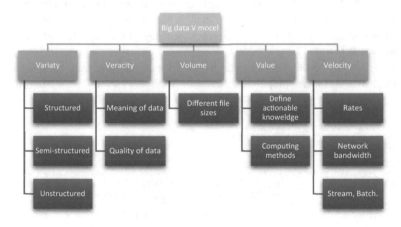

Fig. 2 Big data V model

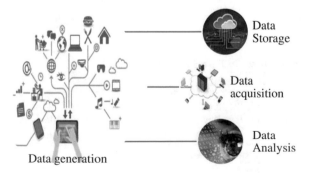

Fig. 3 Big data value chain

storage challenges raised as it describes how can available technology and storage devices handle storing such a huge amount of data. After storing data, it needs to acquiesce. Data acquisition is responsible for collecting and integrating data to be passed to the analysis stage.

The expanding data flooding acquired numerous challenges in all operations applied to that collected data to separate useful information. Challenges can be divided into three main categories data challenges, processing challenges, and management challenges.

Big data is being involved in most of the fields like learning, communication, health, and many other fields. The Internet has a great impact on the learning process whatever you want to learn you find many tutorials that could help. Researchers give great attention to use big data in learning resources and discuss the influence of using it on implementing and supporting learning management via extracting and modifying information to fit the environment of online learning resources (OLR). As data differs in size and types 'variety, velocity and volume. Their work based on big data analysis to provide better online learning resources according to the learners' interests. They proposed a big data emerging technology in supporting resources for the online learning framework to enhance the teaching and learning process [13].

In communication technologies, the spread of using mobile devices not only as a communication way but as a data sharing and storing one also. Most of the users use short-range wireless in sharing data between others because of its high accuracy in delivering and the free cost. Offloading mobile traffic and disseminating content among mobile users by integrating big data technologies are raising up, a device to device framework is proposed to enhance this sharing process in terms of time and number of users. This framework mainly implemented using Hadoop, Spark, and MLlib [14]. Making good use of medical big data (MBD) by analyzing the patients' health information using data mining frameworks [15]. DNA plays a great role in disease diagnoses and the changing of copy number is a type of genetic variation in the human genome which reflects cancer's stage. Using big data analytics and machine learning algorithms with DNA to detect these changes earlier detection of changing increases the chances of medicament [16–19].

In data science, monitoring, visualizing, securing, and reducing the processing cost of high-volume real-time data events and the prediction of future based on big data analysis process results is still a critical issue [20]. Mining algorithms that can be applied to special data, streaming and real-time current infrastructures, dealing with large quantities and varieties and its direct impact on the industry [21]. And whiten social life, using big data analytics helps in a better understanding of the social environment and social change in general. Converting social data to information and information to knowledge helps in better understanding of social problems and implement effective solutions [22]. The social behavior of humans affects the economic systems analyzing the citizens' behaviors' and identifying their needs opens new business areas [23].

3 Big Data Analytics

Term big data analytics can be defined as the application of advanced analytic techniques including data mining, statistical analysis, predictive analytics, etc. [12]. It refers to the processes of examining and analyzing huge amounts of data with variable types to draw a conclusion by uncovering hidden patterns and correlations, trends, and other business valuable information and knowledge, in order to increase business benefits, increase operational efficiency and explore new market and opportunities. Big data analytics permits users to capture, store, and analyze the huge amount of data, from internal as well as external of the organization, from multiple sources such as the corporate database, sensor-captured data such as RFID, mobile-phone records and locations, and internet in order to understand the meaningful insights.

Data analysis introduced to help analyze them using data analytics frameworks that have been built to scale out parallel executions for the purpose of constructing valuable information from big data [24]. As a result, it affords the potential for shifting the collective behavior toward alternative techniques, such as a focus on precise parameter estimation because traditional significance testing is often uninformative in very large datasets [25]. Also, it is defined to be a channel of acquisition, extraction, cleaning, integration, aggregation, and visualization, analysis and modeling, and interpretation.

The analysis process involves five main steps data integration, data management, data preprocessing, data mining and knowledge presentation that are independent of the application field analytics used in [3]. Big data analytic concepts can be classified into three main stages descriptive analytics, predictive analytics, and prescriptive analytics as shown in Fig. 4 [26]. Descriptive analytics helps in understanding what already happened to the data, predictive analytics helps in anticipating what will happen to it and prescriptive analytics helps in responding now what and so what. Advanced analytics combines predictive and prescriptive analytics to provide a

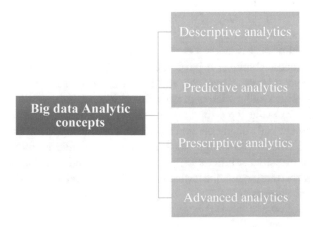

Fig. 4 Big data analytics concepts

forward-looking perspective and make use of techniques encompassing a wide range of disciplines including simulation, learning, statistics, machine, and optimization [27]. Deploying and working with big data system analysis faces challenges that can be classified as data collection, data analytics, and analytic systems issues.

Data collection and management, a massive amount of data is collected which needs to be represented by removing redundant data and unwanted ones in the preprocessing stage. Data analytics, real-time data stream makes data analysis more difficult so the main dependence on an approximation which has two notations approximation accuracy and the output of this approximation, but the data is complex so the need for using sophisticated technologies like machine learning raises. Systems issues as an analytic process deal with large distributed amounts of data that confront some common issues like energy management, scalability, and collaboration. Energy management, the energy consumption of large-scale computing systems has attracted greater concern from economic and environmental perspectives. Scalability, all the components in big data systems must be capable of scaling to address the ever-growing size of complex datasets. Collaboration, dealing with these large amounts of data require specialists from multiple professional fields collaborating to mine hidden values that save time and effort [28].

3.1 Big Data Analytics Methods and Architecture

Big data analysis methods can be classified into classical methods and modern ones. Classical methods are text analytics, audio analytics, and video analytics. Text analytics (text mining) refers to methods that extract information from textual data. Social network feeds, emails, blogs, online forums, survey responses, corporate documents, news, and other data sources. Audio analytics analyze and extract information from unstructured audio data [29]. When applied to human spoken language, audio analytics is also referred to as speech analytics. Speech analytics follows two common technological approaches to the transcript-based approach and the phonetic-based approach. Large-vocabulary continuous speech recognition (LVCSR) systems follow two-phase process indexing, searching. Phonetic-based systems work with sounds or phonemes. Video analytics, known as video content analysis (VCA), includes a variety of techniques to monitor, analyze, and extract significant information from video streams. Modern data analysis methods such as random forests, artificial neural networks, and support vector machines are ideally suited for identifying both linear and nonlinear patterns in the data [30].

Big data analytics architecture is rooted in the concept of data lifecycle framework that starts with data capture, proceeds via data transformation, and culminates with data consumption. Big data analytics capability refers to the ability to manage a huge volume of disparate data to allow users to implement data analysis and reaction. Big data analytics capability is defined as the ability to gather an enormous variety of data—structured, unstructured and semi-structured data—from current and former customers to gain useful knowledge to support better decision making, to predict customer behavior via predictive analytics software, and to retain valuable customers by providing real-time offers.

Big data analytics capability is defined as "the ability to utilize resources to perform a business analytics task, based on the interaction between IT assets and other firm resources [31]. Best practice big data analytics architecture that is loosely comprised of five major architectural layers as shown in Fig. 5: (1) data, (2) data aggregation, (3) analytics, (4) information exploration, and (5) data governance [30]. The data layer contains all the data sources necessary to provide the insights required to support daily operations and solve business problems [31]. Data is classified into structured data, semi-structured data, and unstructured ones. The data aggregation layer is responsible for handling data from various data sources. In this layer, data is intelligently digested by performing three steps, data acquisition to read data provided from various communication channels, frequencies, sizes, and formats, transformation cleaning, splitting, translating, merging, sorting, and validating data, and storage loaded into the target databases such as Hadoop distributed file systems (HDFS) or in a Hadoop cloud for further processing and analysis [32].

The analytics layer is responsible for processing all kinds of data and performing appropriate analyses. The information exploration layer generates outputs such as various visualization reports, real-time information monitoring, and meaningful business insights derived from the analytics layer to users in the organization. The data governance layer is comprised of master data management (MDM), data life-cycle management, and data security and privacy management. The analytics operation goes as shown in Fig. 6.

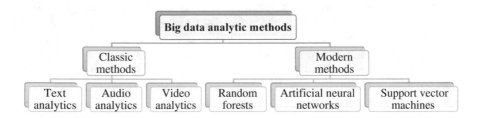

Fig. 5 Big Data analytics methods

Fig. 6 Big data analytics

Target data collected from their sources then stored in distrusted database or clouds, so all are ready to start analyzing this done by different analyzing tools to get useful information and data filtering come at the end of the process to extract useful data according to the target requirement [7]. Researchers pay great attention to the analysis process and its role with scientific and life issues.

3.2 Big Data Analysis Frameworks and Platforms

Many platforms for large-scale processing data were introduced to support big data analysis. Platforms can be divided into (processing, storage, analytic) as shown in Fig. 7. Analytic platforms should be scalable to adapt to the increased amount of data from border perspective like Hadoop, Map Reduce, Hive, PIG, WibiData, Platfora, and Rapidminer [33]. Analytics tools are scalable in two directions horizontal scaling and vertical scaling. Vertical scaling helps in installing more processors, more memory and faster hardware, typically, within a single server and Involves only a single instance of an operating system. Horizontal scaling distributes the workload across many servers. Typically, multiple instances of the operating system are running non-separate machines [34].

Figure 8 summarizes the analytics platforms from a scaling point of view. Mapreduce, Hadoop, Spark, and Apache Flink are examples of big data analytic horizontal scaling platforms [35]. Hadoop MapReduce is an open-source Java-based framework that is designed to provide shared storage and analysis infrastructure and used to handle extensive scale Web applications. Spark is a next-generation paradigm for big data processing developed by researchers at the University of California at Berkeley. It is an alternative to Hadoop which is designed to overcome the disk I/O limitations and improve the performance of earlier systems. Apache Flink is a

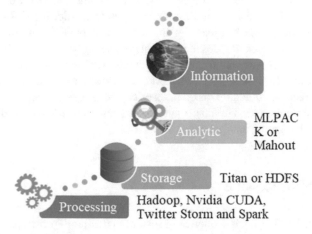

Fig. 7 Platform basic architecture

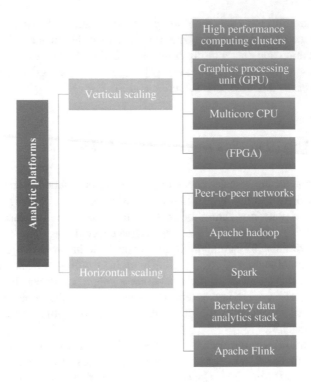

Fig. 8 Scaling analytics platforms

distributed processing framework that concentrates on data stream processing that helps in solving the problem derived from micro-batch processing like in spark. Hadoop is designed to provide shared storage and analysis infrastructure HDFS. Its cluster can store data and perform parallel computation across a large computer cluster having a single master and multiple nodes to extract the information [36]. Hadoop architecture. Spark consists of three main components Driver program and cluster manager and worker nodes that help in processing stream data efficiently [36]. Apache Flink is an opensource framework that has two main operators for describing iterations masters (job manager) and workers iterator (task manager) as shown in Fig. 14. Job manager schedule tasks, determine coordinate checkpoints, coordinate recovery on failures, etc. The task manager executes the tasks of a dataflow, and buffer and exchange the data streams [37].

4   Data Preprocessing

The unbounded and ordered sequence of instances that land over time is called data-stream. Since most real-world databases are highly influenced by negative elements the data streams accompanied by many blemishes such as inconsistencies, missing values, noise and/or redundancies. Streams also come with potentially infinite size, thus it is impossible to store all these incoming floods in the memory, so Extracting knowledge from data involves some major steps as shown in Fig. 8. Preprocessing is a must as it improves the quality of the data and to ensure that the measurement provided is as accurate as possible. Data preprocessing is one of the major phases within the knowledge discovery process which involves more effort and time within the entire data analysis process.

It may be considered as one of the data mining techniques [6]. The traditional data preprocessing methods (e.g., compression, sampling, feature selection, and so on) are expected to operate effectively in the big data age [38]. Data preprocessing for data mining is a set of techniques used prior to the application of a data mining method. It is supposedly the most time-consuming of the whole knowledge discovery phase (anatomy) [39]. Also, it is the conversion of the original useless data into a new and cleaned data prepared for further mining analysis [2].

The preprocessing concept involves data preparation, integration, cleansing, normalization and transformation of data and data reduction tasks [40]. The preprocessing stage can be summarized into two main parts data preparation and data reduction as shown in Fig. 9. Data Preparation is considered as a mandatory step that includes techniques such as integration, normalization, cleaning, and transformation. Data reduction performs simplification by selecting and deleting redundant and noisy features and/or instances or by discretizing complex continuous feature spaces [41]. Preprocessing complexity depends on the data source use [2]. Figure 10 summarizes the most common preprocessing stages and the techniques used in each stage according to the data state [42, 43].

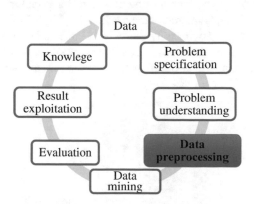

Fig. 9 Extracting knowledge process

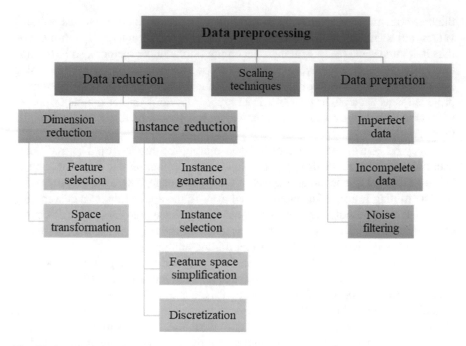

Fig. 10 Preprocessing stages

4.1 Data Preprocessing Related Concepts and Challenges

Preprocessing operation requires an understanding of popular concepts like data streams, concept drift, and data reduction as shown in Fig. 11 [2]. Data streams, Big data comes with a high rate not only the volume of data but also the notion of its velocity. In many real-time situations, we cannot assume that we will deal with a static set of instances. Persistent data flooding leads to the unbounded and ever-growing

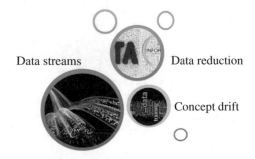

Fig. 11 Preprocessing concepts

dataset. It will expand itself over time and new instances will arrive continuously in batches or one by one. Concept Drift defined as changes in distributions and definitions of learned concepts over time which affects the underlying properties of classes that the learning system aims to discover. Data reduction is the main step task, especially when dealing with large data sets that aims at obtaining an accurate, fast and adaptable model that at the same time is characterized by low computational complexity to quickly respond to incoming objects and changes [44]. It is preferred to apply reduction techniques to elements online or in batch-mode without making any assumptions about data distribution in advance. Data reduction can be applied according to many proposals like data reduction, instance reduction and feature selection space simplification [6].

4.1.1 Concept Drift

Precarious data streams lead to occurrences of the phenomenon called concept drift, the statistical characteristics of the incoming data may change over time [45]. Concept drift locators are external tools used in the aid with the classification module to measure various properties of a data stream, such as standard deviation, predictive error, instance distribution, or stability [6]. A detection signal pings the learning system that the current degree of changes is severe and the old classifier should be replaced by a new one. This solution is also known as explicit drift handling.

Concept drift problem can be detected in many forms, most of the drift detectors work in a two-stage setting, the most common of them are shown in Fig. 12. Concept drift can be detected in the presence of phenomena like sliding windows and online learners and assembler learners as shown in Fig. 13 [6, 46]. Detecting gives the highest importance to the newest ensemble components, and the more sophisticated solutions allow increasing weights of classifiers which results in the best performance, we can say that this algorithm is good if it strikes a balance between the power and memory consumption, recovery and decision time. *Concept drift can be detected from the data distribution of time. According to the speed of change, concept drift can be classified into sudden, incremental, recurring and gradual drift [47].*

4.1.2 Data Reduction

Data mainly collected from IoT sensors, so the data streams could contain noisy ones. Removing noise from data is important. More than 50% of data scientists' work time is spent on data preparation [50]. Dealing with continues incoming data streams is not a straightforward operation that can be done easily due to the following aspects [6]. First, each instance may be only accessed a predetermined number of times and then discarded to best utilize the memory and storage space usage. Second, nowadays real-time is a must, so each Instance must be processed within its limited

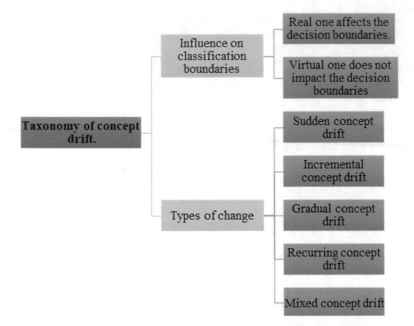

Fig. 12 Concept drift taxonomy

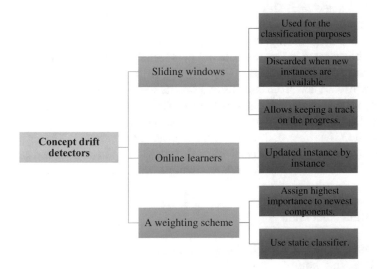

Fig. 13 Concept drift detectors

amount of time to offer real-time responsiveness and keep away from data queuing [48]. Third, accessing genuine class labels is limited due to a high cost of label query for each incoming instance. Fourth, accessing these genuine labels may be overdue as well, in many cases, they are available after a long period. Finally, the statistical characteristics of instances arriving from the stream may be subject to changes over time. So, the need for data reduction raised to transform raw data into high-quality ones which probably fit mining algorithms which mainly done by selecting and deleting redundant and noisy features and/or instances, or by discretizing complex continuous feature spaces. Reduction strategies facilitate processing workload of data and they vary from pure reduction methods to compression ones [43].

Data preprocessing techniques face many challenges like scaling preprocessing techniques that accommodate the current huge amount of data. Data reduction is an important phase in big data preprocessing process that there are many issues that still need more research to be solved or at least enhanced like that all data should be in a balanced state to make benefit of these data, also dealing with the incomplete data streams still an urgent issue as these incomplete ones may have important information which needed to be used in the next phases to get the knowledge of data which is the main goal of big data. Data preparation techniques and new big data learning paradigms need more development that could help in solving data concept drift issues [2].

4.2   Preprocessing Techniques, Models and Technologies

The are many data preprocessing techniques like data cleaning, data imputation data integration, data reduction, and data transformation. Data cleaning corrects inconsistent data and removes noise [40]. Data integration combines data coming from different resources using techniques like data warehouse techniques [40]. Data reduction eliminates or removes redundant or irrelevant properties or samples of data. Data transformation scales data into a certain range to be processed more easily. Each preprocessing technology mainly based on a programming model. Table 3 lists the most common programming models and by which framework they are used.

Investment in developing better algorithms, technologies, and hardware that helps to have preprocessing steps for data as shown in Fig. 14. MLlib is a powerful machine learning library that enables the use of Spark in the data analytics field. This library is formed by two packages Mllib and machine learning [49]. Discretization and normalization, Discretization transforms continuous variables using discrete intervals, whereas normalization performs an adjustment of distributions [2]. Feature extraction used to combine the original set of features to obtain a new set of less-redundant variables. Feature selection tries to select relevant subsets of relevant features without incurring much loss of information. Feature indexers and encoders convert features from one type to another using indexing or encoding techniques [50].

Table 3 Preprocessing programming models

Technique	Facility	Scheme	How it works	Application
MapReduce programming model	Allows distributed	Map and reduce	Map to split data for processing Reduce to rearrange and collect the results	Hadoop
Directed Acyclic Graph Parallel processing	computations in a transparent way for the programmer	Single Instruction Multiple Datasets (SIMD)/ .	Organize jobs by splitting them into a smaller set of atomic tasks Instructions are duplicate and sent from the master to the slave nodes for parallel execution	Spark
Bulk Synchronous Parallel processing	Provides fault tolerance, automatic data partition and management, and automatic	Single Instruction Multiple Datasets	Input data is the starting point. From the start to the end series of steps are applied to portioned data to get the result	Bulk synchronous parallel computing framework for big data applications

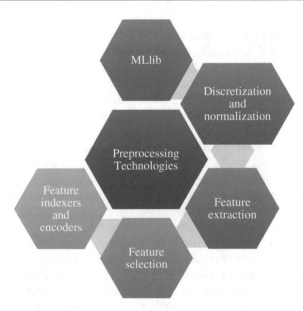

Fig. 14 Preprocessing technologies

Preprocessing technologies' performance and usefulness differ from each other from distinct perspectives like effectiveness, time and memory performance and reduction rate. Effectiveness is a relevant factor in measuring, measured as the number of correctly classified instances divided by the total number of instances in the training set (accuracy). Time and memory performance: measured as the total time spent by the algorithm in the reduction/discretization phase. Usually performed before the learning phase. The reduction rate is measured as the amount of reduction accomplished with respect to the original set (in percentage). The best algorithm is the one with high efficiency and low time and memory utilization and low high reduction rate without affecting the consistency of data [6].

5 Conclusion

Big data is an interesting open research point that involved with many applications and industries. Big data represents any set of data that require large storage, processing time and scalable units to be stored and processed. Big data analytics is one of the major steps that are carried out or applied to data to uncover hidden patterns to gain information that can be used in many applications [51]. Preprocessing data is considered as one of the mining techniques to jewel data via removing noise, representing it in a suitable form to save processing time and storage and provide the best utilization for the resources with high efficiency. It is a classic concept that still being used in many fields that concerned with getting the best benefit from the data like machine learning and artificial. Preprocessing can follow many programming models according to the technology used to do this task. Maching learning algorithms are integrated with big data preprocessing architectures to do the preprocessing job. Preprocessing is not an easy task when the data get more complex or bigger. Problems with data streaming and concept drift are popular research issues that needs more effort.

References

1. Kaisler, S., Armour, F., Espinosa, J.A. Money, W.: Big data: issues and challenges moving forward. In: Proceedings of Annual Hawaii International Conference on System Sciences, pp. 995–1004 (2013)
2. García, S., Ramírez-Gallego, S., Luengo, J., Benítez, J.M., Herrera, F.: Big data preprocessing: methods and prospects. Big Data Anal. 1(1), 1–22 (2016)
3. Durak, U., Becker, J., Hartmann, S., Voros, N.S.: Big data and data analytics in aviation Gerrit. Springer International Publishing (2018)
4. Palominos, F., Díaz, H., Cañete, L., Durán, C., Córdova, F.: A solution for problems in the organization, storage and processing of large data banks of physiological variables. Int. J. Comput. Commun. Control 12(2), 276–290 (2017)
5. Raghupathi, W., Raghupathi, V.: Big data analytics in healthcare: promise and potential. Heal. Inf. Sci. Syst. 2, 3 (2014)

6. Ramírez-Gallego, S., Krawczyk, B., García, S., Woźniak, M., Herrera, F.: A survey on data preprocessing for data stream mining: Current status and future directions. Neurocomputing **239**, 39–57 (2017)
7. Symon, P.B., Tarapore, A.: Defense intelligence analysis in the age of big data. JFQ Jt. Force Q. **79**, 4–11 (2015)
8. Amini, S., Gerostathopoulos, I., Prehofer, C.: Big data analytics architecture for real-time traffic control. In: 5th IEEE International Conference on Models and Technologies for Intelligent Transportation Systems, MT-ITS 2017 - Proceedings, vol. Tum Llcm, pp. 710–715 (2017)
9. Chen, M., Mao, S., Zhang, Y., Leung, V.C.M
10. Hashem, I.A.T., Yaqoob, I., Anuar, N.B., Mokhtar, S., Gani, A., Ullah Khan, S.: The rise of 'big data' on cloud computing: review and open research issues. Inf. Syst. **47**, 98–115 (2015)
11. Addo-Tenkorang, R., Helo, P.T.: Big data applications in operations/supply-chain management: a literature review. Comput. Ind. Eng. **101**, 528–543 (2016)
12. Bhadani, A.K., Jothimani, D.: Big data: challenges, opportunities, and realities. In: IGI Global International Publishing Information Science Technology Research, pp. 1–24 (2016)
13. Huda, M., et al.: Big data emerging technology: Insights into innovative environment for online learning resources. Int. J. Emerg. Technol. Learn. **13**(1), 23–36 (2018)
14. Wang, X., Zhang, Y., Leung, V.C.M., Guizani, N., Jiang, T.: Wireless big data: technologies and applications D2D big data: content deliveries over wireless device-to-device sharing in large-scale mobile networks. IEEE Wirel. Commun. **25**(February), 32–38 (2018)
15. Liao, H., Tang, M., Luo, L., Li, C., Chiclana, F., Zeng, X.J.: A bibliometric analysis and visualization of medical big data research. Sustainability **10**(1), 1–18 (2018)
16. Manogaran, G., Vijayakumar, V., Varatharajan, R., Kumar, P.M., Sundarasekar, R., Hsu, C.H.: Machine learning based big data processing framework for cancer diagnosis using Hidden Markov Model and GM clustering. Wirel. Pers. Commun. **102**(3), 2099–2116 (2018)
17. Ayyad, S.M., Saleh, A.I., Labib, L.M.: A new distributed feature selection technique for classifying gene expression data. J. Biomath. **12**, 1950039 (2019)
18. Ayyad, S.M., Saleh, A.I., Labib, L.M.: Classification techniques in gene expression microarray data. Int. J. Comput. Sci. Mob. Comput. **7**(11), 52–56 (2018)
19. Ayyad, S.M., Saleh, A.I., Labib, L.M.: Gene expression cancer classification using modified K-Nearest Neighbors technique. BioSystems **176**, 41–51 (2019)
20. Negi, A., Bhatnagar, R., Parida, L.: Distributed computing and internet technology, vol. 3347, pp. 295–300 (2005)
21. Lv, Z., Song, H., Basanta-Val, P., Steed, A., Jo, M.: Next-generation big data analytics: state of the art, challenges, and future research topics. IEEE Trans. Ind. Inform. **13**(4), 1891–1899 (2017)
22. Tseng, F.-M., Harmon, R.: The impact of big data analytics on the dynamics of social change. Technol. Forecast. Soc. Change **130**, 56 (2018)
23. Blazquez, D., Domenech, J.: Big data sources and methods for social and economic analyses. Technol. Forecast. Soc. Change **130**, 99–113 (2018)
24. Aggarwal, V.B., Bhatnagar, V., Kumar, D. (eds.) Advances in Intelligent Systems and Computing. Big Data Analytics, vol. 654 (2015)
25. Maxwell, S.E., Kelley, K., Rausch, J.R.: Sample size planning for statistical power and accuracy in parameter estimation. Annu. Rev. Psychol. **59**, 537–563 (2008)
26. Sivarajah, U., Kamal, M.M., Irani, Z., Weerakkody, V.: Critical analysis of big data challenges and analytical methods. J. Bus. Res. **70**, 263–286 (2017)
27. Ramannavar, M., Sidnal, N.S.: A proposed contextual model for big data analysis using advanced analytics. Adv. Intell. Syst. Comput. **654**, 329–339 (2018)
28. Han, H., Yonggang, W., Tat-Seng, C., Xuelong, L.: Toward scalable systems for big data analytics: a technology tutorial. IEEE Access **2**, 652–687 (2014)
29. Vashisht, P., Gupta, V.: Big data analytics techniques: a survey. In: Proceedings of 2015 International Conference Green Computing and Internet of Things, ICGCIoT 2015, pp. 264–269 (2016)

30. Gandomi, A., Haider, M.: Beyond the hype: big data concepts, methods, and analytics. Int. J. Inf. Manage. **35**(2), 137–144 (2015)
31. Wang, Y., Kung, L.A., Byrd, T.A.: Big data analytics: understanding its capabilities and potential benefits for healthcare organizations. Technol. Forecast. Soc. Change **126**, 3–13 (2018)
32. Dumka, A., Sah, A.: Smart ambulance system using concept of big data and internet of things. Elsevier Inc. (2018)
33. Praveena, A., Bharathi, B.: A survey paper on big data analytics. In: 2017 International Conference on Information Communication and Embedded Systems, ICICES 2017 (2017)
34. Singh, D., Reddy, C.K.: A survey on platforms for big data analytics. J. Big Data **2**(1), 1–20 (2015)
35. Fu, C., Wang, X., Zhang, L., Qiao, L.: Mining algorithm for association rules in big data based on Hadoop. In: AIP Conference Proceedings, vol. 1955 (2018)
36. Acharjya, D.P., Ahmed, K.: A survey on big data analytics: challenges, open research issues and tools. Int. J. Adv. Comput. Sci. Appl. **7**(2), 511–518 (2016)
37. Apache Flink® Stateful Computations over Data Streams. https://flink.apache.org/
38. Furht, B., Villanustre, F.: Big Data Technologies and Applications, vol. 2, no. 21 (2016)
39. García, S., Luengo, J., Herrera, F.: Data Preprocessing in Data Mining, vol. 72 (2015)
40. García, S., Luengo, J., Herrera, F.: Data preparation basic models. Intell. Syst. Ref. Libr. **72** (2015)
41. Russom, P.: Big data analytics - TDWI best practices report introduction to big data analytics. Tdwi Res. **1**, 3–5 (2011)
42. Di Martino, B., Aversa, R., Cretella, G., Esposito, A., Kołodziej, J.: Big data (lost) in the cloud. Int. J. Big Data Intell. **1**(1/2), 3 (2014)
43. ur Rehman, M.H., Liew, C.S., Abbas, A., Jayaraman, P.P., Wah, T.Y., Khan, S.U.: Big data reduction methods: a survey. Data Sci. Eng. **1**(4), 265–284 (2016)
44. García, S., Luengo, J., Herrera, F.: Tutorial on practical tips of the most influential data preprocessing algorithms in data mining. Knowl.-Based Syst. **98**, 1–29 (2016)
45. Mangat, V., Gupta, V., Vig, R.: Methods to investigate concept drift in big data streams. Knowl. Comput. Appl. Knowl. Manip. Process. Tech. **1**, 51–74 (2018)
46. Polikar, R.: Ensemble Machine Learning. Springer, Boston (2012)
47. Nagendran, N., Sultana, H.P., Sarkar, A.: A comparative analysis on ensemble classifiers for concept drifting data streams. In: Soft Computing and Medical Bioinformatics. Springer, Singapore, pp 55–62 (2019)
48. Khamassi, I., Sayed-Mouchaweh, M., Hammami, M., Ghédira, K.: Discussion and review on evolving data streams and concept drift adapting. Evol. Syst. **9**(1), 1–23 (2018)
49. Chang, Y.S., Lin, K.M., Tsai, Y.T., Zeng, Y.R., Hung, C.X.: Big data platform for air quality analysis and prediction. In: 2018 27th Wireless and Optical Communication Conference (WOCC), pp. 1–3 (2018)
50. Zhao, L., Chen, Z., Hu, Y., Min, G., Jiang, Z.: Distributed feature selection for efficient economic big data analysis. IEEE Trans. Big Data **4**(2), 164–176 (2016)
51. Ghani, N.A., Hamid, S., Hashem, I.A.T., Ahmed, E.: Social media big data analytics: a survey. Comput. Human Behav. **101**, 417–428 (2019)
52. Tayal, V., Srivastava, R.: Challenges in mining big data streams. In: Data and Communication Networks, pp. 173–183 (2019)

Artificial Intelligence-Based Plant Diseases Classification

Lobna M. Abou El-Maged, Ashraf Darwish, and Aboul Ella Hassanien

Abstract Machine learning techniques are used for classifying plant diseases. Recently, deep learning (DL) is applied in the classification process of image processing. In this chapter, convolutional neural network (CNN) is used to classify plant diseases images. However, CNN suffers from the hyper parameters problem which can affect the proposed model. Therefore, Gaussian optimization method is used to overcome the hyper parameters problem in CNN. This chapter proposed an artificial intelligence model for plants diseases classification based on convolutional neural network (CNN). The proposed model consists of three phases; (a) preprocessing phase, which augmented the data and balanced the dataset; (b) classification and evaluation phase based on pre-train CNN VGG16 and evaluate the results; (c) optimize the hyperparameters of CNN using Gaussian method. The proposed model is tested on the plant's images dataset. The dataset consists of nine plants with thirty-three cases for diseased and healthy plant's leaves. The experimental results before the optimization of pre-trained CNN VGG16 achieve 95.87% classification accuracy. The experimental results improved to 98. 67% classification accuracy after applied the Gaussian process for optimizing hyperparameters.

Keywords Plants diseases · Deep learning · Convolution neural network · Hyper parameter optimization · Gaussian process

L. M. Abou El-Maged (✉)
Computer Science Department, MET High Institute, Mansoura, Egypt
e-mail: lobna_acd@hotmail.com
URL: http://www.egyptscience.net/

A. Darwish
Faculty of Science, Helwan University, Helwan, Egypt

A. E. Hassanien
Faculty of Computers and Artifical Intellgence, Cairo University, Giza, Egypt

1 Introduction

The agricultural sector is one of the most important sectors of the economy for all countries, therefore it is one of the elements of sustainable development. Plant diseases significantly affect crop productivity and agricultural wealth. The economic importance of plant diseases in agricultural production in terms of their impact on the national economy is one of the main factors that affect the quantity of the crop.

Parasitic diseases are the most common diseases among plants, and there are many pathogens of the plant including fungi, viruses and bacteria. The simple farmer can't easily identify plant diseases, but he goes to an expert in plant diseases, which in turn may be specialized only in diseases of fruits or vegetables or even in one type of them. Therefore, in this chapter an intelligent system was built to classify different types of plant diseases of vegetables and fruits. This system may be a substitute for the human expert, which achieves the speed and accuracy in the identification and treatment of plant diseases quickly before the spread of the disease. In this chapter the researchers try to get optimal performance for the identification of the plants' diseases.

DL is an artificial neural network architecture which contains many processing layers. DL have been used in many fields like; image recognition, voice recognition, and also other huge applications that deal with the analysis of large volumes of data, like, e.g., self-driving cars and natural language processing systems [1–3]. The DL is differ than the traditional machine learning in how features are extracted. Traditional machine learning approaches use feature extraction algorithms as a preceding stage for the classifier, but in DL the features are learned and represented hierarchically in multiple levels. So DL is better for machine learning than the traditional machine learning approaches [4]. There are many types of DL tools, the most commonly used are the Convolutional Neural Networks (CNN) [3].

Various researches have been done in the field of agriculture, specifically in the identification of plants and the identification of plant diseases using CNN. In [1] two well-known architectures of CNNs are compared in the identification of 26 plant diseases, using an open database of leaves images of 14 different plants. For the corn specifically the result was for Corn (maize) Zea is 26.1%, Cercospora leaf spot is 35.2%, Common rust is 73.9%, and Northern Leaf Blight is 100%.

Two well-known and established architectures of CNNs AlexNet and GoogLeNet are compared in [19] in the identification of 26 plant diseases, using an open database of leaves images of 14 different plants. In [16] developed a similar methodology for plant disease detection through leaves images using a similar amount of data available on the Internet, which included 13 diseases and 5 different plants accuracy of their models, were between 91% and 98%, depending on the testing data. More recently, in [20] compared the performance of some conventional pattern recognition techniques with that of CNN models, in plants identification, using three different databases of images of either entire plants and fruits, or plant leaves, concluding that CNNs drastically outperform conventional methods. In [21] the researcher developed CNN models for the detection of 9 different tomato diseases and pests, with satisfactory performance (36,573 no of plants). Finally in [1] perform plant disease detection and diagnosis through deep learning methodologies. Training of the models was performed with the use of an open database of 87,848 images, containing 25 different plants in a set of 58 distinct classes, he use several model architectures.

The CNN has many architectures that concerning the identification of plant diseases; AlexNet [5], Overfeat [6], AlexNetOWTBn [7], GoogLeNet [8], and VGG [9]. CNN has a number parameters called hyper parameters, these parameters determine the network structure, these parameters are variables such as; the number of hidden layers and the learning rate, the hyper parameters values are set before training the network [10]. In this chapter, Gaussian Optimization method has been used with CNN to find the optimal hyper parameters in a fine- tuning CNN.

The rest of this chapter is present as follows. Section 2 describes the basics and background. Section 3 presents the Materials and methods. Section 4 the results and discussion. Finally; Sect. 5 presents the conclusion.

2 Preliminaries

This section presents the basic information about CNN, VGG16 & fine-tuning and Gaussian optimization.

2.1 Convolutions Neural Network

CNN is an evolution of traditional artificial neural networks and it is network architecture for deep learning [1, 4, 11]. They are used to object recognition and image classifications. A CNN is like a multilayer neural network, it consists of convolutional layers (may be one or more), and then followed by fully connected layers (may be one or more). Each layer generates a successively abstraction of the input data with high level, called a feature map, which save essential unique information. Amazing performance could to be achieve in modern CNNs by employing a very deep hierarchy of layers [11]. The numbers of layers used in DL range from five to more than a thousand. Generally there are many CNN architectures like; AlexNet, AlexNetOWTBn, GoogLeNet, Overfeat, and VGG [1, 5–9].

In the image analysis tasks, the CNN can be trained to do image classification, object detection, image segmentation and finally image processing [12, 13].

The CNN architecture as shown in Fig. 1 are consist of two main layers; layers for feature extraction and layers for classification [4]. The feature extraction layers are convolution layer and max-pooling layer:

- **Convolution Layer**
  The learnable kernels are used for convolving feature maps from previous layers. The output of the kernels go through a linear or non-linear activation function

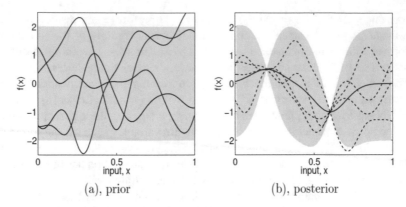

(a), prior   (b), posterior

Fig. 1 a. The prior distribution. **b.** Shows the situation after two data points have been observed

specified as a sigmoid, inflated tangent, Softmax, rectified additive, and sameness functions to make the feature maps. Each of the output feature maps can be conjunctive with much than one input feature map.

$$\text{Suppose}: \quad LR_i^l = f\left(\sum_{j \in Mi} LR^{l-1} * k_{ji}^l + B_i^l\right) \tag{1}$$

Where LR_i^l the product of the present layer is, LR_i^{l-1} is product of the previous layer, k_{ji}^l is the kernel for the present layer, and B_i^l are biases for the present layer. F_i represents a selection of input maps. An additive bias B is given for each output map.

- **max-pooling Layer**
  This bed perfect distribution the dimension of the feature maps depending on the filler of the land distribution mask [4].

The classification layer uses the extracted features from a previous convolutional layer to classify each class. The layers are fully connected in the classification layer. The final layer are represented as vectors with scalar values which are passed to the fully connected layers. The fully connected feed-forward neuronal layers are misused as a soft-max classification layer [4]. Still, in most cases, two to quaternion layers feature been observed in assorted architectures including LeNet, AlexNet, and VGG Net [4].

Soft-max function calculates the probabilities distribution of different events. The soft-max function calculates the probabilities for each class over all possible classes. The calculated probabilities determining the target class for the present inputs [13].

CNN has a number of hyper parameters. The hyper parameters determine the structure of the network such as the number of hidden layers and the learning rate, the hyper parameters are set before the network training [10]. We need to optimize the hyper parameter to reduce the effort of the human necessary for applying machine learning and to enhance the performance of machine learning algorithms. The hyper parameter can be real value such as "learning rate", integer value such "number of layers", binary, or categorical such as "choice of optimize" [14].

For enhancing the deep networks training and to avoid the over-fitting, a lot of techniques were proposed. Typical entirety allows pot normalization (BN), and dropout [12]. To enhance the expressive commonwealth of CNNs, non-linear activation functions were well-studied, like as ReLU, ELU and SELU. Also, a lot of realm peculiar techniques were also mature to add fine-tunes networks on special application [12].

2.2 VGG16 Architecture

VGG16 is the CNN model proposed by Simonyan and Zisserman [9]. The model operated on a dataset contains over 14 million images categories of 1,000 classes. It achieves 92.7% as test accuracy in ImageNet dataset. VGG16 achieves improvement over AlexNet by replacing large kernel-sized filters with 3×3 kernel-sized filters one after another. NVIDIA Titan Black GPU's was used to train VGG16 for weeks.

In VGG16, the image with fixed size 224×224 RGB is an input to the convolution layer. There are stack of convolution layers where the image pass through them using filters with a very small receptive field: 3×3. In one of the configurations, it also utilizes 1×1 convolution filters, which can be seen as a linear transformation of the input channels then followed by non-linearity transformation. The stride is set to 1 pixel; the spatial padding of convolution. The input of the layer is such that the spatial resolution is preserved after convolution, i.e. the padding is 1-pixel for 3×3 convolution layers. After that the spatial pooling is achieved by 5 max-pooling layers, which follow some of the convolution layers not all ones. Max-pooling is performed over a 2×2 pixel window, with stride 2. After the stack of convolutional layers, three Fully-Connected (FC) layers are come. The first two FC layers have 4096 channels each, the third one performs 1000-way ILSVRC classification, so it contains 1000 channels one for each class. The final layer is the soft-max layer. The fully connected layers' configuration is the same all networks. All hidden layers are prepared by the rectification (ReLU) non-linearity. It is also noted that none of the networks (except for one) contain Local Response Normalization (LRN), such normalization does not improve the performance on the ILSVRC dataset, but leads to more complexity for time and memory [9, 15].

The CNN architecture can be reused for new datasets which is differ than its own dataset, the new datasets may vary in size or nature of images. So; we can use fine-tuning deal with this situation. Fine-tuning tries to improve the effectiveness or efficiency of a process or function by making small modifications in the used CNN architecture to improve or optimize the outcome [16].

2.3 Hyper Parameters

The hyper parameters are the variables which determine the network structure such as the number of hidden layers and the variables which determine how the network is trained such the learning rate), the Hyper parameters are set before training the network [10]. Automated hyperparameters optimization has several important use cases; it can reduce the human effort necessary for applying machine learning. It used to improve the performance of machine learning algorithms.

Let A denote a machine learning algorithm with N hyperparameters. We denote the domain of the n-th hyperparameters by Sn and the overall hyperparameters configuration space as $S = s_1 \times s_2 \times \ldots s_N$. A vector of hyperparameters is denoted by $\lambda \in S$, and A with its hyperparameters instantiated to λ is denoted by $A\lambda$. The domain of a hyperparameters can be real-valued (e.g., learning rate), integer-valued (e.g., number of layers), binary (e.g., whether to use early stopping or not), or categorical (e.g., choice of optimizer) [14].

Given a data set DS, our goal is to find

$$\lambda^* = \arg min_{\lambda \in S} E_{(DS_{train}, DS_{valid}) \sim DS} V(L, A_\lambda, DS_{train}, DS_{valid}) \tag{2}$$

where $V(L, A_\lambda, DS_{train}, DS_{valid})$ measures the loss of a model generated by algorithm A with hyperparameters λ on training data DS_{train} and evaluated on validation data DS_{valid}. In practice, it is only access to finite data DS ~ DS and thus need to approximate the expectation in Eq. 2 Popular choices for the validation protocol $V(\cdot, \cdot, \cdot, \cdot)$ are the holdout and cross-validation error for a user-given loss function (such as misclassification rate).

2.4 Gaussian Optimization Method

The optimization method is an iterative algorithm has two main parts, the probabilistic surrogate model and the acquisition function to determine which next point to be evaluated. The alternate modelling is fitted to all observations of the reference use made so far in each iteration. Then the acquisition function uses the prophetic dispersion of the probabilistic copy to determine the utility of various candidate points, trading off exploration and exploitation. Compared to evaluating the dear operate, the acquisition role is to compute and can thence be good optimized [14, 17]. The expected improvement (EI):

$$E[I(\lambda)] = E[\max(f^*_{min} - Y, 0)] \tag{3}$$

EI can be computed in closed form if the model prediction Y at configuration λ according to a normal distribution:

$$E[I(\lambda)] = \left(f_{min}^* - \mu(\lambda)\right) \Phi\left(\frac{f_{min}^* - \mu(\lambda)}{\sigma}\right) + \sigma \emptyset\left(\frac{f_{min}^* - \mu(\lambda)}{\sigma}\right) \tag{4}$$

where $\phi(.)$ is the standard normal density, $\Phi(.)$ is the distribution function, and f_{min}^* is the optimal observed value [14, 17].

Assume that we are then supposition a dataset, the proper random functions showed in Fig. 1a, in Fig. 1b. The broken lines pretense have functions which are reconciled with D, and the solidified line depicts the tight quantity of much functions. Respond how the uncertainness is low nestled to the observations. The combining of the prior and the data leads to the tooth posterior distribution over functions [17].

3   Materials and Methods

3.1   Plant's Image Dataset

The dataset of the plant's diseases images is a "kaggle-dataset" [18], it consists 159.984 images of diseases and healthy leaves' plants present in Fig. 2; 128,028 of plants leaves images for training and 31,956 images for testing. The images are present diseased and healthy plant's leaves. The dataset is characterized by distinguishing images which may be taken at different angles and different backgrounds. Table 1 shows that the data set contains thirty-three categories of nine plants leaves; apple, cherry, corn, grape, peach, Pepper, potato, Strawberry and tomato.

3.2   The Proposed Plant's Diseases Classification Model

The AI proposed model is consisting of three main phases (a) preprocessing phase, (b) classification and evaluation phase and (c) hyperparameter optimization using Gaussian process phase. Each phase is composed of a number of steps as presented in Fig. 3. In the AI proposed model; the image of the plants are gained from the data set, after executing the preprocessing phase, the Boolean variable "optimize" is set true, then process classification and evaluation phase executed, the output of this phase is the evaluation of the results. If the results are satisfying and the value of the "optimize" variable is "false", the algorithm will end otherwise, the optimization phase will execute a number of n times. The detail of each process is given below.

(i) **Preprocessing phase.**

Preprocessing phase is used for processing and formatting images for use in the next stages, it consisting of two processes; data augmentation and balanced the dataset as in Fig. 3. The detail of each process is given below.

Data Augmentation The data was augmented to avoid overfitting and present more varieties in the data set. The augmentation for data was done in five different

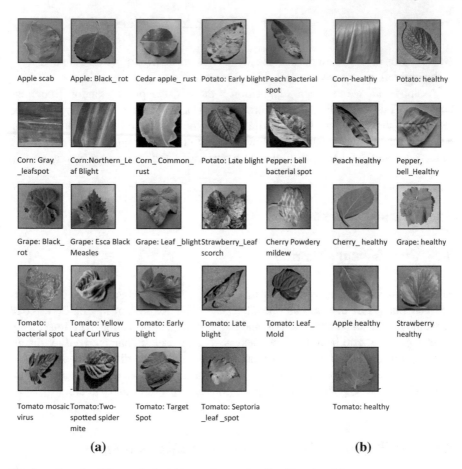

Fig. 2 **a.** Sample of diseased plants' leaves **b.** sample of healthy plants' leaves

methods. The augmentation methods are; rotate right 30°, rotate left 30 & +90°, flip horizontally about Y-axis and shear.

Balance Dataset: The dataset showed in the Table 1 is imbalance dataset, the imbalanced lead to inaccurate results, so we calculate the class weight for the categories in the dataset to use it later for our building model. The class weight will be, for example, the first weight is '1.92135642' which presents 2017 images for the apple scab is less than third weight '4.40165289' presents 880 images for the cedar-apple rust.

(ii) **classification and evaluation phase**

The classification and evaluation phase is consists of 4 processes; hyperparameter setting, proposed CNN architecture, CNN Training, and evaluation the results. The detail of each step is given below.

Table 1 Data set description

#	Plant	Disease	Train samples	Test samples
1	Apple	Scab	2017	505
2		Black rot	1988	496
3		Cedar apple rust	880	202
4		Healthy	5264	1316
5	Cherry	Healthy	2736	680
6		Powdery mildew	3368	840
7	Corn	Cercospora leaf spot Gray leaf spot	1644	408
8		Common rust	3816	952
9		Healthy	3720	920
10		Northern Leaf Blight	3125	788
11	Grape	Black rot	3776	944
12		Esca_(Black Measles)	4428	1104
13		Healthy	1356	336
14		Leaf blight_(Isariopsis_Leaf_Spot)	3444	860
15	Peach	Bacterial_spot	7352	1837
16		Healthy	1152	228
17	Pepper	Bacterial_spot	3193	796
18		Healthy	4725	1180
19	Potato	Early_blight	3200	800
20		Healthy	488	120
21		Late_blight	3200	800
22	Strawberry	Healthy	1460	364
23		Leaf scorch	3552	884
24	Tomato	Bacterial spot	6808	1700
25		Early blight	3200	800
26		healthy	5089	1272
27		Late blight	6109	1524
28		Leaf Mold	3048	760
29		Septoria leaf spot	5668	1416
30		Two-spotted spider mite	5364	1340
31		Target Spot	4496	1120
32		Mosaic virus	1196	296
33		Yellow Leaf Curl Virus	17,144	2484
Total			128,028	38,836

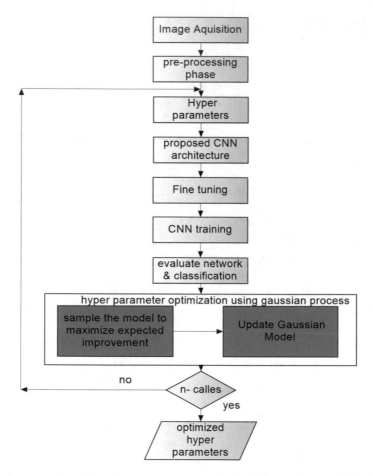

Fig. 3 The proposed model for hyper parameter optimization in fine-tuning CNN using Gaussian process

Hyperparameter Setting: This process is responsible for setting values of the hyper-parameters, this work uses three types of hyperparameters; learning rate, drop_out value, and activation function type.

Proposed CNN Architecture: The proposed architecture depends on the usage of a pre-train CNN VGG16 architecture which is a very convincing way since it has been used previously and has given good results. The proposed architecture is consists of two main steps; the first one is to create a CNN VGG16 then the second one is to make some adaptation in the architecture to fit the dataset. Figure 4 presents the proposed CNN architecture.

As seen in the early section, the VGG16 is a CNN architecture used to classify 1000 classes, it consists of 16 layers. In the proposed architecture, the convolutions layers in the VGG16 is used, then modifying the rest architecture by adding two

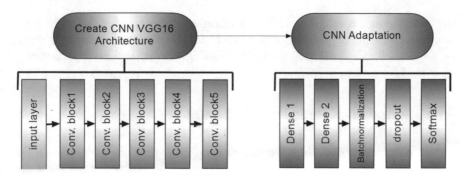

Fig. 4 CNN proposed architecture

dense layers, Batch Normalization layer, Dropout layer to prevent overfitting, and finally the classification layer.

The CNN Training Process: In the training process the fine-tuning is used Fig. 5. The Fine-Tuning is optimized both the weights of the new classification layers that have been added and also all of the layers from the VGG16 model [16]. The training process is based on the hyper-parameter values which sat in the previous process

Evaluation the Results: This process is responsible for evolution the accuracy of the proposed AI model, the evaluation results used to decide a road chart of the model as seen in Fig. 3.

(iii) Hyper Parameter Optimization for CNN using Gaussian process phase.

The optimization method is used for two reasons; the first one is to get the optimal hyper_ parameters' values and the other is to provide the human effort in trying to get the optimal hype_ parameters' values by doing the system with this task.

The proposed model setting parameters used the learning rate = 1e−6 with Adam optimizer, dropout rate = 0.2 and activation function is 'relu' as initial values. In the Gaussian method; the model is sampled with expected improvement, then update

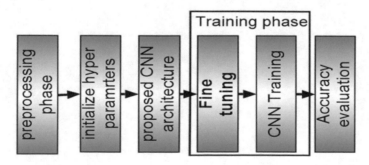

Fig. 5 Proposed model with fine—tuning

the Gaussian model according to the classification accuracy, after that check if the number of iteration calls is maximum or not to decide whether re-execute the model or not. After finishing the n calls iteration, it will give optimized hyperparameters.

4 Experiments Results and Discussion

The experiments are done using tensor flow and Keras with GPU google colab. The experiments are done in three experiments. In the first experiment is implementing the proposed model without optimization. In the second experiment, perform the optimization process using the Gaussian method. In the third experiment, implement the proposed model after optimization.

4.1 Experiment (I): Without Optimization

In this experiment, the structure of VGG16 is modified by adding two dense layers, adding a BatchNormalization layer, adding a Dropout layer, and finally a classification layer. In this experiment, we improved both the weights of the new classification layers and all layers of the VGG16 model.

The hyper_parameters used are learning rate = $6-1e$, drop_out = 0.2 and the activation function is "relu". Figure 6 shows the Test-set classification accuracy: 95.87%.

Fig. 6 Training history of CNN network before hyperparameter optimization

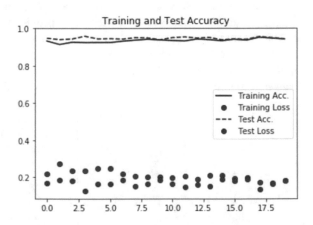

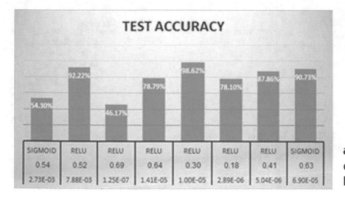

Fig. 7 Test accuracies during optimization process

4.2 Experiment (II): Hyper_ Parameter Optimization Using Gaussian Process

In this experiment, Gaussian process optimization is used with EI and with minimum 11 iteration calls. The hyper-parameters are 'learning_rate' within range (from low = 1e−6 to high = 1e−2), 'drop_out' within range (from low = 0.1 to high = 0.9) and finally 'activation' with categories ('relu' and 'sigmoid'). The fitness function is the function that creates and trains a neural network with the given hyper-parameters, and then evaluates its performance on the data set. The function then returns the so-called fitness, which is the negative classification accuracy on the validation set. It is negative because the performance minimization instead of in Fig. 7.

For example test accuracy is about 91% when learning rate = 6.8.99, Drop_out = 0.63 and activation function is "sigmoid". It is illustrated that the best values are; learning rate = 1e−5 with droup out = 0.3 and activation function is 'relu'.

4.3 Experiment (III): Hyperparameters Optimization

In this experiment, the CNN architecture is run with new parameters such as learning rate = 1e−5, drop out = 0.3 and activation function = 'relu'. It gives Test-set classification accuracy: 98.67%. Figure 8 shows the historical training.

The results as shown in Table 2 shows that the total test accuracy is improved with 2.8%. At the plants level, test accuracy for most plants are improved but test accuracy for a small number of plants doesn't change such as cherry and corn. It is noted that the plant, whose accuracy has not changed, has fairly high accuracy.

Fig. 8 Training history of
CNN network of proposed
AI model

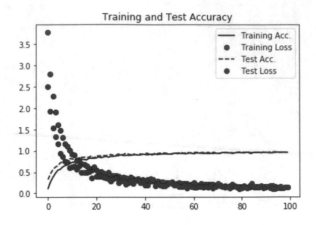

5   Conclusion and Future Work

Agriculture plays an important role in the economic development for many countries
so classification the plant's diseases are very important. Deep learning mimics human
thinking, it has been used extensively in the last decade. VGG16 is an architecture
of CNN. Using fine-tuning CNN VGG16 model by adding several dense layers,
Batch Normalization layer, Dropout layer and finally the classification layer, this
lead to accuracy 95.87%. Automated hyperparameters optimization is important to
help the human by reducing his effort necessary for applying machine learning. It
used to enhance the performance of the algorithm for machine learning. The Gaussian
method can be used for hyperparameters optimization. It uses the predictive distribution of the probabilistic model. The Gaussian method identifies the importance of
different candidate points, trading off exploration and exploitation, this function is
cheap. Using Gaussian process optimization with EI illustrated that the best values
are; learning rate = 1e−5 with drop_ out = 0.3 and activation function is 'real'.
Using fine-tuning with new parameters: learning rate = 1e−5, drop_out = 0.3 and
activation function = 'relu', It gives accuracy 98.76%. I.e. Gaussian process optimization helps in improving accuracy from 95.87% to 98.67%. In future work, we
will use the swarm algorithm for hyperparameters optimization as a trial to enhance
accuracy.

Table 2 Accuracy test results

#	Plant	Disease	Test accuracy before hyper parameters optimization (%)	Test accuracy after hyper parameters optimization (%)
1	Apple	Scab	91.47	94.44
2		Black rot	96.58	100.00
3		Cedar apple rust	97.33	99.55
4		Healthy	99.69	100.00
5	Cherry	Healthy	98.21	98.21
6		Powdery mildew	98.68	98.68
7	Corn	Cercospora leaf spot Gray leaf spot	95.15	98.04
8		Common rust	99.79	100.00
9		Healthy	92.89	96.70
10		Northern Leaf Blight	99.89	99.89
11	Grape	Black rot	97.88	97.88
12		Esca_(Black Measles)	96.24	99.73
13		Healthy	97.93	100.00
14		Leaf blight_(Isariopsis_Leaf_Spot)	99.41	100.00
15	Peach	Bacterial _spot	99.89	99.89
16		Healthy	92.91	95.49
17	Pepper	Bacterial _spot	97.86	98.99
18		Healthy	98.74	99.75
19	Potato	Early _blight	98.27	99.38
20		healthy	93.20	98.24
21		Late _blight	90.08	98.33
22	Strawberry	Healthy	99.55	99.66
23		Leaf scorch	96.15	100.00
24	Tomato	Bacterial spot	95.49	99.00
25		Early blight	86.90	99.59
26		Healthy	96.72	98.62
27		Late blight	91.31	98.42
28		Leaf Mold	97.30	98.80
29		Septoria leaf spot	97.52	99.55
30		Two- spotted spider mite	90.63	98.21
31		Target Spot	97.97	99.11
32		Mosaic virus	84.80	92.23
33		Yellow Leaf Curl Virus	97.22	99.84

(continued)

Table 2 (continued)

#	Plant	Disease	Test accuracy before hyper parameters optimization (%)	Test accuracy after hyper parameters optimization (%)
Total			95.87	98.67

References

1. Ferentinos, K.P.: Deep learning models for plant disease detection and diagnosis. Comput. Electron. Agric. **145**, 311–318 (2018)
2. Zhang, X., et al.: Identification of Maize leaf diseases using improved deep convolutional neural networks. IEEE Xplore Digital Library, Digital Object Identifier (2018). https://doi.org/10.1109/ACCESS.2018.2844405
3. Barbedo, J.G.A.: Factors influencing the use of deep learning for plant disease recognition. Biosyst. Eng. **72**, 84–91 (2018)
4. The History Began from AlexNet: A comprehensive survey on deep learning approaches. https://arxiv.org/abs/1803.01164
5. Krizhevsky, A., Sutskever, I., Hinton, G.E.: ImageNet classification with deep convolutional neural networks. In: Proceedings of 25th International Conference on Neural Information Processing Systems, Lake Tahoe, Nevada, 03–06 December 2012, vol. 1, pp. 1097–1105 (2012)
6. Sermanet, P., et al.: Overfeat: integrated recognition, localization and detection using convolutional networks. arXiv:1312.6229 (2013)
7. Krizhevsky, A.: One weird trick for parallelizing convolutional neural networks. arXiv:1404.5997 (2014)
8. Szegedy, C., Liu, W., Jia, Y., Sermanet, P., Reed, S., Anguelov, D., et al.: Going deeper with convolutions. In: Proceedings of the IEEE Conference on Computer Vision and Pattern Recognition (2015)
9. Simonyan, K., Zisserman, A.: Very deep convolutional networks for large-scale image recognition. arXiv:1409.1556 (2014)
10. https://towardsdatascience.com/what-are-hyperparameters-and-how-to-tune-the-hyperpara meters-in-a-deep-neural-network-d0604917584a. Accessed May 2019
11. Yann, L., et al.: Gradient based learning applied to document recognition. In: Proceedings of the IEEE, November 1998
12. Wang, T., Huan, J., Li, B.: Data dropout: optimizing training data for convolutional neural networks. arXiv:1809.00193v2 (2018)
13. Islam, M.S., et al.: InceptB: a CNN based classification approach for recognizing traditional Bengali games. In: 8th International Conference on Advances in Computing and Communication (ICACC 2018). Elsevier (2018)
14. Feurer, A., Hutter, F.: Hyperparameter optimization, Chap 1. In: Automated Machine Learning: Methods, Systems, Challenges. Springer (2019)
15. https://neurohive.io/en/popular-networks/vgg16/. Accessed March 2019
16. Hussain, M., Bird, J.J., Faria, D.R.: Advances in computational intelligence systems. In: UKCI 2018. AISC, vol. 840, pp. 191–202 (2019)
17. Rasmussen, C.E., Williams, C.K.I.: Gaussian Processes for Machine Learning. The MIT Press, Massachusetts Institute of Technology (2006). www.GaussianProcess.org/gpml, ISBN 026218253X
18. https://storage.googleapis.com/kaggle-datasets/53092/100927/plantdisease.zip
19. Hughes, D., Salathé1, M.: Using deep learning for image-based plant disease detection Sharada Prasanna Mohanty. Front. Plant Sci. **7**, 1419 (2016)

20. da Silva Abade, A., et al.: Plant diseases recognition from digital images using multichannel convolutional neural networks. http://www.insticc.org/Primoris/Resources/PaperPdf.ashx?idPaper=73839VISAPP_2019_144_CR.pdf. Accessed March 2019
21. Fuentes, A., Im, D.H., Yoon, S., Park, D.S.: Spectral analysis of CNN for tomato disease identification. In: ICAISC 2017, Part I, LNAI, vol. 10245, pp. 40–51 (2017)

Artificial Intelligence in Potato Leaf Disease Classification: A Deep Learning Approach

Nour Eldeen M. Khalifa, Mohamed Hamed N. Taha, Lobna M. Abou El-Maged, and Aboul Ella Hassanien

Abstract Potato leaf blight is one of the most devastating global plant diseases because it affects the productivity and quality of potato crops and adversely affects both individual farmers and the agricultural industry. Advances in the early classification and detection of crop blight using artificial intelligence technologies have increased the opportunity to enhance and expand plant protection. This paper presents an architecture proposed for potato leaf blight classification. This architecture depends on deep convolutional neural network. The training dataset of potato leaves contains three categories: healthy leaves, early blight leaves, and late blight leaves. The proposed architecture depends on 14 layers, including two main convolutional layers for feature extraction with different convolution window sizes followed by two fully connected layers for classification. In this paper, augmentation processes were applied to increase the number of dataset images from 1,722 to 9,822 images, which led to a significant improvement in the overall testing accuracy. The proposed architecture achieved an overall mean testing accuracy of 98%. More than 6 performance metrics were applied in this research to ensure the accuracy and validity of the presented results. The testing accuracy of the proposed approach was compared with that of related works, and the proposed architecture achieved improved accuracy compared to the related works.

Keywords Potato leaf blight · Classification · Deep convolutional neural network

N. E. M. Khalifa (✉) · M. H. N. Taha
Faculty of Computers and Artificial Intelligence, Cairo University, Giza, Egypt
e-mail: nourmahmoud@cu.edu.eg
URL: http://www.egyptscience.net

L. M. Abou El-Maged
Faculty of Computers and Information, Mansoura University, Mansoura, Egypt

N. E. M. Khalifa · M. H. N. Taha · L. M. Abou El-Maged · A. E. Hassanien
Scientific Research Group in Egypt (SRGE), Giza, Egypt

© The Author(s), under exclusive license to Springer Nature Switzerland AG 2021
A. E. Hassanien et al. (eds.), *Machine Learning and Big Data Analytics Paradigms: Analysis, Applications and Challenges*, Studies in Big Data 77,
https://doi.org/10.1007/978-3-030-59338-4_4

1 Introduction

Current agricultural practices are incredibly challenging. The agricultural sector has matured into an extremely competitive and international industry in which farmers and other actors must deliberate local climatic and environmental aspects as well as world-wide ecological and political factors to ensure their economic subsistence and sustainable production [1–3].

Potato is considered to be an efficient crop because it produces more protein, dry matter and minerals per unit area compared to cereals. However, potato production is vulnerable by numerous diseases, leading to yield losses and/or a decreasing in tuber quality and causing a rise in potato price. Many diseases especially parasitic diseases affect potato crops, leading to large crop yield reductions and significant economic losses for farmers and producers [4] Traditional methods such as visual inspections, are commonly used to detect and diagnose plant diseases. However, these methods have several disadvantages; they are expensive because they require continual monitoring by experts and time consuming because experts are not always available locally [5, 6].

Recent advances in artificial intelligence have highlighted and accelerated agriculture using various types of Artificial Intelligence essential technologies, such as mobile devices, independent agents (devices) operating in unrestrained environments, independent and collaborative scenarios, robotics, sensing, computer vision, and interactions with the environment. Integrating multiple partners and their heterogeneous information sources has led to the application of semantic technologies [7, 8]. The continuous interest in creating reliable predictions for planning purposes and for controlling agricultural activities requires interdisciplinary cooperation with field specialists for example, between the agricultural research and artificial intelligence domains [7].

Many studies have investigated the problem of classifying plant leaf diseases using computer algorithms. These trials have included different leaf types, including apple, peach, grapevine, cherry, orange, cotton and others [9–11]. In [9], authors proposed an architecture for to detect leaf diseases by converting images from RGB to HSV domain to detect the diseases spots in leaves. In [10], another model was introduced to detect cotton diseased leaf using particle swarm optimization (PSO) [12] algorithm and feed forward neural network and it achieved testing accuracy 95%. In [11], more than 4483 images for different leaves types of fruits which included pear, cherry, peach, apple, and grapevine leaves have been used in a deep learning architecture to classify the images into 13 different classes and achieved accuracy from 91% and 98%.

The focus of this research is potato leaf blight classification; this problem was also addressed in [13, 14]. In [13], the presented algorithms extracted more than 24 (colour, texture, and area) features. The texture features were extracted from the grey level co-occurrence matrix, and a backpropagation neural network-based classifier was used to recognize and classify unknown leaves as either healthy or diseased. The model achieved an overall testing accuracy of 92%.

In [14], the proposed system involved three main phases: segmentation, feature extraction, and classification. Image segmentation was conducted using the k-means algorithm. Then, three types of features were extracted from the segmented images: colour, texture, and shape. Finally, the extracted features were input into a feed-forward neural network for classification. The overall testing accuracy was 95.30%.

The main contributions of this research are a proposed deep neural network archi-tecture with data augmentation which led to a significant improvement in testing accuracy 98% for potato leaf blight classification. Moreover, the proposed achieved a competitive result if the is compared to related works. The rest of this paper is organized as follows. Section 2 introduces artificial intelligence and deep learning. Section 3 describes the dataset. Section 4 presents the proposed CNN architecture, and Sect. 5 introduces the augmentation techniques. The experimental results and environment are reported in Sect. 6. Finally, Sect. 7 summarizes the main findings of this paper.

2 Artificial Intelligence and Deep Learning

In 1947, Alan Turing foreseen that intelligent computers might ascend by the end of the century, and in his classic 1950 article "Can a machine think?" he proposed a test for evaluating whether a machine is intelligent. At the beginning of the 1960s, neural network research became widespread in the world's most prominent laboratories and universities [15]. After that year, AI field has been boosted for the next 50 years in research and industry [16–18].

Artificial intelligence is a broad field that includes both machine learning and deep learning, as illustrated in Fig. 1. Machine learning is a subdomain of artificial intelligence, and deep learning is a subdomain of machine learning [19].

2.1 Deep Learning Activation Functions

Activation functions comprise the non-linear layers and are mixed with other layers to accomplish a non-linear transformation from input to output [20]. Therefore, better feature extraction can be reached by choosing appropriate activation functions [21, 22]. The choice of activation function is critical for complex real-world infor-mation such as text, image, and video to make neural network learn in adequate matter. If a linear activation function was selected to learn from images will lead to losing important features as linear functions are only single-grade polynomials, while non-linear functions is the common presentation for features for images because of they are multi-grade polynomials and multi-layered deep neural networks can learn meaningful features from data [23]. There are several common activation functions, denoted by f. Here are some examples:

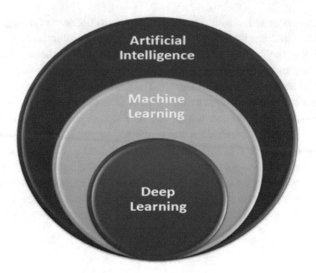

Fig. 1 Artificial intelligence, machine learning, and deep learning

- **Sigmoid function:** Changes variables to values fluctuating from 0 to 1 as shown in Eq. (1) and is frequently used as a Bernoulli distribution [24]:

$$f(a) = \frac{1}{1 + e^{-a}},\tag{1}$$

where a is the input from the front layer. An example of the Bernoulli distribution is shown below:

$$\tilde{f} = \begin{cases} 0 \text{ if } f(a) \leq 0.5 \\ 1 \text{ if } f(a) > 0.5 \end{cases}.\tag{2}$$

- **Hyperbolic tangent:** As shown in Eq. (3), the derivative of f is calculated as $f' = 1 - f^2$, to be applied into back propagation algorithms:

$$f(a) = \tan h(a) = \frac{e^a - e^{-a}}{e^a + e^{-a}}.\tag{3}$$

- **Softmax:** SoftMax is the most frequent layer and is used in the last fully connected layer and calculated as follows:

$$f(a) = \frac{e^{a_i}}{\sum_j e^{a_j}} \tag{4}$$

- **Rectified linear unit (ReLU):** As shown in Eq. (5), the variants of this function and it is original one show superior performance in many situations; thus, ReLU is currently the most widespread activation function in deep learning [25–27].

$$f(a) = \max(0, a). \tag{5}$$

Softplus: As shown in Eq. (6), Softplus is one of the ReLU variants that represents a smooth estimate of ReLU.

$$f(a) = \log(1 + e^a). \tag{6}$$

2.2 Neural Networks

Deep learning architectures depends on Feed-forward neural network architectures. These architectures consist of multiple layers. In these architectures, the neurons in one layer are linked to all the neurons in the succeeding layer. The hidden layers are between the input layers and output ones [28].

In most artificial neural networks, Artificial neurons are represented by mathematical equations that model biological neural structures [29]. Let $x = (x_1, x_2, x_3 \ldots, x_n)$ be an input vector for a given neuron, $w = (w_1, w_2, w_3 \ldots, w_n)$ be a weight vector, and b be the bias. The output of the neuron is presented in Eq. (7):

$$Y = \sigma(w.x + b), \tag{7}$$

where σ represents one of the activation functions presented in the previous section.

2.3 Convolutional Neural Network

The convolutional neural networks (CNN) are deep artificial neural networks that have been applied to image classification and clustering, and object identification within images and video scenes. More specifically, they have also been used to categorize tumours, faces, individuals, street signs and many other types visual data. Recently, CNNs have become popular in computer vision fields, and medical image analysis and applications mainly depend on CNNs.

CNNs do not perceive images as humans do. Their main components typically consist of convolutional layers and pooling layers in its first stages [30]. The layers in a CNN are trained in a robust manner.

Convolutional Neural Networks are the most successful type of architecture for image classification and detection to date. A single CNN architecture contains many different layers of neural networks that work on classifying edges and simple/complex features on shallower layers and more complex deep features in deeper layers. An image is convolved with filters (kernels) and then pooling is applied, this process may go on for some layers and at last recognizable feature are obtained [31]; however, the same combination of filters is shared among all neurons within a feature map. Mathematically, the sum of the convolutions is used instead of the dot product in Eq. (8). Thus, the k-th feature map is calculated by

$$y^k = \sigma\left(\sum_m w_m^k * x_m + b^k\right), \tag{8}$$

where the set of input feature maps are summed, $*$ is the convolution operator, and w_m^k represents the filters.

A pooling layer works on reduce the spatial dimensions of the representation output by a convolutional layer; this operation reduces the number of parameters and the amount of computation within the network. Pooling works self-sufficiently on its input at every depth and has a stride parameter similar to that of a filter in a convolutional layer. Max pooling is most commonly applied.

3 Potato Leaf Blight Dataset

The potato leaf blight dataset was introduced in the Kaggle competition in 2016 [32]. The dataset consists of 1,722 images classified into three categories: healthy potato leaves (122 images), early blight potato leaves (800 images) and late blight potato leaves (800 images). Figure 2 provides some example images from each category.

4 Proposed Neural Network Architecture

This study conducted numerous experimental trials before proposing the following architecture. While similar experiments have been conducted in previous studies [31, 33–35], the resulting test accuracy was unacceptable. Therefore, there was a need to design a new architecture. Figure 3 shows simple view of the proposed architecture for the proposed deep potato leaf blight classification.

The proposed deep learning structure in Fig. 4 consists of 14 layers, including two convolutional layers for feature extraction, with convolution window sizes of

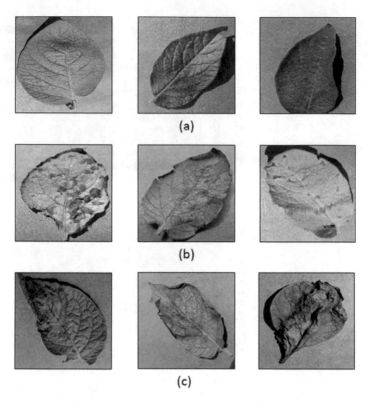

Fig. 2 Sample images from the potato leaf blight dataset, (**a**) healthy leaf category, (**b**) early blight leaf category and (**c**) late blight leaf category

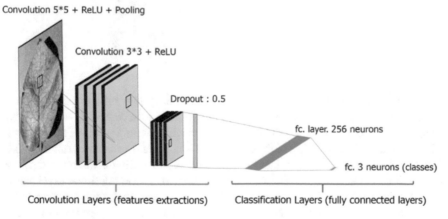

Fig. 3 The proposed deep learning structure

Fig. 4 Layer details of the proposed deep neural architecture

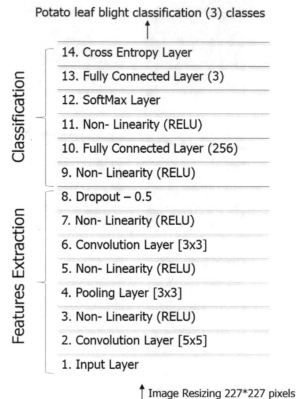

Potato leaf blight classification (3) classes

↑

Classification	14. Cross Entropy Layer
	13. Fully Connected Layer (3)
	12. SoftMax Layer
	11. Non- Linearity (RELU)
	10. Fully Connected Layer (256)
	9. Non- Linearity (RELU)
Features Extraction	8. Dropout – 0.5
	7. Non- Linearity (RELU)
	6. Convolution Layer [3x3]
	5. Non- Linearity (RELU)
	4. Pooling Layer [3x3]
	3. Non- Linearity (RELU)
	2. Convolution Layer [5x5]
	1. Input Layer

↑ Image Resizing 227*227 pixels

5×5 and 3×3 pixels, correspondingly, followed by two fully connected layers for classification. The first layer is the input layer, which accepts a 227×227-pixel image. The second layer is a convolutional layer with a window size of 5×5 pixels. The third layer is a ReLU, which is used as the nonlinear activation function. The ReLU is followed by intermediate pooling with sub-sampling using a window size of 3×3 pixels and another ReLU layer (layer five). A convolutional layer with a window size of 3×3 pixels and another ReLU activation function layer form layers six and seven. A dropout layer and ReLU is performed in layers eight and nine. Layer ten is a fully connected layer with 256 neurons and a ReLU activation function, and the final fully connected layer has 3 neurons to categorize the results into 3 classes for potato leaf blight classification. The model uses a softmax layer as layer number fourteen to determine the class membership, which predicts whether a potato-leaf image belongs to the healthy, early or late blight category.

5 Data Augmentation

The proposed architecture has an enormous amount of learnable parameters compared to the amount of images available in the training set. The original dataset contains 1722 images representing the 3 classes of potato leaf blight. Because of the large difference between the learnable parameters and the number of images in the training set, the model is highly likely to suffer from overfitting. Deep learning architectures achieve better accuracy when large training datasets are offered. One increasingly popular way to make such datasets larger is to conduct data augmentation (also sometimes called jittering) [36]. Data augmentation can increase the size of a dataset by up to 10 or 20 times its original size. The additional training data helps the model avoid overfitting when training on small amounts of data. Thus, data augmentation assists in building simpler, more robust, and more generalizable models [37]. This section introduces the common techniques for overcoming overfitting.

5.1 Augmentation Techniques

Augmentation Techniques are applied on order to overcome overfitting. This is done by increasing the number of images used for training. Data augmentation schemes are applied to the training set, and they can make the resulting model more invariant to reflection, zooming and small noises in pixel values. Images in the training data are transformed by apply augmentation techniques. The equation used for transformation is as follows:

Reflection X: Each image is flipped vertically as shown in Eq. (9):

$$\begin{bmatrix} x' \\ y' \end{bmatrix} = \begin{bmatrix} 1 & 0 \\ 0 & -1 \end{bmatrix} \cdot \begin{bmatrix} x \\ y \end{bmatrix}. \tag{9}$$

Reflection Y: Each image is flipped as shown in Eq. (10).

$$\begin{bmatrix} x' \\ y' \end{bmatrix} = \begin{bmatrix} -1 & 0 \\ 0 & 1 \end{bmatrix} \cdot \begin{bmatrix} x \\ y \end{bmatrix}. \tag{10}$$

Reflection XY: Each image is flipped horizontally and vertically as shown in Eq. (11).

$$\begin{bmatrix} x' \\ y' \end{bmatrix} = \begin{bmatrix} -1 & 0 \\ 0 & -1 \end{bmatrix} \cdot \begin{bmatrix} x \\ y \end{bmatrix}. \tag{11}$$

Zoom: The content of each image is magnified in the training phase by first cropping the image from 0,0 to 150,150 and then scaling the result to the original image size (277 * 277 pixels) using Eq. (12):

$$\begin{bmatrix} x' \\ y' \end{bmatrix} = \begin{bmatrix} \text{Xscale} & 0 \\ 0 & \text{Yscale} \end{bmatrix} \cdot \begin{bmatrix} x \\ y \end{bmatrix} \tag{12}$$

Gaussian noise is considered additive noise, and it generally interferes with the grey values in digital images. The probability density function with respect to the grey values is shown in Eq. (13) [38]:

$$P(g) = \left(\sqrt{\frac{e}{2\pi\sigma^2}} \right)^{\frac{-(g-\mu)^2}{2\sigma^2}}, \tag{13}$$

where g is the grey value, σ is the standard deviation and μ is the mean. As illustrated in Fig. 5, in terms of the probability density function (PDF), the mean value is zero, the variance is 0.1 and there are 256 grey levels.

The data augmentation technique mentioned above were applied to the potato leaf dataset, increasing the number of images fivefold, from 1,722 to 9,822 images. This increase leads to a significant improvement during neural network training. Additionally, it reduces overfitting in the proposed design and makes it more robust

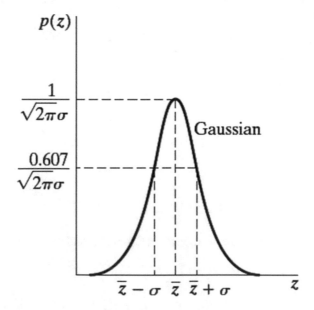

Fig. 5 The probability density function for Gaussian noise

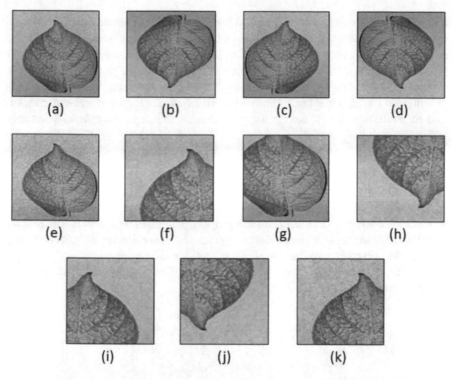

Fig. 6 Samples of augmented images: (**a**) original image, (**b**) reflected image around the X axis, (**c**) reflected image around the Y axis, (**d**) reflected image around the X-Y axes, (**e**) Gaussian noise applied to the original image, (**f**) zoomed image from [0,0] to [150,150], (**g**) zoomed image from [50,50] to [200,200], (**h**) zoomed image (**f**) reflected around the X axis, (**i**) zoomed image from (**f**) reflected around the Y axis, (**j**) zoomed image from (**f**) reflected around the X-Y axis and (**b**) zoomed image from (**f**) with applied Gaussian noise

during the testing and verification phases. Figure 6 shows the result of applying the described data augmentation techniques to a sample image from the dataset.

6 Experimental Results

The proposed architecture was implemented using MATLAB, and GPU specific. All trials were finalised on a computer equipped with an E52620 Intel Xeon processor (2 GHz) and 96 GB of memory.

To measure the accuracy of the proposed architecture for potato leaf blight classification using the proposed deep convolutional neural network, the dataset was divided into 2 sets of 80% and 20%; the 80% portion was used for training, while the remaining 20% was used for testing. The average overall testing accuracy for the original (non-augmented) dataset was 94.8%, while the average overall testing accuracy for the augmented dataset was 98%. A confusion (or error) matrix is considered a quantitative approach for characterizing image classification accuracy because it provides concrete evidence about the performance of a model. The confusion matrix for one of the testing accuracy trials on the original (non-augmented) dataset is presented in Fig. 7, while Fig. 8 shows a confusion matrix for the augmented dataset. The augmentation process has been applied into the training set which contains only 80% of the original dataset.

There are three types of accuracy scores [39]

- **Producer's Accuracy:** The ratio of images properly classified into class X with respect to the number of the images observed to be class X. This metric reflects the model's accuracy from the model's perspective and is equivalent to sensitivity.
- **User's Accuracy:** The ratio of images properly classified into class X with respect to the total number of images predicted as class X. This metric reflects the accuracy from the perspective of the model's user and shows its positive predictive power.
- **Overall Accuracy:** The ratio of properly classified images with respect to the total number of images.

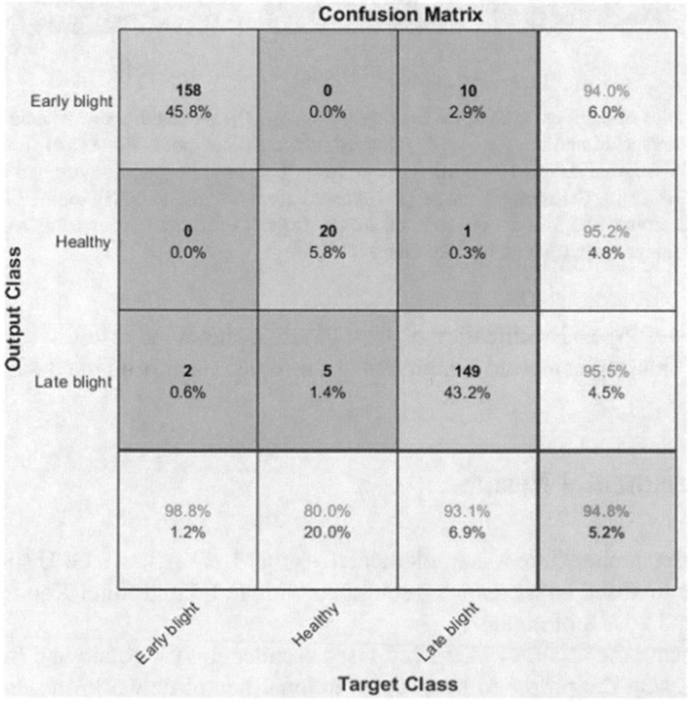

Fig. 7 Confusion matrix for one of the trials on the original (non-augmented) dataset

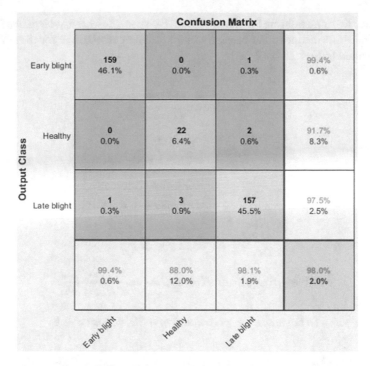

Fig. 8 Confusion matrix for one of the trials on the augmented dataset

Tables 1 and 2 present the producer's and user's accuracy for the original and augmented datasets, respectively.

Performance Evaluation and Discussion

To fully evaluate the performance of the proposed architecture, more performance measures must be investigated in this study. The most common performance metrics

Table 1 Producer's and user's accuracy for the proposed architecture on the original (non-augmented) dataset

	Early blight leaf (%)	Late blight leaf (%)	Healthy leave (%)
Producer accuracy	94.0	95.5	95.2
User accuracy	98.8	93.1	80.8

Table 2 Producer and user accuracy for the proposed architecture on the augmented dataset

	Early blight leaf (%)	Late blight leaf (%)	Healthy leave (%)
Producer accuracy	99.4	97.5	91.7
User accuracy	99.4	98.1	88.0

in the deep learning field are Precision, Recall, F1-score, Selectivity, Negative Predictive Value, Informedness and Markedness [40]. The calculations for these metrics are presented in Eqs. (14)–(20).

$$Precision = \frac{TP}{(TP + FP)} \tag{14}$$

$$Recall = \frac{TP}{(TP + FN)} \tag{15}$$

$$F1 - score = 2.\frac{Precision \cdot Recall}{(Precision + Recall)} \tag{16}$$

$$Selectivity = \frac{TN}{TN + FP} \tag{17}$$

$$Negative\ Predictive\ Value = \frac{TN}{TN + FN} \tag{18}$$

$$Informedness = Precision + Selectivity - 1 \tag{19}$$

$$Markedness = Recall + Negative\ Predictive\ Value - 1 \tag{20}$$

where TP is the number of true positive samples, TN is the count of true negative samples, FP is the count of false positive samples, and FN is the count of false negative samples from a confusion matrix.

Table 3 presents the performance measures for the proposed deep learning architecture on the datasets with/without the augmentation process, clearly showing that the performance measure results are better on the augmented dataset in terms of the achieved accuracies. The adopted augmentation techniques increased the dataset size, reduced overfitting, and finally, helped the model achieve better performance metrics and improved its overall testing accuracy.

Table 4 illustrates comparative results for related works and the proposed architecture. All the related works presented in Table 2 were applied to potato leaf blight datasets—not the same dataset used in our research, but the images are highly similar to those in our dataset. The results reveal that the proposed architecture achieved the highest overall testing accuracy (98%). Moreover, the proposed architecture is more robust and immune overfitting due to the augmentation process adopted in this study.

Table 3 Performance measures for the proposed deep learning architecture on the datasets with and without augmentation

	Without augmentation (%)	With augmentation (%)
Precision	83.71	94.75
Recall	88.44	93.22
F1-score	85.96	93.98
Selectivity	94.86	98.59
Negative predictive value	96.66	98.13
Informedness	78.58	93.35
Markedness	85.11	91.35

Table 4 Comparison results of related works and the proposed architecture

Related work	Year	Description	Accuracy (%)
[13]	2016	K-means clustering + segmentation based on colour, texture, and shape + backpropagation neural network	92.00
[14]	2017	K-means clustering + segmentation based on colour, texture, and shape + feed-forward neural network	95.30
Proposed architecture	2019	Deep convolutional neural networks	98.00

7 Conclusions and Future Work

Crop diseases are a common threat to food security in all nations, but with the artificial intelligence advances in detection and classification allow these threats to be maintained and eliminated at early stages rapidly and accurately. This paper presented a proposed deep learning architecture for potato leaf blight classification. The proposed architecture consists of 14 layers: two main convolutional layers for feature extraction with different convolution window sizes followed by two fully connected layers for classification. Augmentation processes were applied to increase the number of dataset images from 1722 to 9822, resulting in a substantial improvement in the overall testing accuracy. The proposed architecture achieved an overall mean testing accuracy of 98%. A confusion matrix showing all the different accuracy types has been presented in this research, and the performance measures are calculated and presented in this research. Finally, the results of a testing accuracy comparison between the proposed architecture and other related works were presented. The proposed architecture achieved better results than the related works in terms of overall testing accuracy. One prospect for future work is to apply transfer learning based on more advanced pre-trained deep neural network architectures such as AlexNet, VGG-16, and VGG-19. Using pre-trained architectures may reduce the computation time in the training phase and may lead to better testing accuracy.

References

1. Pretty, J.: Agricultural sustainability: concepts, principles and evidence. Philos. Trans. R. Soc. B: Biol. Sci. **363**, 447–465 (2008)
2. Tilman, D., Cassman, K.G., Matson, P.A., Naylor, R., Polasky, S.: Agricultural sustainability and intensive production practices. Nature **418**, 671–677 (2002)
3. Gavhale, M.K.R., Gawande, P.U.: An overview of the research on plant leaves disease detection using image processing techniques. IOSR J. Comput. Eng. **16**, 10–16 (2014)
4. Briggs, G.E.: Advances in plant physiology. Nature **169**, 637 (1952)
5. Balodi, R., Bisht, S., Ghatak, A., Rao, K.H.: Plant disease diagnosis: technological advancements and challenges. Indian Phytopathol. **70**, 275–281 (2017)
6. Martinelli, F., et al.: Advanced methods of plant disease detection. A review. Agron. Sustain. Dev. **35**, 1–25 (2015)
7. Dengel, A.: Special issue on artificial intelligence in agriculture. KI - Künstliche Intelligenz **27**, 309–311 (2013)
8. Bonino, D., Procaccianti, G.: Exploiting semantic technologies in smart environments and grids: emerging roles and case studies. Sci. Comput. Program. **95**, 112–134 (2014)
9. Kumar, R.: Feature extraction of diseased leaf images. J. Signal Image Process. **3**, 60–63 (2012)
10. Revathi, P., Hemalatha, M.: Identification of cotton diseases based on cross information gain_deep forward neural network classifier with PSO feature selection. Int. J. Eng. Technol. **5**, 4637–4642 (2013)
11. Sladojevic, S., Arsenovic, M., Anderla, A., Culibrk, D., Stefanovic, D.: Deep neural networks based recognition of plant diseases by leaf image classification. Comput. Intell. Neurosci. (2016)
12. Particle Swarm Optimization: Particle Swarm Optimization Introduction. Optimization (2007)
13. Athanikar, G., Badar, P.: Potato leaf diseases detection and classification system. Int. J. Comput. Sci. Mob. Comput **5**, 76–78 (2016)
14. El Massi, I., Es-saady, Y., El Yassa, M., Mammass, D., Benazoun, A.: Automatic recognition of vegetable crops diseases based on neural network classifier. Int. J. Comput. Appl. **158**(4), 48–51 (2017)
15. Atkinson, J., Solar, M.: Artificial intelligence and intelligent systems research in Chile. In: Lecture Notes in Computer Science (Including Subseries Lecture Notes in Artificial Intelligence and Lecture Notes in Bioinformatics) (2009)
16. Gevarter, W.B.: Introduction to artificial intelligence. Chem. Eng. Prog. (1987)
17. Saygin, A.P., Cicekli, I., Akman, V.: Turing test: 50 years later. Minds Mach. **10**, 463–518 (2000)
18. Stone, A.T.P.: Artificial intelligence and life in 2030. One Hundred Year Study Artif. Intell. Rep. 2015-2016 Study Panel (2016)
19. Johnson, M.R.: Artificial intelligence. In: Procedural Generation in Game Design (2017)
20. LeCun, Y., Bengio, Y., Hinton, G.: Deep learning. Nature **521**(7553), 436–444 (2015)
21. Krizhevsky, A., Sutskever, I., Hinton, G.E.: ImageNet classification with deep convolutional neural networks. In: ImageNet Classification with Deep Convolutional Neural Networks (2012)
22. Singh, R.G., Kishore, N.: The impact of transformation function on the classification ability of complex valued extreme learning machines. In: 2013 International Conference on Control, Computing, Communication and Materials, ICCCCM 2013 (2013)
23. Nwankpa, C., Ijomah, W., Gachagan, A., Marshall, S.: Activation functions: comparison of trends in practice and research for deep learning, pp. 1–20 (2018)
24. Gooch, J.W.: Bernoulli distribution. In: Encyclopedic Dictionary of Polymers (2011)
25. Bengio, Y.: Practical recommendations for gradient-based training of deep architectures. Lecture Notes on Computer Science (including Subseries Lecture Notes on Artificial Intelligence Lecture Notes Bioinformatics) (2012)
26. Maas, A.L., Hannun, A.Y., Ng, A.Y.: Rectifier nonlinearities improve neural network acoustic models. In: ICML 2013 (2013)

27. Nair, V., Hinton, G.: Rectified linear units improve restricted Boltzmann machines. In: Proceedings of the 27th International Conference on Machine Learning (2010)
28. Liu, W., Wang, Z., Liu, X., Zeng, N., Liu, Y., Alsaadi, F.E.: A survey of deep neural network architectures and their applications. Neurocomputing **234**, 11–26 (2017)
29. Staub, S., Karaman, E., Kaya, S., Karapınar, H., Güven, E.: Artificial neural network and agility. Procedia - Soc. Behav. Sci. **195**, 1477–1485 (2015)
30. Gu, J., et al.: Recent advances in convolutional neural networks. Pattern Recognit. **77**, 354–377 (2018)
31. Khalifa, N.E., Taha, M.H., Hassanien, A.E., Selim, I.: Deep galaxy V2: robust deep convolutional neural networks for galaxy morphology classifications. In: 2018 International Conference on Computing Sciences and Engineering, ICCSE 2018 – Proceedings, pp. 1–6 (2018)
32. Mallah, C., Cope, J., Orwell, J.: Plant leaf classification using probabilistic integration of shape, texture and margin features (2013)
33. Khalifa, N.E.M., Taha, M.H.N., Hassanien, A.E.: Aquarium family fish species identification system using deep neural networks (2018)
34. Khalifa, N.E.M., Taha, M.H.N., Hassanien, A.E., Selim, I.M.: Deep galaxy: classification of galaxies based on deep convolutional neural networks (2017)
35. Khalifa, N., Taha, M., Hassanien, A., Mohamed, H.: Deep iris: deep learning for gender classification through iris patterns. Acta Inform. Medica **27**(2), 96 (2019)
36. Xie, Q., Dai, Z., Hovy, E., Luong, M.-T., Le, Q.V.: Unsupervised data augmentation. In: NIPS Submission (2019)
37. Lemley, J., Bazrafkan, S., Corcoran, P.: Smart augmentation learning an optimal data augmentation strategy. IEEE Access **5**, 5858–5869 (2017)
38. Boyat, A.K., Joshi, B.K.: A review paper: noise models in digital image processing. Signal Image Process. Int. J. **6**(2), 63–75 (2015)
39. Patil, G.P., Taillie, C.: Modeling and interpreting the accuracy assessment error matrix for a doubly classified map. Environ. Ecol. Stat. **10**, 357–373 (2003)
40. Goutte, C., Gaussier, E.: A probabilistic interpretation of precision, recall and f-score, with implication for evaluation (2010)

Granules-Based Rough Set Theory for Circuit Breaker Fault Diagnosis

Rizk M. Rizk-Allah and Aboul Ella Hassanien

Abstract This chapter presents a new granules strategy integrated with rough set theory (RST) to extract diagnosis rules for redundant and inconsistent data set of high voltage circuit breaker (HVCB). In this approach, the diagnostic knowledge base is performed by the granules of indiscernible objects based on tolerance relation in which the objects are collected based on permissible scheme. This permissible scheme is decided by the opinion of the expert or the decision maker. In addition, a topological vision is introduced to induce the lower and upper approximations. Finally, the validation and effectiveness of the proposed granules strategy are investigated through a practical application of the high voltage circuit breaker fault diagnosis.

Keywords Circuit-breaker · Granular · Rough set theory · Topological space

1 Introduction

HIGH-voltage circuit breakers are very important switching equipment in the electrical power system, and have the double function of control and protection in the electrical network. The equipment fault in electric power system effects the operation of systems, which may cause great losses and lead to series of safety and economic problems, in some serious cases it even results in system failure. Therefore, the early fault diagnosis for the high voltage circuit breakers are one of the important means to ensure the system operation safe.

R. M. Rizk-Allah (✉)
Department of Basic Engineering Science, Faculty of Engineering, Menoufia University, Shebin El-Kom, Egypt
e-mail: rizk_masoud@yahoo.com

A. E. Hassanien
Scientific Research Group in Egypt, Cairo, Egypt
URL: http://www.egyptscience.net

© The Author(s), under exclusive license to Springer Nature Switzerland AG 2021
A. E. Hassanien et al. (eds.), *Machine Learning and Big Data Analytics Paradigms: Analysis, Applications and Challenges*, Studies in Big Data 77,
https://doi.org/10.1007/978-3-030-59338-4_5

Rough set theory proposed by Pawlak [1], is a powerful and versatile tool for dealing with inexact, uncertain and insufficient information. It has been successfully applied but not limited to various fields, such as [2–6].

In the theory of rough sets, a pair of lower and upper approximation operators is constructed based on an equivalence relation on the universe of discourse. Equivalence classes of equivalence relation form a partition of the universe; partition is a basic concept in Pawlak rough set model. In order to relax the restriction of partition, many authors try to replace the partition with coverings [7], neighborhood systems [8] and abstract approximation space [9].

Granular computing is an emerging field of study focusing on structured thinking, structured problem solving and structured information processing with multiple levels of granulation [10, 11]. In the viewpoint of granular computing, knowledge of universe can be described by granular structure. Xu et al. [12] construct a generalized approximation space with two subsystems, as there may be no connection between two subsystems, so the corresponding lower approximation and upper approximation are not dual to each other.

In this chapter, we propose a new reduction approach based on a granular structure for fault diagnosis of circuit-breaker. Since there are instabilities in measuring process and fluctuations of the information system, a new relation is introduced. This relation involves a certain tolerance whose possible values may be assigned by the expert. Also, this relation is aided for forming the partitions. In addition, granular structure based on the topological vision is employed to establish the lower and the upper approximation. The lower approximation is obtained from the set system that contains U (universe of discourse), subsets of U and the empty set. The upper approximation is obtained from the dual set system, consists of complements of all elements of the set system. Finally, the correctness and effectiveness of this approach are validated by the result of practical fault diagnosis example.

The rest of the chapter is organized as follows. In Sect. 2 we describe some basic concepts regarding rough set theory. In Sect. 3, granular reduction approach is proposed for solving high voltage circuit breaker. Section 4 presents the experimentation and the results. The conclusion and future work are included in Sect. 5.

2 Basic Concepts

In this section, we will review some basis concepts related to rough set theory. Throughout this chapter, we suppose that the universe U is a finite nonempty set.

An information system (IS) [1] is denoted as a pair $I = (U, A)$, where U, called universe, is a nonempty set of finite objects; A is a nonempty finite set of attributes such that $a : U \rightarrow V_a$ for every $a \in A$; V_a is the value set of a. In a decision system, $A = C \cup D$, where C is the set of condition attributes and D is the set of decision attributes. For an attribute set $P \subseteq A$, there is an associated indiscernibility relation $\text{IND}(P)$ as in (1):

$$\text{IND}(P) : \left\{ (x, y) \in U^2 \mid \forall a \in P, \ a(x) = a(y) \right\} \tag{1}$$

If $(x, y) \in \text{IND}(P)$, then x and y are indiscernible by attributes from P. The family of all equivalence classes of $\text{IND}(P)$, i.e., the partition determined by P, is denoted as U/P. An equivalence class of $\text{IND}(P)$, i.e., the block of the partition U/P, containing x is denoted by $[x]_P$. The indiscernibility relation is the mathematical basis of rough set theory.

In rough set theory, the lower and upper approximations are two basic operations. Given an arbitrary set $X \subseteq U$, the P-lower approximation of X, denoted as $\underline{P}X$, is the set of all elements of U which can be certainly classified as elements of X based on the attribute set P. The P-upper approximation of X, denoted as $\overline{P}X$, is the set of all elements of U, which can be possibly classified as elements of X based on the attribute set P. These two definitions can be expressed as in (2) and (3) respectively as follows:

$$\underline{P}X = \{x \in U \mid [x]_P \subseteq X\} \tag{2}$$

$$\underline{P}X = \{x \in U \mid [x]_P \cap X \neq \phi\} \tag{3}$$

The P-boundary region of X is: $PN(X) = \overline{P}X - \underline{P}X$. Where, X is said to be roughly if and only if $\overline{P}X \neq \underline{P}X$ and boundary region $\neq \phi$, X is said to be definable if and only if $\overline{P}X = \underline{P}X$ and boundary region $=\phi$.

Another issue of great practical importance is that of "superfluous" data in a data table. Superfluous data can be eliminated, in fact, without deteriorating the information contained in the original table [1]. Let $P \subseteq A$ and $a \in P$. It is said that attribute a is superfluous in P if $\text{IND}(P) = \text{IND}(P - \{a\})$ otherwise, a is indispensable in P.

The set P is independent (orthogonal) if all its attributes are indispensable. The subset P' of P is a reduct of P (denotation RED (P)) if P' is independent and $\text{IND}(P) = \text{IND}(P')$.

More than one reduct of P may exist in a data table. The set containing all the indispensable attributes of P is known as the core. Formally as in (4).

$$Core(P) = \cap RED(P) \tag{4}$$

3   The Proposed Granular Structure and Reduction

The condition of equivalence relation in the approximation space limits the range of applications, which absolutely does not permit any overlapping among its granules, seems to be too restrictive for real world problems. The purpose of this article is to use a granular structure to overcome this limitation that is, the partition determined by P does permit small degree of overlapping that assigned by the expert, i.e., $[x_i]_P \cap [x_j]_P \neq \phi$. Then the classification which is formed by this approach is reformed as a general topological method, science it does not use any distance measures and depends on a structure with minimal properties.

3.1 Knowledge Base

The approximation generalized space is a pair of (U, R), where U is a non-empty finite set of objects (states, patients, digits, cars, ... etc.) called a universe and R is an equivalence relation over U which makes a partition for U, i.e. a family $C = \{X_1, X_2, \ldots, X_n\}$ such that $X_i \subseteq U, X_i \cap X_j \neq \phi \vee X_i \cap X_j = \phi$ for $i \neq j, i, j = 1, 2, \ldots, n$ and $\cup X_i = U$, the class C is called the knowledge base of (U, R). Thus the introduced relation is defined as follows:

Definition 3.1 Let $\delta = (U, R)$ be a generalized approximation space, then $R(x) = \{y \in U : xRy\}$ is called R-related element of x as in (5):

$$x R_P y = \sum_{a \in P} |a(x_i) - a(y_j)|/|P| \leq \lambda, \quad i, j = 1, 2, \ldots, |U| \tag{5}$$

where λ is user defined that assigned by expert and the collection of all R-related element in δ is denoted by $\chi = \{R(x) : x \in U\}$.

The following is an example of an approximation space and its associated knowledge base.

Example 3.1 Let $U = \{1.3, 2.1, 1.05, 2.3, 1.5\}$ and relation R which partition U into classes with permissible degree, i.e. $\lambda \leq 0.25$, then knowledge base is:

$$U/R = \{\{1.3, 1.05, 1.5\}, \{2.1, 2.3\}, \{1.3, 1.05\}, \{1.3, 1.5\}\}$$
$$X_1 = \{1.3, 1.05, 1.5\}, X_2 = \{2.1, 2.3\}, X_3 = \{1.3, 1.05\}, X_4 = \{1.3, 1.5\},$$
$$X_1 \cap X_2 = \phi, X_1 \cap X_3 \neq \phi, X_1 \cap X_4 \neq \phi, X_2 \cap X_3 = \phi, X_2 \cap X_4 = \phi,$$
$$X_3 \cap X_4 \neq \phi \text{ and } \cup X_i = U$$

3.2 Topological Granular Space

A topological space [1] is a pair (U, τ) consisting of a set U and family τ of subset of U satisfying the following conditions:

(T1) $\phi \in \tau$ and $U \in \tau$.
(T2) τ is closed under arbitrary union.

(T3) τ is closed under finite intersection.

The pair (U, τ) is called a space, the elements of U are called points of the space, the subsets of U belonging to τ are called open set in the space, and the complement of the subsets of U belonging to τ are called closed set in the space; the family τ of open subsets of U is also called a topology for U.

It often happens that some of subsets, τ, comes from a real-life issue, these subsets can be used as known knowledge about the universe. In some cases it is very complicated to satisfy the conditions (T2) and (T3). In this case, the complements of subsets of τ can also be understandable in general, thus they can be used as another type of known knowledge. Consequently, topological granular approximation method is proposed as follows:

Definition 3.2 Let (U, S) consisting of a set U and family S of subset of U, denoted by a set system on U, satisfying the condition T1. Then the triplet (U, S, S^c) is called a topological granular space. For any $X \subseteq U$, lower and upper approximations of X in (U, S, S^c) are defined as in (6) and (7) respectively by

$$\underline{S}X = \cup \{A \in S | A \subseteq X\} \tag{6}$$

$$\overline{S}X = \cap \{A \in S^c | X \subseteq A\} \tag{7}$$

where $\underline{S}X$, $\overline{S}X$ and $\overline{S}X - \underline{S}X$ are called the positive domain, negative domain and boundary domain of X in (U, S, S^c), respectively. For any $X \subseteq U, X$ is called lower-definable if $\underline{S}X = X$, X is called upper-definable if $\overline{S}X = X$. If X is lower-definable and upper definable, then X is called definable.

Example 3.2 Let topological granular space (U, C, C^c) is defined as

$$U = \{1, 2, 3, 4, 5, 6, 7\},$$
$$S = \{U, \phi, \{1, 3\}, \{4, 7\}, \{3, 6, 7\}, \{2, 5\}, \{1, 3, 5\}, \{3, 5, 7\}\},$$
$$S^c = \{U, \phi, \{2, 4, 5, 6, 7\}, \{1, 2, 3, 5, 6\}, \{1, 2, 4, 5\},$$
$$\{1, 3, 4, 6, 7\}, \{2, 4, 6, 7\}, \{1, 2, 4, 6\}\},$$
$$X = \{1, 3, 5, 7\},$$

Then we can compute the lower and upper approximation using Definition 3.2 as follows:

$$\underline{S}X = \{1, 3, 5, 7\}, \quad \overline{S}X = \{U\}.$$

Remark It should be pointed out that the notions of the lower and upper approximation proposed in Definition 3.2 are generalization of Pawlak rough approximations. In Pawlak approximation space (U, R), taking $S = U/R$, it is not difficult to check that $\underline{S}X = \underline{R}X$ and $\overline{S}X = \overline{R}X$ for all $X \in P(U)$.

3.3 Reduction of Knowledge

Let $B \subseteq A, a \in B$ then a is superfluous attributes in B if the partitions is described as in (8):

$$U/IND(B) = U/IND(B - \{a\}) \tag{8}$$

The set M is called a minimal reduct of B if:

$$U/IND(M) = U/IND(B) \ \& \ U/IND(M) \neq U/IND(M - \{a\}), \quad \forall a \in M.$$

For example let the reduct of B denoted by $RED(B)$ = $\{a, b, c\}, \{a, b\}, \{a, b, d\}, \{c, d\}$. But the minimal reduct is $M = \{\{a, b\}, \{c, d\}\}$. Then the core is the set of all characteristic of knowledge, where cannot be eliminated from knowledge at reduct of knowledge as in (9).

$$CORE(B) = \cap RED(B) = \{a, b\} \cap \{c, d\} = \phi \tag{9}$$

4   Experimentation and Comparison of Results

In order to certify the correctness of the fault diagnosis method, take ZN42-27.5kV indoor vacuum circuit breaker fault diagnosis as an example. High voltage circuit breaker uses operating actuator's control contact to break or close the circuit. An action includes three stages-energy storage, closing operation and opening operation. We monitor its current signal of opening and closing operating coil with Hall sensor. The current waveform of typical opening/closing operating coil is illustrated in Fig. 1.

Fig. 1 The current waveform of opening/closing operating coil

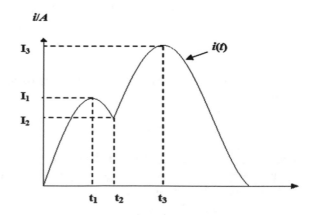

Table 1 The original decision table

U	I_1	I_2	I_3	D
x_1	1.61	1.11	2.23	1
x_2	1.62	1.17	2.18	0
x_3	1.61	1.21	2.28	0
x_4	1.61	1.13	2.21	0
x_5	1.64	1.09	2.27	1
x_6	1.625	1.13	2.26	1
x_7	1.63	1.15	2.21	0

Taking t_0 as time zero, we measured three characteristic parameters of I_1, I_2 and I_3. These three values can respectively reflect the information of supply voltage and coil resistant, as well as electromagnet core's movement. So, we chose these three values as condition attribute and took it that whether has "closed core jamming at the beginning (HKS)" fault as decision attribute. To select seven groups of data to build the original decision table, while D represents that the sample whether has any fault. If 1 means yes, while 0 means no. The original decision table is shown in Table 1.

This section is devoted to analyze the fault samples data of high voltage circuit breaker with regarding the proposed approach method and get diagnosis rules from it. The obtained rules show the proposed approach is practical and feasible regarding two cases, the normal operation case and abnormal operation case. The distinguished feature of the proposed approach is to overcome the drawbacks of classical rough set theory such as discretization of continuous attributes and disjoint classes.

The proposed approach is implemented in details regarding all attributes, $P = \{I_1, I_2, I_3\}$. Let the universe $U = \{x_1, x_2, x_3, x_4, x_5, x_6, x_7\}$ represents the sample date of the circuit breaker and $X = \{x_2, x_3, x_4, x_7\}$ represents the decision at normal operation of circuit breaker that means no fault. Then, the equivalence classes are determined by groping that objects that satisfy R-related element by using Definition 3.1 as shown in Table 2.

The relation R which partition U into classes with permissible degree, i.e. $\lambda \leq 0.02$, then knowledge base is:

Table 2 Demonstrates the relation matrix

P	x_1	x_2	x_3	x_4	x_5	x_6	x_7
x_1	0	0.04	0.05	0.0133	0.03	0.0217	0.0267
x_2	0.04	0	0.05	0.0267	0.0633	0.0417	0.02
x_3	0.05	0.05	0	0.05	0.0533	0.0383	0.05
x_4	0.0133	0.0267	0.05	0	0.0433	0.0217	0.0133
x_5	0.03	0.0633	0.0533	0.0433	0	0.0217	0.0433
x_6	0.0217	0.0417	0.0383	0.0217	0.0217	0	0.025
x_7	0.0267	0.02	0.05	0.0133	0.0433	0.025	0

$$U/IND(P) = \{\{x_1, x_4\}, \{x_2, x_7\}, \{x_3\}, \{x_1, x_4, x_7\}, \{x_5\}, \{x_6\}, \{x_2, x_4, x_7\}\}$$
$$S = \{\phi, \{x_1, x_4\}, \{x_2, x_7\}, \{x_3\}, \{x_1, x_4, x_7\}, \{x_5\}, \{x_6\}, \{x_2, x_4, x_7\}, U\}$$
$$\{S\}^c = \{\phi, \{x_2, x_3, x_5, x_6, x_7\}, \{x_1, x_3, x_4, x_5, x_6\}, \{x_1, x_2, x_4, x_5, x_6, x_7\},$$
$$\{x_2, x_3, x_5, x_6, \}, \{x_1, x_2, x_3, x_4, x_6, x_7\},$$
$$\{x_1, x_2, x_3, x_4, x_5, x_7\}, \{x_1, x_3, x_5, x_6\}, U\}$$

The lower and upper approximations are obtained using the Definition 3.2 as follows:

$$\underline{P}X = \{x_2, x_3, x_4, x_7\} \text{ and } \overline{P}X = \{x_1, x_2, x_3, x_4, x_7\}.$$

For attribute reduction the superfluous attribute is checked to get the minimal reduction as follows:

$$U/IND(P - \{I_1\}) = \{\{x_1, x_4\}, \{x_2\}, \{x_3\}, \{x_1, x_4, x_7\}, \{x_5\}, \{x_6\}, \{x_4, x_7\}\}$$
$$S = \{\phi, \{x_1, x_4\}, \{x_2\}, \{x_3\}, \{x_1, x_4, x_7\}, \{x_5\}, \{x_6\}, \{x_4, x_7\}, U\}$$
$$\{S\}^c = \{\phi, \{x_2, x_3, x_5, x_6, x_7\}, \{x_1, x_3, x_4, x_5, x_6, x_7\},$$
$$\{x_1, x_2, x_4, x_5, x_6, x_7\}, \{x_2, x_3, x_5, x_6\}, \{x_1, x_2, x_3, x_4, x_6, x_7\},$$
$$\{x_1, x_2, x_3, x_4, x_5, x_7\}, \{x_1, x_2, x_3, x_5, x_6\}, U\}.$$

$$U/IND(P - \{I_2\}) = \{\{x_1, x_4, x_7\}, \{x_2, x_4, x_7\}, \{x_3, x_5, x_6\}, \{x_1, x_2, x_4, x_7\},$$
$$\{x_3, x_5, x_6\}, \{x_3, x_5, x_6\}, \{x_1, x_2, x_4, x_7\}\}$$
$$S = \{\phi, \{x_1, x_4, x_7\}, \{x_2, x_4, x_7\}, \{x_3, x_5, x_6\}, \{x_1, x_2, x_4, x_7\},$$
$$\{x_3, x_5, x_6\}, \{x_3, x_5, x_6\}, \{x_1, x_2, x_4, x_7\}, U\}$$
$$\{S\}^c = \{\phi, \{x_2, x_3, x_5, x_6\}, \{x_1, x_3, x_5, x_6\}, \{x_1, x_2, x_4, x_7\}, \{x_3, x_5, x_6\}$$
$$\{x_1, x_2, x_4, x_7\}, \{x_1, x_2, x_4, x_7\}, \{x_3, x_5, x_6\}, U\}.$$

$$U/IND(P - \{I_3\}) = \{\{x_1, x_4, x_6\}, \{x_2, x_7\}, \{x_3\}, \{x_1, x_4, x_6, x_7\}, \{x_5\},$$
$$\{x_1, x_4, x_6, x_7\}, \{x_2, x_4, x_6, x_7\}\}$$
$$S = \{\phi, \{x_1, x_4, x_6\}, \{x_2, x_7\}, \{x_3\}, \{x_1, x_4, x_6, x_7\}, \{x_5\},$$
$$\{x_1, x_4, x_6, x_7\}, \{x_2, x_4, x_6, x_7\}, U\}$$
$$\{S\}^c = \{\phi, \{x_2, x_3, x_5, x_7\}, \{x_1, x_3, x_4, x_5, x_6\},$$
$$\{x_1, x_2, x_4, x_5, x_6, x_7\}, \{x_2, x_3, x_5\}$$
$$\{x_1, x_2, x_3, x_4, x_6, x_7\}\{x_2, x_3, x_5\}, \{x_1, x_3, x_5\}, U\}$$

We found that

$$U/IND(P - \{I_1\}) = U/IND(P)$$
$$U/IND(P - \{I_2\}) \neq U/IND(P)$$

Table 3 Represents the reduct for $\{I_2, I_3\}$

U	I_2	I_3	D
x_1	1.11	2.23	1
x_2	1.17	2.18	0
x_3	1.21	2.28	0
x_4	1.13	2.21	0
x_5	1.09	2.27	1
x_6	1.13	2.26	1
x_7	1.15	2.21	0

$$U/IND(P - \{I_3\}) \neq U/IND(P)$$

Then attribute I_1 is superfluous attribute and we get:

$$U/IND(P - \{I_1\}) = U/IND(P)$$

The minimal reduct of P is $M = \{I_2, \ I_3\}$ which is shown in Table 3.

The proposed approach also is implemented in case of fault occurrence. Let the universe $U = \{x_1, x_2, x_3, x_4, x_5, x_6, x_7\}$ represents the sample date of the circuit breaker and $X = \{x_1, x_5, x_6\}$ represents the decision at abnormal operation of circuit breaker. Then, the equivalence classes as discussed earlier.

$$U/IND(P) = \{\{x_1, x_4\}, \{x_2, x_7\}, \{x_3\}, \{x_1, x_4, x_7\}, \{x_5\}, \{x_6\}, \{x_2, x_4, x_7\}\}$$
$$S = \{\phi, \{x_1, x_4\}, \{x_2, x_7\}, \{x_3\}, \{x_1, x_4, x_7\}, \{x_5\}, \{x_6\}, \{x_2, x_4, x_7\}, U\}$$
$$\{S\}^c = \{\phi, \{x_2, x_3, x_5, x_6, x_7\}, \{x_1, x_3, x_4, x_5, x_6\},$$
$$\{x_1, x_2, x_4, x_5, x_6, x_7\}, \{x_2, x_3, x_5, x_6, \}, $$
$$\{x_1, x_2, x_3, x_4, x_6, x_7\}, \{x_1, x_2, x_3, x_4, x_5, x_7\}, \{x_1, x_3, x_5, x_6\}, U\}$$

The lower and upper approximations are obtained using the Definition 3.2 as follows:

$\underline{P}X = \{x_5, x_6\}$ and $\overline{P}X = \{x_1, x_5, x_6\}$.

For attribute reduction the superfluous attribute is checked. In addition, the lower and upper approximation is obtained as follows:

When I_1 is eliminated, we obtain

$$U/IND(P - \{I_1\}) = \{\{x_1, x_4\}, \{x_2\}, \{x_3\}, \{x_1, x_4, x_7\}, \{x_5\}, \{x_6\}, \{x_4, x_7\}\}$$
$$S = \{\phi, \{x_1, x_4\}, \{x_2\}, \{x_3\}, \{x_1, x_4, x_7\}, \{x_5\}, \{x_6\}, \{x_4, x_7\}, U\}$$
$$\{S\}^c = \{\phi, \{x_2, x_3, x_5, x_6, x_7\}, \{x_1, x_3, x_4, x_5, x_6, x_7\},$$
$$\{x_1, x_2, x_4, x_5, x_6, x_7\}, \{x_2, x_3, x_5, x_6\}, \{x_1, x_2, x_3, x_4, x_6, x_7\},$$
$$\{x_1, x_2, x_3, x_4, x_5, x_7\}, \{x_1, x_2, x_3, x_5, x_6\}, U\}$$
$$\text{and } \underline{P}X = \{x_5, x_6\} \text{ and } \overline{P}X = \{x_1, x_5, x_6\}.$$

For elimination I_2 we have

$$U/IND(P - \{I_2\}) = \{\{x_1, x_4, x_7\}, \{x_2, x_4, x_7\}, \{x_3, x_5, x_6\},$$
$$\{x_1, x_2, x_4, x_7\}, \{x_3, x_5, x_6\}, \{x_3, x_5, x_6\}, \{x_1, x_2, x_4, x_7\}\}$$
$$S = \{\phi, \{x_1, x_4, x_7\}, \{x_2, x_4, x_7\}, \{x_3, x_5, x_6\}, \{x_1, x_2, x_4, x_7\},$$
$$\{x_3, x_5, x_6\}, \{x_3, x_5, x_6\}, \{x_1, x_2, x_4, x_7\}, U\}$$
$$\{S\}^c = \{\phi, \{x_2, x_3, x_5, x_6\}, \{x_1, x_3, x_5, x_6\}, \{x_1, x_2, x_4, x_7\},$$
$$\{x_3, x_5, x_6\}, \{x_1, x_2, x_4, x_7\}, \{x_1, x_2, x_4, x_7\}, \{x_3, x_5, x_6\}, U\}$$
$$\text{and } \underline{P}X = \phi \text{ and } \overline{P}X = \{x_1, x_3, x_5, x_6\}.$$

Finally, when I_3 is omitted we get

$$U/IND(P - \{I_3\}) = \{\{x_1, x_4, x_6\}, \{x_2, x_7\}, \{x_3\}, \{x_1, x_4, x_6, x_7\},$$
$$\{x_5\}, \{x_1, x_4, x_6, x_7\}, \{x_2, x_4, x_6, x_7\}\}$$
$$S = \{\phi, \{x_1, x_4, x_6\}, \{x_2, x_7\}, \{x_3\}, \{x_1, x_4, x_6, x_7\}, \{x_5\},$$
$$\{x_1, x_4, x_6, x_7\}, \{x_2, x_4, x_6, x_7\}, U\}$$
$$\{S\}^c = \{\phi, \{x_2, x_3, x_5, x_7\}, \{x_1, x_3, x_4, x_5, x_6\}, \{x_1, x_2, x_4, x_5, x_6, x_7\},$$
$$\{x_2, x_3, x_5\}, \{x_1, x_2, x_3, x_4, x_6, x_7\}\{x_2, x_3, x_5\}, \{x_1, x_3, x_5\}, U\}$$
$$\text{and } \underline{P}X = \{x_5\} \text{ and } \overline{P}X = \{x_1, x_4, x_5, x_6\}.$$

We found that

$$U/IND(P - \{I_1\}) = U/IND(P)$$
$$U/IND(P - \{I_2\}) \neq U/IND(P)$$
$$U/IND(P - \{I_3\}) \neq U/IND(P)$$

Then attribute I_1 is superfluous attribute and we get:

$$U/IND(P - \{I_1\}) = U/IND(P).$$

The minimal reduct of P is $M = \{I_2, I_3\}$ which is shown in Table 3.

The condition of equivalence relation in the approximation space limits the range of applications, which absolutely does not permit any overlapping among its granules, seems to be too restrictive for real world problems. The purpose of this article is to use a granular structure to overcome this limitation that is, the partition determined by P does permit small degree of overlapping that assigned by the expert, i.e., $[x_i]_P \cap [x_j]_P \neq \phi$. Then the classification which is formed by this approach is reformed as a general topological method, science it does not use any distance measures and depends on a structure with minimal properties.

5 Conclusion and Future Work

This chapter has presented an effective and efficient approach to extract diagnosis rules from inconsistent and redundant data set of high voltage circuit breaker using a granular reduction approach. The extracted diagnosis rules can effectively reduce space of input attributes and simplify knowledge representation for fault diagnosis. The motivation of the proposed granular reduction approach is to overcome the drawbacks of traditional rough set theory which is not only suitable for continuous attributes, but also suppress the universe to disjoint equivalence classes. The proposed granular reduction has two characteristic features. Firstly, the fault diagnosis decision table is treated as continuous attributes instead of discrete attributes, where dealing with continuous attributes increases the computational complexity. In addition, the partitions of the universe are built by considering a general relation, which permits a small degree of overlapping that assigned by the expert. Secondly, the lower and upper approximations are established based on topological vision. Finally, the proposed approach represents a good promotion to build the fault diagnose expert system by considering the no fault case and the occurring fault case. Therefore, the correctness and effectiveness of this approach are validated by the result of practical fault diagnosis example. For future work, we attempt to construct a program language for large scale information system in the real-world applications.

References

1. Pawlak, Z.: Rough sets. Int. J. Comput. Inf. Sci. **11**, 341–356 (1982)
2. AngiuJli, F., Pizzuti, C.: Outlier mining in large high-dimensional data sets. IEEE Trans. Knowl. Data Eng. **17**(2), 203–215 (2005)
3. Jensen, R., Shen, Q.: Semantics-preserving dimensionality reduction: rough and fuzzy-rough-based approaches. IEEE Trans. Knowl. Data Eng. **16**(12), 1457–1471 (2004)
4. Pal, S., Mitra, P.: Case generation using rough sets with fuzzy representation. IEEE Trans. Knowl. Data Eng. **16**(3), 292–300 (2004)
5. Su, C.T., Hsu, J.H.: An extended Chi2 algorithm for discretization of real value attributes. IEEE Trans. Knowl. Data Eng. **17**(3), 437–441 (2005)
6. Hassanien, A.E., Suraj, Z., Slezak, D., Lingras, P.: Rough Computing: Theories, Technologies and Applications. IGI (2008)
7. Bonikowski, Z., Bryniariski, E., Skardowska, V.W.: Extension and intensions in the rough set theory. Inf. Sci. **107**, 149–167 (1998)
8. Yao, Y.Y.: Relational interpretations of neighborhood operators and rough set approximation operators. Inf. Sci. **111**, 239–259 (1998)
9. Cattaneo, G.: Abstract approximation spaces for rough theories. In: Polkowski, L., Skowron, A. (eds.) Rough Sets in Data Mining and Knowledge Discovery, pp. 59–98. Physica, Heidelberg (1998)
10. Bargiela, A., Pedrycz, W.: Granular Computing: An Introduction. Kluwer Academic Publishers, Boston (2002)

11. Polkowski, L., Semeniuk-Polkowska, M.: On foundations and applications of the paradigm of granular rough computing. Int. J. Cogn. Inf. Nat. Intell. **2**, 80–94 (2008)
12. Xu, F.F., Yao, Y.Y., Miao, D.Q.: Rough set approximations in formal concept analysis and knowledge spaces. In: An, A. (ed.) ISMIS 2008, LNAI 4994, pp. 319–328. Springer, Berlin (2008)

SQL Injection Attacks Detection and Prevention Based on Neuro—Fuzzy Technique

Doaa E. Nofal and Abeer A. Amer

Abstract A Structured Query Language (SQL) injection attack (SQLIA) is one of most famous code injection techniques that threaten web applications, as it could compromise the confidentiality, integrity and availability of the database system of an online application. Whereas other known attacks follow specific patterns, SQLIAs are often unpredictable and demonstrate no specific pattern, which has been greatly problematic to both researchers and developers. Therefore, the detection and prevention of SQLIAs has been a hot topic. This paper proposes a system to provide better results for SQLIA prevention than previous methodologies, taking in consideration the accuracy of the system and its learning capability and flexibility to deal with the issue of uncertainty. The proposed system for SQLIA detection and prevention has been realized on an Adaptive Neuro-Fuzzy Inference System (ANFIS). In addition, the developed system has been enhanced through the use of Fuzzy C-Means (FCM) to deal with the uncertainty problem associated with SQL features. Moreover, Scaled Conjugate Gradient algorithm (SCG) has been utilized to increase the speed of the proposed system drastically. The proposed system has been evaluated using a well-known dataset, and the results show a significant enhancement in the detection and prevention of SQLIAs.

Keywords SQL injection attacks · Neuro-fuzzy · ANFIS · FCM · SCG · Web security

D. E. Nofal (✉)
Information Technology Department, Institute of Graduate Studies and Research, Alexandria University, Alexandria, Egypt
e-mail: Doaa_ean@yahoo.com

A. A. Amer
Computer Science and Information System Department, Sadat Academy for Management and Sciences, Alexandria, Egypt
e-mail: abamer_2000@yahoo.com

© The Author(s), under exclusive license to Springer Nature Switzerland AG 2021
A. E. Hassanien et al. (eds.), *Machine Learning and Big Data Analytics Paradigms: Analysis, Applications and Challenges*, Studies in Big Data 77,
https://doi.org/10.1007/978-3-030-59338-4_6

1   Introduction

In just a few decades, the Internet has become the biggest network of connections known to users. However, with the rising dominance of the Internet in our daily lives, and the broadened use of its web-based applications, threats have become more and more prominent. SQLIA is a widely common threat, and it has been rated as the number one attack in the open web application security project (OWASP) list of top ten web application threats [1] as shown in Fig. 1. SQLIA is a class of code injection attacks that take advantage of a lack of validation of user input [2, 3]. There are several SQLIA techniques performed by hackers to insert, retrieve, update, and delete data from databases; shut down an SQL server; retrieve database information from the returned error message; or execute stored procedures [4, 5]. Generally, there are several classifications of SQLIA, such as tautologies, illegal/logically incorrect queries, union query, piggy-backed queries, stored queries, inference, and alternate encodings [6–11].

The main problem of SQLIAs and other security threats is that developers did not previously consider structured security approaches and dynamic and practical policy framework for addressing threats. Moreover, when such approaches are taken into consideration, attackers aim to develop new ways that can bypass the defenses designed by developers; they began to use different techniques to perform the SQLIA [12, 13]. The rising issue is that SQLIA techniques have become more and more complex. Thus, most of the current defense tools cannot address all types of attacks. Furthermore, there is a large gap between theory and practice in the field nowadays; some existing techniques are inapplicable in real, operating applications. Some of the used techniques also need additional infrastructures or require the modification of the web application's code [4, 12]. Lack of flexibility and scalability is another challenge, as some existing techniques solve only a subset of vulnerabilities that lead to SQLIAs [14, 15]. The lack of learning capabilities is another significant hurdle. In the last few

OWASP Top 10 – 2013 (Previous)	OWASP Top 10 – 2017 (New)
A1 – Injection	A1 – Injection
A2 – Broken Authentication and Session Management	A2 – Broken Authentication and Session Management
A3 – Cross-Site Scripting (XSS)	A3 – Cross-Site Scripting (XSS)
A4 – Insecure Direct Object References - Merged with A7	A4 – Broken Access Control (Original category in 2003/2004)
A5 – Security Misconfiguration	A5 – Security Misconfiguration
A6 – Sensitive Data Exposure	A6 – Sensitive Data Exposure
A7 – Missing Function Level Access Control - Merged with A4	A7 – Insufficient Attack Protection (NEW)
A8 – Cross-Site Request Forgery (CSRF)	A8 – Cross-Site Request Forgery (CSRF)
A9 – Using Components with Known Vulnerabilities	A9 – Using Components with Known Vulnerabilities
A10 – Unvalidated Redirects and Forwards - Dropped	A10 – Underprotected APIs (NEW)

Fig. 1   OWASP top list [1]

years, machine learning techniques were adapted to overcome the aforementioned problems [13, 15, 16]. However, most existing machine learning techniques suffer from high computational overhead. Furthermore, a number of existing solutions do not have the capability to detect new attacks [14]. Uncertainty is a common phenomenon in machine learning, which can be found in every stage of learning [15, 17, 18].

One of the most important machine learning techniques is the neural network (NN) model. The main characteristic of NN is the fact that these structures have the ability to learn through input and output samples of the system. The advantages of the fuzzy systems are the capacity to represent inherent uncertainties of the human knowledge with linguistic variables; simple interaction of the expert of the domain with the engineer designer of the system; easy interpretation of results, which is achieved due to the natural rules representation; and easy extension of the base of knowledge through the addition of new rules [19]. Nevertheless, an interpretation of the fuzzy information in the internal representation is required to reach a more thorough insight into the network's behavior. As a solution to such problem, neuro-fuzzy systems are employed to find the parameters of a fuzzy system (i.e., fuzzy sets and fuzzy rules) by utilizing approximation techniques from NN. By using a supervised learning algorithm, the neuro-fuzzy systems can construct an input-output mapping based on either human knowledge or stipulated input-output data pairs. Therefore, it is rendered a powerful method that addresses uncertainty, imprecision, and non-linearity [20–22].

Web applications have become a crucial method for daily transactions. Web applications are often vulnerable to attacks, in which attackers intrude easily to the application's underlying database [6]. SQLIAs are increasing continuously and expose the Web applications to serious security risks as attackers gain unrestricted access to the underlying database [10]. Programming practices such as defensive programming and sophisticated input validation techniques can prevent some of the vulnerabilities mentioned but attackers continue to find new exploits that can avoid the restrictions employed by the developers [11]. The various types of injections at different levels require a solution that can deal with such changes. This research work focuses on the analysis and finding a resolution of the problem of SQLIA, in order to protect Web applications.

This paper presents a modified approach for the detection and prevention of SQLIA. This modified approach is proposed to address the problems of uncertainty, adaptation, and fuzziness that are associated with existing machine learning techniques in the field of SQLIA. An ANFIS has the ability to construct models solely based on the target system's sample data. The FCM has been utilized in this work as a clustering method to enhance system performance by solving the fuzziness and uncertainty problems of the input data. Moreover, the SCG algorithm has been used to speed up and, thus, improve the training process. Finally, a malicious SQL statement has been prevented from occurring in the database. To the best of our knowledge, there is no previous work that uses ANFIS-FCM for detecting and preventing SQLIAs.

The subsequent parts of this paper are organized as follows: Sect. 2 provides a background and a literature survey of works related to SQLIA detection and prevention systems, an overview of the proposed is explained in Sect. 3, the experimental result and evaluation of the proposed system are discussed in Sect. 4, and Sect. 5 presents the conclusion of the work and the directions that could be taken in future works.

2 Literature Review and Related Work

There are four main categories of SQL injection attacks against databases, namely SQL manipulation, code injection, function call injection, and buffer overflow [23]. SQLI detection and prevention techniques are classified into the static, dynamic, combined analysis, and machine learning approaches, as well as the hash technique or function and black box testing [3, 24–27]. Static analysis checks whether every flow from a source to a sink is subject to an input validation and/or input sanitizing routine [28]; whereas dynamic analysis is based on dynamically mining the programmer's intended query structure on any input and detects attacks by comparing it against the structure of the actual query issued [29]. The existing machine learning models deal with uncertain information, such as input, output, or internal representation. Hence, the proposed system can be classified as a machine learning technique.

The analysis for monitoring and NEutralizing SQLIAs, known as the AMNESIA technique, suggested in [3] is a runtime monitoring technique. This approach has two stages: a static stage, which automatically builds the patterns of the SQL queries that an application creates at each point of access to the database, and a dynamic stage, in which AMNESIA intercepts all the SQL queries before they are sent to the database and checks each query against the statically built patterns. If the queries happen to violate the approach, then this technique prevents executing the queries on the database server. This tool limits the SQLIAs when successfully building the query models in the static analysis, but it also has some limitations, particularly in preventing attacks related to stored procedures and in supporting segmented queries. A. Moosa in [5] investigates Artificial Neural Networks (ANN) in SQL injection classification—a technique that is an application layer firewall, based on ANN, that protects web applications against SQL injection attacks. This approach is based on the ability of the ANN concept to carry out pattern recognition when it is suitably trained. A set of malicious and normal data is used to teach the ANN during the training phase. The trained ANN is then integrated into a web application firewall to protect the application during the operational phase. The key drawback of this approach is that the quality of a trained ANN often depends on its architecture and the way the ANN is trained. More importantly, the quality of the trained ANN also depends on the quality of the training data used and the features that are extracted from the data. With relatively limited sets of training data, the resulting ANN seems to be sensitive to content that has an SQL keyword. Another work related to ANN-based SQLI detection is introduced in [30, 31]. It depends on limited SQL patterns for training, which renders it susceptible to generate false positives.

Shahriar and Haddad in [32] introduced a fuzzy logic-based system (FLS) to assess the risk caused by different types of code injection vulnerabilities. Their work identified several code-level metrics that capture the level of weakness originating from the source code. These metrics, along with essential MF, can be used to define the subjective terms in the FLS. There are three FLS systems, each of which has three general steps: defining linguistic terms (fuzzifying crisp inputs), defining rule sets and evaluating the rules based on inputs, and, finally, aggregating the rule outputs and defuzzifying the result. This approach can effectively assess high risks present in vulnerable applications, which is considered as an advantageous aspect. Neverthe-less, fuzzy logic-based computation is flexible and tolerates inaccuracy while spec-ifying rule and membership functions. The main problem of this technique is the overhead on the system caused by web code scanning and training. In [33], Joshi and Geetha designed an SQL injection attack detection method based on Naïve-Bayes machine learning algorithm combined with role-based access control mechanism. The Naïve-Bayes classifier is a probabilistic model, which assumes that the value of a particular feature is unrelated to the presence or absence of any other feature. The Naïve Bayes algorithm detects an attack with the help of two probabilities, the prior probability and the posterior probability. The drawback of this system is that small datasets are used in testing and evaluating the system's efficiency. Said datasets, based on and used in the test cases, are mostly derived from only three SQLIA attacks: comments, union, and tautology. Therefore, this technique cannot detect the other types of SQLI.

The main objective of this paper is to propose an adaptive approach that solves and bypasses the limitations of ANN and machine learning approaches. It also aims to reduce the occurrence of errors in the stage of producing the output of the adaptive network and improve the accuracy of the system in detecting and preventing SQLIs. This shows the ability of the system to resolve uncertainty and fuzziness problems founded in SQLI attack statement. Moreover, the proposed system can avoid the overhead problem that is encountered in neuro-fuzzy approaches by applying the SCG learning algorithm.

3 Proposed System

This paper introduces an ANFIS with FCM cluster method for detecting and preventing SQLIs, which is presented in Fig. 2. As previously mentioned in the intro-duction, the reason behind choosing machine learning and ANFIS lies in ANFIS's ability to avoid the issues of uncertainty and lack of flexibility, and machine learning's ability to handle and overcome the problems that arose in previous techniques. In addition, to overcome the overhead problem that initially faced the proposed system, the SCG algorithm was used instead of the BP algorithm to enhance the proposed

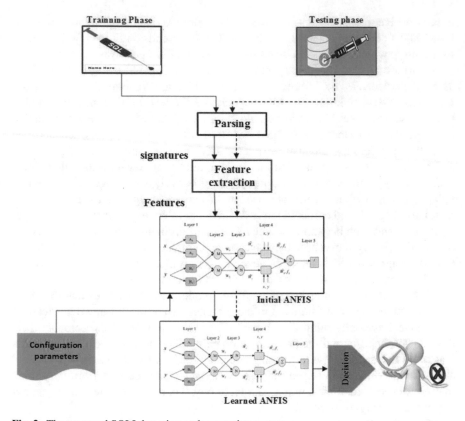

Fig. 2 The proposed SQLI detection and prevention system

system's training speed [34]. The following subsections describe, in detail, the steps of the proposed system as mentioned in the following:

- Parsing and Extracting SQLI Features from Dataset
- SQL Signatures Fuzzification
- Clustering
- Building an initial and learned ANFIS
- Prevention of SQLIAs

3.1 Parsing and Extracting SQLI Features from Dataset

In the proposed system, the SQL statement is treated as features, which characterize the SQLI attack keywords. Keywords that include create, drop, alter, where, table, etc., are the most well-known keywords in the SQL language, and they are used to perform operations on the tables in underlying database. The proposed system has investigated the most common features that have been used in numerous approaches

Table 1 SQLI signatures and keywords

Signature	Keywords
Punctuations(PU)	;, and, , , or, ', $+$, $-$, !, , ;, Existence of statements that always result in true value like "$1 = 1$" or "$@ = @$" or "$A = A$"
Dangerous words(DW)	delete, drop, create, wait, rename, exec, shutdown, sleep, load_file, userinfo, information_schema, if, else, convert, xp_cmdshell, sp_, ascii, hex, execchar(\w*\w*), xp_, sp_,
Combination words(CW)	\\, //, $\gg$, $<<$, &#, &#x, */, /*, *, *\, %, @@, (,), {, }, [,], *, $-$
SQL keyword(K)	Union, union all, select, from, insert, where, table, into, update, set, alter, like, revoke, truncate, having, union select, join, group

related to machine learning and NN dealing with the problem of SQLIAs [3, 15, 16, 35]. These features are categorized into four signatures, namely punctuation signatures (PU), dangerous words signatures (DW), combination word signatures (CW), and SQL keyword signatures (K). Each query statement, whether it is a normal or an attack query, is converted into a vector of numerical values; each number inside the vector represents a signature. The value of a signature is calculated by the addition of the frequency of each feature in the signature. A query statement is thus parsed into a list of signatures, as illustrated in Table 1. There are more than 70 features that have been used in the feature extraction stage. Each feature belonged to one of the following signatures: the SQL keyword signature, a punctuations signature, a dangerous words signature, or a combination of words signature that appear in the content of the query statement.

The features are chosen because of their ability to identify most of the SQLIA types, such as tautologies, union, piggybacked, illegal/logically incorrect, inference, alternate encodings, and stored procedures. They also work on increasing the ability of the system to detect a new malicious code, and other features can be added. Undoubtedly, the appropriate selection of the system features is the most critical step in establishing strong and practical SQLI detection and prevention systems. Basically, the feature extraction stage can be described as follows: if a keyword is discovered in the sentence, its corresponding signature will increase the value by 1. Therefore, if a feature appears more than once corresponding, its signature's value will increase by the number of occurrences.

3.2 SQL Signatures Fuzzification

Due to its low computational requirement and capability of modelling human perception, FL is probably the most efficient and flexible method available for managing degrees of uncertainty in the detection of dangerous attacks. FL is a theory that allows the natural descriptions, in linguistic terms, of problems to be solved rather than having to use numerical values. Therefore, after the stage of feature extraction, the numerical signatures that represent SQL statement features are converted into linguistic terms. A linguistic variable is defined as a variable whose values are words or sentences in a natural or synthetic language. For example, 'frequency of dangerous words' can be a linguistic variable that takes the fuzzy sets "low", "medium" and "high" as its linguistic term.

3.3 Clustering Methods

Before a final optimal model can be derived, the initial fuzzy model in the proposed system can be determined based on the fuzzy rules formed by either using the Subtractive Clustering (SC), Grid Partitioning (GP) or FCM clustering method, as SC and GP are two main clustering methods that are commonly used with ANFIS. GP is the most frequently used input partitioning method for ANFIS [34]. In this paper, we will explore the suitability of FCM as a powerful data clustering method; in which each data point has a membership degree between 0 and 1 to each fuzzy subset [36]. As aforementioned, the data point resembles a vector of five numbers; FCM partitions a collection of n vectors x_i, $i = 1, 2, ..., n$ into fuzzy groups, and determine a cluster center for each group $i = 1, 2, ..., c$ that are arbitrarily selected from the n points. Where n is the total number of training data set and c is the cluster center. The steps of the FCM method are explained briefly [36]:

1- The centers of each cluster c_i, $i = 1, 2,..., c$ are randomly selected from the n data patterns (training data set) $\{x_1, x_2, x_3, ..., x_n\}$.
2- The membership matrix (μ) is computed with the following equation.

$$\mu_{ij} = \frac{1}{\sum_{k=1}^{c} \left(\frac{d_{ij}}{d_{kj}}\right)^{2/m-1}}, \tag{1}$$

where μ_{ij} is the degree of membership of object j (new signature value; punctuation signature or dangerous words or combination words) in cluster I, m is the degree of fuzziness determined by the user; $m = 2$ is initially chosen, m is important because it significantly influences the fuzziness of the resulting partition and $d_{ij} = \|c_i - x_j\|$ is The Euclidean distance between c_i and x_j

3- The objective function is calculated with the following equation:

$$J(U, c_1, c_2, ..., c_c) = \sum_{i=1}^{c} \sum_{j=1}^{n} \mu_{ij}^m d_{ij}^2 \quad 1 \leq m < \infty \tag{2}$$

where U is the partition matrix that contains all the data points and the computed cluster center in each cluster.

4- The new c fuzzy cluster c_i, $i = 1, 2, ..., c$ is calculated using the following equation [36].

$$c_i = \frac{\sum_{j=1}^{n} \mu_{ij}^{m} x_j}{\sum_{j=1}^{n} \mu_{ij}^{m}}.$$

(3)

By applying the FCM clustering to each class of **data** individually, a set of rules for identifying each class of data has been obtained. Due to this, the system must be pre-configured with the numbers of clusters that is determined by the user. The individual sets of rules are then combined to form the rule base of the classifier. Furthermore, according to the Sugeno model, each rule has a number of consequence parameters in FIS output part; where each MF is determined by two parameters c and σ (σ is used to determine the width of the MF and c is the center point of the Gaussian MF).

3.4 Building an Initial and Learned ANFIS

ANFIS is basically a graphical network representation of Sugeno-type fuzzy systems endowed with the neural learning capabilities [34, 36]. Inputs for the proposed ANFIS for SQLI detection and prevention are the numerical values (the four signatures). The output of the system has been configured in such manner that it is equal to 2 if there is an attack, and 1 otherwise. Gaussian MF has been used for fuzzy set due to its nonlinear, smooth and continuous derivatives [35]. An ANFIS presented in Fig. 3 it is functionally equivalent to the fuzzy inference system (as seen in Fig. 4). Figure 3 illustrates the reasoning mechanism for the Sugeno model where it is not only the basis of ANFIS model; but also, it is simple in computation and easy to be combined with optimizing and self-adapting methods [17]. Subsequently, the corresponding equivalent ANFIS architecture is as shown in Fig. 4, where nodes of the same layer have similar function [33–35].

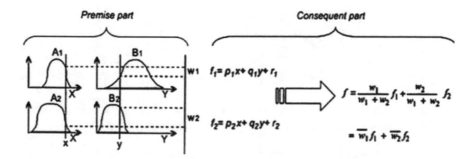

Fig. 3 First-order Sugeno fuzzy model

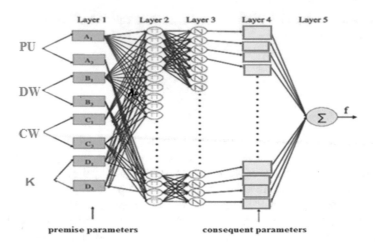

Fig. 4 ANFIS structure

1. ANFIS Layers

The input data and output data were fed into the ANFIS model to extract the rules. The 'fuzzification' layer is set and adapted according to the parameters for the chosen membership. After that, the strength firing layer represents the IF conditions to set the rules. The output of the firing strength is normalized in the normalization layer. Before the final layer, there is another adaptation layer that works as a 'defuzzification' layer of the rules, where the consequent model parameters are tuned to derive the best matching between input and output [36]. Here, a four-input and single-output fuzzy system has been used. three fuzzy variables including 'low', "medium" and 'high' have been used to describe the features. Their respective MFs (μ_A) are Gaussian function that introduces the fuzzification operation of input parameters.

Layer 1: The output of this layer represents the membership grade of the inputs; for example, the MF for PU signature can be parameterized as given in the following equation [33–35]

$$\mu_{Ai}(PU) = \exp\left[-\left(\frac{PU - c_i}{2\sigma_i}\right)^2\right] \tag{4}$$

$$O_{1,i} = \mu_{A_i}(PU), \text{ for } i = 1 \tag{5}$$

where $\{\sigma_1, c_i\}$ is the premise parameter set that changes the shape of the membership function; and $O_{1,i}$ is the output of the i-th node of layer 1.

Layer 2: Every node in this layer represents the firing strength of a rule:

$$O_{2,i} = w_i = \mu_{A_i}(PU)\mu_{B_i}(DW)\mu_{c_i}(CW)\mu_{D_i}(K), \ i = 1, 2 \tag{6}$$

where A_i, B_i, C_i, D_i are the fuzzy sets for each signature. Each rule is assigned a firing strength that measures the degree to which the rule matches the inputs.

Layer 3: Each node in this layer calculates the ratio of the *ith* rule's firing strength to the sum of all rules' firing strengths. The outputs of this layer are called normalized firing strengths.

$$O_{3,i} = \bar{w}_i = \frac{w_i}{\sum_i w_i}, \quad i = 1, 2 \tag{7}$$

Layer 4: Every node in this layer is an adaptive node with a node function, where $\bar{w}_i$ is a normalized firing strength from layer 3 and $\{a_i, q_i, t_i, e_i r_i\}$ is the parameter set of this node. Parameters in this layer are referred to as consequent parameters. Consider a first order Sugeno type of fuzzy system having the rule base.

Rule 1 : **If** PU is *low* and DW is *low* and CW is *low* and K is *low*
then *output* $f1 = 0.0229 * PU + 0.0101 * DW + 0.266 * CW$
$+ 0.0392 * K + 1.5112$
Rule 2 : **If** PU is *high* and DW is *high* and CW is *high* and K is *high*
then *output* $f2 = 0.2212 * PU + 0.1280 * DW + 0.0574 * CW$
$+ 0.0301 * K + 0.9644$
Rule 3 : **If** PU is *low* and DW is *high* and CW is *high* and K is *high*
then *output* $f1 = -0.0372 * PU + 0.0583 * DW + 0.0647 * CW$
$+0.0886 * K + 0.816$
Rule 4 : **If** PU is *low* and DW is *high* and CW is *low* and K is *high*
then *output* $f1 = 0.3932 * PU + 0.3844 * DW - 0.0456 * CW$
$+0.0688 * K + \{1\}.1169$
$$\vdots$$
$$\vdots$$
$$\vdots$$

$$\tag{8}$$

Rule 8

Layer 5: The output of the fuzzy system in the SQLI detection system is linear and the single node in this layer computes the overall output as the summation of all incoming signals.

$$O_{5,1} = \sum_i \bar{w}_i f_i = \frac{\sum_i w_i f_i}{\sum_i w_i} \tag{9}$$

2. Learning Algorithm

In this paper, the ANFIS has been utilized which is a fuzzy inference system implemented in the framework of adaptive networks. ANFIS has been utilized in

Table 2 Hybrid learning process

Type	Path forwards	Path backwards
Premise parameter	Fixed	Back propagation or SCG
Consequent parameter	Least Squares estimation	Fixed

this paper because it is proficient in constructing input-output mapping accurately based predetermined input output data pairs. In addition, it is dealing with fuzzy and uncertain input data in SQLI attack problem. The advantage of ANFIS is that it obtains the membership functions and set the rules by itself adaptively using the training data. ANFIS uses different types of learning algorithms of NN like back propagation algorithm (BP), least mean squares algorithm etc..... [22, 37]. ANFIS is a NN that is functionally the same as a Takagi–Sugeno type inference model. That employing learning method works similarly to that of NN to update the parameters of the Takagi–Sugeno type inference model.

In ANFIS, parameters that determine MF shapes of each input in the first layer which is nonlinear while the fourth layer contains linear consequent parameters. All these parameters are adjusted by BP algorithm or a combination of least squares estimation and BP algorithm. In the hybrid method, premise parameters are adjusted by BP algorithm while consequent parameters are adjusted by least squares estimation. There are two parts of a hybrid learning algorithm, namely the forward path and backward path [21, 37] as shown in Table 2. In the forward path, the premises parameters must be in a steady state. A recursive least square estimator method was applied to repair the consequent parameter in the fourth layer. As the consequent parameters are linear, then least square estimator method can be applied to accelerate the convergence rate in hybrid learning process. Next, after the consequent parameters are obtained, input data is passed back to the adaptive network input, and the output generated will be compared with the actual output. While backward path is run, the consequent parameters must be in a steady state. The error occurred during the comparison between the output generated with the actual output is propagated back to the first layer [21, 37]. There is a number of training algorithms that can be used in training the premise parameters of the ANFIS systems. In this system, two training algorithms have been evaluated, the BP and SCG algorithms. SCG algorithm has been presented as one of the algorithms that enhance the processing time [22, 37]. Generally, the SCG algorithm shows great performance over a wide variety of problems [21]. Therefore, in this paper, ANFIS has been improved with the SCG learning algorithm to speed up its training process [38].

3.5 Prevention of SQLIA

After ANFIS model has been trained with the input and output training data, the system that has been tested using the testing dataset and SQLIA is detected;

this system prevents the malicious SQL statement from accessing the database by converting it into a comment. As it is well-known, the comment statements do not execute in the database engine.

4 Experimental Results and Discussion

All these experiments were carried out on windows 10 (64-bit) operating system with i7 processor and 8 GB RAM. To evaluate the performance of the proposed system, all methods and training functions that are used are coded in MATLAB. The dataset is downloaded from Testbed [39], which is then used to evaluate the Amnesia approach in [3]. The testbed has two sets of inputs: "legit" set, which consists of legitimate inputs for the application, and "attack" set, which consists of attempted SQLIAs. All types of attacks were represented in this set with the exception of the multi-phase attacks. The multi-phase attacks include inference attacks and illegal/logically incorrect queries, such attacks require human intervention and interpretation. The testbed includes seven folders; four of which are used for the training and the remainder of the three sets are used for testing. At first, to compare the efficiency of the different clustering methods GP, SC and FCM, the popular measure, RMSE (root mean squared error), was employed for performance evaluation according to the next formula:

$$MSE = \frac{\sum_{i=1}^{N} (y_i - o_i)^2}{N} \qquad (10)$$

$$RMSE = (MSE)^{1/2} \qquad (11)$$

where y_i is the target value, o_i the observed output, and n is the number of data set [38].

The results are shown in Table 3; in which the RMSE of the testing dataset

Table 3 The RMSE for the clustering methods in different epochs based on BP learning algorithm

Epochs No.	Clustering method	RMSE of the test data set
40	GP	0.1632
	SC	0.1342
	FCM	0.0468
60	GP	0.15838
	SC	0.1205
	FCM	0.0402
100	GP	0.1307
	SC	0.0984
	FCM	0.0402

displays the error between the target output and the observed output in the testing dataset. According to the results, it is verified that with the use of 60 epochs, the FCM cluster has achieved the minimum value of the RMSE among the others. The GP and the SC clustering behaviours did not achieve the minimal error, such as the case with FCM. The reason for such results is that the interpretation of m (fuzziness degree) is different than the case of FCM, where values of m increase the sharing of points among all the clusters which will lead to better performance; they also lead to the reduction of the objective function of the dissimilarity measure. Therefore, this improvement in FCM-ANFIS is related to its ability to manage uncertainty and the fuzziness degree in dangerous attacks' statements. Moreover, any value with more than 60 epochs resulted in the overtraining of the model.

Figure 5 shows the waveforms of RMSE for the ANFIS-FCM system based on BP learning algorithm in details. It is obvious that RMSE waveforms start descending before 50 epochs. After 60 epochs, the RMSE curve tends to be stabilized with very small variation (overfitting). In this case, the network parameters are saturated as the network output matches the target; any additional epochs will decrease the accuracy performance inside NN as the reason of overtraining [40].

Figure 6 indicates that the suggested system is improved by the SCG Algorithm and reaches the same RMSE in less epochs. With 60 epochs, the utilized BP learning algorithm archives 0.0402 RMSE and this value is obtained through the utilization of the SCG learning algorithm in only 5 epochs. As a result, the SCG learning algorithm needs lower computational time as SCG involves twice as much calculation work per iteration when compared with BP. As stated in [41], one iteration in SCG needs the calculation of two gradients, and in addition to this it requires one call to the error

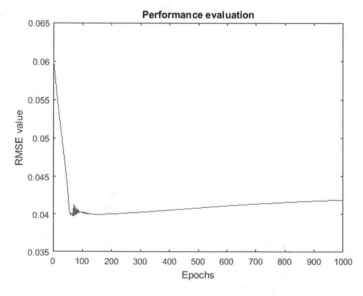

Fig. 5 RMSE curve for the ANFIS-FCM system based on BP learning algorithm

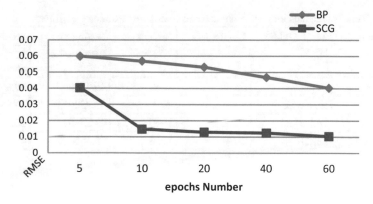

Fig. 6 RMSE curves for ANFIS based on BP and SCG learning algorithm

function, while one iteration in standard BP needs the computation of one gradient and one call to the error function.

In the third experiment, the performance of the suggested system under SCG learning algorithm with different number of epochs is discussed. As shown in Fig. 7, the RMSE curve starts descending before 20 epochs. After 20 epochs, the RMSE curve tends to be stabilized; this is due to the error decreasing in monotonic towards zero, which is characteristic for SCG. In this case, an error increase is not allowed and second order information (second derivatives) of global error function hasn't been

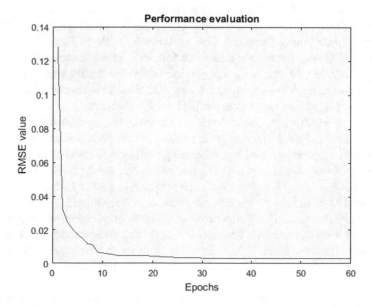

Fig. 7 RMSE curve for the ANFIS-FCM system based on SCG learning algorithm

Table 4 Time per seconds for testing phase based on different learning algorithms

Training folder subject	Testing folder subject	No of attack URLs	No of legit URLs	Testing Time per seconds	
				BP algorithm	SCG algorithm
Bookstore	Employee	6530	1268	45.821000	39.00562
Checkers	Employee	6939	2019	58.950023	47.99056
Classifieds	Office talk	6958	1000	43.85002	38.70112
Events	Portal	5970	1980	55.560023	45.98002

Table 5 Comparison of the processing time between BP and SCG for training and testing phases

Learning algorithm	Time per minutes
BP	11.3481206
SCG	7.99533351

positively definite, only in the beginning of the minimization. This is not surprising because the closer the current point is to the desired minimum the bigger is the possibility that global error function is positive definite [41].

The fourth experiment compares between the two learning algorithms in terms of the computational cost (processing time) under different attack folders for each a testing phase and all phases respectively. Table 4 reveals that using SCG as a learning algorithm inside the proposed system requires less time as compared with BP in all attack folders (about 20% reduction in time) in the testing phase. For the whole system (training and testing phases), as it was expected, the SCG learning algorithm achieves reduction in the running time with 30% as shown in Table 5. This improvement is due to that SCG does not require a line search at each iteration step, unlike the other training algorithms. In another words, the used step size scaling mechanism avoids the time consuming line search per learning iteration. This mechanism makes the SCG learning algorithm faster than the BP learning algorithm.

In the fifth experiment, the accuracy of the proposed system that employs the SCG algorithm and the FCM clustering method for the detection and prevention of SQLI, this is compared with the algorithm suggested by N. Sheykhkanloo in [30] that utilized traditional neural network. The traditional NN model has 10 hidden layers and 32 input features. Generally, the correct response of the NN system depends on the number of hidden layers that are commonly determined by the user. From the illustrated results in Fig. 8, the proposed system outperforms the other one by 3.08%. Based on the research findings, the proposed system for SQLIA detection and prevention trained by the SCG training algorithm with four inputs and one output achieved high accuracy, less time, and avoided the computational complexity of the network.

In the last experiment presented in Table 6, the accuracy of the proposed system compared with the algorithm suggested by C. Basta in [15] shows 98.4% accuracy with the duration being 217 s for 13,079 attacks, i.e. 16.6 ms for one attack; while

Fig. 8 Comparative study

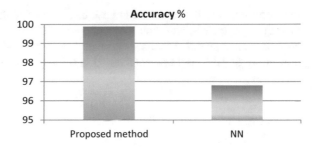

Accuracy %

Proposed method   NN

Table 6 Comparative Study of the accuracy and processing time

Technique	Accuracy	Time per sec for one attack
Proposed system	99.8	14.7
Genetic fuzzy classifier	98.4	16.6

the proposed system shows 99.8% accuracy with the duration being 180.920011 s for 12,241 attacks, i.e. 14.7 ms for one attack.

5 Limitations

The solution has some limitations such as the dependability on the quality of the training data and the features extracted from it. With a relatively limited set of training data, the resulting classifier can be sensitive to the features of benign queries. Despite the success of the technique overall, an acceptable amount of overhead still takes place due to the learning phase.

6 Conclusions and Future Work

In this paper, a new approach based on the ANFIS system for SQLIA detection and prevention has been proposed. The proposed model includes two main elements: URL parser and feature extraction and the training ANFIS classifier. The proposed system has been implemented with the benign and malicious URLs depending on the extracted features. Furthermore, the FCM is employed as a clustering method in the first layer of ANFIS to enhance the system's performance by dealing with fuzzy and uncertain inputs. Additionally, SCG learning algorithm is employed instead of BP learning algorithm for enhancing the system's processing time. The approach that is based on the ANFIS classifier has the advantages of learning through patterns (input and output system data), and the easy interpretation of its functionality. Future work includes upgrading the system to detect XSS attacks with the ANFIS classifier.

Furthermore, the learning algorithm can be replaced with another appropriate one to fine tune parameters, and then the results should be evaluated to deduce the better optimization approach. Finally, employing a new algorithm to encrypt data query for preventing SQLIA could be considered in future work.

References

1. O.W.A.S.P.Top10Vulnerabilities.https://www.owasp.org/index.php/Top_10_2013-Top_10. Accessed 1 Dec 2016\
2. Shegokar, A., Manjaramkar, A.: A survey on SQL injection attack, detection and prevention techniques. Int. J. Comput. Sci. Inf. Technol. **5**(2), 2553–2555 (2014)
3. Halfond, W., Orso, A.: AMNESIA: analysis and monitoring for neutralizing SQL-injection attacks. In: 20th IEEE/ACM International Conference on Automated Software Engineering, pp. 174–183, USA (2005)
4. Bhagat, M., Mane, V.: Protection of web application against SQL injection attack. Int. J. Sci. Res. Publ. **3**(10), 1–5 (2013)
5. Moosa, A.: Artificial neural network based web application firewall for SQL injection. Int. J. Comput. Electr. Autom. Control Inf. Eng. **4**(4), 12–21 (2010)
6. Nithya, V., Regan, R.: A survey on SQL injection attacks, their detection and prevention techniques. Int. J. Eng. Comput. Sci. **2**(4), 886–905 (2013)
7. Tajpour, A., Massrum, M., Heydari, Z.: Comparison of SQL injection detection and prevention techniques. In: 2nd International Conference on Education Technology and Computer, China, pp. 174–179 (2010)
8. Gomaa, Y., El Aziz Ahmed, A., Mahmood, M., Hefny, H.: Survey on securing a querying process by blocking SQL injection. In: 3rd World Conference on Complex Systems, pp. 1–7, Morocco (2015)
9. Som, S., Sinha, S., Kataria, R.: Study on SQL injection attacks: mode, detection and prevention. Int. J. Eng. Appl. Sci. Technol. **1**(8), 23–29 (2016)
10. Verma, N.: A detailed study on prevention of SQLI attacks for web security. Int. J. Comput. Appl. Technol. Res. **1**(4), 308–311 (2015)
11. Kumar, P., Pateriya, K.: A survey on SQL injection attacks, detection and prevention techniques. In: 3rd International Conference on Computing Communication & Networking Technologies, India, pp. 1–5 (2012)
12. Al-Khashab, E., Al-Anzi, F., Salman, A.: PSIAQOP: preventing SQL injection attacks based on query optimization process. In: The 2nd Kuwait Conference on e-Services and e-Systems, Kuwait, pp. 1–10 (2011)
13. Valeur, F., Mutz, D., Vigna, G.: A learning-based approach to the detection of SQL attacks. In: International Conference on Detection of Intrusions and Malware, and Vulnerability Assessment, Austria, pp. 123–140 (2005)
14. Wu, X., Chan, P.: SQL injection attacks detection in adversarial environments by K-centers. In: International Conference on Machine Learning and Cybernetic, China, pp. 406–410 (2012)
15. Basta, C., Elfatatry, A., Darwish, S.: Detection of SQL injection using a genetic fuzzy classifier system. Int. J. Adv. Comput. Sci. Appl. **7**(6), 129–137 (2016)
16. Komiya, R., Paik, I., Hisada, M.: Classification of malicious web code by machine learning. In: 3rd International Conference on Awareness Science and Technology, China, pp. 406–411 (2011)
17. Wang, X., Zhai, J.: Learning with Uncertaintey. CRC Press, pp. 1–227 (2016). ISBN 9781498724128 - CAT# K25713
18. Hammer, B., Villmann, T.: How to process uncertainty in machine learning? In: European Symposium on Artificial Neural Networks, Belgium, pp. 79–90 (2007)

19. Toosi, A., Kahani, M.: A novel soft computing model using adaptive neuro-fuzzy inference system for intrusion detection. In: The 2007 IEEE International Conference on Networking, Sensing and Control, UK, pp. 15–17 (2007

20. Wu, X., Zhu, X., Li, X., Yu, H.: Realization of an improved adaptive neuro-fuzzy inference system in DSP. In: The International Symposium on Neural Networks, pp. 170–178. Springer (2007)

21. Ghaffari, A., Abdollahi, H., Khoshayand, M., Bozchalooi, I., Dadgar, A., Rafiee-Tehrani, M.: Performance comparison of neural network training algorithms in modeling of bimodal drug delivery. Int. J. Pharm. **327**(1), 126–138 (2006)

22. Falas, T., Stafylopatis, A.: Implementing temporal-difference learning with the scaled conjugate gradient algorithm. Neural Process. Lett. **22**(3), 361–375 (2005)

23. Tajpour, A., Ibrahim, S., Masrom, M.: SQL injection detection and prevention techniques. Int. J. Adv. Comput. Technol. **3**(7), 82–91 (2011)

24. Wassermann, G., Su, Z.: An analysis framework for security in Web applications. In: The FSE Workshop on Specification and Verification of Component-Based Systems, pp. 70–78 (2004)

25. Buehrer, G., Weide, B., Sivilotti, P.: Using parse tree validation to prevent SQL injection attacks. In: The 5th International Workshop on Software Engineering and Middleware, pp. 106–113 (2005)

26. Bandhakavi, S., Bisht, P., Madhusudan, P., Venkatakrishnan, V.: CANDID: preventing SQL injection attacks using dynamic candidate evaluations. In: The 14th ACM Conference on Computer and Communications Security, pp. 12–24 (2007)

27. Singh, S., Tripathi, U., Mishra, M.: Detection and prevention of SQL injection attack using hashing technique. Int. J. Mod. Commun. Technol. Res. **2**(9), 27–30 (2014)

28. Shar, L., Tan, H.: Defeating SQL injection. Comput. Softw. Eng. **46**(3), 69–77 (2013)

29. Tajpour, A., JorJor Zade Shooshtari, M.: Evaluation of SQL injection detection and prevention techniques. In: 2nd IEEE International Conference on Computational Intelligence, Communication Systems and Networks, UK, pp. 216–221 (2010)

30. Sheykhkanloo, N.: SQL-IDS: evaluation of SQLI attack detection and classification based on machine learning techniques. In: The 8th International Conference on Security of Information and Networks, USA, pp. 258–266 (2015)

31. Sheykhkanloo, N.: Employing neural networks for the detection of SQL injection attack. In: The 7th International Conference on Security of Information and Networks, UK, pp. 318–323 (2014)

32. Shahriar, H., Haddad, H.: Risk assessment of code injection vulnerabilities using fuzzy logic-based system. In: The 29th Annual ACM Symposium on Applied Computing, Korea, pp. 1164–1170 (2014)

33. Joshi, A., Geetha, V.: SQL injection detection using machine learning. In: The International Conference on Control, Instrumentation, Communication and Computational Technologies, India, pp. 1111–1115 (2014)

34. Othman1, M., Yau, T.: Neuro fuzzy classification and detection technique for bioinformatics problems. In: The First Asia International Conference on Modelling & Simulation (AMS'07), pp. 375–380 (2007)

35. Batista, L., et al. Fuzzy neural networks to create an expert system for detecting attacks by SQL injection. Int. J. Forensic Comput. Sci. **13**(1), 8–21 (2018)

36. Abdulshahed, A., Longstaff, A., Fletcher, S., Myers, A.: Thermal error modelling of machine tools based on Anfis with fuzzy C-means clustering using a thermal imaging camera. Appl. Math. Model. **39**(7), 1837–1852 (2015)

37. Sharma, B., Venugopalan, K.: Comparison of neural network training functions for hematoma classification in brain CT images. J. Comput. Eng. **16**(1), 31–35 (2014)

38. Hager, W., Zhang, H.: A survey of nonlinear conjugate gradient methods. Pacific J. Optim. **2**(1), 35–58 (2006)

39. Halfond, W.: Testbed. http://wwwbcf.usc.edu/~halfond/testbed.html

40. Nasr, M., Mahmoud, A., Fawzy, M., Radwan, A.: Artificial intelligence modeling of cadmium biosorption using rice straw. Appl. Water Sci. **7**(2), 823–831 (2017)
41. Sehga, P.: Prerana: comparative study of GD, LM and SCG method of neural network for thyroid disease diagnosis. Int. J. Appl. Res. **1**(10), 34–39 (2015)

Convolutional Neural Network with Batch Normalization for Classification of Endoscopic Gastrointestinal Diseases

Dalia Ezzat, Heba M. Afify, Mohamed Hamed N. Taha, and Aboul Ella Hassanien

Abstract In this paper, an approach for classifying gastrointestinal (GI) diseases from endoscopic images is proposed. The proposed approach is built using a convolutional neural network (CNN) with batch normalization (BN) and an exponential linear unit (ELU) as the activation function. The proposed approach consists of eight layers (six convolutional and two fully connected layers) and is used to identify eight types of GI diseases in version two of the Kvasir dataset. The proposed approach was compared with other CNN architectures (VGG16, VGG19, and Inception-v3) using five elements (number of convolutional layers, number of total parameters of the convolutional layers, number of epochs, validation accuracy and test accuracy). The proposed approach achieved good results compared to the compared architectures. It achieved a validation accuracy of 88%, which is superior to other architectures and a test accuracy of 87%, which outperforms the Inception-v3 architecture. Therefore, the proposed approach has less trained images and less computational complexity in the training phase.

Keywords Endoscopic gastrointestinal (GI) images · Kvasir dataset · Convolutional neural network (CNN) · Batch normalization (BN) · Exponential linear unit (ELU)

1 Introduction

The American Society for Gastrointestinal Endoscopy (ASGE) supports the analysis of endoscopic images in gastrointestinal (GI) tract to assist clinicians in making correct decisions [1]. Endoscopic imaging technology has refined the diagnostic and

D. Ezzat (✉) · H. M. Afify · M. H. N. Taha · A. E. Hassanien
Scientific Research Group in Egypt, Giza, Egypt
e-mail: dalia.Azzat@yahoo.com
URL: http://www.egyptscience.net/

H. M. Afify
Systems and Biomedical Engineering Department, Higher Institute of Engineering in El-Shorouk City, Cairo, Egypt

© The Author(s), under exclusive license to Springer Nature Switzerland AG 2021 113
A. E. Hassanien et al. (eds.), *Machine Learning and Big Data Analytics Paradigms: Analysis, Applications and Challenges*, Studies in Big Data 77,
https://doi.org/10.1007/978-3-030-59338-4_7

therapeutic purposes that can be used as alternative techniques by which patients can avoid biopsy and surgical procedures [2]. Explorations of the digestive system based on endoscopic images are performed using gastroscopy for the upper GI tract and colonoscopy for the lower GI tract. Based on statistical analyses of GI disorders, various different diseases exist such as oesophageal, stomach and colorectal cancer [3], that may result in death. The most common cancers in the GI tract are colorectal cancer representing 1.80 million cases, and stomach cancer representing 1.03 million cases. In the United States, approximately 862, 000 colorectal cancer and 783 000 stomach cancer deaths occur each year [4]. The factors related to GI diseases, include environmental factors (Helicobacter pylori infection, a wrong diet, food storage), treatment factors (using antibiotics to kill a specific bacterium, poorly qualified gastroenterologists), genetic factors (inherited cancer genes), and unknown factors. In some cases, optical diagnoses by endoscopic imaging examination suffer from endoscopist errors, lengthy procedures and poor quality images [5]. Therefore, computer-assisted diagnosis systems for GI images can affect accurate and rapid classification by discriminating between normal and diseased GI tract and reducing the mortality level for GI diseases [6].

In general, endoscopic images for the GI tract are considered to be biomedical images. It is essential to create deep learning algorithms to process these huge images before the disease diagnosis. A major challenge in biomedical images is to perform classification for low-level visual images obtained from imaging devices. The deep convolutional neural network (CNN) is a common learning algorithm that has achieved success in medical images classification [7]. For example, CNNs have been efficiently applied for polyp detection in colonoscopy videos [8], for lung images classification [9], for pancreas segmentation in CT images [10], and for brain tumour segmentation in magnetic resonance imaging (MRI) scans [11]. Additionally, CNN frameworks running on accelerated hardware have been utilized for medical image retrieval [12] and for medical image segmentation ration [13]. Thus, the CNN architecture has encouraged rapid automated classification for large number of medical images.

This paper demonstrates a CNN model for classifying GI diseases from endoscopic images. The remainder of this paper is structured as follows. Related works are discussed in Sect. 2, especially from the perspective of previous CNN architectures when using BN [14] and when ELU is used as the activation function [15]. Section 3 presents an explanation of the image dataset used in this paper. In Sect. 4, the proposed methodology is explained. The experimental outcomes are notified in Sect. 5. Lastly, a few concluding comments are estimated in Sect. 6.

2   Related Works

CNNs have been used extensively to solve issues related to computer vision, such as image identification [16] because a CNN is one of the most effective ways to extract features for non-trivial tasks [17]. Numerous variants of CNN architectures can be found in the literature that have advanced results in different image classification task, for example, VGG16 and VGG19 [18], which won the runner-up award in the ILSVRC-2014. VGG16 is a 16-layer network containing 13 convolutional layers, three fully-connected layers, and five max-pooling layers, while VGG19 is a 19-layer network containing 16 convolutional layers, three fully-connected layers, and five max-pooling layers. Despite the successes of these architectures, one of their drawbacks is that they are difficult to train [19].

In addition, a wide range of techniques have been developed to improve the performance or facilitate the training of CNNs, such as incorporating BN or using an ELU as the activation function. BN is a technique introduced by Ioffe and Szegedy for accelerating deep network training. BN has become a typical element in modern better performing CNN designs such as Inception V3 [20], which achieved the lowest error rate (3.08%) in the ImageNet challenge. BN helps the network to train faster, achieve higher accuracy, stabilize the distribution and reduce the internal covariate shift [14, 21, 22].

The experimental results in [15] indicated that the ELU activation function accelerates learning in deep neural networks, leads to higher classification accuracies that achieve better generalization performance than other activations function such as rectified units (RELUs), and using ELU with BN outperforms RELU with BN. According to the previous work, a CNN model incorporating the BN technique and using ELU as an activation function accelerates GI diseases identification from endoscopic images.

3   Dataset Description

The Kvasir dataset [23] has two versions. Deep learning methods were implemented using version one in [24, 25]; however, version two, which was released in 2017, has not been used until now in any previous studies in this field. Therefore, in this paper the proposed model is applied to version two of the Kvasir database. Kvasir version two has a size of 2.3 GB and contains 8,000 images with 720×576 pixels. These data are divided into eight classes with 1,000 images for each class. This Kvasir dataset was created from endoscopic images of GI tract diseases. The descriptions of the eight classes are listed in Table 1. These data consist of three types: anatomical landmarks, pathological findings, and polyp removal, as shown in Fig. 1.

4   The Proposed Approach

The proposed approach involves four phases: images preprocessing, data augmentation, feature extraction, and classification. The feature extraction phase includes convolutional layers, BN layers, ELU layers, and max-pooling layers, while the classification phase contains fully connected layers, BN layer, ELU layer, dropout layer,

Table 1 Descriptions of the 8 of endoscopic images classes in the GI tract

Class	Types	Description
Dyed and lifted-polyps	Polyps Removal = endoscopic mucosal resection (EMR)	A polyp detected by injection of saline injection and indigocarmine. The presence of non-lifted areas indicates malignancy
Dyed and resection margins	Polyps Removal = endoscopic mucosal resection (EMR)	Polyp indicator is either completely removed or not. Residual polyp tissue leads to continued growth and malignancy development
Normal caecum	Anatomical landmarks	The nearest part of the large bowel. This is used as an indicator for colonoscopy and the appendiceal orifice
Normal pylorus	Anatomical landmarks	The region around the opening from the stomach into the first part of the small bowel (duodenum). It appears as a dark circle encircled by homogeneous pink stomach mucosa
Normal Z-line	Anatomical landmarks	The transition place between the oesophagus and the stomach. It is used as a reference mark when explaining pathology in the oesophagus
Polyps	Pathological findings	Lesions in the bowel noticeable as mucosal outgrows. The polyps are flat, elevated or pedunculated, and are distinguishable from normal mucosa by color and surface shape. Most bowel polyps are safe, but some have the possibility to become cancerous
Ulcerative colitis	Pathological findings	A chronic inflammatory illness impacting on the large bowel. The level of inflammation changes among none, mild, moderate and severe

(continued)

Table 1 (continued)

Class	Types	Description
Esophagitis	Pathological findings	It acted as an inflammation of the esophagus visible as a break in the Esophageal mucosa in relation to the Z-line. It used as indictor for gastric acid flows back into the esophagus as gastroesophageal reflux, vomiting or hernia

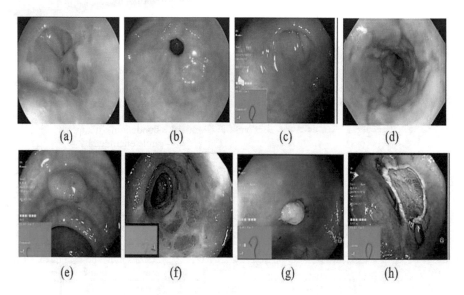

Fig. 1 Endoscopic images of gastrointestinal (GI) tract for anatomical landmarks, pathological findings, and polyps removal: (**a**) Z-line, (**b**) pylorus, (**c**) caecum, (**d**) oesophagitis, (**e**) polyps, (**f**) ulcerative colitis, (**g**) dyed and lifted polyps, (**h**) dyed resection margins

and a softmax layer. Figure 2 demonstrates a structural representation of the proposed approach. In the proposed approach, the RMSProp optimizer [26] with a learning rate of 1e−4, categorical cross-entropy as the loss function [27], a batch size of 32 and 115 epochs were used as shown in Table 2.

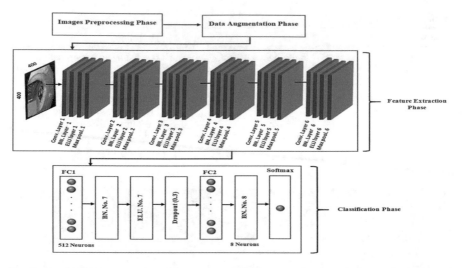

Fig. 2 Graphical representation of the proposed CNN approach. Conv. Layer = Convolutional layer, BN. Layer = Batch normalization layer, ELU layer = Exponential Linear Unit layer, FC1 = the first fully connected layer, FC2 = the second fully connected layer

Table 2 Hyper-parameters values of the proposed CNN approach. FC1 = the first fully connected layer, FC2 = the second fully connected layer

Hyper-parameters	Layer 1	Layer 2	Layer 3	Layer 4	Layer 5	Layer 6	FC1	FC2	General
Number of filters	64	64	64	64	128	128	–	–	–
Kernel size	3×3	3×3	3×3	3×3	3×3	3×3	–	–	–
Stride	2	2	2	2	2	2	2	2	–
Padding	Same	Same	Same	Same	Same	Same	–	–	–
Number of neurons	–	–	–	–	–	–	512	8	–
Batch size	–	–	–	–	–	–	–	–	32
Number of epochs	–	–	–	–	–	–	–	–	115
Dropout rate	–	–	–	–	–	–	–	–	0.3
Learning rate	–	–	–	–	–	–	–	–	1e−4

4.1 Images Preprocessing Phase

The dataset was split into three separate file groups. The first file group comprised the training set, which included 700 images of each class; each class was stored in a separate file. The second file group comprised the validation set, which included 150 images of each class, and each class was stored in a separate file. The third file group was the test set, which included 150 images for each class, and each class was stored in a separate file.

Table 3 Data augmentation techniques and their corresponding values

Transformation type	Value
Rotation	20
Width shift	0.2
Height shift	0.2
Shear	0.2
Zoom	0.2
Horizontal flip	True
Fill mode	Nearest

Before loading the images into the proposed approach, all the images in the training, validation and test sets were resized to a resolution of 400 × 400 to decrease the computational time and normalized by dividing the colour value of each pixel by 255 to achieve values in the range 0, 1.

4.2   Data Augmentation Phase

Data augmentation techniques increase the amount of training data available, which is crucial when training a deep learning model from scratch [28]. Data augmentation was used in this paper to overcome the overfitting that can result from small training dataset sizes. The data augmentation has a lot of techniques, such as rotation, width shift, height shift, shear, zoom, horizontal flip and fill mode. These techniques were used in this paper to apply various transformation to the images as listed in Table 3.

4.3   Feature Extraction Phase

The convolutional layers, BN layers, ELU layers, and Max pooling layers were used to extract important features from the images.

- **Convolutional Layers:** the proposed approach involves six convolutional layers [29]. All convolutional layers contain 64 filters except for the last two layers (layer five and layer six) which each contained 128 filters. A kernel size of 3 × 3, a stride of 2 and the same padding were used in all convolutional layers (see Table 2).
- **Batch Normalization** is a recent approach for accelerating deep neural network training that normalizes each scalar feature independently by making it have a mean of zero and unit variance, as shown in step one, two and three in Algorithm 1. Then, the normalized value for each training mini-batch is scaled and shifted by the scale and shift parameters γ and β as shown in step four in Algorithm 1. This conversion confirms that the input distribution of each layer remains unchanged

within different mini-batches; thus, BN reduces the internal covariate shift and the number of iterations required for convergence and simultaneously improves the final performance. BN maintains non-trainable weights (the mean and variance vectors) that are updated via layer updates instead of through back propagation [30]. The BN can be considered as another layer that can be inserted into the model architecture, similar to a convolutional layer, an activation layer or a fully connected layer [31]. The proposed CNN approach includes eight BN layers in which six are used in the feature extraction phase and two are used in the classification phase. In the proposed approach, the BN layers were added before each activation function layer.

Algorithm 1. Batch normalization Transform applied to activation x over a mini-batch.

Input: Values of x over a mini-batch: $B = \{x_{1....m}\}$;

Parameters to be learned: γ and β

output: $\{y_i = BN_{\gamma,\beta}(x_i)\}$

$$\mu_B \leftarrow \frac{1}{m} \sum_{i=1}^{m} x_i \qquad \text{# Step one: the mean of the mini batch}$$

$$\sigma_B^2 \leftarrow \frac{1}{m} \sum_{i=1}^{m} (x_i - \mu_B)^2 \quad \text{# Step two: the variance of the mini batch}$$

$$\hat{x}_i \leftarrow \frac{x_i - \mu_B}{\sqrt{\sigma_B^2 + \epsilon}} \qquad \text{# Step three: normalize}$$

$$y_i \leftarrow \gamma \hat{x}_i + \beta \equiv BN_{\gamma,\beta}(x_i) \qquad \text{# Step four: scale and shift}$$

- **An Exponential Linear Unit (ELU)** is the activation function used in the proposed approach and given in [32] as

$$elu(x) = \begin{cases} \alpha(\exp(x) - 1) & if\ x \leq 0 \\ x & if\ x > 0 \end{cases} \tag{1}$$

in which the gradient w.r.t. the input is

$$\frac{d}{dx} elu(x) = \begin{cases} elu(x) + \alpha & if\ x \leq 0 \\ 1 & if\ x > 0 \end{cases} \tag{2}$$

where $\alpha = 1$.

The proposed CNN approach involves seven ELU layers in which six are used in the feature extraction phase and one is used in the classification phase. Each ELU layer was implemented after each BN layer as shown in Fig. 2.

Max-Pooling aims to down-sample the input representation in the feature extraction phase [33]. In this paper, six max-pooling layers of size (2×2) were used and these layers were implemented after each ELU layer.

4.4 Classification Phase

The classification phase classifies the images after flattening the output of the feature extraction phase [34] using two fully connected layers (FC), in which the first (FC1) contains 512 neurons and the second (FC2) contains 8 neurons, a BN layer, an ELU layer and a dropout layer [35] with a dropout rate of 0.3 to prevent overfitting. Finally, a softmax layer was added [36].

4.5 Checkpoint Ensemble Phase

When training a neural network model, the checkpoint technique [37] can be used to save all the model weights to obtain the final prediction or to checkpoint the neural network model improvements to save the best weights only and then obtain the final prediction, as shown in Fig. 3. In this paper, checkpointing was applied to save the

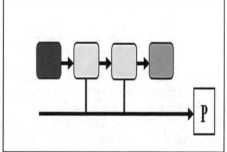

Fig. 3 The rounded boxes going from left to right represent a model's weights at each step of a particular training process. The lighter shades represent better weight. In checkpointing neural network model, all the model weights are saved to obtain the final prediction P. In checkpointing neural network improvements, only the best weights are saved to obtain the final prediction P

best network weights (those that maximally reduced the classification loss of the validation dataset).

5 Experimental Results

The architecture of the proposed approach was built using Keras library [38] using Tensorflow [39] as the backend. The Keras library's ImageDataGenerator function [40] was used to normalize the images during images preprocessing phase and to perform the data augmentation techniques as shown in Fig. 4, while the ModelCheckpoint function was used to perform the checkpoint ensemble phase. The proposed approach has 2,702,056 total parameters of which 2,699,992 are trainable parameters and 2,064 are non-trainable parameters that come from using the BN layers.

The proposed approach was tested using traditional metrics such as accuracy (Table 4), precision, recall, F1 score (Table 5), and a confusion matrix (Table 6).

Accuracy measures the ratio of correct predictions to the total number of instances evaluated and is calculated by the following formula [41]:

$$Accuracy = \frac{Number\ of\ correct\ prediction}{Total\ number\ of\ prediction} \tag{3}$$

Precision measures the ability of a model to correctly predict values for a particular category and is calculated as follows:

$$Precision = \frac{particular\ category\ predicted\ correctly}{all\ category\ predictions} \tag{4}$$

Recall measures the fraction of positive patterns that are correctly classified and is calculated by the following formula:

$$Recall = \frac{Correctly\ Predicted\ Category}{All\ Real\ Categories} \tag{5}$$

The F1 score is the weighted average of the precision and recall. Additionally, the confusion matrix is a matrix that maps the predicted outputs across actual outputs [42]. Additionally, the pyplot function of matplotlib [43] was used to plot the loss and accuracy of the model over the training and validation data during training to ensure that the model did not suffer from overfitting, as shown in Fig. 5.

To evaluate the proposed approach, transfer learning and fine-tuning techniques [44] with data augmentation technique were applied to the VGG16, VGG19 and Inception-v3 architectures using the same dataset and batch size used in the proposed approach. Transfer learning was conducted first by replacing the fully connected layers with new two fully connected layers, where FC1 contains 512 neurons and FC2 contains 8 neurons, and one added dropout layer after the flatten layer with

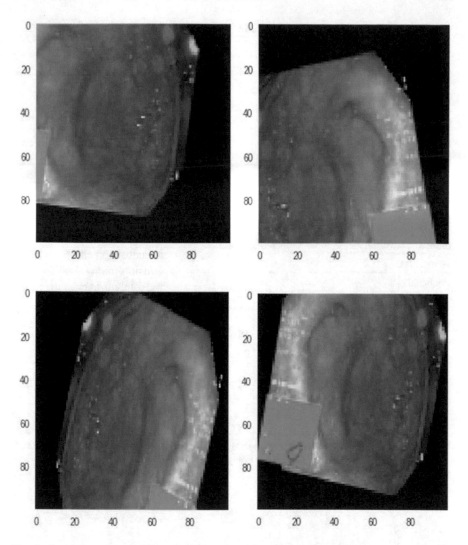

Fig. 4 Generation of normal caecum picture via random data augmentation

Table 4 Accuracy and loss values of the proposed approach

	Accuracy (%)	Loss
Training data	88	0.37
Validation data	88	0.38
Test data	87	0.38

Table 5 Classification results of the proposed approach

Classes	Precision	Recall	F1-score	Support
Dyed-lifted-polyps	0.81	0.93	0.87	150
Dyed-resection-margins	0.94	0.79	0.86	150
Oesophagitis	0.75	0.93	0.83	150
Normal-caecum	0.94	0.98	0.96	150
Normal-pylorus	0.97	1.00	0.99	150
Normal-z-line	0.91	0.71	0.81	150
Polyps	0.92	0.82	0.87	150
Ulcerative-colitis	0.89	0.93	0.91	150
Avg/total	0.89	0.89	0.89	1200

Table 6 Confusion matrix of the proposed approach

Actual class									
Predicted class		A	B	C	D	E	F	G	H
	A	140	7	0	0	0	0	2	1
	B	31	118	0	1	0	0	0	0
	C	0	0	139	0	0	11	0	0
	D	1	0	0	147	0	0	0	2
	E	0	0	0	0	150	0	0	0
	F	0	0	43	0	0	107	0	0
	G	0	0	1	8	4	0	123	14
	H	0	0	1	1	0	0	8	140

A = Dyed-lifted-polyps, B = Dyed-resection-margins, C = Oesophagitis, D = Normal-caecum, E = Normal-pylorus, F = Normal-z-line, G = Polyps, H = Ulcerative-colitis

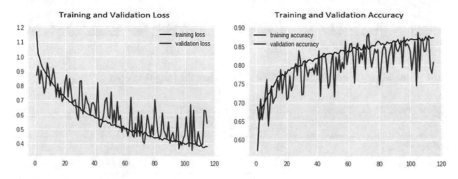

Fig. 5 Accuracy and loss values of the proposed approach on the training and validation data

Table 7 Comparative results for GI diseases identifications

Model	E1	E2	E3	E4 (%)	E5 (%)
VGG16	13	14,714,688	50	87	87
VGG19	16	20,024,384	50	87	87
Inception-v3	94	21,802,784	50	81	80
Proposed approach	**6**	**336,067**	**115**	**88**	**87**

E1 = Number of convolutional layers, E2 = Total parameters of the convolutional layers, E3 = Number of epochs, E4 = Validation accuracy, E5 = Test accuracy

a dropout rate of 0.3 in VGG architectures and a dropout rate of 0.5 in inception-v3 architecture to prevent overfitting. During the transfer learning the RMSProp optimizer with a learning rate of 1e−4 was used, and the models were trained for 15 epochs. Then, the top convolutional block of the VGG16 and VGG19 architectures and the top two blocks of Inceptionv3 architecture were fine-tuned, with a small learning rate of 1e−5 and trained for 35 epochs.

The elements used for comparison were the number of convolutional layers, the total number of parameters of the convolutional layers, the number of epochs, validation accuracy and test accuracy, as shown in Table 7 which shows as comparison of the models' results when identifying GI diseases. The first and second comparative elements are the number of convolution layers and the number of parameters of the convolutional layers. The proposed approach includes the fewest convolutional layers and parameters compared to the other architectures, which reduces its computational complexity in the training phase. The third comparative element is the number of epochs in the training phase. The proposed approach has the maximum number of epochs; however, the great number of epochs was expected because unlike the other architectures, the proposed approach network was not pre-trained; thus, more epochs are required to achieve a stable accuracy. As shown by the validation accuracy comparison, the proposed approach obtained an accuracy of 88%, which is better than the accuracy of the compared models. Regarding test accuracy, the proposed approach achieved an accuracy of 87%, similar to VGG16 and VGG19, while Inception-v3 achieved an accuracy of only 80%.

6 Conclusion

Automatic classification of GI diseases from imaging is increasingly important; it can assist the endoscopists in determining the appropriate treatment for patients who suffer from GI diseases and reduce the costs of disease therapies. In this paper, the proposed approach is introduced for this purpose. The proposed approach consists of a CNN with BN and ELU. The results of comparisons show that the proposed approach although it has low trained images and low computational complexity in training phase, outperforms the VGG16, VGG19, and Inception-v3 architectures regarding validation accuracy and outperforms the Inception-v3 model in test accuracy.

References

1. Subramanian, V., Ragunath, K.: Advanced endoscopic imaging: a review of commercially available technologies. Clin. Gastroenterol. Hepatol. **12**, 368–376 (2014)
2. Pereira, S.P., Goodchild, G., Webster, G.J.M.: The endoscopist and malignant and non-malignant biliary obstruction. BBA – Mol. Basis Dis. **2018**, 1478–1483 (1864)
3. Ferlay, J., et al.: GLOBOCAN 2012 v1.0, cancer incidence and mortality worldwide: IARC CancerBase No. 11. International Agency for Research on Cancer (2013). http://globocan.iarc.fr
4. American Cancer Society: Cancer Facts & Figures 2018. American Cancer Society, Atlanta, Ga (2018). https://www.cancer.org/
5. Repici, A., Hassan, C.: The endoscopist, the anesthesiologists, and safety in GI endoscopy. Gastrointest. Endosc. **85**(1), 109–111 (2017)
6. Kominami, Y., Yoshida, S., Tanaka, S., et al.: Computer-aided diagnosis of colorectal polyp histology by using a real-time image recognition system and narrow-band imaging magnifying colonoscopy. Gastrointest. Endosc. **83**, 643–649 (2016)
7. Khalifa, N.E.M., Taha, M.H.N., Ali, D.E., Slowik, A., Hassanien, A.E.: Artificial intelligence technique for gene expression by tumor RNA-Seq data: a novel optimized deep learning approach. IEEE Access **8**, 22874–22883 (2020)
8. Tajbakhsh, N., Gurudu, S.R., Liang, J.: A comprehensive computer aided polyp detection system for colonoscopy videos. In: Information Processing in Medical Imaging. Springer, pp. 327–338 (2015)
9. Cai, W., Wang, X., Chen, M.: Medical image classification with convolutional neural network. In: 13th International Conference on Control Automation Robotics & Vision (ICARCV) (2014)
10. Roth, H.R., Farag, A., Lu, L., Turkbey, E.B., Summers, R.M.: Deep convolutional networks for pancreas segmentation in CT imaging. In: SPIE Medical Imaging. International Society for Optics and Photonics, pp. 94 131G (2015)
11. Havaei, M., Davy, A., Warde-Farley, D., Biard, A., Courville, A., Bengio, Y., Pal, C., Jodoin, P.-M., Larochelle, H.: Brain tumor segmentation with deep neural networks. Med. Image Anal. **35**, 18–31 (2017)
12. Qayyum, A., Anwar, S.M., Awais, M., Majid, M.: Medical image retrieval using deep convolutional neural network. Neurocomputing **266**, 8–20 (2017)
13. Vardhana, M., Arunkumar, N., Lasrado, S., Abdulhay, E., Ramirez-Gonzalez, G.: Convolutional neural network for bio-medical image segmentation with hardware acceleration. Cognit. Syst. Res. **50**, 10–14 (2018)
14. Ioffe, S., Szegedy, C.: Batch normalization: accelerating deep network training by reducing internal covariate shift (2015). arXiv preprint arXiv:1502.03167v3
15. Clevert, D.A., Unterthiner, T., Hochreiter, S.: Fast and accurate deep network learning by exponential linear units (ELUs). In: ICLR Conference (2016)
16. Ezzat, D., Taha, M.H.N., Hassanien, A.E.: An optimized deep convolutional neural network to identify nanoscience scanning electron microscope images using social ski driver algorithm. In: International Conference on Advanced Intelligent Systems and Informatics, pp. 492–501. Springer, Cham (2019)
17. Albeahdili, H.M., Alwzwazy, H.A., Islam, N.E.: Robust convolutional neural networks for image recognition. Int. J. Adv. Comput. Sci. Appl. (ijacsa) **6**(11) (2015). http://dx.doi.org/10.14569/IJACSA.2015.061115

18. Simonyan, K., Zisserman, A.: Very deep convolutional networks for large-scale image recognition. In: Proceedings of the International Conference on Learning Representations, San Diego, CA, USA, 7–9 May 2015 (2015)
19. Srivastava, R.K., Greff, K., Schmidhuber, J.: Training very deep networks. In: Proceedings of the 28th International Conference on Neural Information Processing Systems, Montreal, Canada, pp. 2377–2385 (2015)
20. Szegedy, C., Vanhoucke, V., Ioffe, S., Shlens, J., Wojna, Z. Rethinking the inception architecture for computer vision. In: 2016 IEEE Conference on Computer Vision and Pattern Recognition (CVPR). https://doi.org/10.1109/cvpr.2016.308
21. He, K., Zhang, X., Ren, S., Sun, J.: Deep residual learning for image recognition. In: 2016 IEEE Conference on Computer Vision and Pattern Recognition (CVPR) (2016)
22. Ioffe, S.: Batch renormalization: towards reducing minibatch dependence in batch-normalized models. CoRR, abs/1702.03275 (2017)
23. http://datasets.simula.no/kvasir/
24. Asperti, A., Mastronardo, C.: The effectiveness of data augmentation for detection of gastroin testinal diseases from endoscopical images. In: Proceedings of the 11th International Joint Conference on Biomedical Engineering Systems and Technologies, KALSIMIS, Funchal, Madeira, Portugal, vol. 2, pp. 199–205 (2018)
25. Pogorelov, K., et al.: Kvasir: a multi-class image dataset for computer aided gastrointestinal disease detection. In: Proceedings of the 8th ACM on Multimedia Systems Conference, MMSys 2017, New York, NY, USA, pp. 164–169. ACM (2017)
26. Bello, I., Zoph, B., Vasudevan, V. and Le, Q.V.: Neural optimizer search with reinforcement learning. In: International Conference on Machine Learning, pp. 459–468 (2017)
27. Santosa, B.: Multiclass classification with cross entropy-support vector machines. Procedia Comput. Sci. **72**, 345–352. https://doi.org/10.1016/j.procs.2015.12.149
28. Ezzat, D., Hassanien, A.E., Taha, M.H.N., Bhattacharyya, S., Vaclav, S.: Transfer learning with a fine-tuned CNN model for classifying augmented natural images. In: International Conference on Innovative Computing and Communications, pp. 843–856. Springer, Singapore (2020)
29. Albawi, S., Mohammed, T.A., Al-Zawi, S.: Understanding of a convolutional neural network. In: International Conference on Engineering and Technology (ICET) (2017). https://doi.org/10.1109/icengtechnol.2017.8308186
30. Guo, Y., Wu, Q., Deng, C., Chen, J., Tan, M.: Double forward propagation for memorized batch normalization. AAAI (2018)
31. Schilling, F.: The effect of batch normalization on deep convolutional neural networks. Master thesis at CSC CVAP (2016)
32. Pedamonti, D.: Comparison of non-linear activation functions for deep neural networks on MNIST classification task. (2018). arXiv preprint arXiv:1804.02763v1
33. Darwish, A., Ezzat, D., Hassanien, A.E.: An optimized model based on convolutional neural networks and orthogonal learning particle swarm optimization algorithm for plant diseases diagnosis. Swarm Evol. Comput. **52**, 100616 (2020)
34. Passalis, N., Tefas, A.: Learning bag-of-features pooling for deep convolutional neural networks. In: IEEE International Conference on Computer Vision (ICCV) (2017). https://doi.org/10.1109/iccv.2017.614
35. Wu, H., Gu, X.: Towards dropout training for convolutional neural networks. Neural Netw. **71**, 1–10 (2015). https://doi.org/10.1016/j.neunet.2015.07.007
36. Yuan, B.: Efficient hardware architecture of softmax layer in deep neural network. In: 2016 29th IEEE International System-on-Chip Conference (SOCC). https://doi.org/10.1109/socc.2016.7905501
37. Chen, H., Lundberg, S., Lee, S.-I.: Checkpoint ensembles: ensemble methods from a single training process (2017). arXiv preprint arXiv:1710.03282v1
38. Chollet, F.: Keras: deep learning library for theano and tensorflow. 2015. https://keras.io/. Accessed 19 Apr 2017
39. Zaccone, G.: Getting Started with TensorFlow. Packt Publishing Ltd. (2016)

40. Ayan, E., Unver, H.M.: Data augmentation importance for classification of skin lesions via deep learning. In: 2018 Electric Electronics, Computer Science, Biomedical Engineerings Meeting (EBBT). IEEE, pp. 1–4 (2018). https://doi.org/10.1109/EBBT.2018.8391469
41. Hossin, M., Sulaiman, M.N.: A review on evaluation metrics for data classification evaluations. Int. J. Data Min. Knowl. Manag. Process 5(2), 01–11 (2015). https://doi.org/10.5121/ijdkp.2015.5201
42. Maria Navin, J.R., Balaji, K.: Performance analysis of neural networks and support vector machines using confusion matrix. Int. J. Adv. Res. Sci. Eng. Technol. 3(5), 2106–2109 (2016)
43. Jeffrey, W.: The Matplotlib basemap toolkit user's guide (v. 1.0.5). Matplotlib Basemap Toolkit documentation. Accessed 24 Apr 2013
44. Wang, G., Sun, Y., Wang, J.: Automatic image-based plant disease severity estimation using deep learning. Comput. Intell. Neurosci. 1–8 (2017). https://doi.org/10.1155/2017/2917536

A Chaotic Search-Enhanced Genetic Algorithm for Bilevel Programming Problems

Y. Abo-Elnaga, S. Nasr, I. El-Desoky, Z. Hendawy, and A. Mousa

Abstract In this chapter, we propose chaotic search-enhanced genetic algorithm for solving bilevel programming problem (BLPP). The proposed algorithm is a combination between enhanced genetic algorithm based on new selection technique named effective selection technique (EST) and chaos searching technique. Firstly, the upper level problem is solved using enhanced genetic algorithm based on EST. EST enables the upper level decision maker to choose an appropriate solution in anticipation of the lower level's decision. Then, lower level problem is solved using genetic algorithm for the upper level solution. Secondly, local search based on chaos theory is applied for the upper level problem around the enhanced genetic algorithm solution. Finally, lower level problem is solved again using genetic algorithm for the chaos search solution. The incorporating between enhanced genetic algorithm supported by EST and chaos theory increases the search efficiency and helps in faster convergence of the algorithm. The performance of the algorithm has been evaluated on different sets of test problems linear and nonlinear problems, constrained and unconstrained problems and low-dimensional and high-dimensional problems. Also, comparison between the proposed algorithm results and other state-of-the-art algorithms is introduced to show the effectiveness and efficiency of our algorithm.

Keywords Bi-level optimization · Evolutionary algorithms · Genetic algorithm · Chaos theory

Mathematics Subject Classification 68T20 · 68T27 · 90C26 · 90C59 · 90C99

Y. Abo-Elnaga
Department of Basic Science, Higher Technological Institute, 10th of Ramadan City, Egypt

S. Nasr (✉) · I. El-Desoky · Z. Hendawy · A. Mousa
Department of Basic Engineering Science, Faculty of Engineering, Menoufia University, Shebin El-Kom, Egypt
e-mail: sarah.nasr.eid@gmail.com

A. Mousa
Department of Mathematics and Statistics, Faculty of Sciences, Taif University, Taif, Saudi Arabia

© The Author(s), under exclusive license to Springer Nature Switzerland AG 2021 129
A. E. Hassanien et al. (eds.), *Machine Learning and Big Data Analytics*
Paradigms: Analysis, Applications and Challenges, Studies in Big Data 77,
https://doi.org/10.1007/978-3-030-59338-4_8

1 Introduction

In the common form of optimization problems, there are just one decision maker seeks to find an optimal solution [1, 2]. On the contrary, many optimization problems appeared in real world in engineering design, logistics problems, traffic problems, economic policy and so on consist of a hierarchical decision structure and have many decision makers [3, 4]. This hierarchical decision structure can modeled as "k" levels nested optimization problems. The first level is the upper level and its decision maker is the leader. Next levels are lower levels and its decision makers are the followers. Any decision taken by leader to optimize his/her problem is affected by the followers' response. The hierarchical process starts by taking the upper level decision maker his/her decisions. Then, the lower level decision makers take its own decisions upon the leader's decisions [5–7]. BLPP solution requirements produce difficulties in solving such as disconnectedness and non-convexity even to the simple problems [8, 9].

Since 1960s, the studies of bilevel programming models have been started and since that many strategies have been appeared to solve it. The methods developed to solve BLPP can be classified into two main categories: classical methods and evolutionary methods. The classical methods include the K-th best algorithm, descent direction method, exact penalty function, etc. [10, 11]. Classical methods can treat only with differentiable and convex problem. Furthermore, the BLPP solution requirements produce disconnectedness and non-convexity even to the simple problems. Lu et al. proposed a survey on multilevel optimization problems and its solution techniques [12], they proposed detailed comparisons between BLPP solution methods. They accentuated that classical methods have been used to solve a specified BLPP and have not been used to solve various multilevel programming problems especially for large-scale problems. The evolutionary methods are conceptually different from classical methods. It can treat with differentiable and convex optimization problem. Evolutionary methods are inspired by natural adaptation as the behavior of biological, molecular, swarm of insects, and neurobiological systems. Evolutionary methods include genetic algorithms, simulated annealing particle swarm optimization ant colony optimization and neural-network-based methods... etc. [13–16].

In this chapter, we propose a combination between genetic algorithm based on a new effective selection technique and chaos search denoted by CGA-ES. Genetic algorithm is well suited for solving BLPP and chaos theory is also represented as one of new theories that have attracted much attention applied to many aspects of the optimization [17]. Using genetic algorithm supported with our new effective selection technique and chaos theory enhances the search performance and improves the search efficiency. The search using chaos theory not only enhances the search characteristics but also it helps in faster convergence of the algorithm and avoids trapping the local optima.

The performance of the algorithm has been evaluated on different sets of test problems. The first set of problems is constrained problems with relatively smaller number of variables named TP problems [18, 19]. The second set is SMD test set [20]. SMD problems are unconstrained high-dimensional problems that are recently proposed. The proposed algorithm results have been analyzed to show that our proposed algorithm is an effective strategy to solve BLPP. Also, comparison between the proposed algorithm results and other state-of-the-art is introduced to show the effectiveness and efficiency of the proposed algorithm.

This chapter is organized as follows. In Sect. 2, formulation of bilevel optimization problem is introduced. Section 3 provides our proposed algorithm to solve BLPP. Numerical experiments are discussed in Sect. 4. Finally, Sect. 5 presents our conclusion.

2 Bilevel Programming Problems

BLPP consists of two nested optimization problems with two decision makers. The two decision makers are constructed in hierarchical structure, leader and follower. The leader problem is to take a decision that make the follower take his/her decision in the interest of leader's problem. Contrariwise, the follower doesn't have a problem. He/She make an optimal decision upon the leader's decision. Thus, the lower-level problem is parameterized by the upper-level decision and appears as a constraint in the upper-level problem. Bilevel Programming Problems can be formulated as follows:

$$BLPP : \min_{x,y} F(x, y)$$
$$\text{s.t. } G(x, y) \leq 0, \quad \text{(Upper level)}$$

where for each given x obtained by the upper level, solves y

$$\min_{y} f(x, y)$$
$$\text{s.t. } g(x, y) \leq 0 \quad \text{(Lower level)}$$

where $F(x, y)$, $f(x, y) : R^{n1} \times R^{n2} \to R$ are object functions of the upper and lower level problems. $G(x, y) : R^{n1} \times R^{n2} \to R^p$, $g(x, y) : R^{n1} \times R^{n2} \to R^q$ are the constraint functions of the upper and lower level problems. $x \in R^{n1}$, $y \in R^{n2}$ are the decision variables controlled by the upper and lower level problems, respectively. $n1, n2$ are the dimensional decision vectors for the upper and lower level problems. A vector $v^{\cdot} = (x^{\cdot}, y^{\cdot})$ is an feasible solution if it satisfies all upper and lower level constrains $G(x, y)$, $g(x, y)$ and if vector $y^{\cdot}$ is an optimal solution for the lower level objective function $f(x, y)$ solved for the upper level vector solution $x^{\cdot}$ [21].

| Check chromosome in constrains and repair out constrain values |
| Evaluate chromosome in the objective function |
| Do: |
| Children population [Select parents from population and recombine parents (Crossover and mutation operators)] |
| Evaluate children in the objective function |
| Construct best population of parents and children population |
| While satisfactory solution has been found |

Fig. 1 The pseudo code of the general GA algorithm

3 The Proposed Algorithm (CGA-ES)

Our proposed algorithm for BLPP is based on concepts of one of evolutionary algorithms, genetic algorithm, and chaos theory. In this section, we first introduce basic concepts of genetic algorithm and chaos theory and then we explain the proposed algorithm in details.

3.1 Basic Concepts of Genetic Algorithm

Genetic algorithm (GA) is a powerful search technique inspired by the biological evolution process. GA was proposed originally by Holland [21] in the early 1970s. Since then it has been widely used for solving different types of optimization problem. GA starts with a set of random individuals (chromosomes) that are generated in feasible boundaries. Then, the individuals move through a number of generations in search of optimal solution. During each generation, genetic search operators such as selection, mutation and crossover are applied one after another to obtain a new generation of chromosomes. New generation of chromosomes are expected to be better than its previous generation. The evaluation of chromosomes is according to the fitness function. This process is repeated until the termination criterion is met, and the best obtained chromosome is the solution. Figure 1 shows the pseudo code of the general GA algorithm. Generate an initial population.

3.2 Chaos Theory for Optimization Problems

Chaos theory studies the behavior of systems that follow deterministic laws but appear random and unpredictable. Chaos theory was initially described by Henon and summarized by Lorenz. Chaos being radically different from statistical randomness, especially the inherent ability to search the space of interest efficiently, could improve the performance of optimization procedure. It could be introduced into the optimization strategy to accelerate the optimum seeking operation and find the global optimal solution [22]. Chaos theory distinguishes from other optimization techniques by its inherent ability to search the space of interest efficiently. The chaotic behavior is described using a set of different chaotic maps. Some of chaotic maps are discrete time parameterized or continuous time parameterized. The basic idea of chaos search for solution is to transform the variable of problems from the solution space to chaos space and then perform search for solution. The transformation starts with determining the range of chaos search boundary. Then, chaotic numbers are generated using one of chaotic maps. The transformation finishes by mapping the chaotic numbers into the variance range search. Logistic map is known that is more convenient to be used and increase the solution quality rather than other chaotic maps. Here we offer some well-known chaotic maps found in the literature [23].

- **Chebyshev map**
  Chebyshev map is represented as:

$$x_{t+1} = \cos(t \cos^{-1}(x_t));\qquad(1)$$

- **Circle map**
  Circle map is defined as the following representative equation:

$$x_{t+1} = x_t + b - (a - 2\pi)\sin(2\pi x_t) \bmod (1);$$
$$x_{t+1} = \cos(t \cos^{-1}(x_t));\qquad(2)$$

where $a = 0.5$ and $b = 0.2$.

- **Gauss/mouse map**
  The Gauss map consists of two sequential parts defined as:

$$x_{t+1} = \begin{cases} 0 & \text{if } x_t = 0 \\ 1/x_t \text{ else } \bmod (1) \end{cases};\qquad(3)$$

where $\frac{1}{x_t} \bmod (1) = \frac{1}{x_t} - \left\lfloor \frac{1}{x_t} \right\rfloor$.

- **Intermittency map**
  The intermittency map is formed with two iterative equations and represented as:

$$x_{t+1} = \begin{cases} \varepsilon + x_t + cx_t^n & \text{if } 0 < x_t \le p \\ \frac{x_t - p}{1-p} & \text{elseif } p < x_t < 1 \end{cases};\qquad(4)$$

where $c = \frac{1-\varepsilon-p}{p^2}$, $n = 2.0$ and ε is very close to zero.

- **Iterative map**

  The iterative chaotic map with infinite collapses is defined with the following as:

  $$x_{t+1} = \sin\left(\frac{a\pi}{x_t}\right); \tag{5}$$

  where $a \in (0, 1)$.

- **Liebovitch map**

  The proposed chaotic map can be defined as:

  $$x_{t+1} = \begin{cases} \alpha x_t & 0 < x_t \le p_1 \\ \frac{p_2 - x_t}{p_2 - p_1} & p_1 < x_t \le p_2 \\ 1 - \beta(1 - x_t) & p_2 < x_t \le 1 \end{cases}; \tag{6}$$

  where $\alpha = \frac{p_2(1 - (p_2 - p_1))}{p_1}$ and $\beta = \frac{((p_2 - 1) - p_1(p_2 - p_1))}{p_2 - 1}$.

- **Logistic map**

  Logistic map demonstrates how complex behavior arises from a simple deterministic system without the need of any random sequence. It is based on a simple polynomial equation which describes the dynamics of biological population.

  $$x_{t+1} = cx_t(1 - x_t); \tag{7}$$

  where $x_0 \in (0, 1)$, $x_0 \notin \{0.0, 0.25, 0.50, 0.75, 1.0\}$ and when $c = 4.0$ a chaotic sequence is generated by the Logistic map.

- **Piecewise map**

  Piecewise map can be formulated as follows:

  $$x_{t+1} = \begin{cases} \frac{x_t}{p} & 0 < x_t < p \\ \frac{x_t - p}{0.5 - p} & p \le x_t < 0.5 \\ \frac{(1 - p - x_t)}{0.5 - p} & 0.5 \le x_t < 1 - p \\ \frac{(1 - x_t)}{p} & 1 - p < x_t < 1 \end{cases}; \tag{8}$$

  where $p \in (0, 0.5)$ and $x \in (0, 1)$.

- **Sine map**

  Sine map can be described as:

  $$x_{t+1} = \frac{a}{4}\sin(\pi x_t); \tag{9}$$

  where $0 < a < 4$.

- **Singer map**

  One dimensional chaotic Singer map is formulated as

  $$x_{t+1} = \mu\left(7.86x_t - 23.31x_t^2 + 28.75x_t^3 - 13.302875x_t^4\right); \tag{10}$$

where $\mu \in (0.9, 1.08)$.

- **Sinusoidal map**
  Sinusoidal map is generated as the following equation:

$$x_{t+1} = ax_t^2 \sin(\pi x_t); \qquad (11)$$

where $a = 2.3$.

- **Tent map**
  Tent map is defined by the following iterative equation:

$$x_{t+1} = \begin{cases} x_t/0.07 & x_t < 0.7 \\ \frac{10}{3}(1.0 - x_t) & x_t \geq 0.7 \end{cases} \qquad (12)$$

3.3 CGA-ES for BLPP

The proposed algorithm to solve the BLPP is incorporation between an enhanced genetic algorithm and chaos search. Firstly, the upper level problem using genetic algorithm based on our new effective selection technique. Then, lower level problem is solved for the obtained upper level solution. Secondly, another search based on chaos theory is applied to the obtained solution.

Effective selection technique: The new effective selection technique operates on enabling the upper level decision maker to choose an appropriate solution in anticipation of the lower level's decision. After the individuals are evaluated by upper level objective function and are selected according to its fitness. Then, these selected solutions are evaluated by lower level problem objective function and are selected according to its fitness. The steps of the proposed algorithm are listed in details as follows:

Step 1. Solve upper level problem using genetic algorithm based on effective selection technique with constrain handling.
Step 1.1 Constrains handling
Any solution for the upper level is feasible solution for the upper level only if it is feasible and optimal solution for the lower level. Constrains can be handled by adding constrains of the lower level problem to upper level constrains.
Step 1.2 Initial population
The population vectors are randomly initialized and within the search space bounds [23].
Step 1.3 Obtaining reference point
At least one feasible reference point is needed to enter the process of repairing infeasible individuals of the population and to complete the algorithm procedure.

Step 1.4 Repairing
When the problem is constrained, some of generated individuals don't satisfy
constrains and become infeasible individuals. The proposed of repairing process
is to transform the infeasible individuals to be feasible individuals. The reader
can refer to [24].

Step 1.5 Evaluation
The individuals are evaluated using both upper level objective function and lower
level objective function.

Step 1.6 Create a new population
In this step, a new population is generated by applying a new effective selection
technique as mentioned above, crossover operator and mutation operator [25].

- **Ranking**: Ranks individuals according to their fitness value, and returns a
  column vector containing the corresponding individual fitness value, in order
  to establish later the probabilities of survival that are necessary for the selection
  process.
- **Selection**: There are several techniques of selection. The commonly used techniques for selection of individuals are roulette wheel selection, rank selection,
  steady state selection, stochastic universal sampling, etc. Here we will use
  Stochastic Universal Sampling (SUS) [26] where, the most important concern
  in a stochastic selection is to prevent loss of population diversity due to its
  stochastic aspect.
- **Crossover**: In GAs, crossover is used to vary individuals from one generation to
  the next; where it combines two individuals (parents) to produce a new individuals (offspring) with probability (Pc). There are several techniques of crossover,
  one-point crossover, two-point crossover, cut and splice, uniform crossover and
  half uniform crossover, etc. Here, we will use One-point crossover involving
  splitting two individuals and then combining one part of one with the other
  pair. This method performs recombination between pairs of individuals and
  returns the new individuals after mating, and gives offspring the best possible
  combination of the characteristics of their parents.
- **Mutation**: Premature convergence is a critical problem in most optimization
  techniques, which occurs when highly fit parent individuals in the population breed many similar offspring in early evolution time. Mutation is used to
  maintain genetic diversity from one generation of a population to the next. In
  addition, Mutation is an operator to change element in a string which is generated after crossover operator. In this study, we will use real valued mutation;
  which means that randomly created values are added to the variables with a
  low probability (Pm).

Step 1.7 Migration
In this step, the best individuals of the new generation and old generation are
migrated for the new generation.

Step 1.8 Termination test for GA
Terminated of algorithm is achieved either when the maximum number of generations is achieved, or when the individuals of the population convergence occur. Otherwise, return to step 1.6.

Step 2. Solve lower level problem using genetic algorithm as upper level steps for the solution of the upper level decision variables.

Step 3. Evaluate the solution obtained from solving lower level problem using upper level objective function.

Step 4. Search based on chaos theory for the solution obtained in step 3.
In this step, chaos search is applied to the solution obtained in step 3 (x_i^*, y_i^*)[23]. The detailed description of chaotic search is described as follows:

Step 4.1 Determine chaotic search range
The range of chaotic search is $[a, b]$ for the upper level variables and $[c, d]$ for the lower level variables. The range is determined by the following equation

$$x_i^* - \varepsilon_1 < a_i, \ x_i^* + \varepsilon_1 > b_i \tag{13}$$

$$y_i^* - \varepsilon_2 < c_i, \ y_i^* + \varepsilon_2 > d_i \tag{14}$$

where $\varepsilon_1, \varepsilon_2$ is specified radiuses of chaos search for the upper level and lower level variables receptively.

Step 4.2 Generate chaotic number using logistic map
We choose the logistic map because it is more convenient to use and increase the solution quality rather than other chaotic maps [23]. Chaotic random numbers z^k is generated by the logistic map by the following equation

$$z^{k+1} = \mu z^k (1 - z^k), \ z^0 \in (0, 1), \ z^0 \notin \{0.0, 0.25, 0.50, 0.75, 1.0\}, k = 1, 2, \ldots \tag{15}$$

Step 4.3 Map the chaos variable into the variance range
Chaos variable z^k is mapped into the variance range of optimization valuable $[a, b]$ and $[c, d]$ by:

$$x_i = x_i^* - \varepsilon_1 + 2\varepsilon_1 z^k \quad \forall i = 1, \ldots, n \tag{16}$$

$$y_i = y_i^* - \varepsilon_2 + 2\varepsilon_2 z^k \quad \forall i = 1, \ldots, m \tag{17}$$

Step 4.4 Update the best value
Set chaotic iteration number as $k = 1 \to$ Do

$$x_i^k = x_i^* - \varepsilon + 2\varepsilon z \quad \forall i = 1, \ldots, n \tag{18}$$

$$y_i^k = y_i^* - \varepsilon_2 + 2\varepsilon_2 z^k \quad \forall i = 1, \ldots, m \tag{19}$$

If $F(x^k, y^k) < F(x^*, y^*)$ then set $(x^*, y^*) = (x^k, y^k)$.
Else if $F(x^k, y^k) \geq F(x^*, y^*)$ then give up the K-th iterated
Result (x^*, y^*).
Loop runs until $F(x^*, y^*)$ is not improved after k searches.

Step 4.5 Update the boundary
Update the boundary value $[a, b]$ and $[c, d]$ of the new optimal point (x^*, y^*) as the new chaos search range. Use Eqs. (4) and (5) to map the chaos variables into the new search range then go to 4.2.

Step 4.6 Stopping Chaos search
Stop chaos search for, K-th for the specified iterations and put out (x^*, y^*) as the best solution.

Step 5. Solve lower level problem using genetic algorithm for the solution after chaos search then evaluate the solution obtained from solving lower level problem.
In this step, repeat step 2 and 3 for the solution obtained from chaos search (x^*, y^*).

Step 6. Stopping algorithm
Compare the result before chaos search at step 3 and after chaos search at step 5. The best result is the algorithm solution. The flow chart of the proposed algorithm is shown in Fig. 2.

4   Numerical Experiments

Our proposed algorithm has been evaluated on several sets of well-known multi-modal test problems including linear and nonlinear problems, constrained and unconstrained problems, low-dimensional and high-dimensional problems. The first set of problems is 10 constrained problems with relatively smaller number of variables named TP problems [18, 19]. TP problems have been extensively used test-suite for comparing different algorithms that have been solved BLPP. The second set is SMD test set constructed in [20]. The used set of SMD problems are un-constrained high-dimensional problems that are recently proposed, which contains problems with controllable complexities. SMD test problems cover a wide range of difficulties associated with BLPP as non-linearity, multi-modality and conflict in upper and lower level objectives.

CGA-ES is coded in matlab 7.8 and the simulations have been executed on an Intel core (TM) i7-4510u cpu @2.00GHZ 2.60 GHz processor. CGA-ES involves a number of parameters that affect the performance of algorithm. The applied GA parameters for both upper and lower level problem are:

(a) Generation gap	0.9
(b) Crossover rate	0.9

(continued)

(continued)

(c)	Mutation rate	0.7
(d)	Selection operator	Stochastic universal sampling
(e)	Crossover operator	Single point
(f)	Mutation operator	Real-value
(g)	GA generation	200–1000
(h)	Chaos generation	10,000
(i)	Specified neighborhood radius	1E−3

Table 1 proposes a standard test problems from TP1 to TP10. The problem dimensions (n, m) are given in the first column, and the problem formulation is defined in the second column. Dimensions (n) denotes to the upper level dimensions. Dimensions (m) denotes to the lower level dimensions.

4.1 Results and Analyses

In this section, we first introduce analyses of our results for the chosen test set. We compare our results and other algorithms results solved same chosen test set. The results and comparisons reveal of our proposed algorithm feasibility and efficiency to solve BLPP and that it has better ability and precision than other proposed methods in literature.

1. **Results analyses for TP test set**

Sinha et al. presented an improved evolutionary algorithm based on quadratic approximations of lower level named as BLEAQ. In Table 2, TP problems result is presented. The results for upper level and lower level objective functions are introduced in second column. The best solutions of upper level and lower level objective functions according to the reference of the problems are in the third column [18, 19].

As indicated in Table 1, our proposed algorithm find solutions better than the best solutions according Sinha for four problems TP3, TP4, TP5, TP6 and reach to same solutions as best known solutions for both levels TP7 and reach to same solutions as best known solutions for upper levels for TP2. The remainder problems have also a small difference from best known solutions.

2. **Results analyses for SMD test set**

Table 2 presents results for SMD problems. In the second column our proposed results for upper level and lower level objective functions. The optimal solution of upper level and lower level objective functions SMD problems are in the third column [20]. From Table 3, our proposed algorithm reach to optimal solutions for two problems SMD1, SMD3. The remainder problems have also a small error.

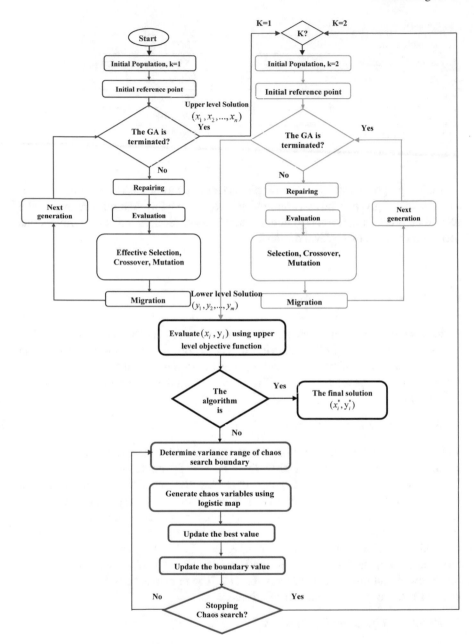

Fig. 2 The flow chart of the proposed algorithm CGA-ES

The simulation results of various numerical studies have been demonstrated the superiority of CGA-ES to solve BLPP. CGA-ES superior on other algorithms proposed in literature solved same test problems. It can produce better solutions and performs well. CGA-ES can manage the difficulties resulting from the increasing dimension.

Table 1 BLPP formulation for problems TP1–TP10

Problem	Problem dimensions (n, m)	Problem formulation
TP1	$n = 2$ $m = 2$	$Minimize\ F(x,\ y) = (x_1 - 30)^2 + (x_2 - 20)^2 - 20y_1 + 20y_2$ $s.t$ $y \in \underset{(y)}{\arg\min} \begin{cases} Minimize\ f(x,\ y) = (x_1 - y_1)^2 + (x_2 - y_2)^2 \\ s.t. \\ 0 \le y_i \le 10 \quad i = 1, 2 \end{cases}$ $x_1 + 2x_2 \ge 30,$ $x_1 + x_2 \le 25,$ $x_2 \le 15$
TP2	$n = 2$ $m = 2$	$Minimize\ F(x,\ y) = 2x_1 + 2x_2 - 3y_1 - 3y_2 - 60$ $s.t$ $y \in \underset{(y)}{\arg\min} \begin{cases} Minimize\ f(x,\ y) = (y_1 - x_1 + 20)^2 + (y_2 - x_2 + 20)^2 \\ s.t \\ x_1 - 2y_1 \ge 10 \\ x_2 - 2y_2 \ge 10 \\ -10 \le y_i \le 20 \quad i = 1, 2 \end{cases}$ $x_1 + x_2 + y_1 - 2y_2 \le 40,$ $0 \le x_i \le 50 \quad i = 1, 2$
TP3	$n = 2$ $m = 2$	$Minimize\ F(x,\ y) = -x_1^2 - 3x_2^2 - 4y_1 + y_2^2$ $s.t$ $y \in \underset{(y)}{\arg\min} \begin{cases} Minimize\ f(x,\ y) = 2x_1^2 + y_1^2 - 5y_2 \\ s.t \\ x_1^2 - 2x_1 + x_2^2 - 2y_1 + y_2 \ge -3 \\ x_2 + 3y_1 - 4y_2 \ge 4 \\ 0 \le y_i \quad i = 1, 2 \end{cases}$ $x_1^2 + 2x_2 \le 4,$ $0 \le x_i \quad i = 1, 2$

(continued)

Table 1 (continued)

Problem	Problem dimensions (n, m)	Problem formulation
TP4	$n = 2$ $m = 3$	$Minimize\ F(x,\ y) = -8x_1 - 4x_2 + 4y_1 - 40y_2 - 4y_3$ $s.t$ $y \in \underset{(y)}{arg\ min} \begin{cases} Minimize\ f(x,\ y) = x_1 + 2x_2 + y_1 + y_2 + 2y_3 \\ s.t \\ y_2 + y_3 - y_1 \le 1 \\ 2x_1 - y_1 + 2y_2 - 0.5y_3 \le 1 \\ 2x_2 + 2y_1 - y_2 - 0.5y_3 \le 1 \\ 0 \le y_i \quad i = 1, 2 \end{cases}$ $0 \le x_i \quad i = 1, 2$
TP5	$n = 2$ $m = 2$	$Minimize\ F(x,\ y) = rt(x)x - 3y_1 + 4y_2 + 0.5t(y)y$ $s.t$ $y \in \underset{(y)}{arg\ min} \begin{cases} Minimize\ f(x,\ y) = 0.5t(y)hy - t(b(x))y \\ s.t \\ -0.333y_1 + y_2 - 2 \le 0 \\ y_1 - 0.333y_2 - 2 \le 0 \\ 0 \le y_i \quad i = 1, 2 \end{cases}$ $where,\ h = \begin{pmatrix} 1 & 3 \\ 3 & 10 \end{pmatrix}, b(x) = \begin{pmatrix} -1 & 2 \\ 3 & -3 \end{pmatrix} x, r = 0.1$ $t(.)\ denotes\ transpose\ of\ a\ vector$
TP6	$n = 1$ $m = 2$	$Minimize\ F(x,\ y) = (x_1 - 1)^2 + 2y_1 - 2x_1$ $s.t$ $y \in \underset{(y)}{arg\ min} \begin{cases} Minimize\ f(x,\ y) = (2y_1 - 4)^2 + (2y_2 - 1)^2 + x_1 y_1 \\ s.t \\ 4x_1 + 5y_1 + 4y_2 \le 12, \\ 4y_2 - 4x_1 - 5y_1 \le -4 \\ 4x_1 - 4y_1 + 5y_2 \le 4, \\ 4y_1 - 4x_1 + 5y_2 \le 4 \\ 0 \le y_i \quad i = 1, 2 \end{cases}$ $x_1 \ge 0$

(continued)

Table 1 (continued)

Problem	Problem dimensions (n, m)	Problem formulation
TP7	$n = 2$ $m = 2$	$Maximize\ F(x, y) = -\dfrac{(x_1 + y_1)(x_2 + y_2)}{1 + x_1 y_1 + x_2 y_2}$ $s.t$ $y \in \underset{(y)}{\arg\min} \left\{ \begin{array}{l} Maximize\ f(x, y) = \frac{(x_1 + y_1)(x_2 + y_2)}{1 + x_1 y_1 + x_2 y_2} \\ s.t \\ 0 \le y_i \le x_i \quad i = 1, 2 \end{array} \right\}$ $x_1^2 + x_2^2 \le 100,$ $x_1 - x_2 \le 0$ $0 < x_i, \quad i = 1, 2$
TP8	$n = 2$ $m = 2$	$Minimize\ F(x, y) = |2x_1 + 2x_2 - 3y_1 - 3y_2 - 60|$ $s.t$ $y \in \underset{(y)}{\arg\min} \left\{ \begin{array}{l} Minimize\ f(x, y) = (y_1 - x_1 + 20)^2 + (y_2 - x_2 + 20)^2 \\ s.t \\ 2y_1 - x_1 + 10 \ge 0 \\ 2y_2 - x_2 + 10 \ge 0 \\ -10 \le y_i \le 20 \quad i = 1, 2 \end{array} \right\}$ $x_1 + x_2 + y_1 - 2y_2 \le 40,$ $0 \le x_i \le 50 \quad i = 1, 2$
TP9	$n = 10$ $m = 10$	$Minimiyze\ F(x, y) = \sum_{i=1}^{10} (|x_i - 1| + |y_i|)$ $s.t$ $y \in \underset{(y)}{\arg\min} \left\{ \begin{array}{l} Minimize\ f(x, y) = e^{(1 + \frac{1}{4000} \sum_{i=1}^{10} (y_i)^2 - \prod_{i=1}^{10} \cos(\frac{y_i}{\sqrt{i}}) \sum_{i=1}^{10} (x_i)^2}} \\ s.t \\ -\pi \le y_i \le \pi \quad i = 1, 2, \ldots, 10 \end{array} \right\}$
TP10	$n = 10$ $m = 10$	$Minimiyze\ F(x, y) = \sum_{i=1}^{10} (|x_i - 1| + |y_i|)$ $s.t$ $y \in \underset{(y)}{\arg\min} \left\{ \begin{array}{l} Minimize\ f(x, y) = e^{(1 + \frac{1}{4000} \sum_{i=1}^{10} (y_i x_i)^2 - \prod_{i=1}^{10} \cos(\frac{y_i}{\sqrt{i}}))} \\ s.t \\ -\pi \le y_i \le \pi \quad i = 1, 2, \ldots, 10 \end{array} \right\}$

Table 2 CGA-ES results and best known solutions for TP1–TP10

Problem	CGA-ES results		Best solutions in [18, 19]	
	$F(x^\cdot, y^\cdot)$	$f(x^\cdot, y^\cdot)$	$F(x^*, y^*)$	$f(x^*, y^*)$
TP1	−225.0001	99.9999	225.0000	100.0000
TP2	0	200.0000	0	100.0000
TP3	−18.9365	−1.1563	−18.6787	−1.0156
TP4	−29.3529	3.008	−29.2000	3.2000
TP5	−3.9014	−2.0300	−3.6000	−2.0000
TP6	−1.2520	8.0708	−1.2091	7.6145
TP7	−1.9600	1.9660	−1.9600	1.9600
TP8	8.7468E−05	199.9989	0	100.0000
TP9	8.6845E−06	2.7183	0	1.0000
TP10	1.23793E−04	2.7183	0	1.0000

Table 3 CGA-ES results and best known solutions for SMD1-SMD6

Problem	CGA-ES results		Optimal solutions	
	$F(x^\cdot, y^\cdot)$	$f(x^\cdot, y^\cdot)$	$F(x^*, y^*)$	$f(x^*, y^*)$
SMD1	0	0	0	0
SMD2	4.7659e−06	2.2190e−06	0	0
SMD3	0	0	0	0
SMD4	5.5692E−12	3.4094E−11	0	0
SMD5	1.1324e−09	1.1324e−09	0	0
SMD6	9.3428E−11	9.3428e−11	0	0

5 Conclusions

In this chapter, we proposed a genetic algorithm based on a new effective selection technique and chaos theory to solve bilevel programming problems. To verify the performance of CGA-ES, extensive numerical experiments based on a suite of multimodal test have applied. A careful observation will reveal the following benefits of the CGA-ES:

- CGA-ES is presented as a powerful global feasible and efficient technique to solve BLPP.
- CGA-ES overcomes the difficulties associated with BLPP as non-linearity, multimodality and conflict in upper and lower level objectives.
- The combination between genetic algorithm technique and chaos search offer the advantages of both genetic algorithm as a powerful global searching technique and chaos search as an efficient and fast searching technique.

- The new effective selection technique enables the upper level decision maker to choose an appropriate solution in anticipation of the lower level's decision.
- The new effective selection technique helps in dispensing of solving lower level problem for every generation that fast convergence to the optimal solution.

References

1. Kalashnikov, V., Dempe, S., Pérez-Valdés, A., Kalashnykova, I., Camacho-Vallejo, J.: Bilevel programming and applications. Math. Probl. Eng. **2015**, 1–16 (2015)
2. Ruusk, S., Miettinen, K., Wiecek, M.: Connections between single-level and bilevel multiobjective optimization. J. Optim. Theory Appl. **153**, 60–74 (2012)
3. Shuang, M.; A nonlinear bilevel programming approach for product portfolio management. SpringerPlus **5**(1), 1–18 (2016)
4. Gaspar, I., Benavente, J., Bordagaray, M., Jose, B., Moura, L., Ibeas, A.: A bilevel mathematical programming model to optimize the design of cycle paths. Transp. Res. Procedia **10**, 423–443 (2015)
5. Aihong, R., Yuping, W., Xingsi, X.: A novel approach based on preference-based index for interval bilevel linear programming problem. J. Inequalities Appl. **2017**, 1–16 (2017)
6. Birla, R., Agarwal, V., Khan, I., Mishra, V.: An alternative approach for solving bi-level programming problems. Am. J. Oper. Res. **7**, 239–247 (2017)
7. Osman, M., Emam, M., Elsayed, M.: Interactive approach for multi-level multi-objective fractional programming problems with fuzzy parameters. J. Basic Appl. Sci. **7**(1), 139–149 (2018)
8. Migdalas, M., Pardalos, M., Värbrand, P: Multilevel Optimization: Algorithms and Applications, 4th edn, pp. 149–164. Kluwer, USA (1998)
9. Bard, J.: Some properties of the bilevel programming problem. J. Optim. Theory Appl. **68**(2), 371–437 (1991)
10. Bard, J., Moore, J.: A branch and bound algorithm for the bi-level programming problem. SIAM J. Sci. Stat. Comput. **11**(2), 281–292 (1990)
11. Bard, J., Falk, J.: An explicit solution to the multi-level programming problem. Comput, Oper. Res. **9**(1), 77–100 (1982)
12. Jie, L., Jialin, H., Yaoguang, H., Guangquan, Z.: Multilevel decision-making: a survey. Inf. Sci. **346–347**, 463–487 (2016)
13. Wang, G., Wan, Z., Wang, X., Yibing, L.: Genetic algorithm based on simplex method for solving linear-quadratic bilevel programming problem. Comput. Math. Appl. **56**, 2550–2555 (2008)
14. El-Desoky, I., El-Shorbagy, M., Nasr, S., Hendawy, Z., Mousa, A.: A hybrid genetic algorithm for job shop scheduling problem. Int. J. Adv. Eng. Technol. Comput. Sci. **3**(1), 6–17 (2016)
15. Hosseini, E.: Solving linear tri-level programming problem using heuristic method based on bi-section algorithm. Asian J. Sci. Res. **10**, 227–235 (2017)
16. Carrasqueira, P., Alves, M., Antunes, C.: Bi-level particle swarm optimization and evolutionary algorithm approaches for residential demand response with different user profiles. Inf. Sci. **418**, 405–420 (2017)
17. Wang, L., Zheng, D., Lin, Q.: Survey on chaotic optimization methods. Comput. Technol. Autom. **20**(1), 1–5 (2001)
18. Sinha, A., Malo, P., Kalyanmoy, D.: An improved bilevel evolutionary algorithm based on quadratic approximations. In: 2014 IEEE Congress on Evolutionary Computation, Beijing, China (July 2014)
19. Sinha, A., Malo, P., Kalyanmoy, D.: Test problem construction for single-objective bilevel optimization. Evol. Comput. **22**(3) (2014)

20. Sinha, A., Malo, P., Deb, K.: Unconstrained scalable test problems for single-objective bilevel optimization. In: Proceedings of the 2012 IEEE Congress on Evolutionary Computation, Brisbane, Australia (June 2012)
21. Zhongping, W., Guangmin, W., Bin, S.: A hybrid intelligent algorithm by combining particle swarm optimization with chaos searching technique for solving nonlinear bi-level programming problems. Swarm Evol. Comput. **8**, 26–32 (2013)
22. Wang, Y., Jiao, Y., Li, H.: An evolutionary algorithm for solving nonlinear bilevel programming based on a new constraint-handling scheme. IEEE Trans. Syst. Man Cybern. **35**, 221–232 (2005)
23. El-Shorbagy, M., Mousa, A., Nasr, S.: A chaos-based evolutionary algorithm for general nonlinear programming problems. Chaos, Solitons Fractals **85**, 8–21 (2016)
24. Mousa, A., El-Shorbagy, M., Abd-El-Wahed, W.: Local search based hybrid particle swarm optimization algorithm for multiobjective optimization. Swarm Evol. Comput. **3**, 1–14 (2012)
25. Mousa, A., Abd-El-Wahed, W., Rizk-Allah, R.: A hybrid ant optimization approach based local search scheme for multiobjective design optimizations. Electr. Power Syst. Res. **81**, 1014–1023 (2011)
26. Baker, E.: Reducing bias and inefficiency in the selection algorithm. In: Erlbaum, L. (ed.) Proceedings of the Second International Conference on Genetic Algorithms, pp. 14–21. Morgan Kaufmann Publishers, San Mateo, CA (1987)

Bio-inspired Machine Learning Mechanism for Detecting Malicious URL Through Passive DNS in Big Data Platform

Saad M. Darwish, Ali E. Anber, and Saleh Mesbah

Abstract Malicious links are used as a source by the distribution channels to broadcast malware all over the Web. These links become instrumental in giving partial or full system control to the attackers. To overcome these issues, researchers have applied machine learning techniques for malicious URL detection. However, these techniques fall to identify distinguishable generic features that are able to define the maliciousness of a given domain. Generally, well-crafted URL's features contribute considerably to the success of machine learning approaches, and on the contrary, poor features may ruin even good detection algorithms. In addition, the complex relationships between features are not easy to spot. The work presented in this paper explores how to detect malicious Web sites from passive DNS based features. This problem lends itself naturally to modern algorithms for selecting discriminative features in the continuously evolving distribution of malicious URLs. So, the suggested model adapts a bio-inspired feature selection technique to choose an optimal feature set in order to reduce the cost and running time of a given system, as well as achieving an acceptably high recognition rate. Moreover, a two-step artificial bee colony (ABC) algorithm is utilized for efficient data clustering. The two approaches are incorporated within a unified framework that operates on the top of Hadoop infrastructure to deal with large samples of URLs. Both the experimental and statistical analyses show that improvements in the hybrid model have an advantage over some conventional algorithms for detecting malicious URL attacks. The results demonstrated that the suggested model capable to scale 10 million query answer pairs with more than 96.6% accuracy.

S. M. Darwish
Department of Information Technology, Institute of Graduate Studies and Research, Alexandria University, 163 Horreya Avenue, El-Shatby, P.O. Box 832, Alexandria 21526, Egypt

A. E. Anber (✉)
Faculty of Computers and Information, Damanhour University, Damanhour, Egypt
e-mail: ali.anber@damanhour.edu.eg

S. Mesbah
Arab Academy for Science, Technology and Maritime Transport, Alexandria, Egypt

© The Author(s), under exclusive license to Springer Nature Switzerland AG 2021
A. E. Hassanien et al. (eds.), *Machine Learning and Big Data Analytics Paradigms: Analysis, Applications and Challenges*, Studies in Big Data 77,
https://doi.org/10.1007/978-3-030-59338-4_9

Keywords Machine learning · Malicious URLs · Passive DNS · Big data ·
Hadoop · Genetic algorithm · Artificial bee colony

1 Introduction

The web has become a platform for supporting a wide range of criminal enterprises
such as spam-advertised commerce, financial fraud, and malware propagation. The
security group has responded to this by creating blacklisting tools. The distribution
channels use harmful URLs as a medium to spread malware throughout the Internet.
These relationships help to give the attackers, who use systems for various cyber-
crimes, partial or comprehensive control over the system. Systems with the ability to
detect malicious content should be quick and precise to detect such crimes [1]. DNS
data analysis has several advantages compared to other methods like the blacklist
of compromised domains as the DNS traffic has a considerable number of useful
features to classify domain names affiliated with fraudulent activities [2].

Detection of malicious domains through the analysis of DNS data has a number
of benefits compared to other approaches such as blacklists [2, 3]. First, DNS data
constitutes only a small fraction of the overall network traffic, which makes it suitable
for analysis even in large-scale networks that cover large areas. Moreover, caching,
being an integral part of the protocol, naturally facilitates further decrease in the
amount of data to be analyzed, allowing researchers to analyze even the DNS traffic
coming to top level domains. Second, the DNS traffic contains a significant amount
of meaningful features to identify domain names associated to malicious activities.
Third, many of these features can further be enriched with associated information,
such as autonomous system number, domain owner, and so on, providing an even
richer space exploitable for detection. The large amount of features and the vast
quantity of traffic data available have made DNS traffic a prime candidate for experimentation with various machine-learning (ML) techniques applied to the context
of security. Fourth, although the solutions to encrypt DNS data exist, still a large
fraction of DNS traffic remains unencrypted, making it available for the inspection
in various Internet vantage points. Fifth, sometimes researchers are able to reveal
attacks at their early stages or even before they happen due to some traces left in the
DNS data.

Detecting malicious URLs is an essential task in network security. Despite many
exciting advances over last decade for malicious URL detection using machine
learning techniques, there are still many open problems and challenges which are critical and imperative, including but not limited to the following [4]: (1) **High volume
and high velocity**: The real-world URL data is obviously a form of big data with
high volume and high velocity. It is almost impossible to train a malicious URL
detection model on all worlds' URL data using machine learning. (2) **Difficulty in
collecting features**: collecting features for representing a URL is crucial for applying
machine learning techniques. In particular, some features could be costly (in terms

of time) to collect, e.g., host-based features. (3) **Feature representation**: In addition to high volume and high velocity of URL data, another key challenge is the very high-dimensional features. Some commonly used learning techniques, such as feature selection, dimension reduction and sparse learning, have been explored, but they are far from solving the challenge effectively. Besides the high dimensionality issue, another more severe challenge is the evolving high dimensional feature space, where the feature space often grows over time when new URLs and new features are added into the training data. This again poses a great challenge for a clever design of new machine learning algorithms which can adapt to the dynamically changing feature spaces. (4) **Concept drifting and new emerging challenges**: Another challenge is the concept drifting where the distribution of malicious URLs may change over time due to the evolving behaviors of new threats and attacks. This requires machine learning techniques to be able to deal with concept drifting whenever it appears. Besides, another recent challenge is due to the popularity of URL shortening services, which take a long URL as input and produce a short URL as an output.

A variety of approaches have been attempted to tackle the problem of Malicious URL Detection. According to the fundamental principles, these approaches can be broadly grouped into three major categories: (i) Blacklisting, (ii) Heuristics, and (iii) Machine Learning approaches. The key principles of each category are briefly described in [4]. Despite many exciting advances over last decade for malicious URL detection using machine learning techniques, there are still many open problems and challenges which are critical and imperative, including but not limited to the following [3]: high volume and high velocity, difficulty in acquiring labels, difficulty in collecting features, feature representation, and concept drifting [1]. In recent times, improvements in technology and infrastructure have led to creating the problem of big data. Accordingly, it is necessary for cyber security professionals to design and implement novel methods to efficiently and effectively mitigate cyber security threats in a big data world. Hadoop, an open-source distributed storage platform that can run on commodity hardware, has been utilized to better accommodate the big data storage requirements of massive volume and fast-speed processing criteria of potentially very complex, heterogeneous data structures [5].

1.1 Problem Statement and Research Motivation

Compromised websites that are used for cyber-attacks are named as malicious URLs. In fact, nearly one third of all websites were identified as potentially malicious in nature, showing that malicious URLs are used widely in order to commit cyber-crimes [3]. To overcome the problem to maintain and update a blacklist, some systems inspect the web page content and analyze the behavior after visiting the Web page. Unfortunately, this increases the run time overhead and affects the user experience. With the development of machine learning, several classification-based methods, using the features of Web page content and URL text, are also used to detect malicious

URLs. However, the attackers adjust their strategies accordingly and invent new kinds of attacks [6]. This work is motivated by the observation that attackers tend to abuse certain domain features during this process. By basing a detection system on such features, we can effectively detect and blacklist the malicious web pages.

1.2 Contribution and Methodology

This paper addresses the problem of detecting malicious Web sites using machine learning over URL-based features. The challenge is to construct a system that is accurate, scalable, and adaptive. To this end, the contributions of this paper are the following: (1) Show that better classification is possible by extracting more meaningful URL data characteristics that outweigh the advantages that a skilled learner makes. For this reason, a bio-inspired reduction process is applied that adopts GA to refine the lists of features (optimal features), with the goal of building a robust and efficient learning model. (2) Adapt a two-step artificial bee colony (ABC) algorithm for efficient Malicious URL clustering.

The rest of this paper is organized as follows: Sect. 2 presents a review of the recent related works. Section 3 introduces the proposed malicious URL detection model. Section 4 exhibits the experimental results to evaluate the performance study of the proposed model. Finally, we conclude the paper and give future directions in Sect. 5.

2 Related Work

Because the suggested technique is based on passive domain name-based features of the URL, more emphasis in surveying related work that incorporates those features are placed. Ma et al. [6] presented an approach for classifying URLs automatically as either malicious or benign based on supervised learning across both lexical and host-based features. Another approach was suggested by Zhang et al. in 2007 [7] that presented the design and evaluation of Carnegie Mellon Anti-phishing and Network Analysis Tool (CANTINA), a novel content-based approach for detecting phishing web sites. CANTINA uses the well-known Term Frequency/Invest Document Frequency (TF-IDF) algorithm and applies it to anti-phishing, to robust hyperlinks, the idea of overcoming problems with page.

Kan and Thi [8] introduced the notion of bag-of-words representation for classifying URLs. Concurrently, in 2009, Guan et al. [9] had examined the aspect of instant messaging (IM) for classifying URLs. Although they used several URL-based features, they also take advantage of a number of IM-specific features such as message timing and content. Yet, this algorithm needs more URL message samples to make their experiments more accurate and convincible. Watkins et al. [3] a strategy focused on an integral feature of Big Data was launched in 2017: the overwhelming majority of network processing in a historically guarded business (i.e., using defense-in-depth) is non-malicious. The core objective of Bilge et al. [10] work is to build an exposure system that is designed to detect such domains in real time by applying 15 unique features grouped into four categories. Although URLs cyber security models have been studied for nearly many decades, there is still room to make it more efficient and practical in the real application.

According to the aforementioned review, it can be found that past studies were primarily devoted to (1) Blacklisting, which cannot predict the status of previously URLs or systems based on site content or behavior assessments that require visits to potentially risky sites. (2) Not addressing the issues related to the selection of optimal feature set from the pool of extracted features. In general, with appropriate classifiers, it is feasible to automatically shift through comprehensive feature sets (i.e., without requiring domain expertise) and identify the most predictive features for classification. However, to best of my knowledge, little attention has been paid to devising a new bio-inspired feature selection technique for a malicious URL detections system that relies on a big number for training samples (big data environment). Most of the current bio-inspired optimization techniques for malicious URL detection depend on combining two or more algorithms to enhance the exploration and exploitation fitness of the basic algorithm. The next Section discusses in detail the suggested model that integrates the Hadoop framework to handle big data with a bio-inspired artificial Bee colony algorithm for URL classification.

3 The Proposed Model

The work presented in this paper explores how to detect malicious Web sites from passive DNS based features. This issue naturally leads to new algorithms in which biased features may be chosen while Malicious URLs are constantly being spread. So, the suggested model adapts a bio-inspired feature selection technique to choose an optimal feature set in order to reduce the cost and running time of a given system, as well as achieving an acceptably high recognition rate. Moreover, a two-step artificial bee colony (ABC) algorithm is utilized for efficient data clustering. The two approaches are incorporated within a unified framework that operates on the top of the Hadoop infrastructure to deal with large samples of URLs. Herein, a modified representation learning model based on genetic algorithm is proposed to select the most representative features that keep the classification accuracy as high of the state of the art models that use hundreds or thousands of features, allowing possible embedded programs to run fast looking for the characteristics that match malicious URL behavior. Moreover, this model requires large datasets to train and to tune the learning prediction algorithm. To address this problem, Apache Hadoop is employed as a distributed computing platform. Figure 1 shows the main components of the suggested prediction model, and how these components are linked together and the following subsections discuss its steps in detail.

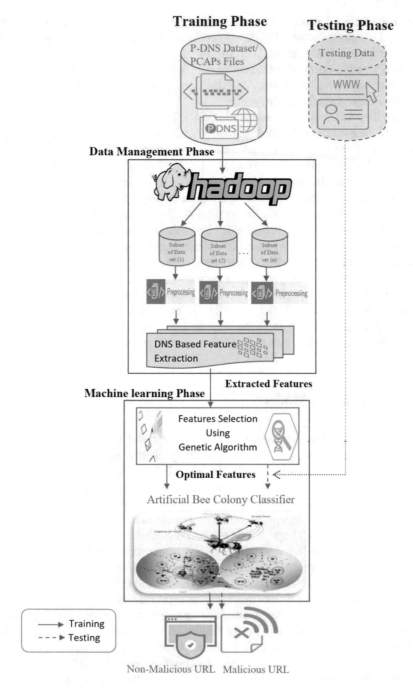

Fig. 1 The proposed malicious URL detection model

3.1   Training Phase

Step 1: Passive DNS Dataset. The challenges of access to DNS data faced by the research community lie in two aspects [2]: (1) first is the data collection phase; the peculiarity of many existing DNS-based malicious domain detection techniques is that they work best in big data scenarios. Thus, they may not be able to produce meaningful results on datasets collected in small networks. Meanwhile, integrating data from DNS servers belonging to different organizations would often face significant bureaucratic/legal obstacles, due to the sensitive nature of DNS logs. (2) Even a bigger challenge lies in data sharing. Unfortunately, security related data are notoriously sensitive and hard to share. In general, Passive DNS data collection happens through the installation of sensors to DNS servers or the connection to DNS server logs for the purpose of obtaining real DNS queries. Furthermore, passive DNS data are linked to the behavior of individual users, so passive DNS data could be used to detect malicious domains with techniques that rely on user-level features (e.g., temporal statistics of user queries).

Step 2: Data Management Phase. Traditional computing storage platforms like relational databases do not scale effectively against the onslaught of big data challenges posed by malicious URL detection. There should be only two types of headings. The headings of the lower level stay unnumbered and formatted as run-in headings. To address this problem, some authors suggested using distributed computing platforms such as Apache Hadoop. Hadoop, an open-source distributed storage platform that can run on commodity hardware, has been utilized to better accommodate Big data processing requires of massive volume and high speed along with heterogeneous data structures theoretically very complex. Hadoop provides a software framework for distributed storage and distributed computing. It divides a file into the number of blocks and stores it across a cluster of machines. It does distributed processing by dividing a job into a number of independent tasks. These tasks run in parallel over the computer cluster.

Hadoop MapReduce includes several stages [11]: In the first step, the program locates and reads the «input file» containing the raw data. As the file format is arbitrary, there is a need to convert data into something the program can process. The «InputFormat» and «Record Reader» does this job. InputFormat uses the InputSplit function to split the file into smaller pieces. Then the Record Reader transforms the raw data for processing by the map. It outputs a list of key-value pairs. Once the Mapper processes these key-value pairs, the result goes to «OutputCollector». There is another function called «Reporter» which intimates the user when the mapping task finishes. In the next step, the Reduce function performs its task on each key-value pair from the Mapper. Finally, Output Format organizes the key-value pairs from Reducer for writing it on HDFS.

Step 3: Data Preprocessing. Data pre-processing is an important phase in machine learning, since the quality of the data and its useful information affects the capacity of the proposed model to learn directly; therefore, it is extremely important that the data are preprocessed before feeding it into the model [12]. Data preprocessing is the process of simply transforming raw data into an understandable format. Preprocessing involves various steps that help to convert raw data into a processed and sensible format such as data cleaning, data integration, data transformation, and data reduction. The main use of the cleaning step is based on detecting incomplete, inaccurate, inconsistent and irrelevant data and applying techniques to modify or delete this useless data.

Step 4: Feature Extraction. In machine learning, feature extraction starts from an initial set of measured data and builds derived values (features) intended to be informative and non-redundant, facilitating the subsequent learning and generalization steps, and in some cases leading to better human interpretations. Feature extraction is a dimensionality reduction process, where an initial set of raw variables is reduced to more manageable groups (features) for processing, while still accurately and completely describing the original data set [6]. Content-features usually require downloading the web-page, which would affect the feature collection time. In general, the success of a machine learning model critically depends on the quality of the training data, which hinges on the quality of feature representation. Given a URL $u \in \mathbb{U}$ where $\mathbb{U}$ denotes a domain of any valid URL strings, the goal of feature representation is to find a mapping $g : \mathbb{U} \to \mathbb{R}^d$ such that $g(\mathbb{U}) \to X$ where $X \in \mathbb{R}^d$ is a d-dimensional feature vector, that can be fed into machine learning models. The process of feature representation can be further broken down into two steps:

- **Feature Collection**: This phase is engineering oriented, which aims to collect most if not all relevant information about the URL.
- **Feature Preprocessing**: In this phase, the unstructured information about the URL is appropriately formatted and converted to a numerical vector so that it can be fed into machine learning algorithms.

The suggested model relies on DNS Answer-based, TTL Value-based, and Domain-Name-based features. See [3] for more details.

Step 5: Feature Selection Using Genetic Algorithm. It is essential to select a subset of those features which are most relevant to the prediction problem and are not redundant. Heuristic search is an intelligent search process through an extremely wide range of solutions to detect a satisfactory solution. No exhaustive sequential selection process can generally be guaranteed for the optimal subset; any ordering of the error probabilities of each of the 2^n feature subsets is possible. In this case, an instance of a GA-feature selection optimization problem can be described in a formal way as a four-tuple (R, Q, T, f) defined as [13, 14]:

- R is the solution space (initial population—a combination of 16 feature vector per URL—a matrix n × 16) where n represents the number of URL samples. Each bit is signified as a gene that represents the absence or existence of the feature within the vector. Every feature vector is represented as a chromosome.
- Q is the feasibility of predicate (different operators—selection, crossover, and mutation). The crossover is the process of exchanging the parent's genes to

produce one or two offspring. The purpose of mutation is to prevent falling into a locally optimal solution of the solved problem [14]. A uniform mutation is employed for its simple implementation. The selection operator retains the best fitting chromosome of one generation and selects the fixed numbers of parent chromosomes. Tournament selection is probably the most popular selection method in genetic algorithm due to its efficiency and simple implementation.

- ζ is the set of feasible solutions (new generation populations). With these new generations, the fittest chromosome will represent the URL feature vector with a set of salient elements. This vector will specify the optimal feature combination explicitly according to the identification accuracy.
- f is the objective function (fitness function). The individual that has higher fitness will win to be added to the predicate operators mate. Herein, the fitness function is computed based on accuracy Acc value that shows the difference between the real URL's classification, and it's computed one.

$$Acc = (True\ Positive + True\ Negative)/(no.\ Positive + no.\ Negative)$$
(1)

Accuracy (Acc) is the ratio of the correctly identified domains to the whole size of the test set. The higher the value is, the better ($Acc \in [0, 1]$). True Positive (TP) is the correctly identified malicious domains, True Negative (TN) is the correctly identified benign domains, P is the total number of malicious domains, and N is the total number of benign domains.

Step 6: Artificial Bee Colony (ABC) Classifier. The final step includes employed an artificial bee colony to classify the URLs malicious or benign based on the training dataset that contains the best feature vector of each URL. In this case, to accelerate the convergence rate and maintaining the balance between exploration and exploitation, in this research work two-step ABC algorithm is utilized to improve the ABC algorithm for clustering problems by using K-means algorithm [15]. The combination of ABC and K-means (named ABCk) uses the merits of the k-means and ABC algorithms for solving the problem of malicious URL classification. In this case, the suggested model uses sensitivity as a fitness function. Sensitivity is defined in Eq. 2 as the ratio of the True Positives to the sum of the True Positives and the False Negatives. The True Positives are the correctly identified malicious domains, and the False Negatives are the domains that are malicious but were incorrectly identified as non-malicious.

$$\text{Sensitivity} = \frac{True\ Positive}{True\ positive + False\ negative}$$
(2)

3.2   Testing Phase

In this step, given the unknown URL, the model starts with extracting the features vector for this URL that follows the indices of the best features vector learned from the training stage. This extracted feature vector is then classified according to its similarity to the final cluster centers generated from applying the artificial bee colony classifier in the training phase.

4   Experimental Analysis and Results

In this section, many experiments are conducted to validate the performance of the suggested malicious URL detection model and compare it with some common detection techniques. The performance is validated in terms of precision, false positive rates, Accuracy, True Negative rate, Recall, and F-Measure based. The experiment was carried out in Intel Xeon E5-2620v3 @ 2.4 GHz (12 CPUs) processor with 32.00 GB RAM implemented in Java. The experiments are conducted using a benchmark dataset that is comprised of captured passive DNS data, which are answers (e.g., IP address, time to live (TTL), record counts) from authoritative DNS servers given domain name queries from the browsers of users. A big dataset totaling 184 million rows is extracted.

- **Experiment 1: (The significance of features selection)**

    Aim: To validate the benefits of employing a feature selection module within the suggested model; this experiment implements the suggested model using both full features vector and optimal features vector to investigate the difference between the two runs in terms of detection accuracy and time.
    Observations: The results in Figs. 2 and 3 reveals that the use of optimal features achieves an increase in accuracy in terms of True Positive, True Negative, False Positive and False Negative of approximately 1% compared to using full feature vector. Although this increase is relatively small, the benefit is to reduce the testing time for each URL from approximately one second to 500 ms.
    Discussions: As the proposed model tries to select of the most prominent features that contain the URLs characteristics which is able to distinguish Web sites either malicious or benign, so this features vector as expected yields increasing in detection accuracy. One possible explanation of these results is that the feature selection module able to remove redundant features (high correlated features) and discards features leading to mislabelled based on fitness function.

- **Experiment 2: (Classifier evaluation)**

    Aim: Since there is a wide range of supervised classification algorithms, this set of experiments is conducted to assess a sample of the collected dataset

Fig. 2 Confusion matrix
using optimal features

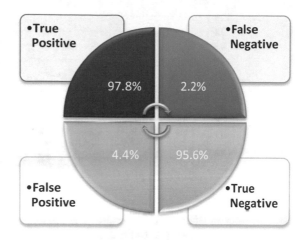

Fig. 3 Confusion matrix
using all features

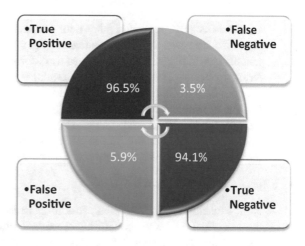

according to three classifiers using Weka [16] as well as ABC and two-step
ABC classifiers. The three classifiers were tested covering tree-based (Random
Forest, C4.5) and function-based (SVM). The classification was made without
parameters tuning through a ten-fold cross-validation as a first step to select
the most promising approach.

Observations: Results for accuracy, true positives and true negatives are given
in Fig. 4 for each classifier. Among the tested classifiers, SVM yields the
worst accuracy (86.31%) while being efficient in identifying legitimate URLs
(93.1%). Tree-based classifiers have approximately the same performance
(around 90%) with disproportionate true positives and true negatives. The ABC
classifier has approximately the same performance of random forest classifier.
The best performer is two-step ABC classifier, correctly classifying 96.6% of
URLs, being the best.

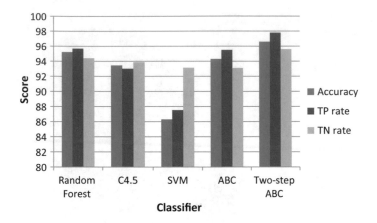

Fig. 4 Classification results for 6 classifiers

Discussions: ABC classifier takes a long time because of its stochastic nature. It is also observed that the position of bees and food sources is identified randomly initially which takes more time for optimizing, especially in clustering problems. The solution search equation of ABC is good at exploration, but poor at exploitation, which results in the poor convergence. It is also noted that the convergence speed of ABC algorithm is decreased as dimensions of the problem are increased. In two-step ABC algorithm, the initial positions of food sources are identified using the K-means algorithm instead of random initialization. So, it yields more accurate classification.

- **Experiment 3: (Concept drift)**

    Aim: Phishing tactics and URL structures keep changing continuously over time as attackers come up with novel ways to circumvent the existing filters. As phishing URLs evolve over time, so must the classifier's trained model to improve its performance. In general, retraining algorithms continuously with new features is crucial for adapting successfully to the ever-evolving malicious URLs and their features. An interesting future direction would be to find the effect of variable number of features using online algorithms in detecting phishing URLs. Herein, the whole passive DNS data set is divided into 15 patches; each patch contains approximately 1,200,000 samples with different numbers of benign and malicious URLs. The suggested model was trained with each patch separately to validate its classification error rates using each patch as a training set.

    Observations: Fig. 5 shows the classification error rates for the suggested classifier after training using different patches. The x-axis shows the patch number in the experiment. The y-axis shows the error rates on testing the suggested classifier. Figure 5 reveals that the error rate fluctuates between 2.2

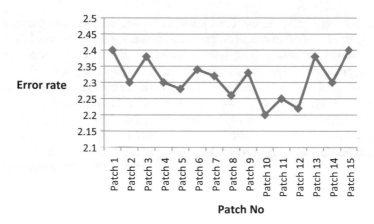

Fig. 5 Error rates of the suggested classifier after training them on different patches

and 2.4% for all patches. So, the suggested model demonstrates the stability regarding error rate despite the changing the dataset used.

Discussions: one possible explanation for these results is that the suggested model is built based on optimal features vector that has discriminative ability to distinguish between benign and malicious URLs. This feature vector minimizes the inter-class similarity while maximizes intra-class similarity. The inter-class cluster show the distance between data point with cluster centre, meanwhile intra-class cluster show the distance between the data point of one cluster with the other data point in another cluster.

5 Conclusion

DNS data carry rich traces of the Internet activities, and are a powerful resource to fight against malicious domains that are a key platform to a variety of attacks. To design a malicious domain detection scheme, one has to consider the following major questions that hinder the advances of the field: (1) data sources: what types of DNS data, ground truth and auxiliary information are available; (2) features and data analysis techniques: how to derive features to match intuitions of malicious behaviors, and what types of detection techniques the malicious domain discovery problem can be mapped to; (3) evaluation strategies and metrics: how to evaluate the robustness of a technique given the adaptive nature of attackers, and what metrics to use for these purposes.

The work presented in this paper proposes a new model to discover malicious domains by analyzing passive DNS data. This model takes advantage of the dynamic nature of malicious domains to discover strong associations among them, which are further used to infer malicious domains from a set of existing known malicious ones. The central finding of the work in this paper is that machine learning techniques can alleviate the disadvantages of blacklists, heavyweight (use more features and so have a higher accuracy) and lightweight (useless features and consumes the features from the browser) classifiers for detecting malicious URLs. This paper demonstrated that the approach of using an optimal number of features (middleweight) extracted from the original features vector had clear advantages over larger feature sets. One of the major contributions of the work in this paper was to explore a hybrid machine learning technique (K-mean and ABC) that used selected discriminative features and the use of the Hadoop framework to handle the big size of URLs data. The results demonstrated that the suggested model capable of scaling 10 million query answer pairs with more than 96.6% accuracy. The major limitation of the suggested model is that it cannot take an arbitrary given domain and decide whether it is potentially malicious or not. Similarly, if a domain never shares IPs with other domains, it will not appear in the domain graph, and the suggested model is not applicable to such domain either. Future endeavors in this regard include improving the level of filtering of non-malicious data to provide the analysts with an even smaller set of candidate data to investigate.

References

1. Sayamber, A., Dixit, A.: Malicious URL detection and identification. Int. J. Comput. Appl. **99**(17), 17–23 (2014)
2. Zhauniarovich, Y., Khalil, I., Yu, T., Dacier, M.: A survey on malicious domains detection through DNS data analysis. ACM Comput. Surv. **51**(4), 1–36 (2018)
3. Watkins, L., Beck, S., Zook, J., Buczak, A., Chavis, J., Mishra, S.: Using semi-supervised machine learning to address the big data problem in DNS networks. In: Proceedings of the IEEE 7th Annual Computing and Communication Conference (CCWC), pp. 1–6, USA (2017)
4. Sahoo, D., Liu, C., Hoi, S.: Malicious URL Detection Using Machine Learning: A Survey. arXiv preprint arXiv:1701.07179, pp. 1–21 (2017)
5. Antonakakis, M., Perdisci, R., Lee, W., Vasiloglou, N., Dagon, D.: Detecting malware domains at the upper DNS hierarchy. In: Proceedings of the 20th USENIX Conference on Security (SEC'11), pp. 1–16, USA (2011)
6. Ma, J., Saul, L., Savage, S., Voelker, G: Beyond blacklists: learning to detect malicious web sites from suspicious URLs. In: Proceeding of the 15th ACM SIGKDD International Conference on Knowledge Discovery and Data Mining, pp. 1245–1254, France (2009)
7. Zhang, Y., Hong, J., Cranor, L.: CANTINA: a content-based approach to detecting phishing web sites. In: Proceedings of the 16th International Conference on World Wide Web, pp. 639–648, Canada (2007)
8. Kan, M.-Y., Thi, H.: Fast webpage classification using URL features. In: Proceedings of the 14th ACM International Conference on Information and Knowledge Management, pp. 325–326, Germany (2005)
9. Guan, D., Chen, C., Lin, J.: Anomaly based malicious URL detection in instant messaging. In: Proceedings of the Joint Workshop on Information Security, Taiwan (2009)
10. Bilge, L., Sen, S., Balzarotti, D., Kirda, E., Kruegel, C.: EXPOSURE: a passive DNS analysis service to detect and report malicious domains. ACM Trans. Inf. Syst. Secur. **16**(4), 1–28 (2014)
11. Manikandan, S., Ravi, S.: Big data analysis using Apache Hadoop. In: Proceedings of the International Conference on IT Convergence and Security (ICITCS), pp. 1–4, China (2014)
12. Figo, D., Diniz, P., Ferreira, D., Cardoso, J.: Preprocessing techniques for context recognition from accelerometer data. Pers. Ubiquit. Comput. **14**(7), 645–662 (2010)

13. El-Sawy, A., Hussein, M., Zaki, E., Mousa, A.: An introduction to genetic algorithms: a survey, a practical issues. Int. J. Sci. Eng. Res. **5**(1), 252–262 (2014)
14. Sivanandam, S., Deepa, S.: Introduction to Genetic Algorithms. Springer, USA (2007)
15. Kumar, Y., Sahoo, G.: A two-step artificial bee colony algorithm for clustering. Neural Comput. Appl. **28**(3), 537–551 (2015)
16. Veček, N., Liu, S., Črepinšek, M., Mernik, M.: On the importance of the artificial bee colony control parameter 'Limit'. Inf. Technol. Control **46**(4), 566–604 (2017)

Machine Learning and Applications

TargetAnalytica: A Text Analytics Framework for Ranking Therapeutic Molecules in the Bibliome

Ahmed Abdeen Hamed, Agata Leszczynska, Megean Schoenberg, Gergely Temesi, and Karin Verspoor

Abstract Biomedical scientists often search databases of therapeutic molecules to answer a set of molecule-related questions. When it comes to drugs, finding the most specific target is a crucial biological criterion. Whether the target is a gene, protein, and cell line, target specificity is what makes a therapeutic molecule significant. In this chapter, we present TargetAnalytica, a novel text analytics framework that is concerned with mining the biomedical literature. Starting with a set of publications of interest, the framework produces a set of biological entities related to gene, protein, RNA, cell type, and cell line. The framework is tested against a depression-related dataset for the purpose of demonstration. The analysis shows an interesting ranking that is significantly different from a counterpart based on drugs.com's popularity factor (e.g., according to our analysis Cymbalta appears only at position #10 though it is number one in popularity according to the database). The framework is a crucial tool that identifies the targets to investigate, provides relevant specificity insights, and help decision makers and scientists to answer critical questions that are not possible otherwise.

Keywords Text analytics · Ranking framework · Therapeutic molecules · Target specificity

A. Abdeen Hamed (✉)
Norwich University, Northfield, VT, USA
e-mail: ahamed@norwich.edu

A. Abdeen Hamed · M. Schoenberg
Merck and Co., Inc., Boston, MA, USA

A. Leszczynska · G. Temesi
MSD, Prague, Czech Republic

K. Verspoor
The University of Melbourne, Melbourne, Australia

1 Introduction

The problem of ranking certain biological entities according to their association with other entities is well-established. Such a problem is manifested in various studies. For instance, gene ranking is a crucial task in bioinformatics for making biotarget predictions and has been previously investigated. (1) Bragazzi et al presented an algorithmic approach for gene ranking that leads to a list of few strong candidate genes when studying complex and multi-factorial diseases [1], and (2) Winter et al conducted a study that ranks gene entities using a modified version of Google's PageRank, which they called NetRank [2]. Protein ranking is another problem that has also been researched; Weston et al demonstrated the importance of protein ranking in a network constructed from a sequence database [3]. Their algorithm proved more significant than rankings generated from a local network search algorithm such as PSI-BLAST. Other studies have ranked biological entites as sets to provide answers to various scientific questions. Wren et al., mined PubMed for biological entities (genes, proteins, phenotypes, chemicals, and diseases), grouped them in sets, and ranked them to satisfy scientific needs [4]. While some researchers addressed the ranking algorithmically [5], others addressed it from a biological networks point of view. Using centrality measures for the purpose of identifying the most influential elements in the network is also common [6, 7].

From a drug discovery point of view, ranking molecules in a biological network holds great promise in the efforts of expanding the domain of druggable targets, especially one that integrates network biology and polypharmacology. Integrating network biology and polypharmacology holds the promise of expanding the current opportunity space for druggable targets. The authors shows how such networks offers great implications for tackling the two major sources of attrition in drug development–efficacy and toxicity [9]. In this chapter we focus our effort on exploring the efficacy part of drug development from a drug specificity point of view. For a drug to be effective, it must a target specific gene, protein, or a cell type etc. TagetAnalytica is designed to study this aspect of the research based on two foundations (1) the very large network of biological entities we extracted from the biomedical literature, and (2) querying the network as a knowledge graph to answer a given question that is corresponding to a molecule, using a SPARQL query. The ranking analytics generated here are essentially a post-processing phase that takes place over the query result of the given question being asked. To date, graph databases do not provide ranking based on the specificity notion that is required for ranking the molecules. Such a complex approach: (mining $\rightarrow$ transforming $\rightarrow$ persisting $\rightarrow$ querying then $\rightarrow$ ranking) is worthwhile because it is flexible and can answer an unlimited number of question for various purposes (drug discovery, drug design, decision support, hypothesis testing, etc). An example of a query can be: list all the molecules related to Breast Cancer and BRCA1 gene and HMECs and Adipocytes cell type. What we mean by *specificity* is not by forcefully limiting a given molecule to be associated with one specific target, but rather show how a few arbitrary entities present themselves to be specific enough to cure a certain disease(s) with minimal side effects as it will be described in Fig. 1 in a future section.

In this chapter, we show that starting with a disease-related set of publications of interest (e.g., depression), an entity recognition step is performed to identify the biological entities associated with the drugs of study. TargetAnalytica empowers analytics framework to deriving ranking insights when specificity is key. In the heart of the framework is a network centrality algorithm that generated the ranking analytics utilizing the biological specificity notion. When it comes to drugs, specificity is a crucial biological criterion in the search when querying the entities that are specific to a gene, protein, RNA, etc. Therefore, specificity it has inspired the design of our algorithm to identify top-k ranked drugs. Experiments comparing our ranking algorithm with counterpart centrality measures (i.e., degree) expressed consistent correlations (both positive and negative) that ranged from [0.55–0.60]. The results highlight the promise that TargetAnalytica offers in two folds: (a) a fully mature framework that is capable for processing biomedical abstracts and generate biological entities. The entities are recognized by means of Natural Language Processing using Machine Learning algorithms and and Ontology-based feature extraction tools, (b) an interesting network analysis algorithm that generates ranking analytics [8] that we also demonstrate its importance in deriving decisions related to drug discovery and design.

1.1 Biological Specificity

Specificity is one of fundamental notions in biochemistry. The ability of enzymes to catalyze biochemical reactions and transform substrate molecules into new products depends on its specificity and catalytic power. When the enzyme and substrate bind, they form an enzyme-substrate complex, which enables the substrate to be catalytically converted into a product molecule [11]. Specificity of an enzyme towards a particular substrate by performing kinetic measurements. This can be done by testing a series of varied concentrations to study a particular substrate and applying non-regression analysis to obtain the specificity constants for comparison. The specificity constant (also called kinetic efficiency), is a measure of how efficiently an enzyme converts substrates into products. A comparison of specificity constants can help determine the preference of an enzyme for different substrates (i.e., substrate specificity). The higher the specificity constant, the more the enzyme "prefers" that substrate [12].

Specificity is also important both for drug discovery and for determining toxicity profiles. Most drugs exert their therapeutic effects by binding to specific receptor sites with subsequent change in receptor's conformation, which triggers the transduction of a signal to produce a biological effect. Drug-target specificity plays a crucial role in this process. A drug's specificity to a particular receptor warrants that it will only bind to a particular enzyme, thus exerting a specific biological effect. In experience, the majority of drugs will act on more than one receptor site once they reach an appropriately high concentration [13]. Conversely, nonspecific drug molecules can bind to different target enzymes in human body, exerting wide range of biological reactions, some of which are manifested in occurrence of adverse effects [14]. In the process of pre-clinical drug development specificity of binding to target molecule or receptor constitutes one of the desired properties for drug candidate [15].

Target specificity is a critical consideration for any small-molecule tool [16]. As more and more drugs are developed to inhibit particular molecular targets, it seems reasonable that extreme specificity of a potential drug for the intended target will be important [17]. An effective drug molecule often must bind with high specificity to its intended target in the body; lower specificity implies the possibility of significant binding to unintended partners, which could instigate deleterious side effects [18]. Specificity of the new molecule's binding with the target protein is one of the crucial characteristics evaluated in the process of early drug discovery [19]. In the case of new antitumor drug development efforts, specificity of new molecule's binding with certain Deoxyribonucleic acid (DNA) sequence (typically encoding for the oncogenes) is a subject to detailed investigation and a crucial success factor [20]. The overall process of drug discovery is thus facilitated by experiments with highly specific compounds. And, of course, a highly specific drug would be expected to have fewer side effects, simplifying the process of evaluating the results of a clinical trial [17]. Some studies explored how Binding Specificity may involve proteins, DNA, and Ribonucleic acid(RNA) when investigating a given disease [21].

This brief introduction constitutes the heuristic and the computational motivation for measuring specificity algorithmically as a network analysis measure. To illustrate the concept, we present the following simple mockup scenario. For a given disease, suppose we find that a given number of molecules are mentioned in publications in conjunction with the following biological entity types (DNA, RNA, protein and cell type). Typically, some molecules exhibit stronger binding with a given DNA, RNA, or a protein instance than others.

As mentioned above, the stronger the binding, the stronger the specificity. Here we presented a weighted network that models the binding specificity aspects and captures its essence. Figure 1 shows how the five hypothetical molecules (MK1, MK2, MK3, MK4, and MK5) are linked to specific instances of genes, proteins, RNA, and cell types. In a real example, a molecule may bind with an RNA instance, while another may not.

Considering the figure, one can observe MK3 is binding to exactly one entity instances, {tRNA, breast, BRAC1, EML4} for RNA, cell type, DNA, and protein entities respectively. On the other hand, MK5 is binding with two DNA instances {BRCA1, and HNF1A}. This suggests that MK3 is more specific in terms of binding with a DNA entity than MK5. Another observation, it appears that MK4 is binding with two protein instances {p53, and Leptin}, but is missing information on binding with cell type and genes. When comparing MK3 with MK4, we can clearly that MK3 is specific and has more knowledge than MK2. The missing gene and cell type information makes MK2 more risky. Therefore, MK3 would be ranked higher.

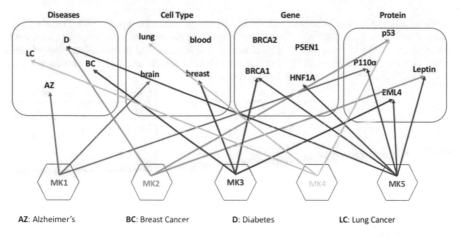

Fig. 1 Motivating example for explaining the notion of specificity using a limited number of biological entities (e.g. DNA, protein, cell type and RNA), and their association with the therapeutic molecule mentions in a given set of publications. The hexagon shapes at the bottom represents five hypothetical molecules encoded with MK_i where i is a numerical index to distinguish the molecules. Each MK_i is also encoded with a given color to better visualize the outgoing connections to the corresponding biological entities. It is imperative to note that some molecules MK_i have exactly one link associated with a given instance of a gene, cell type, etc such as the case in MK3 in each of the four categories. Others have exactly one instance with some category when info existed but missing any links with some other categories. This is clear in the case of MK1 and MK4. On the extreme, others molecules have several connections to several instances associated with in the same category and and also missing connections to whole categories as in the case of MK2 and MK5. Clearly the case of MK3 is more desirable than MK2 and MK5

When comparing MK2 with MK5 we realize that MK5 is less specific than MK4. This is related to the fact that MK5 is connected with three different proteins namely (P110α, EML4, and Leptin), while MK4 is connected to two protein instances as opposed to three. Despite the fact that MK2 is missing cell type and gene information, it remains more specific than MK5. The risk of being of not being specific enough also increases the possibility of being toxic, therefore, we suppress its ranking analytics even if we have more knowledge (MK5 has gene knowledge while MK2 is missing such knowledge). This addresses the toxicity component that is mentioned above [9]. Both MK1 and MK4 have the same specificity so should be ranked about the same. However, when comparing MK1 or MK4 with MK2 or MK5, we find that MK1 and MK4 have higher ranking insights thank either MK2 and MK5. The full description of the example is also summarized in Table 1.

Table 1 Summary of the preliminary example which is plotted in the figure above. There are four column entities in each molecule. The last column shows the order of ranking analytics for each molecule according to the notion of specificity (to be fully described onward). According to this notion we have the following order: MK3 is ranked#1, then MK1 and MK4 are ranked#2, then MK2 and MK5 are ranked last 3 and 5 respectively

	Disease	Cell type	Gene	Protein	Rank analytics
MK1	Alzheimer's	Brain	–	$P110\alpha$	2
MK2	Diabetes	Blood	–	2 Instances	3
MK3	Breast cancer	Breast	BRCA1	EML4	1
MK4	Lung cancer	Lung	–	p53	2
MK5	Breast cancer	Breast	2 instances	3 instances	4

2 Methods

2.1 Data Gathering, Processing, and Entity Identification

The main focus of this chapter is to present the components of a novel framework its main objective is to analyze a network of entities that we identifying in publications. In order to produce useful results, it is important to collect data items that are highly relevant to the ranking problem we are addressing. This offers a deep level of validation of the final results. It also establishes confidence in the methods and the algorithm covered in this chapter. This is why we searched PubMed for the disease keyword("depression"). This search Query returned exactly 390,000 PubMed abstracts, from which we extracted the publication ID (PMID) and abstracts (AB). Such abstracts represents the raw dataset that we further analyzed to identify the entities. This dataset is very rich with molecules (we use the term molecule and drug interchangeably here). To be sure, we also searched the well-known "drugs.com" database and have downloaded the entire list of depression drugs. The list includes drugs in their prescribed name or the commercial brand-name.

Using means of machine learning and Natural Language Processing (NLP) tools, we extracted the mentions of biological entities of each of the abstracts. The co-occurrences among such entities and the molecules/drugs (prescribed or brand name) offer ranking insights that we further use to bring order to the therapeutic molecules. The types of biological entities varied from molecular, taxonomic, and clinical. For each type, we used appropriate tools and algorithms to extract the entities. As stated above, there is a great need to capture high quality recognized entities so we can establish confidence in the methods and produce beneficial results. This was achieved using a dictionary lookup mechanism empowered by some specialized ontology whenever possible. Ontology offers a comprehensive list of terms along with their semantics. For biological entities (e.g., protein, cell, RNA, etc), we used a tool called A BioMedical Name Entity Recognition(ABNER) [22], which recognizes the instances and their entity types found in the text. We extracted and tagged these entities with

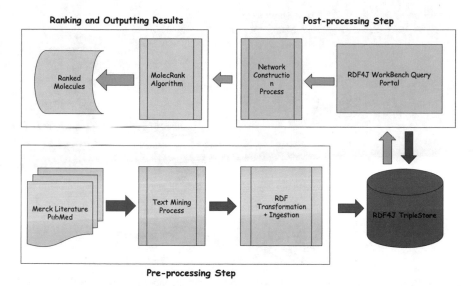

Fig. 2 Workflow of three processes: (1) a pre-processing step, where the literature is mined and entities are recognized, (2) post-processing step, where the triple store is queried, and (3) ranking step, where the ranking is produced

their original type and source of extraction (ABNER, etc). This part will play a significant role in the algorithm we present below. As for the actual drug names, we fed the list downloaded from drugs.com to another well-known information extraction tool known as LingPipe [23] we trained and used against the dataset to identify such drug names.

2.2 From Text to Graph

When we analyzed the abstracts, we identified entities of the following types: biological, chemical, disease, and gene. Representing the entire set of entities as a network offers an attractive theoretical model that can be computationally explored. Representing the resulting entities to Resource Descriptor Framework (RDF)[24] triples offers a mechanism to linking these entities naturally into a graph. Each triple is uniquely described in the information that it communicates (subject →predicate→object). For instance, to express some breast cancer knowledge extracted it can be done in the following factual items: ("breast cancer" is a disease type), ("breast cancer" is mentioned in a PMID:28042876), ("breast cancer" co-occurred with "BT-549"), ("BT-549" is a cell line type), and (PMID:28042876 mentions a Molecule ID). This info can be expressed in triples and visualized Fig. 3 demonstrates.

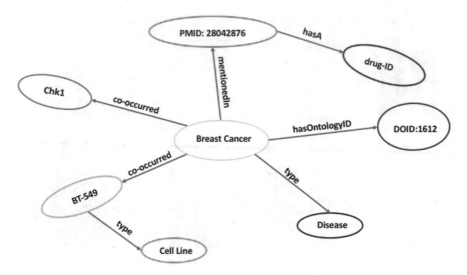

Fig. 3 Graph resulting from the biological entities identified from a set of publication and expressed as RDF triples. Specifically, there is a triple expressing that lung cancer was mentioned in PMID:28042876. Another triple also shows a drug-ID, was also mentioned in the same PMID among other facts demonstrated by the graph diagram

This way of modeling the entities using RDF has the advantage of being queried using Semantic Web Query Languages (e.g., SPARQL). The outcome of the SPARQL queries becomes the input data to the MolecRank analytics algorithm. Figure 2 shows the workflow of how the pre-processing step communicates with post-processing until the ranking is completed.

Listing 10.1 shows the SPARQL query that we issue to calculate the frequencies of molecules and their entities associations.

Listing 10.1 A SPARQL query that calculates the frequencies of the RX Drugs to contribute to the network weight

```
PREFIX xsd: <http://www.w3.org/2001/XMLSchema#>

SELECT ?xrDrug (GROUP_CONCAT(?strength; separator = ", ")
  AS ?strengths)
WHERE {
    SELECT ?xrDrug (concat(?type, " = ", str(COUNT(?pubmed)))
      AS ?strength)
    WHERE {
      GRAPH ?pubmed {
        ?molec xsd:name ?xrDrug .
        ?feature xsd:type ?type .
        FILTER (!sameTerm(?molec, ?feature))
        }
    GROUP BY ?molecNo ?type
    }
}
GROUP BY ?molecNo
```

2.3   The Ranking Algorithm and Heuristics

Here we present the heuristic that empowers our network analysis algorithm to generate molecule ranks of a given search query. This algorithm is designed based on the notion of molecular specificity. The notion starts with a molecule and how it is connected to the various entity types. A molecule is considered specific if it is attached to one instance in each entity type. Contrarily, the more entity instances of the same type a molecule is linked to, the more general and less significant. This specificity notion is not be to confused with the knowledge-gain which entails: the more we know about the molecule (even if it is not specific enough), the more it contributes to the rank. For example: if we have two different molecules A and B, such that A is attached to a single instance of three entity types. Whereas, B is attached to a single instance of each of the four entities types. Then B should be ranked higher than A since we know more knowledge by subscribing to more entities. The notion of specificity has a precedence over the notion of knowledge-gain when there is a tie. Figure 1 shows this notion when examining molec1 vs molec3. In the case of molec1 is subscribes to three feature types (disease, cell and protein) which MK3 subscribes to all four entities (disease, cell, gene and protein). Hence, it is ranked higher as also shown in Table 1.

Depending on the purpose of the ranking, for scientists who are actively searching for molecules for a given purpose, a SPARQL query may include a context of a disease (i.e., depression in this case). Such a query can be asked in the following form: list all molecules that are found associated with depression disease, all proteins instances, all cell types, all chemical entities). We see that both types of ranking are necessary. The algorithm we present here is indeed capable of producing both types of ranking results. On the one hand, the global ranking starts with molecule of type and returns a ranked list of of all of the molecules in the dataset. On the other hand, a specific ranking starts with an instance of any type (e.g., gene, protein, etc) as a starting node. Following are the steps that describe MolecRank, and the corresponding pseudocode in Algorithm 1.

1. Generate the network, resulting from the query result, into feature types clusters.
2. For each molecule, associated with a given feature type, calculate the set of its neighbors.
3. For each molecule and its neighbors, generate two matrices:

    (a) A knowledge-gain matrix: whether a molecule is subscribed to a feature type regardless of its cardinality. This part of the measure was inspired by Shannon's entropy. Particularly, the notion of information-gain for subscribing to various feature types [25]. Here we also keep in mind the knowledge gained by the association with a category. In Shannon's Entropy, this is represented by 1's and 0's, here we do the same by normalize by the actual network weights. This accounts for the value of the feature subscribing to a category and doesn't allow the specificity to be overrun.

(b) A specificity matrix: the cardinality of the global knowledge-gains each fea-
     ture type regardless of number of instances it connects to.

4. Normalize the knowledge-gain matrix (if cardinality is > 0) replace with 1, otherwise it is 0.
5. Normalize the specificity matrix (if cardinality is = 1) replace with 1, otherwise
   it is 0.
6. For each matrix, calculate the sum of each row. This generates two scores for each
   molecule.
7. Merge the scores using a common sense heuristic that has the following characteristics:

   (a) Favoring specificity by multiplying by the sum of total knowledge-gains. This
        guarantees that knowledge-gain is always lower
   (b) Incorporating a noice factor ϵ that is of a value between [0, 1]

8. Sort the score in descending order and return top K matches.

Data: A SPARQL query result in CSV format
Result: Ranked D's based on their computed scores
1 - Construct a network G of drugs and all other feature instances;
while G *not clustered* **do**
  | 2 - Associate each D_i with each feature instance (make bipartite);
end
for $D_i \in G$ **do**
  | 3 - Generate neighbors' adjacency matrices S and C ;
end
for $e \in S$ **do**
  | **if** $e > 1$ **then**
  | | e $\leftarrow$ 0, otherwise e $\leftarrow$ 1 ;
  | **end**
end
for $e \in C$ **do**
  | **if** $e > 0$ **then**
  | | e $\leftarrow$ 1, otherwise e $\leftarrow$ 0 ;
  | **end**
end
4 - Calculate the product of S and the weights of each feature ;
5 - Calculate the row sum of each matrix S and C ;
6 - Calculate specificity Ψ and knowledge-gain Ω ;
7 - Combine Ψ and Ω as in equation: 1 ;
8 - Sort descendingly and return list of D's ;

Algorithm 1: MolecRank Pseudocode: MolecRank: A Molecule Ranking Algorithm for Biomedical Knowledge Networks

2.4   Experiments Using Mockup Data

The example show in Table 1 and Fig. 1 is a good step to demonstrate the idea of specificity in general. However, it lacks some significant details once it is connected to the drug mentions in publications and how they are impacted by their frequencies of appearances. Which is something that the earlier example lacks to show. To remedy this issue and demonstrate a much more closer example, here we show some drugs, and they association with entities along with the frequencies of cach association hypothetically derived from publications. Table 2 captures a list of four drugs (D1 …D4) and their corresponding feature association and weights.

The table displayed in the manner it is because it makes it possible to construct a network which can be further analyzed. The Drug column can be viewed as a source node while the Feature Instance is the target node. This makes the network bipartite in nature where the source node is a Drug and the target is a feature of a specific type. The remaining columns (Feature Frequency and Feature type) act as attributes for edges and nodes. This table can be fed into Python iGraph library as a CSV value to produce the desired network. This explains how step 1 and 2 of the algorithm are accomplished in the algorithm. Figure 4 shows how the final network as weighted bipartite graph. The figure shows the drugs in blue with a bigger nodes than all the other nodes to represent the source nodes. All nodes are labeled with the instance

Table 2   Contrived dataset of drugs (D1, D2,…, D4) and their corresponding neighbors of biological entities, types and counts. The entities types are DNA, RNA, Protein, Cell Type, and Cell Line

Molecule/drug	Feature instance	Feature frequency	Feature type
D1	g1	0.1	DNA
	g24	0.23	DNA
	rna1	0.05	RNA
	g2	0.15	DNA
D2	prt1	0.3	PROT
	prt2	0.15	PROT
	ctype1	0.05	CTYPE
	cline1	0.25	CLINE
	g1	0.1	DNA
D3	g24	0.1	DNA
	rna3	0.1	RNA
	prot13	0.1	PROT
	cline1	0.17	CLINE
	cline23	0.05	CLINE
D4	g24	0.15	DNA
	prt2	0.15	PROT
	cline1	0.15	CLINE
	rna1	0.25	RNA

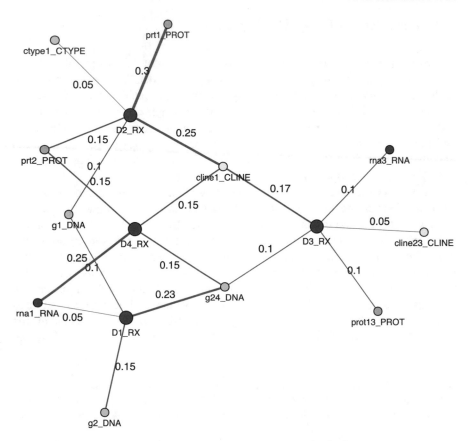

Fig. 4 Visual of a simulated network of molecules and their corresponding entities. Each feature type is colored differently

and type to make it easier for the reader to understand and trace the algorithm steps. This also applies for the edges as they are labeled in addition to having different thickness according to the weight to demonstrate which links are more frequent and which are not.

Calculating the adjacency matrices (specificity and knowledge-gain) is a straight-forward process: Since Drugs are encoded as a source so we query the graph and retrieve all source. For each source, we access it neighbors which are only the feature instances. The source is prohibited to connect to a drug since the graph is bipartite. This is why it is guaranteed to have neighbors of all types except for Drug nodes. Now start with specificity matrix since it is more significant and requires more work. For each drug, a count of how many links to the neighbors for each type is calculated. For each type, if the count is = 1, a value of one is placed in the type field, otherwise, a zero value is placed. For instance D1's adjacency record would be (D1→ DNA:0, RNA:1, PROT:0, CTYPE:0, CLINE:0).

Now, when applying the Feature Frequency values, we get the following considering the feature frequency weights, which captures the specificity strength, it produces (D1→ DNA:0, RNA:0.05, PROT:0, CTYPE:0, CLINE:0). Clearly this is because the RNA Feature Instance "rna1" is specific which has a 1 value in the specificity matrix. When multiplying by the strength of such specificity, presented as a weight, it produces the values of "0.05". The RNA type instance is the only specific feature according to out definition. The remaining entities are not specific, hence their corresponding elements are values of a "0".

Calculating the knowledge-gain is treated differently. The data demonstrates that one drug may have association with more than one instance of the same type. An example of that is actually D1 where it links to three different DNA instances (g1, g2, g24). This makes the D1 is not specific for linking to a given DNA, however, it is information we gained that must be factored in. The algorithm treats such probabilistically as in Eq. 1. This may result in a marginal number but it does address the issue of how much information we gained correctly, which eventually contribute to the final rank of the molecule significantly.

$$\Omega(d_i) = \Pr_{(w_{i1})} \times \Pr_{(w_{i2})} \times \Pr_{(w_{i3})} \tag{1}$$

The full results of computing the specificity and knowledge-gain are shown in Table 3. The two measures (specificity and knowledge-gain) are represented as row vectors respectively. This step produces a single score for each feature type in each vector. In turn, calculating the final ranks using the two vectors for each drug is a mechanical step. The final ranks for each drug drug are: [('D4': 1.4), ('D2', 0.845), ('D3', 0.609), ('D1', 0.103)].

$$\Psi_i = [\forall_{i=1} \mid (\psi_i + \epsilon)] \tag{2}$$

$$\sum_{i=1}^{n} \Theta_i = [\Psi_i + \Omega_i] \tag{3}$$

3 Network Centrality Analysis

Before we share the ranking analytics generated in this chapter, we would like to provide an overview of the most common network centrality measures.

3.1 Centrality Analysis

The Centrality Analysis is a method that is used to rank network elements and identify interesting ones [26]. The interestingness of vertices is based on the connection

Table 3 Shows the both the specificity versus knowledge-gain for each molecule and its corresponding feature. A molecule is represented by two rows, the top row is for specificity, while the bottom is for information gain. Hence the grouping to it easier for the readers to understand the step and make sense to the data

DRUG	DNA	RNA	CLINE	CTYPE	PROT
D1_SPEC	0	0.05	0	0	0
D1_IG	0.003	0.05	0	0	0
D2_SPEC	0.1	0	0.25	0.05	0
D2_IG	0.1	0	0.25	0.05	0.045
D3_SPEC	0.1	0.1	0	0	0.1
D3_IG	0.1	0.1	0.009	0	0.1
D4_SPEC	0.15	0.25	0.15	0	0.15
D4_IG	0.15	0.2	0.15	0	0.15

structure of the given network. Such a ranking is useful in the identification of key players, particularly in biological networks. For example, highly connected vertices in protein interaction networks often signals functionally significant proteins, which in the case of deletion, it leads to lethality [27]. Centrality in the formal terms is a function such that each vertex in the network has a numeric value $C(v)$ [28]. Once assigned, such a value determines whether a vertex a is more important than b if and only if $C(a) > C(b)$.

Various centrality methods are used to express a notion of importance based on a certain characteristic, which leads to an interesting ranking. The simplest form of such ranking is established from the degree of the vertex. The degree is a local value derived from the immediate neighbors connected to this vertex in consideration. This can be formalized as:

$$C_d(u) = \sum_u^N A_{uv} \tag{4}$$

where u is the focal node, v represents all other nodes, and N is the total number of nodes and A is the adjacency matrix such that A_{uv} is the cell that captures whether any two nodes are connected. A value of 1 shows connected and a value of 0 shows otherwise [29, 30]. In biological networks, the degree is often used to co-relate with another variable [31, 32].

Another ranking mechanism is based on the length of the shortest-path within the network (a.k.a. closeness centrality). It is formally defined as the reciprocal sum of the minimal distance, for pairwise strongly connected vertices, to all other vertices in the network. Therefore, closeness only applies to a strongly connected network. The closeness centrality is defined as

$$C_{clo}(u) = \frac{1}{\sum_{v \in V} dist(u, v)} \tag{5}$$

The shortest-path betweenness is another intuitive centrality measure. It is quantified by the counting ability of a vertex to minor communications between two other vertices. The rate of communication can is expressed as

$$\delta_{st}(v) = \frac{\sigma_{st}(v)}{\sigma_{st}} \tag{6}$$

While σ_{st} denotes the number of shortest paths between two vertices s and t, and $\sigma_{st}(v)$ denotes the number of shortest paths between two vertices s and t that use v. The betweenness centrality is defined as :

$$C_{spb}(v) = \sum_{s \neq v \in V} \sum_{t \neq v \in V} \sigma_{st}(v) \tag{7}$$

The contemporary PageRank [33] is a link analysis algorithm that is designed for a directed network. Vertices gain their reputation by being voted by others. Though, originally designed for establishing order for the web, it has been widely used among the major centrality measures. The PageRank formula is expressed as follows:

$$P_r(k) = \sum_{h \to k} \frac{P_r(h)}{o(h)}, \qquad k = 1, 2, \dots, n \tag{8}$$

where $P_r(h)$ is the PageRank of page h, $o(h)$ is the number of out-links of page h and the sum is extended to all Web pages h pointing to page k; n is the number of pages in the Web [34].

3.2 Discussion

We have presented TargetAnalytica a fully mature analytics platform. TargetAnalytica is equipped with Machine Learning and Ontology-based entity recognition tools. The entities of interest are potential targets to answer a specific question (a query). Though the TargetAnalytica tested the hypotheses around gene, protein, cell type and a cell line as possible targets, it is also possible to extend to allows the incorporation of other entities (whether they are exported from ontologies or databases). To demonstrate the potential of TargetAnalytica, we performed a ranking-type analysis against a PubMed dataset related to the Depression disease. The purpose of such analysis was to explore how the specificity of biological targets (genes, protein, cell type) can drive interesting ranking analytics that is very different from their counterparts. Specifically, we compared our ranking with Drugs.com's Popularity measure. We have found out that drugs such as Cymbalta is the known to number #1 in popularity was ranked as #10 according to the ranking algorithm. Clearly, there are other factors such as marketing, and promotions might contribute to the popularity. TargetAnalytica, however, produces more objective ranking that is not biased by commercial

Table 4 Shows the correlations among SpecRank scores for each drug and how how correlation it has with the centrality

Centrality	MR-Deg	MR-Clos	MR-Bet	MR-ECC
Correlation	−0.58	−0.62	−0.59	0.55

marketing and addresses the ranking in a purely biological way. This is indeed more advantageous to scientists and decision makers to work with.

We also compared the ranking analytics to common network centrality measures. According to the correlation analysis it appears that there are some interesting patterns consistent which we believe consistent with the centrality measures. While degree, closeness, and eccentricity demonstrated the highest "negative" correlation of (−0.60), and eccentricity have a positive correlation (0.55). Most of the co-relation values, negative or positive are above 0.50 and they range from (0.55–0.60) regardless of the sign. This can be explained due to the fact that the centrality measures are all about the underlying network structure(edge frequencies and how they are connected). For example, the degree centrality non-selectively counts the neighbors of the each drug. However, specificity selectively count the edges of the neighbors according to the singularity and the strength of the bond. This means the more neighbors connected with edges to the drug the less the specificity. However, the information of the neighbors are not entirely lost since it was captured by the notion of knowledge-gain, which was computed by the multiplications of the probabilities of how the neighbors of connected. Table 4 shows how the various centrality measures are co-related with our ranking algorithm.

Since the ranking algorithm favors the specificity over the knowledge-gain, the specificity always triumph. This explains why there is about 0.60 of negative correlation, which there remaining 0.40 for a perfect correlation is lost due accounting for the knowledge-gain probabilities. The same notion is also applied for closeness and betweenness since the two measures consider the neighbors of the vertices in question (i.e., the drug). As for eccentricity, which is about finding the maximum distance of a drug, which depends on the neighbors.

According the specificity, we selectively look for the singular neighbors, which will affect contribute to the distance and makes is longer. In other words, the higher the specificity values, the farther the maximum distance, which is also explain the positive correlation. It is interesting to see that the absolute values all ranged around 0.55–60 which indeed demonstrates the spirit of the algorithm in capturing the quality of the connection as opposed to the bare frequencies without losing the other connectivity information embedded in the network. The ranking algorithm could potentially be a favorite ranking analytics measure that summarizes all other measures. Clearly, the algorithm can be not only used for ranking molecules but can also be used in a merely theoretical and also all other problems where specificity is in question. Table 5 shows the top-10 drugs and the corresponding ranks in each of the centrality measures (degree, closeness, betweenness, and eccentricity) respectively.

Table 5 Shows the top-10 drugs ranked using the specificity-centrality to rank the drugs and the corresponding order using the various centrality measures

RX_DRUG	SPEC	DEG	CLOS	BET	ECC
Protriptyline	1	40	41	36	26
Ludiomil	2	49	55	51	28
Symbyax	3	56	60	61	20
Amoxapine	4	33	36	34	48
Sinequan	5	63	68	65	8
Parnate	6	44	61	48	21
Methylphenidate	7	19	19	18	53
Trintellix	8	53	56	52	10
Surmontil	9	70	72	70	1
Cymbalta	10	42	48	41	3

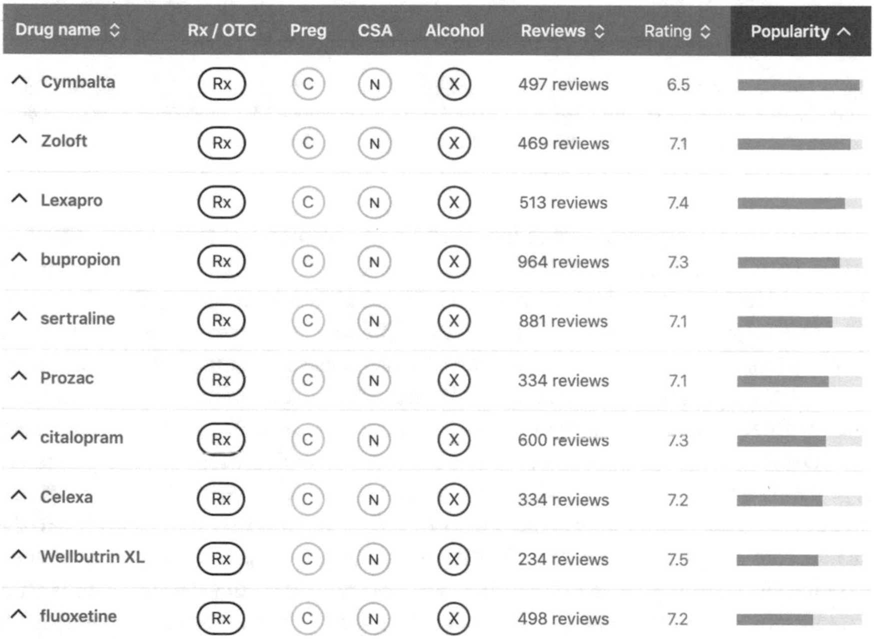

Drug name ⌃	Rx / OTC	Preg	CSA	Alcohol	Reviews ⌃	Rating ⌃	Popularity ⌃
⌃ Cymbalta	Rx	C	N	X	497 reviews	6.5	
⌃ Zoloft	Rx	C	N	X	469 reviews	7.1	
⌃ Lexapro	Rx	C	N	X	513 reviews	7.4	
⌃ bupropion	Rx	C	N	X	964 reviews	7.3	
⌃ sertraline	Rx	C	N	X	881 reviews	7.1	
⌃ Prozac	Rx	C	N	X	334 reviews	7.1	
⌃ citalopram	Rx	C	N	X	600 reviews	7.3	
⌃ Celexa	Rx	C	N	X	334 reviews	7.2	
⌃ Wellbutrin XL	Rx	C	N	X	234 reviews	7.5	
⌃ fluoxetine	Rx	C	N	X	498 reviews	7.2	

Fig. 5 Shows the order derived by the popularity measure generated by drugs.com database. The drugs appear in an ascending order, the most popular drugs comes first

By inspecting the list of the top-10 we see some surprising incidents. Though Cymbalta is ranked number 1 in popularity on the drugs.com database, it claims the 10th position on the MolecRank list. Figure 5 shows the list of the top-10 popular drugs. The popularity notion is calculated using various aggregate statistics from analytics generated from the website. The details of such a measure if currently

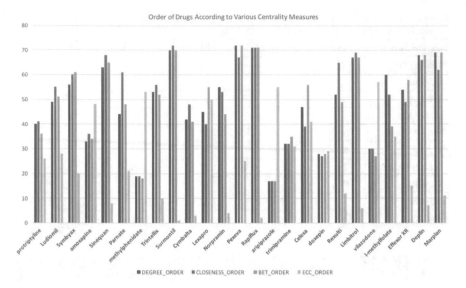

Fig. 6 Shows the order derived by the centrality measures for the the top-10 drugs. We, here, notice that the lower the eccentricity, the higher the other measures. The opposite is also true which confirms the correlation analysis demonstrated and explained above

unpublished as it is proprietary to the database. According to the ranking algorithm, a drug such as Protriptyline comes first. There are a few factors to consider in this analysis, which are yet to be incorporated in the next future phase: (1) The interaction with alcohol, (2) the impact on pregnancy, (3) Controlled Substance Act. Without keeping these factors in mind, it is not directly possible to link the popularity of a drug to its specificity. However, learning that Protriptyline is manufactured by companies such as Pfizer and Sanofi among a few other pharmaceutical companies, one may conclude that the confidence in the drug's ability to treat depression. As for not being as popular as Cymbalta, that could be to various factors such as marketing and media advertisements. Figure 6 shows the order of each of the top-10 drugs according to the various centrality measures above. The figure shows that when degree, betweenness, and closeness are high in the rank, the eccentricity is low, and the opposite is true. This is consistent with what we have observed above in the correlation analysis of specificity-centrality and all other centrality measures.

4 Future Directions

The pharmaceutical expert authors have validated the preprocessing results (entity identifications), their associations with each other and disease, and validated the ranks manually to increase the confidence in the future ranking analytics predictions. This offers a great value to scientists and subject matter experts who are seeking the use of a credible tool for starting exploration of large datasets related to molecules. This motivates the authors to extend the scope of the study in various directions: (2) studying other types of disease (e.g., Alzheimer's, cancer, and diabetes) in addition to depression, (2) naturally the scale of the study will be much bigger which could be millions of publications as opposed to 390,000 for depression, (3) as we are validating the results, we have discovered associations of adverse drug reactions. This is something that needs to be explored further and incorporate such knowledge into the algorithm. We believe the ranks should be penalized by the number of side effects associated with a drug, (4) the algorithm delves into the realm of intractability as new entities are added or search results are abundant. An interesting possibility is to redesign the algorithm and make it a hybrid between classical and quantum employing Grover's algorithm. We have only explored a small fraction of a 30M publications and expand the entity type to model organisms, biological assays, chemical entities, etc to enrich the network and get more comprehensive analytics.

Acknowledgements The authors would like thank Dr. Mark Schreiber and Dr. Ramiro Barrantes for their valuable discussions. The authors also greatly appreciate the tremendous feedback on this work giving by Dr. Barabasi and his lab members.

References

1. Bragazzi, N.L., Nicolini, C.: A leader genes approach-based tool for molecular genomics: from gene-ranking to gene-network systems biology and biotargets predictions. J. Comput. Sci. Syst. Biol. **6**, 165–176 (2013)
2. Winter, C., Kristiansen, G., Kersting, S., Roy, J., Aust, D., Knösel, T., Rümmele, P., Jahnke, B., Hentrich, V., Rückert, F., Niedergethmann, M., Weichert, W., Bahra, M., Schlitt, H.J., Settmacher, U., Friess, H., Büchler, M., Saeger, H.-D., Schroeder, M., Pilarsky, C., Grützmann, R.: Google goes cancer: improving outcome prediction for cancer patients by network-based ranking of marker genes. PLOS Comput. Bio. **8**(5), 1–16 (2012)
3. Weston, J., Elisseeff, A., Zhou, D., Leslie, C.S., Noble, W.S.: Protein ranking: from local to global structure in the protein similarity network. Proc. Nat. Acad. Sci. USA **101**(17), 6559–6563 (2004)
4. Wren, J.D., Garner, H.R.: Shared relationship analysis: ranking set cohesion and commonalities within a literature-derived relationship network. Bioinformatics **20**(2), 191–198 (2004)
5. Chen, J., Jagannatha, N.A., Fodeh, J.S., Yu, H.: Ranking medical terms to support expansion of lay language resources for patient comprehension of electronic health record notes: adapted distant supervision approach. JMIR Med. Inform. **5**(4), e42 (2017)
6. Koschützki, D., Schwöbbermeyer, H., Schreiber, F.: Ranking of network elements based on functional substructures. J. Theoret. Bio. **248**(3), 471–479 (2007)
7. Junker, B.H., Koschützki, D., Schreiber, F.: Exploration of biological network centralities with centibin. BMC Bioinform. **7**(1), 219 (2006)
8. Hamed, A.A., Leszczynska, A., MolecRank, M.S.: A specificity-based network analysis algorithm the international conference on advanced machine learning technologies and applications (AMLTA2019) (2020)
9. Hopkins, A.L.: Network pharmacology: the next paradigm in drug discovery. Nat. Chem. Biol. **4**(11), 682–690 (2008)
10. Bodnarchuk, M.S., Heyes, D.M., Dini, D., Chahine, S., Edwards, S.: Role of deprotonation free energies in p k a prediction and molecule ranking. J. Chem. Theo. Comput. **10**(6), 2537–2545 (2014)

11. Koshland, D.E.: Application of a theory of enzyme specificity to protein synthesis. Proc. Nat. Acad. Sci. **44**(2), 98–104 (1958)
12. Lehninger, A., Nelson, D.L., Cox, M.M.: Lehninger principles of biochemistry. In: Freeman, W.H. 5th edn. (2008)
13. Wood, E.J.: Harper's biochemistry 24th edition. In: Murray, R.K., Granner, D.K., Mayes, P.A., Rodwell, V.W. pp. 868. Appleton & lange, stamford, ct. 1996.£ 28.95 isbn 0-8385-3612-3. Biochem. Edu. **24**(4), 237–237 (1996)
14. Hu, L., Fawcett, J.P., Gu, J.: Protein target discovery of drug and its reactive intermediate metabolite by using proteomic strategy. Acta Pharm. Sinica B **2**(2), 126–136 (2012)
15. Hefti, F.F.: Requirements for a lead compound to become a clinical candidate. BMC Neurosci. **9**(3), S7 (2008)
16. Degterev, A., Maki, J.L., Yuan, J.: Activity and specificity of necrostatin-1, small-molecule inhibitor of rip1 kinase. Cell Death Differ. **20**(2), 366 (2013)
17. Eaton, B.E., Gold, L., Zichi, D.A.: Let's get specific: the relationship between specificity and affinity. Chem. Bio. **2**(10), 633–638 (1995)
18. Radhakrishnan, M.L., Tidor, B.: Specificity in molecular design: a physical framework for probing the determinants of binding specificity and promiscuity in a biological environment. J. Phys. Chem. B **111**(47), 13419–13435 (2007)
19. Strovel, J., Sittampalam, S., Coussens, N.P., Hughes, M., Inglese, J., Kurtz, A., Andalibi, A., Patton, L., Austin, C., Baltezor, M., et al.: Early drug discovery and development guidelines: for academic researchers, collaborators, and start-up companies (2016)
20. Hartley, J.A., Lown, J.W., Mattes, W.B., Kohn, K.W.: Dna sequence specificity of antitumor agents: Oncogenes as possible targets for cancer therapy. Acta Oncol. **27**(5), 503–510 (1988)
21. Timchenko, L.T., Timchenko, N.A., Caskey, C.T., Roberts, R.: Novel proteins with binding specificity for dna ctg repeats and rna cug repeats: implications for myotonic dystrophy. Hum. Mol. Genet. **5**(1), 115–121 (1996)
22. Settles, B.: ABNER: an open source tool for automatically tagging genes, proteins, and other entity names in text. Bioinformatics **21**(14), 3191–3192 (2005)
23. Carpenter, B.: Lingpipe for 99.99% recall of gene mentions. In: Proceedings of the Second BioCreative Challenge Evaluation Workshop, vol. 23, pp. 307–309 (2007)
24. Candan, K.S., Liu, H., Suvarna, R.: Resource description framework: metadata and its applications. SIGKDD Explor. Newsl. **3**(1), 6–19 (2001)
25. Shannon, C.E.: Prediction and entropy of printed english. Bell Labs Tech. J. **30**(1), 50–64 (1951)
26. Koschützki, D., Schreiber, F.: Centrality analysis methods for biological networks and their application to gene regulatory networks. Gene Regul. Syst. bio. **2**, 193 (2008)
27. Jeong, H., Mason, S.P., Barabási, A.-L., Oltvai, Z.N.: Lethality and centrality in protein networks. Nature **411**(6833), 41–42 (2001)
28. Koschützki, D., Lehmann, K.A., Peeters, L., Richter, S., Tenfelde-Podehl, D., Zlotowski, O.: Centrality indices, pp. 16–61. Springer Berlin Heidelberg, Berlin, Heidelberg (2005)
29. Freeman, L.C.: Centrality in social networks conceptual clarification. Soc. Netw. **1**(3), 215–239 (1978)
30. Opsahl, T., Agneessens, F., Skvoretz, J.: Node centrality in weighted networks: generalizing degree and shortest paths. Soc. Netw. **32**(3), 245–251 (2010)
31. Zhou, Q., Womer, F.Y., Kong, L., Wu, F., Jiang, X., Zhou, Y., Wang, D., Bai, C., Chang, M., Fan, G., et al.: Trait-related cortical-subcortical dissociation in bipolar disorder: analysis of network degree centrality. J. Clin. Psychiatry **78**(5), 584–591 (2017)
32. Costenbader, E., Valente, T.W.: The stability of centrality measures when networks are sampled. Soc. Netw. **25**(4), 283–307 (2003)
33. Page, L., Brin, S., Motwani, R., Winograd, T.: The pagerank citation ranking: bringing order to the web. Technical report, Stanford InfoLab (1999)
34. Pretto, L.: A theoretical analysis of google's pagerank. In: String Processing and Information Retrieval. Springer, pp. 125–136 (2002)

Earthquakes and Thermal Anomalies in a Remote Sensing Perspective

Utpal Kanti Mukhopadhyay, Richa N. K. Sharma, Shamama Anwar, and Atma Deep Dutta

Abstract Earthquakes are the sudden tremors of the ground leaving behind damages to life and property, ranging from smaller to massive scale. Since very early times, earthquake prediction has been in the limelight of the scientific community. In the texts of ancient civilizations, earthquake predictions can be found which is based on the position of the planets with respect to the earth. With the advent of real time observations from various data sources many attempts are going on this direction. The present chapter investigates and put forward some facts based on data obtained from satellites for an earthquake which occurred in Imphal, India, in 2016. It studies the thermal anomaly data that took place before the earthquake. MODIS Land Surface Temperature (LST) product was used wherein daily night time images of 6 years have been used for the study. Good quality pixels having maximum information were identified by performing Quality Assurance of the datasets. A change detection technique for satellite data analysis namely Robust Satellite Technique has been used and RETIRA index has been calculated. The study of this RETIRA index has been done for 3 years and it has been found that the RETIRA index is considerably high for the earthquake year. But it cannot be concluded that high value of RETIRA index is a sure indicator for an earthquake and hence it leaves scope for future studies.

Keywords Thermal anomaly · Robust Satellite Technique (RST) · Land Surface Temperature (LST) · Moderate Resolution Imaging Spectro-radiometer (MODIS) · Robust Estimator of TIR Anomalies (RETIRA) index

U. K. Mukhopadhyay
Indian Institute of Technology Kharagpur, Kharagpur, West Bengal, India

R. N. K. Sharma (✉) · S. Anwar · A. D. Dutta
Birla Institute of Technology, Mesra, Ranchi, Jharkhand, India
e-mail: richasharma.ranchi@gmail.com

1   Introduction

Prediction of earthquakes, form in themselves, a specialized study. However, after their occurrences, there are lots of ways and means to study their causes as a lot of data can be reconstructed after its occurrence. Hence, a better understanding about the earthquake phenomenon can be made by analysing the past earthquakes and their characteristics. The most widely acclaimed reasons behind the occurrence of an earthquake among all the geological theories are the tectonic movement and the stress developed in rocks beyond the elastic limit. Researchers, since the last decade are continuously trying to find clues, more technically precursors, that is left by nature before an eminent earthquake. Ultra low frequency (ULF), magnetic field, Total Electron Content (TEC) change in ionosphere, noises in communication signal across wide frequencies which can be observed in mobile network, TV and GPS signals etc., sudden bright light emission from ground, thermal infra-red anomalies, gravity anomalies, abnormal cloud patterns, changes in animal behaviour etc. are some of the previously reported and tested precursory signals in the field of earthquake detection [1]. But the main problem with these precursors is that they are not consistent which means that they do not occur prior to all the earthquakes. Uncertainty involved in time and place of occurrence of earthquake has lured the scientists across the globe but decades of research have failed to reliably predict the time and magnitude of occurrence an earthquake. But, from time to time this topic of earthquake prediction has been rekindled by newer and more sophisticated ways to analyse the precursors before an earthquake and provide important clues for prediction of an earthquake.

Since the last decade, the fluctuations in Earth's thermally emitted radiation, detected by thermal sensors on board satellites have been regarded as an observable precursor prior to earthquakes. Although a large number of studies have been performed since the 1980s but these studies have often been questioned and greeted with scepticism due to insufficiency of validation dataset, scarce importance related to causes other than seismic events and also the poor methodologies deployed to discern the existence of thermal anomalies [2]. Recently, a statistical approach named Robust Satellite Technique (RST) has been extensively used as a suitable method for studying anomalous behaviour of TIR signals prior to and after earthquake events [3]. This chapter uses the RST technique to investigate the presence of thermal anomaly prior to Imphal Earthquake that occurred on 4th January 2016 and understand the spatio-temporal correlation between the earthquake occurrence and the appearance of such transient anomaly in space-time domain. Six years of MODIS Land Surface Temperature data in terms of reference dataset has been used to analyse the thermal behaviour of the study area.

1.1 Application of Thermal Remote Sensing in Detection of Thermal Anomaly: A Brief History

A British geologist Milne in 1931 reported possible air temperature fluctuations in relation to seismic event long before the advent of remote sensing. Attempts were made to measure potential precursory Land Surface Temperature (LST) Change with the implementation of remote sensing in seismology in the 1980s. Land Surface Temperature (LST) is one among the most important parameters in the physical processes of the surface energy and water balance at local through global scales because it acts as the direct driving force in the exchange of long-wave radiation and turbulent heat fluxes at the surface-atmosphere interface [4]. LST is the physical manifestation of surface energy and moisture flux exchange between the atmosphere and the biosphere [5, 6]. International Geosphere and biosphere Program considers LST as one of the high priority parameters in studying climate change, monitoring vegetation, urban climate and environmental research [7].

Land Surface Temperature (LST) changes both changes significantly, in both space and time owing to the diversity of land surface features such as vegetation, soil and topography [8, 9]. Natural reasons contributing to TIR signal variability include local geographical factors such as altitude solar expositions, meteorological, spatial and temporal variations of observational conditions and vertical humidity profiles among others. Several phenomena, other than being natural, are reckoned to instigate the raising of Land Surface Temperature before earthquakes strike. However, such phenomena are still vaguely understood and not widely acknowledged by the scientific community.

Subsequent research was carried out by scientists using satellite data to study the thermal regime of the earth before and after earthquakes that struck India, Iran, China and Japan [10–20]. Diverse methodologies were propounded by researchers for detecting thermal anomalies. Recently, a study carried Zhang et al. [5, 6] reported thermal anomaly, 2 months prior to MS 6.2 Earthquake in Zadio, Qinghai on October 17th 2016 and it was concluded that the earthquake occurred after the anomaly reached its peak. A study by Bhardwaj et al. [21] show that 3–17 days prior to the Nepal Earthquake of 2015, 3–14 K of temperature anomaly had been observed. Rise in temperature diminished the snow cover of the region considerably was also reported from the study. However, not all earthquakes are preceded by thermal anomalies have been presumed by few reports [22–26]. Tronin et al. [10, 27] studied earthquakes in Japan and China. Thermal images of 7 years acquired from NOAA–AVHRR satellite thermal data over China were used as reference dataset to investigate the presence of thermal anomaly. For each image, temporal mean and standard deviation were computed and subsequently, pixels showing ±2 standard deviation from the average were deemed to be a deviation from the normal regime. Fluctuations in the thermal regime was observed 6 days to 24 days prior to the seismic event and persisted for a week after the earthquake. Amplitude of the anomaly was around 3°C. Meteorological and topographical factors were not held responsible for the observed time dependent pattern of the anomaly [28]. Inference was drawn that thermal anomalies were sensitive to the earthquakes greater than 4.7 and can be spatially distributed within 100–200 km from the epicentre [27]. Ouzounov and Freund were the pioneers of the method of differentiation of spatially averaged LST in order to discern thermal anomaly [17]. The earthquake under probe was the Bhuj earthquake that jolted Gujarat on January 26th 2001. Subsequently differentiation of daily night time MODIS LST of the earthquake year and same of the non-earthquake year were carried out. Results claimed that a TIR anomaly with amplitude of 3–4 °C was observed 5 days prior to the earthquake and died out shortly after the event passed.

Saraf et al. [18] also concentrated on Bhuj, Gujarat Earthquake. As the result showed, inferences were drawn by analysing NOAA–AVHRR images over time period from December 2000 to February 2001. Homogeneous images belonging to the same period of the month, time of the day, same sensor data of 2003 were analysed visually in the process for obtaining information about the thermal regime in the non-earthquake year. Thermal anomaly of magnitude 5–7 °C was reported [18].

A study by Blackett et al. [29] challenged the argument led by Saraf et al. [25] for Bhuj, Gujarat Earthquake. It was deduced that LST differencing based in multiple years cannot justly indicate a potential precursory Land Surface Temperature change before the earthquake, since both the year of interest and the temporal mean of the years have considerable effect on the calculations [30]. Blackett et al. [29] ascertained that the presence of missing data owing to cloud cover, mosaicking gap etc. particularly over normally warmer and cooler areas can induce errors in RST calculations.

Saraf et al. [18, 26] focused on the earthquakes of Algeria that occurred in May 2003. Historical datasets were not used in the study. Rather, a spatio-temporal analysis of LST fluctuations was carried out using images of the earthquake year only. The anomalous pixels were premised on the comparison of the concerning pixel values and value of neighbouring pixel. For the anomalous pixels, the LST difference was reported as high as 5–10 °C.

Panda et al. [19] researched the presence of thermal anomalies prior to the Kashmir earthquake (Mw 7.6) that struck on October 8th 2005 and pursued a different approach. Daily MODIS LST products (pertaining to daytime) between September 26th 2005 and October 27th 2005 were used for studying the earthquake year. For retrieving historical thermal background, thermal images of the same time period between 2000 and 2004 were analysed by averaging the daily pixel values of images through 2000–2004 [19]. Subsequently, the daily averages were subtracted from corresponding days in the earthquake year 2005. The results showed that LST anomaly reached 5–10 °C around the epicentre region. Truth be told, the fundamental issues in some of these methods of investigations were the absence of a thorough meaning of Thermal anomaly and the rare consideration paid to factors other than seismic movement (e.g. meteorological) which could be liable for the appearance of thermal infrared anomalies [14, 31].

A Robust Satellite Technique (RST) has been employed to unearth thermal anomalies preceding certain earthquakes. RST is a multi-temporal approach of data analysis that considers every anomaly in the space-time domain as a deviation from an 'unperturbed' state, specific for each place and time of observation [11]. In order to reconstruct the normal behaviour of the signal, long term historical series of satellite records are processed in multi-temporal sequence stacked according to homogeneity criteria [32]. The same RST technique has been used to exploit the thermal regime before and after several earthquakes as for example the Kocaeli Earthquake, Turkey (1999) [11], the Bhuj Earthquake, India (2001) [14], the Abruzzo Earthquake, Italy (2009) [31]. However, depending on the availability of the data, the number of years that has been used as a reference dataset has varied. It is seen that taking more number of years increases the accuracy and reliability of the methodology implemented to detect thermal anomaly but there is no hard and fast rule for the number of years that should be taken for consideration for the computation of historical or reference datasets. It is judicious to use 4–10 years of data for computing the reference dataset [13]. Night time LST data has been preferred in most of the studies because of the fact that it is less sensitive to the soil–air temperature differences which are normally higher during daytime than at night and cloud cover and shadows as well [3].

1.2   The Physical Basis of Thermal Infrared Anomaly

The earth's crust passes through an earthquake preparatory phase before an imminent earthquake. Accumulation of stresses and resultant pressure development prompts the ascent in LST. The enhanced TIR emanation from Earth's surface retrieved by satellites prior to earthquakes is known a thermal anomaly [22–24]. A number of studies have been performed in the past years using satellite data acquired in TIR (thermal infrared radiation band, 8–14 μm) which imply the existence of a relation between anomalous space-time signal fluctuations and earthquake occurrence. Various studies have been performed in the previous years using satellite information procured in TIR which infer the presence the presence of a connection between abnormal space time TIR fluctuations and earthquake event.

Thermal Remote Sensing was a blessing to the scientific community who were intrigued in investigating the thermal regime of earth's surface. Subsequently, researcher postulated theories justifying the reason of occurrence of thermal anomalies before and after earthquake. Yet, theories explaining the physical basis of the occurrence of pre earthquake TIR anomaly are widely debated since the first realization of this anomalous phenomenon [33]. The propounded theories include increased degassing activity particularly for optically active gases like CO_2 and CH_4 [11, 12, 34], near-ground air ionization due to enhanced radon emission leading to the condensation of water vapour and, hence, the release of latent heat [35, 36] and activation of positive-hole pairs in rocks under stress [11, 12, 17, 34, 37].

Pertaining to the first model, it is well known that Earth's degassing activity (and particularly of optically active gases like CO_2 and CH_4) is more intense near seismogenic faults [38]. Abrupt variations of such gases in near-surface atmospheric layers could result in a local greenhouse effect. This may increase the near-surface temperature and, consequently TIR emission [11, 12, 34]. The Thermal IR transients affect areas of several thousand square kilometres and could appear up to several hundred kilometres far away from the Earthquake epicentre zone. This distribution of thermal anomaly in space-time is consistent with atmospheric gas dispersion [11, 12, 34].

Respect to the theory of near-ground air ionization due to enhanced Radon emission, using data analysis in laboratory experiments Freund et al. [22] described that substantial air ionization can take place at the surface of rocks. The build-up of stress within the Earth's crust prior to major Earthquake may correspondingly lead to air ionization. This may evoke ionospheric perturbations and a host of other phenomena. Pulinets et al. [35, 36] put forward an explanation of enhanced TIR emission prior to strong earthquake. Ionizing radiation by Radon gas in the near-surface atmosphere over active tectonic faults and tectonic plate boundaries was held responsible for the enhanced, hence abnormal LST increase prior to earthquakes.

Freund's positive-hole pair activation keeps parity with lab experiments. The theory says that most crustal rocks contain dormant electronic charge carriers in the form of peroxy defects. When rocks are under stress, these peroxy links break. This in turn releases electronic charge carriers, known as positive holes. The positive holes remain in the form of positive-hole pairs (PHP). When rock deformation takes place, the PHP breaks and releases positives holes. When p-holes arrive at the rock surface, they recombine and release energy, which in turn leads to an enhanced IR emission [34].

Another concept is known as Remote sensing rock Mechanics put forward by Geng, Cui and Deng in 1992 in the wake of the emergence of modern thermal remote sensing techniques and requirement of prediction of rock failure [20, 26, 39], has some credibility. Based on several experimental studies, it was inferred that rock infrared radiation energy increased with gradually increasing stress on the rock subjected to loading. A similar phenomenon may be triggering the increased LST prior to earthquakes. These experimental studies indicate that the structure of the rock material changes on the application of stress [20, 39]. Studies also concluded that the infrared radiations descended with the reduction in stress after rock failure.

2 Materials and Methods

2.1 Study Area

The earthquake under investigation is the Imphal Earthquake that jolted Imphal, the capital of the state of Manipur in the north-east India on January 4, 2016 with a moment magnitude of 6.7. This was the largest seismic event in the last six decades. Prior to this event, the region experienced an earthquake of magnitude 7.3 in 1957.

Epicenter of the Imphal Earthquake of 2016 was located at 24.834°N and 93.656°E near Noney Village of Tamenglong district of Manipur. The focal depth of the earthquake was estimated to be 50 km. Eleven people lost their lives, 200 others were injured and various structures were damaged. The shake was widely felt in the eastern and north-eastern India and was even felt in Bangladesh. The entire north-east region of India comes under seismic zone V with regards to the seismic zone map of India. This means that the entire region is very much vulnerable to earthquakes. The study is a post event analysis of this earthquake event (Fig. 1).

2.2 Datasets Used

The MODIS (Moderate Resolution Imaging Spectro-radiometer) LST, namely, MOD11A1 daily night time images of 6 years (2011–2016) was used to examine the ground thermal condition before and after the Imphal Earthquake. MODIS is an instrument aboard Terra and Aqua satellite. Terra orbits earth from North Pole to South Pole across the equator in the morning at 10:30 a.m., whereas, Aqua orbits earth from South Pole to North Pole over the equator in the afternoon at 1:30 p.m. MODIS consists of 36 spectral bands from 0.4 to 14 μm. MODIS also caters eight day and monthly global gridded LST products. The daily LST product (Level 3) has a spatial resolution of 1 km. LST is gridded in the Sinusoidal Projection. A tile consists of 1200 × 1200 grids in 1200 rows and 1200 columns. The exact grid size is 0.928 km by 0.928 km at spatial resolution of 1 km.

The daily MOD11 LST and Emissivity data are acquired by implementing generalized split window algorithm. The split window method corrects for atmospheric effects based on the differential absorption in adjacent infrared bands [40]. The product comprises of daytime/night time LSTs, quality assessment bit flags, satellite view angle, observation time and emissivity which are collectively termed as Sub datasets (SDS) [41]. All of the thermal data are downloaded from USGS Earth Explorer. Information on earthquake as regard to location of epicenter, focal depth, magnitude, time of event and casualty were acquired from The United States Geological Surveys (USGS) website and Indian Meteorological Department (IMD) website.

2.3 Data Pre-processing

In order to extract information from raw data, pre-processing of MOD11A1 night time products were essential. Three adjacent tiles of MODIS data cover the area of interest. Associated tiles were mosaicked and re-projected in ArcMap 10.3.1. Each MODIS HDF file contains Quality Science dataset layers which provide user information regarding usability and usefulness of the data product. Information extracted from these quality sciences SDS are used for analysis. The LST Quality Science Dataset

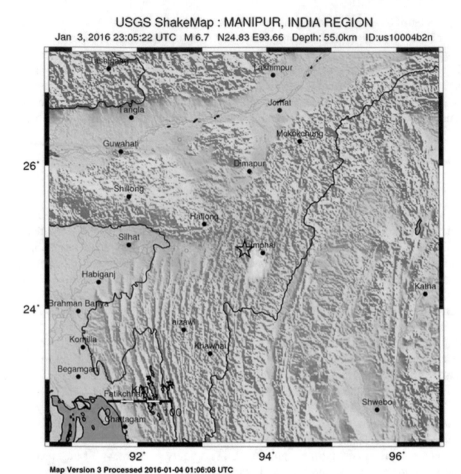

Fig. 1 Shake Map of Imphal Earthquake. (Image downloaded from https://earthquake.usgs.gov/archive/product/shakemap/us10004b2n/atlas/1582928540336/download/intensity.jpg.)

layers are binary encoded meaning each pixel has an integer value that must be converted to bit binary value for cloud masking and interpretation. MODIS LST image contains both good and bad quality pixels. The bad quality pixels are attributed due to cloud or sensor defects.

In order to determine the pixel values in the Quality Control image to be retained as good quality pixels, a number of permutations and combinations were calculated.

Table 1 Defined bit flags for quality assurance scientific dataset QC night/day in MOD11A1 adapted from the Collection 6 MODIS Land Surface Temperature Products Users Guide

Bits	Bit flag name	Key
1 and 0	Mandatory QA flag	00: LST produced, good quality not necessary to examine more detailed QA 01: LST produced, other quality, recommended examination of more detailed QA 10: LST not produced due to cloud effect 11: LST not produced other than due to cloud effect
3 and 2	Data quality flag	00: Good Quality data 01: Other Quality data 10: TBD 11: TBD
5 and 4	Emissivity error flag	00: Average emissivity error ≤ 0.01 01: Average emissivity error ≤ 0.02 10: Average emissivity error ≤ 0.04 11: Average emissivity error > 0.04
7 and 6	LST error flag	00: Average LST error ≤ 1 K 01: Average LST error ≤ 2 K 10: Average LST error ≤ 3 K 11: Average LST error > 3 K

Different combinations and permutations of bit flags compose an 8 bit binary number which has a decimal equivalent in Quality Control (QC) image within 0–255 range. Only those binary values which are acceptable are taken for calculations such as Mandatory QA flag 00 and 01, Data Quality flag 00 and 01, Emissivity error flag 00, 01, 10 and LST error flag 00 and 01. Here the Least Significant Bit (LSB) is 0. The bit flag values involved in analysis are as follows (Table 1):

Next the QC layer is used to create a logical mask. Bits that are "No Data" in QC layer will be "No Data" in the destination, if a pixel is not NA in "QC layer" and has an acceptable bit flag value like 00 or 17 or 25, and then the value of the source is written in the destination.

The no data (NULL) value has to be set to zero. Unfortunately, the value zero also indicates highest quality pixels. In order to preserve pixels with value zero, pixel with a QC value of 0 and a valid LST value are of highest quality, while pixels with a QC value of 0 and an invalid LST value can be set to NULL (Fig. 2).

The no data (NULL) value must be set to zero. Tragically, the value 0 also shows most astounding quality pixels. Keeping in mind the end goal to save good quality pixels, pixels with a QC value equals to 0 and valid LST value are of good, while pixels with a QC value of 0 and an invalid LST value can be set to NULL.

Information on how to convert the digital values of the satellite data can be obtained in the MODIS Land Surface Temperature product user guide [5]. The LST night time dataset is a 16 bit unsigned integer with a valid range of 7500–65,533. It has a scale factor of 0.02 which means that the LST night time data is 50 times magnified. Scale factor is used for a linear DN value rescaling to temperatures in Kelvin scale.

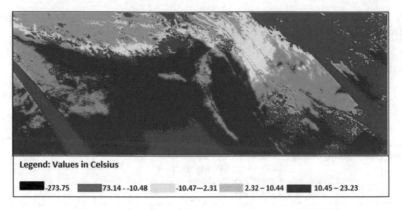

Fig. 2 LST night time SDS converted to Celsius scale

Temperatures in °C can be evaluated with the formula

$$(DN * 0.02 - 273.15).$$

It must be remembered that the no data pixels are ignored in this conversion.

2.4 Robust Satellite Technique (RST)

Robust Satellite Technique is a change detection technique for satellite data analysis used for monitoring natural hazards and is invariant with the utilization of any satellite or sensor. Therefore, RST can be utilized on different satellite data and applied for different events such as earthquakes, volcanoes, floods, forest fires etc. [31]. Beside thermal anomaly detection prior to earthquakes, Robust Satellite Technique has been used for Ash Plume detection and Tracking [32], for oil spill detection and monitoring [42, 43], for monitoring sea water turbidity by RST [44], RST technique for Sahara dust detection and monitoring, RST for pipeline rupture detection [15], RST approach for volcanic hotspot detection and monitoring among others [28] (Fig. 3).

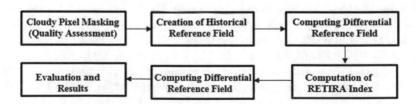

Fig. 3 Flowchart of RETIRA index computation

The methodology is premised on analyzing multi-temporal homogeneous satellite thermal IR images of several years which are co-located in time and space domain. Each pixel of the satellite image is processed in terms of expected and natural variability range of the TIR signal for each pixel in the satellite image. Anomalous TIR signal, therefore an anomalous pixel, is identified as a deviation from its normal or expected value using a unit less index called Robust Estimator of TIR Anomalies (RETIRA) index [22]. Mathematically RETIRA index is as follows

$$\text{RETIRA INDEX} = \frac{\Delta LST(r, t) - \mu \Delta LST(r)}{\sigma \Delta LST(r)}$$

where $\Delta LST(r, t)$ = the difference between the current LST value on the associated image $LST(r, t)$ at the location 'r' $\equiv (x, y)$, at the acquisition time 't' $(t - \tau)$ and its spatial average of the image 'T'(t). $\mu \Delta LST(r)$ and $\sigma \Delta LST(r)$ are the time average and standard deviation of $\Delta LST(r, t)$ respectively, at the location (r) calculated on cloud free pixels.

Time average and temporal mean are evaluated by processing several years of homogeneous cloud free reference datasets. It must be noted that only land pixels are taken into consideration leaving all sea pixels. The numerator part i.e. the difference between $\Delta LST(r, t)$ and $\mu \Delta LST(r)$ can be thought of as the signal to be investigated for possible relation with seismic activity while the standard deviation in the denominator ($\sigma \Delta LST(r)$) represents the noise due to natural and observational causes. That is to say, the computation of the RETIRA index is based on the comparison of the signal and its local variability which is noise. The index depicts the strength of the thermal anomaly. Temporal mean and standard deviation is calculated using cell statistics tool in ArcMap 10.3.1. Temporal mean is the mean LST calculated for every pixel over the reference time period. Standard deviation is the measure of deviation of LST from the mean LST. For each day RETIRA index was calculated for two earthquake years and one non earthquake year. For each day, a high value and a low value of RETIRA index was obtained. This chapter takes under consideration on high values of RETIRA indices for analysis. A baseline value of 3.5 was considered above which RETIRA indices are abnormal.

3 Results and Discussions

3.1 Results of Night Time RETIRA Index Computation for Earthquake Year 2016 (Observation)

On 26th December 2015, 2nd January 2016, 11th January 2016 and 12th January 2016 RETIRA indices obtained on computation were abnormal. In fact 2 days before the earthquake, the RETIRA index was beyond baseline. As the earthquake event passed, the values of RETIRA indices decreased. But a week after the earthquake event on 11th January and 12th January there was a surge in the values of RETIRA Indices (Figs. 4, 5, 6, 7 and 8).

3.1.1   Analysis

In an attempt to identify any correlation between transient temporal thermal anomalies earthquake events, a graphical analysis was carried out for the main earthquake year. For each day under investigation, RETIRA INDICES were plotted. On 26th December 2015, the RETIRA INDEX was abnormal. This was the first instance within the days under investigation when the RETIRA INDEX was observed anomalous. 2 days prior to the main seismic event of magnitude 6.8 on Richter scale i.e. on 02.01.2016, the RETIRA INDEX was again above the baseline value. After the earthquake gradual lowering of RETIRA INDICES were observed. Again, 7 and 8 days after the earthquake, RETIRA INDICES were anomalous. The graph also depicts that the RETIRA INDICES increased gradually before it became anomalous on 11th January 2016 and 12th January 2016. A fitting curve was interpolated from the RETIRA INDICES as a function of time. To achieve a good degree of precision a polynomial of degree 5 was used to interpolate (Figs. 9 and 10).

3.2   Graphical Results of Night-Time RETIRA Index Computation for Secondary Earthquake Year (2014–2015) (Analysis)

With an aim to see if there exists similar kind of trend of RETIRA INDICES in some other earthquake year over the study area, another earthquake year was taken into consideration. Between the 25th December 2014 and 14th January 2015, two earthquakes, one on 28th December 2014 and the other on 6th January 2015, both of magnitude 4.8 were recorded in the north-eastern region of India, within the study area. Hence this period has been considered as another seismically perturbed year. Again, RETIRA INDICES were plotted as a function of time for each day under investigation. As it can be seen from the graph, the RETIRA INDICES were abnormal prior to the first seismic event of magnitude 4.8 on 28.12.14. After the event the value of RETIRA INDICES dropped. Before the second seismic event on 06.01.2015, the RETIRA INDICES once again gradually increased (see graph). Similar trend of RETIRA INDEX was seen for the main earthquake year where prior to the seismic event the indices escalated. For this secondary earthquake year, a fitting curve was interpolated from the RETIRA INDICES as a function of time. Again, to achieve a good degree of precision a 5th order polynomial was used to interpolate. But the curves were inconclusive of a strong relation between the magnitude of earthquakes and the fluctuation of RETIRA INDICES. This is due to the fact that the main earthquake year saw a magnitude 6.8 earthquake whereas the secondary earthquake

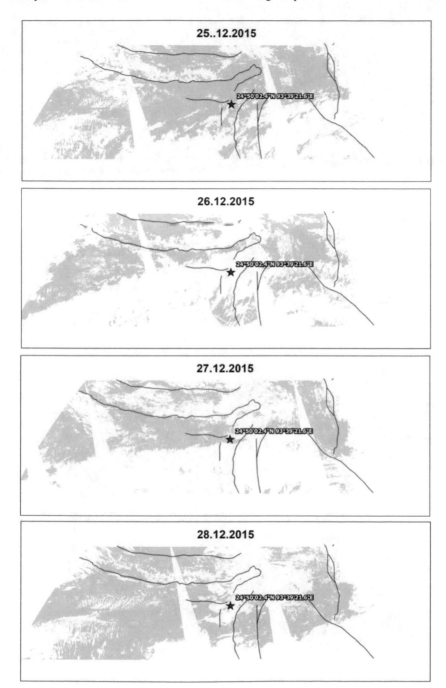

Fig. 4 Results of night-time RETIRA index computation for the main earthquake year (25th Dec 2015–14th Jan 2016)

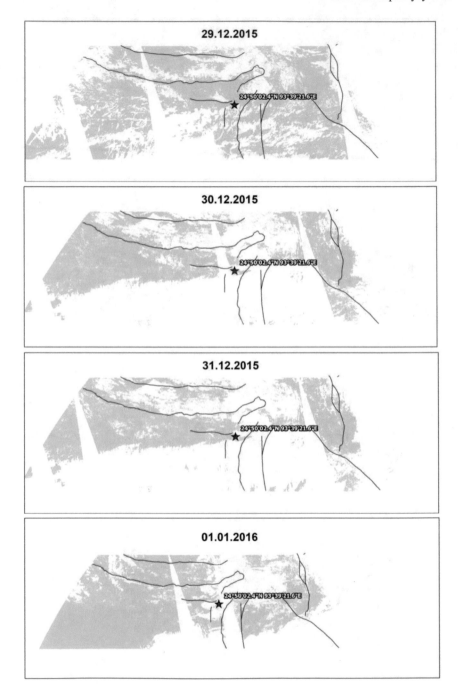

Fig. 4 (continued)

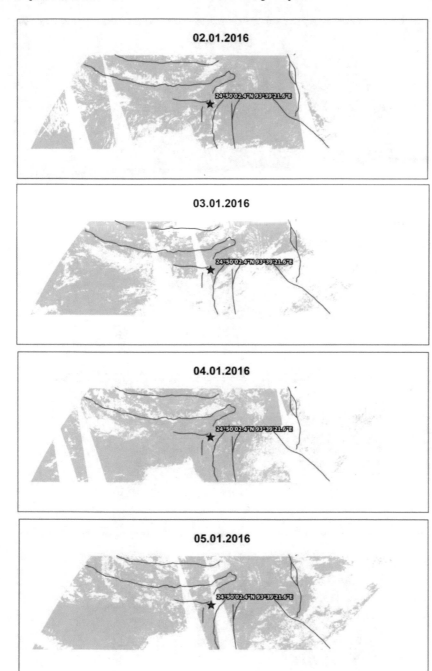

Fig. 4 (continued)

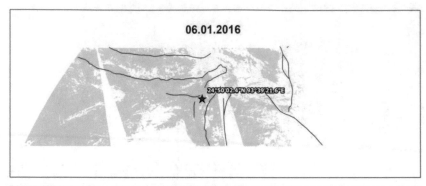

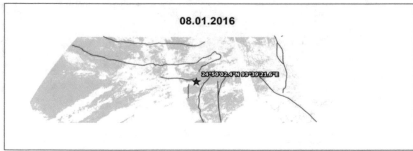

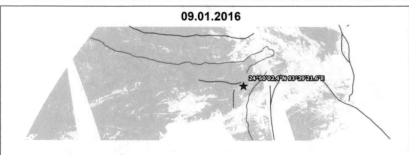

Fig. 4 (continued)

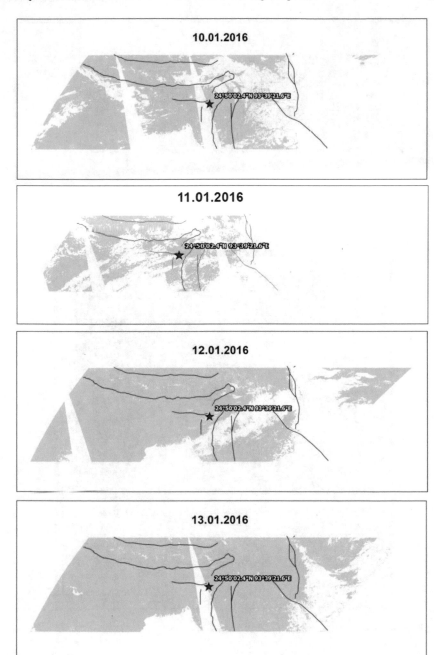

Fig. 4 (continued)

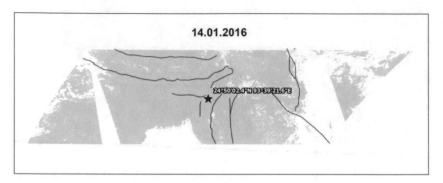

Fig. 4 (continued)

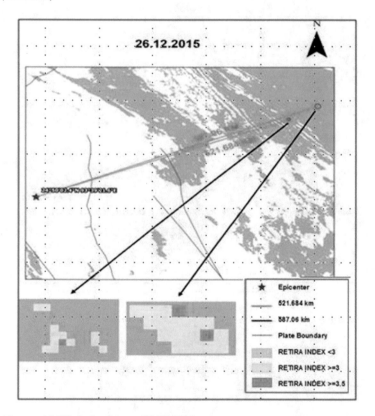

Fig. 5 Abnormal LST pixel values on 26.12.2015

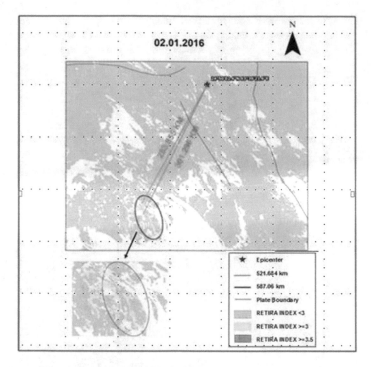

Fig. 6 Abnormal LST pixel values on 02.01.2016

year (2014–15) experienced two earthquakes of relatively smaller magnitude 4.8, but if we closely look at the both the graphs for the two earthquake years, the RETIRA INDICES had higher abnormal values for 2014–2015 (Figs. 11 and 12).

3.3 Results of Night-Time RETIRA Index Computation for Non-earthquake Year (2013–2014) (Observation)

See Fig. 13.

3.3.1 Analysis

Similar graphical analysis was carried out for the non-earthquake year also where RETIRA INDICES were plotted for each day. The year 2013–14 has been considered as a non-earthquake year because of the reason that no earthquake of magnitude greater than 5.0 occurred within the time window under consideration (i.e. 25th December 2013–14th January 2014). Only one seismic event of magnitude 4.2 occurred in the study area during this period. As the graph depicts, there is only one

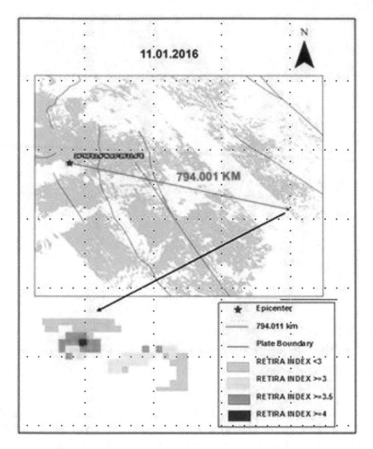

Fig. 7 Abnormal LST pixel values on 11.01.2016

time on 8th January 2014, when the RETIRA INDICES touch the baseline that is exactly 3.5. Above this, the index would be abnormal. A fitting curve was interpolated from the RETIRA INDICES as a function of time. A polynomial of degree 5 was used to interpolate. Degree of order 5 was enforced to achieve a good degree of precision. But the fitting curve never crossed the baseline of 3.5. Analysing the fitting curves reveals that the two earthquake years are similar and a non-earthquake years show quite a different trend (Figs. 14 and 15).

3.4 Pixel Based Analysis for the Main Earthquake Year

Pixels labeled 'No data' hindered our understanding of the evolution of thermal anomaly. Cloud cover, among others is the main reason for such 'no data' pixels. With this aim a pixel based analysis was carried out. The results shown in the table

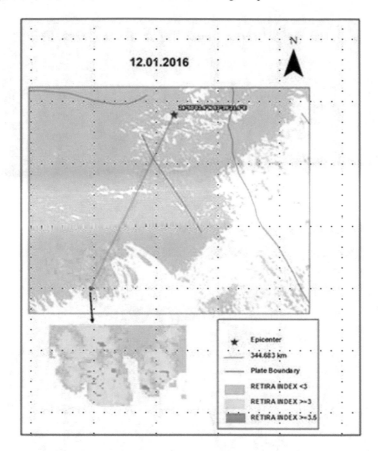

Fig. 8 Abnormal LST pixel values on 12.01.2016

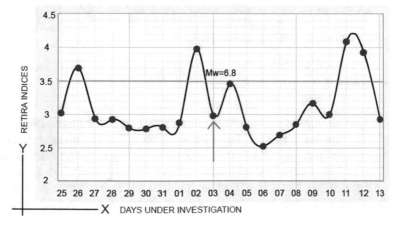

Fig. 9 Graphical analysis of the Earthquake year (2015–16)

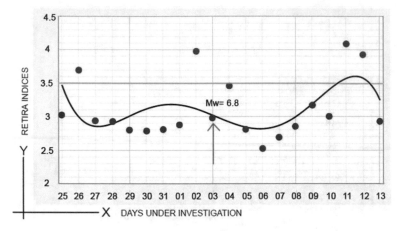

Fig. 10　Graph depicts a fitting curve between 25.12.2015 and 14.01.2016

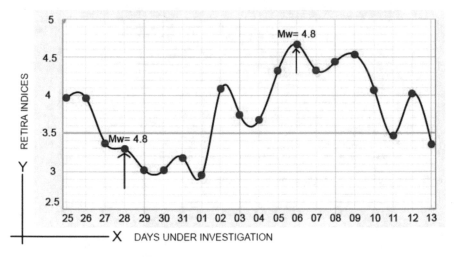

Fig. 11　Graph depicting RETIRA INDICES plotted against the days under investigation for secondary earthquake year (2014–2015)

dictate that the number of pixels with abnormal RETIRA index surged from 3 (nine days prior to the earthquake) to 115 (two days prior to the earthquake). On the day of the earthquake, RETIRA index was normal but 7 and 8 days after the event the RETIRA indices were beyond normal value and number of anomalous pixels increased from 16 to 51 in 1 day. These anomalous pixel counts give us some idea about the evolution of thermal anomalies before and after the event (Table 2).

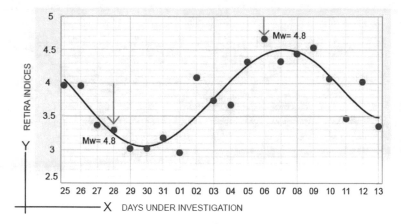

Fig. 12 Graph depicting a fitting curve of RETIRA INDICES plotted against the days under investigation for secondary earthquake year (2014–2015)

4 Conclusion and Future Scope of Day

Till date a valid prediction specifying the date, time, place and magnitude of earthquake have not been standardized [45]. Study of earthquake precursors in many ways facilitate the understanding of their occurrences and thereby lead one step closer to predict earthquakes. Single precursor in pre-seismic monitoring research has its own limitations [46]. Devoting attention to multiple precursors is a promising choice which can raise the reliability and credibility of the relationship between thermal anomalies and imminent seismic events [3]. The region under study did show an anomalous behavior of LST prior to the main earthquake year 2015–2016. The RETIRA index showed thermal anomaly on 26th December 2015, 2nd January 2016, 11th January 2016 and 12th January 2016 for the main earthquake year, before and after the seismic event. In addition the pixel based analysis also reiterated the rise in temperature in the same year. For statistically validating the results of RETIRA index of the main earthquake year, a secondary earthquake year and anon-earthquake year were considered. Although, the present study showed that the RETIRA index has crossed over the baseline in both the earthquake years, yet the earthquake with relatively higher magnitude has relatively lower RETIRA index values whereas the earthquake with low magnitude has a higher RETIRA index above the baseline as in the case of secondary earthquake year (2014–2015). The present study is inconclusive in this aspect and leaves scope for future investigations. However, efforts in investigating precursory earthquake signals had been carried out with Outgoing Long Wavelength Radiation (OLR) [47] pertaining to this Imphal Earthquake of 2016. The authors' result showed that on 27th of December, 2015 an abnormal OLR flux of 303 W/m^2 was recorded against a mean of 279 W/m^2 [47]. Such precursor studies if combined and applied may pave way for further investigations in this direction to strengthen earthquake prediction.

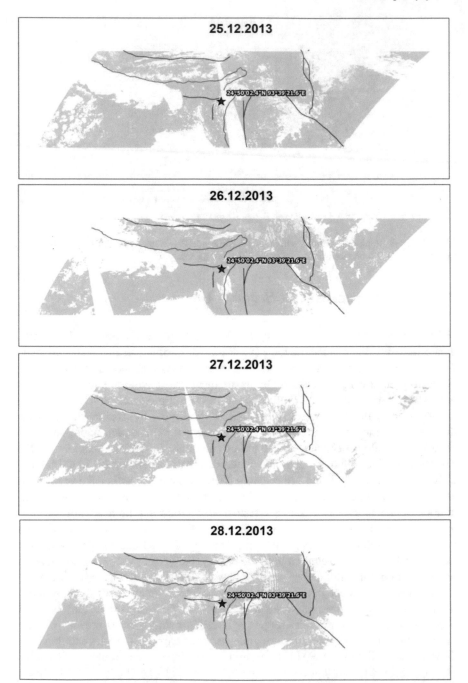

Fig. 13 Results of night-time RETIRA index computation for non-earthquakeyear (25th Dec 2013–14th Jan 2014)

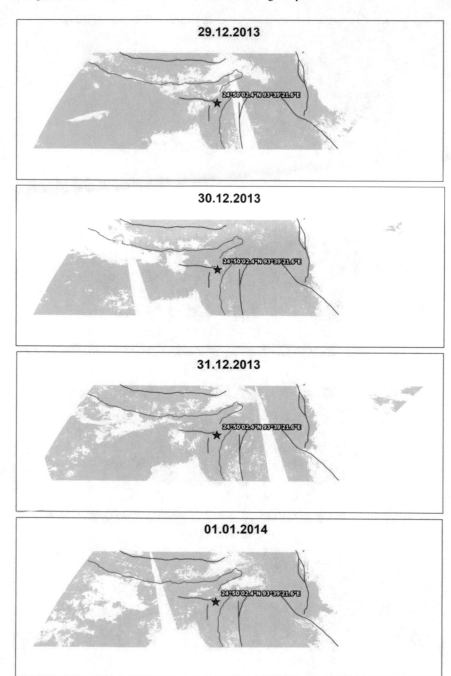

Fig. 13 (continued)

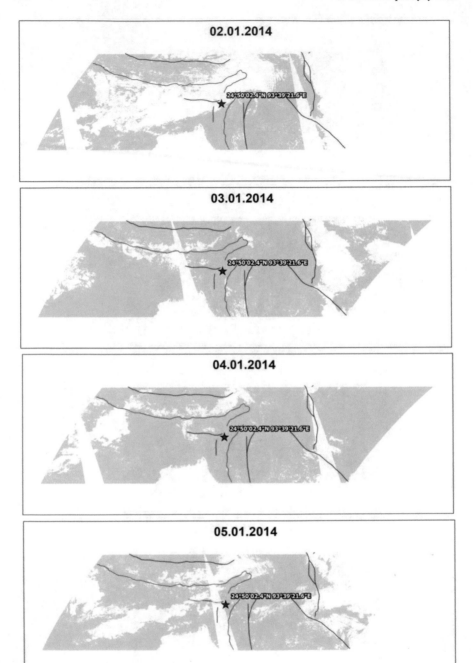

Fig. 13 (continued)

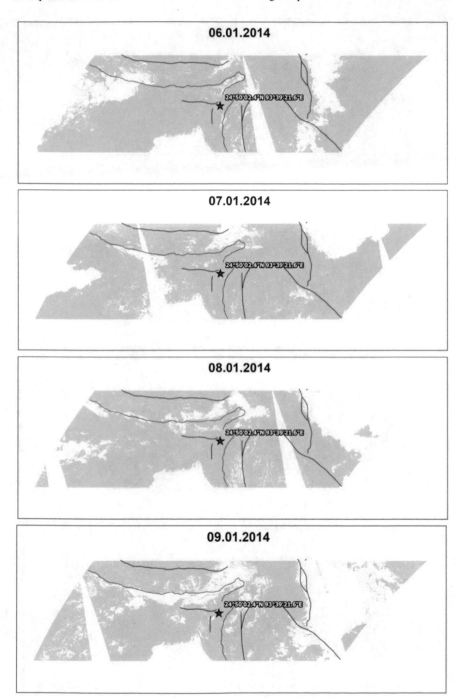

Fig. 13 (continued)

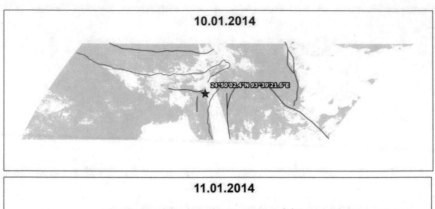

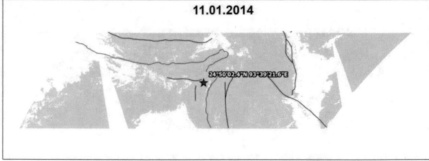

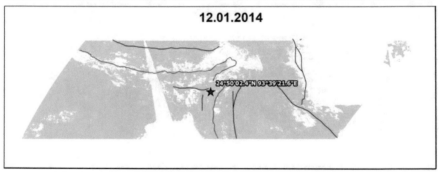

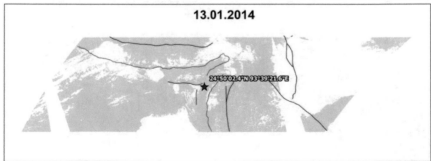

Fig. 13 (continued)

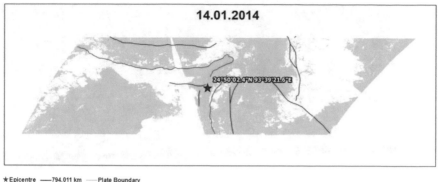

★Epicentre ——794.011 km ——Plate Boundary

▬▬ RETIRA Index <3 ▭ RETIRA Index <=3

▬▬ RETIRA Index >3.5 ▬▬ RETIRA Index >4

Fig. 13 (continued)

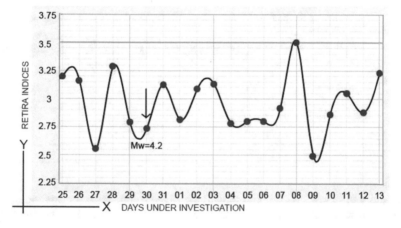

Fig. 14 Graphical analysis of the non-earthquake year (2013–2014)

Forecasting moderate and strong earthquakes in the future, prior to several months and days at the regional and global scale is highly demanded to mitigate disaster. Up to this point, no single quantifiable geophysical variable and information investigation strategy represents impressive potential for an adequately dependable earthquake forecast. Passive microwave remote sensing can be good choice because of its ability to penetrate cloud masking observation of the surface in the cloud cover. Passive microwave remote sensing suffers from one main disadvantage, i.e. its spatial resolution (1–5 kms) unlike thermal infrared imagery (nearly 1 km). The combination of TIR and microwave information proves to be essential viewpoint for understanding Land Surface Temperature changes before earthquakes. Such data combination can exploit the high accuracy and spatial resolution of satellite TIR data and the infiltrating cloud capacity of microwave data. Combined analysis of multiple precursors

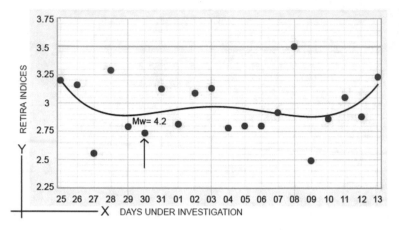

Fig. 15 A fitting curve of the non-earthquake year (2013–2014)

Table 2 Table depicting the number of anomalous pixels recorded on days before and after the earthquake event of 2016

Days under investigation with abnormal RETIRA index values	RETIRA index highest value	No. of pixels with RETIRA index > 3.5
26.12.2015	3.69	3
02.01.2016	3.97	115
04.01.2016 (Mw = 6.8)	2.97	NIL
11.01.2016	4.08	16
12.01.2016	3.92	51

is a feasible option because it can reduce the detection of false thermal anomalies. Because utilizing only a single approach to identify the pre-earthquake anomaly is not statistically robust [21].

In the meantime, ground estimations, for example, air temperature data, geochemical observation, groundwater level changes along faults can provide a superior understanding and supplement the missing data attributed due to cloud cover. New machine learning algorithms can optimize this multi-method scheme. Big data analysis based on long-term series of remote sensing data and global earthquake cases is an important means to establish the statistical relationship between thermal anomalics and earthquake events [46].

Acknowledgements The authors are deeply thankful to the United States Geological Survey for making available the Shake Map used in this study. They also express their thanks for the satellite data and data products made available as freeware on which this research is based.

References

1. Jeganathan, C., Gnanasekaran, G., Sengupta, T.: Analysing the spatio-temporal link between earthquake occurences and orbital perturbations induced by planetary configuration. Int. J. Adv. Remote Sens. GIS Geogr. **3**(2), 123–146 (2015)
2. Pavlidou, E., van der Mejide, M., van der Werff, H. M. A., Ettma, J. Time series analysis of remotely sensed TIR emissions: linking anomalies to physical processes. Presented at AGU Fall Meeting 2013, San Francisco, USA, 9–13 Dec 2013, pp. 1–2
3. Eleftheriou, A., Fizzolla, C., Genzano, N., Lacava, T., Lisi, M., Paciello, R., Pergolla, N., Vallianatos, F., Tramutoli, V.: Long term RST analysis of anomalous TIR sequences in relation with earthquakes occurred in Greece in the period 2004–2013. Pure Appl. Geophys. **173**, 285–303 (2016)
4. Anderson, M.C., Norman, J.M., Kustas, W.P., Houborg, R., Starks, P.J., Agam, N.: A thermal based remote sensing technique for routine mapping of land surface carbon, water and energy fluxes from field to regional scales. Remote Sens. Environ. **112**, 4227–4241 (2008)
5. Zhang, H., Zhang, F., Ye, M., Che, T., Zhang, G.: Estimating daily air temperatures over the Tibetan Plateau by dynamically integrating MODIS LST data. J. Geophys. Res. Atmos. (2016)
6. Zhang, X., Zhang, Y.: Tracking of thermal infrared anomaly before one strong earthquake—in the case of MS 6.2 earthquake in Zadoi, Qinghai on October 17th, 2016. Int. J. Rock Mech. Min. 879–908
7. Townshend, J.R.G., Justice, C.O., Sloke, D., Malingreau, J.P., Cihlar, J., Teillet, P., Sadwoski, F., Ruttenberg, S.: The 1 km resolution global dataset: needs of the International Geosphere Biosphere Programme. Int. J. Remote Sens. 3417–3441 (1994)
8. Vauclin, M., Vieria, R., Bernard, R., Hatfield, J.L.: Spatial variability of surface temperature along two transects of a bare. Water Resour. Res. 1677–1686 (1982)
9. Prata, A.J., Caselles, V., Coll, C., Sobrino, J.A., Ottle, C.: Thermal remote sensing of land surface temperature from satellites: current status and future prospects. Remote Sens. Rev. 175–224 (1995)
10. Tronin, A.A.: Satellite thermal survey—a new tool for study of seismoactive regions. Int. J. Remote Sens. 1439–1455 (1996)
11. Tramutoli, V., Aliano, C., Corrado, R., Filizzola C., Genzano, N., Lisi, M.: Assessing the potential of thermal infrared satellite surveys for monitoring seismically active areas: the case of Kocaeli earthquake, August 17, 1999. Remote Sens. Environ. 409–426 (2005)
12. Tramutoli, V., Di Bello, G., Pergola, N., Piscitelli, S. Robust satellite techniques for remote sensing of seismically active areas. Ann. Geophys. (2001)
13. Lisi, M., Filizzola, C., Genzano, N., Paicello, R., Pergola, N., Tramutoli, V.: Reducing atmospheric noise in RST analysis of TIR satellite radiances for earthquake prone areas satellite monitoring. Phys. Chem. Earth **85–86**, 87–97 (2015)
14. Genzano, N., Aliano, C., Filizzola, C., Pergola, N., Tramutoli, V.: A Robust Satellite Technique for monitoring seismically active areas: the case of Bhuj-Gujarat Earthquake. Tectonophysics **431**, 197–210 (2007)
15. Filizzola, C., Pergola, N., Pietrapertosa, C., Tramutoli, V.: Robust satellite techniques for seismically active areas monitoring: a sensitivity analysis on September 7, 199 Athen's Earthquake. Phys. Chem. Earth **29**, 517–527 (2004)
16. Aliano C., Corrado R., Filzzola C., Pergola N., Tramtoli V.: Robust satellite technique for thermal monitoring of earthquake prone areas: the case of Umbria-Marche, 1997 seismic events. **11**(2), 362–373
17. Ozounov, D., Freund, F.: Mid infrared emission prior to strong earthquakes analyzed by remote sensing data. Adv. Space Res. **33**, 268–273 (2004)
18. Saraf, A.K., Rawat, V., Banerjee, P., Choudhury, S., Panda, S.K., Dasgupta, S., Das, J.D.: Satellite detection, of earthquale thermal infrared precursors in Iran. Nat. Hazards **47**, 119–135 (2008)

19. Panda, S.K., Choudhury, S., Saraf, A.K., Das, J.D.: MODIS land surface temperature data detects thermal anomaly preceeding 8th October 2005 Kashmir Earthquake. Int. J. Remote Sens. **28**, 4587–4596 (2007)

20. Rawat, V.: Applications of thermal remote sensing in earthquake precursor studies. **47**, 119–135 (2008)

21. Bhardwaj, A., Singh, S., Sam, L., Bhardwaj, A., Martin-Torres, F.J., Singh, A., Kumar, R. MODIS based estimates of strong snow surface temperature anomaly related to high altitude earthquakes of 2015. Remote Sens. Environ. **188**, 1–8 (2017)

22. Freund, F.T., Kulachi, I., Cyr, G., Ling, J., Winnick, M., Tregloan, R.: Air ionization at rock surfaces and pre-earthquake signals. J. Atmos. Sol. Terr. Phys. **71**, 1824–1834 (2009)

23. Freund, F., Keefiner, J., Mellon, J.J., Post, R., Teakeuchi, A., Lau, B.W.S., La, A., Ouzounov, D.: Enhanced mid-infrared emission from igneous rocks under stress. Geophys. Res. Abstr. **7**, 09568

24. Freund, F.T.: Pre-earthquake signals: underlying physical processes. J. Asian Earth Sci. **41**, 383–400 (2011)

25. Saraf, A.K., Choudhury, S.: NOAA-AVHRR detects thermal anomaly associated with 26th January, 2001 Bhuj Earthquake, Gujarat, India. Int. J. Remote Sens. **26**, 1065–1073 (2005)

26. Saraf, A.K., Rawat, V., Choudhury S., Dasgupta, S., Das, J.: Advances in understanding of the mechanism for generation of earthquake thermal precursors detected by satellites. Int. J. Appl. Earth Obs. Geoinform. 373–379 (2009)

27. Tronin, A.A., Hayakawa, M., Molchanov, O.A.: Thermal IR satellite data application for earthquake research in Japan and China. J. Geodyn. **519**, 534 (2002)

28. Harris, A.J., Swabey S.E.J., Higgins, J.: Automated threshold 463 of active lava using AVHRR data. Int. J. Remote Sens. **16**, 3681–3686

29. Blackett, M., Wooster, M.J., Malamud, B.D.: Exploring land surface temperature earthquake precursors: a focus on the Gujarat (India) earthquake of 2001. Geophys. Res. Lett. **38**, L15303

30. Okyay, U.: Evaluation of thermal remote sensing for detection of thermal anomalies as earthquake precursors: a case study for Malataya Puturge Doganyol (Turkey) earthquake July 13, 2003. M.S. Thesis, Universitat Jaume-I, Spain

31. Pergola, N., Aliano, C., Coveillo, I., Filzzola, C., Genzano, N., Lavaca, T., Lisi, M., Mazzeo G., Tramutoli, T.: Using RST approach and EOS-MODIS radiances for monitoring seismically active regions: a study on the 6th April 2009 Abruzzo earthquake. Hazards Earth Syst. Sci. **10**, 239–249

32. Marchese. F., Pergola, N., Telesca, L.: Investigating the temporal fluctuations in satellite AVHRR thermal signals measured in the volcanic area of Etna (Italy). Fluct. Noise Lett. 305–316 (2006)

33. Gorny, V.I., Salman, A.G., Tronin, A.A., Shilin, B.B.: The Earth outgoing IR radiation as an indicator of seismic activity. J. Geophys. Res. Solid Earth **83**, 3111–3121

34. Tramutoli, V., Aliano, C., Corrado, R., Filizzola, C., Genzano, N., Lisi, M., Martinelli, G., Pergola, N.: On the possible origin of thermal radiation infrared radiation (TIR) anomalies in Earthquake prone areas observed using Robust Satellite Techniques (RST). Chem. Geol. 157–168

35. Pulinets, S., Ozounov, D.: Lithosphere–Atmosphere–Ionosphere Coupling (LAIC) model—an unified concept for earthquake precursors validation. J. Asian Earth Sci.**41**, 371–382 (2011)

36. Pulinets, S.A., Ouzounov, D., Karelin, A.V., Boyarchuk, K.A., Pokhmelnykh, L.A.: The physical nature of thermal anomalies observed before strong earthquakes. Remote Sens. Rev. **12**, 175–224

37. Ouzounov, D., Bryant, N., Logan, T., Pulinets, S., Taylor, P.: Satellite thermal IR phenomena associated with some of the major earthquakes in 1999–2003. Phys. Chem. Earth **31**, 153–163 (2006)

38. Irwin, W.P., Barnes, I.: Tectonic relations of carbon dioxide discharges and earthquakes. J. Geophys. Res. Solid Earth **85**, 3113–3121

39. Wu., L.X., Cui, C.Y., Geng, N.G., Wang, J.Z.: Remote Sensing rock mechanics (RSRM) and associated experimental studies. Int. J. Rock Mech. Min. 879–888 (2000)

40. Wan, Z., Dozier, J.: A generalized split window algorithm for retrieving land surface temperature from space. IEEE Trans. Geosci. Remote Sens. **34**(4), 892–905
41. Wan, Z.: New refinement and validation of the Collection 6 MODIS land surface temperature/emissivity products. Remote Sens. Environ. 36–45 (2014)
42. Grimaldi, C.S.L., Coviello, I., Lacava, T., Pergola N., Tramutoli, V.: A new RST based approach for continuous oil spill detection in TIR range: the case of the Deepwater Horizon platform in the Gulf of Mexico. In: Liu, Y., MacFadyen A., Ji., Z.G., Weisberg, R.H. (eds.) Monitoring and Modeling the Deepwater Horizon Oil Spill: A Record-Breaking Enterprise, Geophysical Monograpph Series, pp. 19–31. American Geophysical Union (AGU), Washington, DC, USA (2011)
43. Casciello, D., Lacava, T., Pergola, N., Tramutoli, V.: Robust Satellite Technique (RST) for oil spill detection and monitoring. International Workshop on the Analysis of Multi-temporal Remote Sensing Images, 2007, MultiTemp 2007, 18–20 July 2007
44. Lacava, T., Ciancia, E., Coviello, I., Polito, C., Tramutoli, V.: A MODIS based Robust Satellite Technique (RST) for timely detection of oil spilled areas. Phys. Chem. Earth 45–67 (2015)
45. Geller, R.S.: Predicting Earthquakes is impossible. Los Angeles Times, Feb 1997
46. Jiao, Z.H., Zhao, J., Shan, X.: Pre-seismic anomalies from optical satellite observations: a review. Nat. Hazards Earth Syst. Sci. **18**, 1013–1036
47. Venkatanathan, N., Hareesh, V., Venkatesh, W.S.: Observation of earthquake precursors—a study on OLR scenario prior to the Earthquakes of Indian and neighboring regions occurred in 2016. ISSN: 0974–5904, vol. 9, No. 3, pp. 264–268, June 2016

Literature Review with Study and Analysis of the Quality Challenges of Recommendation Techniques and Their Application in Movie Ratings

Hagar El Fiky, **Wedad Hussein**, and **Rania El Gohary**

Abstract During the past few decades, the web-based services and organizations like Amazon, Netflix and YouTube have been raised aggressively. These web services have shown the demand for the recommender systems and their growing place in our lives. More steps deeper, we noticed that the severity of the quality and accuracy of these recommendation systems is very high to match users with same interests. For that reason and for being in competitive position with the most outstanding recommendation web services, these recommendation systems should be always monitored and evaluated from a quality perspective. However, due to the steep growth rate of the available web-based services, new challenges like data sparsity, scalability problem and cold start issue have been burst and threaten the performance and the quality of the predicted recommendations. Accordingly, many data scientists and researchers got excited to figure out ways for these challenges especially if they are in scaled environments and distributed systems. These solutions could be achieved using multiple approaches such as machine learning and data mining.

Keywords System quality · System accuracy · Collaborative filtering · Content based filtering · Rating prediction · Large scale recommendation systems · RMSE · IMDB · Netflix

H. E. Fiky (✉) · W. Hussein · R. E. Gohary
Information Systems, Faculty of Computer and Information Sciences, Ain Shams University, Cairo, Egypt
e-mail: hagar_elfiky@hotmail.com

W. Hussein
e-mail: wedad.hussein@fcis.asu.edu.eg

R. E. Gohary
e-mail: rania.elgohary@cis.asu.edu.eg

© The Author(s), under exclusive license to Springer Nature Switzerland AG 2021   219
A. E. Hassanien et al. (eds.), *Machine Learning and Big Data Analytics Paradigms: Analysis, Applications and Challenges*, Studies in Big Data 77,
https://doi.org/10.1007/978-3-030-59338-4_12

1 Introduction

The intensive expansion of the available and accessible digital data and the number of active users to the internet has a direct impact on the performance and efficiency of predicting the users' interests and preferences in an accurate manner. Many of internet-based applications, require users' interactions to express their satisfaction, opinion, and ratings towards a certain item or product of interest through recommendation systems.

A recommendation system is a system that aims to predict interests based on users' preferences. However, these systems tend to be more accurate only with an enormous amount of data fed into them. The quality of these recommendation systems varies from one domain to another depending on the industry such as entertainment, medical, academic, research, news, tourism, products, and data collection from users within any of these domains. The major differentiating factor between all these systems is the quality of their recommendations, which directly impact the users.

This chapter is focused on illustrating the major techniques of recommender systems followed by the various challenges of recommendation systems. On the other hand, we discussed how the recommendation system got evaluated through common quality measurers. The purpose of this chapter is to represent the latest and most recent strategies of evaluating larger-scale internet-based recommendation systems in order to recommend accurately the right item for the right user especially if its newly registered user. Besides, the exploration of data sparsity issues that are caused by such systems. Movie recommendation systems like such as IMDb, Netflix, Rotten Tomatoes were all built and used for recommendation purposes; however, they are not all of the same reliability, quality, and validity. Readers will be exposed to distinct factors and strategies in evaluating the quality as well as the challenges that affect the performance of such recommendation systems.

The rest of the chapter is organized as follows: Sect. 2 presents an overview of recommendation systems. In Sect. 3 a comparative analysis between the advantages and disadvantages of recommendation systems is introduced. Section 4, we discuss the prediction from the point of view of large-scale systems and their main issues. Section 5 sheds the light on the quality issues and challenges of recommendation systems. In Sect. 6, quality evaluation measures are discussed. Section 7, we have provided a brief explanation of the literature review and some related work on recommendation systems and their applications. Finally, we summarized how IMDb and Netflix distributed movie rating datasets have contributed and helped in solving and improving the quality of the movie rating prediction systems in Sect. 8.

2   Overview of Recommendation Systems

Recommendation systems are systems or applications that attempt to suggest items to significant potential customers, based on the available data and information or based on predefined interests and preferences that these systems asked their customers about [1, 2]. Recommendation systems are not field specific since they can be applied in any area like what articles to read [3], brands to shop, music to listen to or even movies to watch.

These systems got mature by time and by enlarging the training set they continuously deal with. The more mature these systems got the more significant improvement they do in the revenue of companies' electronic commerce and in the facilitation, they offer for the users over the internet. Recommendation and prediction algorithms have been associated with each other due to their common final goal. These recommendation systems have been ultimately utilized in many applications such as Netflix, Amazon, Google news and YouTube.

In the past few years, it has been discovered the plinth relationship between the movie rating prediction and the recommendations later on. These two factors are inter-dependable on each other producing a connected ring. This connected ring operates as follows, the predictive models feed the recommendation systems with certain features of the user's interest as an input and in return, the recommendation systems reply back with items of the same interests and preferences of that user. Therefore, the more the rating prediction was based on the accurate and precise algorithm, the more efficient will be the generated recommendation system in the future.

In [4], the authors showed the following flowchart in Fig. 1 that illustrates the variety of techniques that could be implemented in any recommendation system. A regular recommender system is categorized into one of three distinct techniques; Content-based filtering, Collaborative filtering, and Hybrid filtering.

Collaborative filtering (CF), constructs a model from the user's history such as items viewed or purchased by the user, in addition to equivalent decisions taken by other users. Accordingly, the constructed model is utilized to predict items similar to the users' interests as shown in Fig. 2.

CF can be further subdivided into two filtering techniques which are; Model-based, and memory-based filtering techniques. Model-based filtering technique is based on data mining and machine learning techniques such as clustering, association rules, etc. While the memory-based filtering technique is used to rate items using the correspondence between the users and items.

Content-based filtering (CB), uses a sequence of distinguished features of an item so that further items could be recommended with the same characteristics. Content-based filtering technique depends mainly on the item's characteristics, and the user's persona.

Hybrid filtering technique is a mixture of two or more recommendation approaches in order to dominate the drawbacks of using any approach individually and to improve the overall prediction quality [5–7] and computational performance [8]. An example

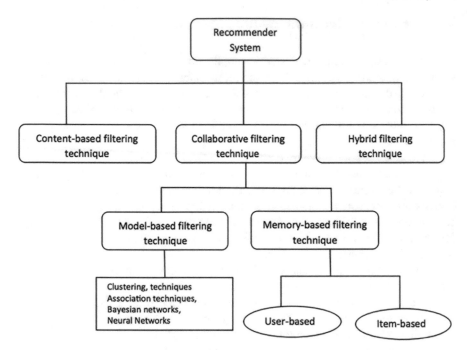

Fig. 1 Diagram of recommendation systems techniques (https://www.sciencedirect.com/science/article/pii/S1110866515000341)

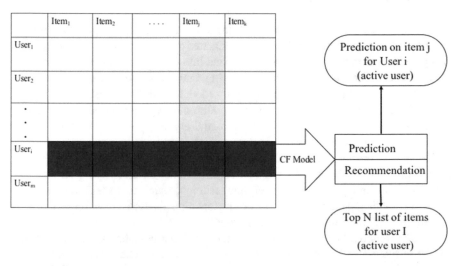

Fig. 2 Collaborative filtering user-item rating matrix (https://www.sciencedirect.com/science/article/pii/S1110866515000341)

of hybrid filtering is combing memory-based and model based techniques. Although this category showed potentials in improving the prediction quality, space and time complexity were highly exceeded in addition to the high cost of implementation using constraint resources.

Social filtering (SF) or community-based recommendation systems is based on the interests of the user's circle of friends. These recommendation systems are built upon the saying of "Tell me who your friends are and I will tell you who you are" [9]. These systems collect social information from social networks such as Facebook and Instagram to generate recommendations based on the votes of the user's friends.

Knowledge-based systems (KB), are usually depending on the certain domains in which they get their knowledge from, to recommend items to users. Knowledge-based systems are also known as case-based systems, in which a similarity equation computes the degree of matching the user's interests with the final recommendations [10].

3 Advantages and Dis-advantages of Recommendation Techniques

After reviewing multiple scientific studies and researches, it was discovered that the disadvantages of one algorithm can be overcome by another algorithm [11]. Consequently, we have determined and analyzed the advantages and disadvantages of recommendation techniques and their impact on these systems' overall quality. Among various types of recommendation techniques, CF has showed a great effectiveness and accuracy even in large range problems. While on the other hand other techniques like CB cannot afford data sparsity, scalability and accuracy [12]. Due to the nature of hybrid filtering techniques, they can overcome each other's weakness in order to gain better efficiency [5–7]. As per the adaptive and learning- based nature of CF and CB they can recommend based on the implicit data with no domain knowledge requirement. Accordingly, their quality of recommendations got improved over time due to dealing with more implicit data. Unlikely the KB techniques made their recommendations depending on their knowledge base so any preference change in it may reflect in the recommendations as they map the user interests to the products features as well as considering the product's value factors like delivery schedule and warranty conditions not only the product's specific features [13].

Tables 1 and 2 are produced to conclude these strengths and weaknesses respectively of various recommendation techniques from different quality perspectives.

3.1 Advantages

See Table 1.

Table 1 Advantages of various recommender systems

Quality perspective	Recommender system				
	CF	CB	Hybrid	SF	KB
Accuracy and efficiency	X		X		
Ignoring domain knowledge	X	X			
Quality improvement over time	X	X			
Preference change sensibility					X
Considering non-product properties					X
Matching users' needs to the recommended item					X
Recommendations sensitivity				X	
Scalability adaptation	X			X	
Overcome latency		X			

3.2 Dis-advantages

See Table 2.

4 Large Scale Recommender Systems

The main challenge for large scale recommendation systems is to enhance the recommendation process for large volume and diversity of data, through the utilization of all the available information to analyze social interaction, ratings and content similarity graphs (similarity between user's profile and item's properties). In addition to

Table 2 Dis-advantages of various recommender systems

Quality perspective	Recommender system				
	CF	CB	Hybrid	SF	KB
Latency	X				
Privacy				X	
Plenty of past data affects the quality	X				
Requires domain knowledge					X
Low performance when limited resource	X				
Recently added user	X	X			
Recently added item	X				

the flexibility of these algorithms towards the dynamically improving graphs and the ability of these algorithms to scale to large graphs [14]. As in [15] the CF- based recommender systems got scaled to large scale ratings graphs using the Lenskit Recommender library and Apache Spark. Approaching the market and industry field, authors in [16] have introduced a solution based on the Hadoop framework to construct a large-scale and distributed advertising recommendation system in which the seller pays for the advertising when it only got clicked by users. These experiments and studies have shown the intervention of recommendations and predictions into large scale systems and how they tried to overcome their main issues. The major large-scale shortcomings are abated in the following challenges.

4 1 Data Diversity

Since the data diversity is one of the major factors in the large-scale recommender system due to the variety and multiplicity of available information. Keeping the user's preferences and personal info from the social networks has helped a lot in solving the poor performance of cold start recommendation of new users [17]. As they keep this information to fill many gaps in user profiles through extracting the hidden user preferences.

However, handling all information in social networks made the latest CF techniques not able to absorb the data volume, variety and complexity of new information. Thus, amendments and extensions of traditional techniques have been introduced like matrix factorization, social based matrix factorization [18, 19] to combine the hidden preferences within the current techniques.

Based on this, social networks have played a double agent role in the domain of data diversity. This means as social networks have helped in filling many gaps in the user's profile, however, it also has introduced the issue of large-scale data.

4.2 Data Size

The data size issue started to appear when the total number of users and items in the system increase exponentially exceeding the performance capabilities of the traditional collaborative filtering methods. At this point, the computational processors became highly demanding to be practically utilized [20]. However, this problem can be solved with dimensionality reduction techniques, data partitioning and parallel, and distributed algorithms.

The scalability of recommendation systems depend on the availability of the resources, system architecture that allows distributed or parallel data management and processing and lastly adaptable, scalable algorithms. Previously, distributed or parallel algorithms were introduced and designed to achieve better recommendation results. However, recent algorithms can achieve comparable results without consuming all of the available data. Users are the first-dimension input of recommendation systems over the web followed by items' properties. Item property contains a lot of high-level semantic meanings that requires a huge amount of computing resources. The most common techniques for this issue are to involve the extraction of high-level metadata to be later on compared with the user's preferences [21].

Incremental Collaborative Filtering Algorithm

In the current recommendation systems $i \times j$ ratings matrix is given as an input where i is the list of users while j is the list of items. In return, the system has to predict which j's are assigned to which i's. This raises multiple problems such as scattering of rating matrixes and noise production resulting in poor recommendation quality. Consequently, latent factor analysis methods (feature analysis methods) such as s Principal Component Analysis (PCA), Latent Dirichlet Analysis (LDA) and Singular Value Decomposition (SVD) are used to compute the correlation on ratings matrixes in order to get more accurate and scalable results [22]. For example, in Netflix prize, they used the latent factor technique to achieve higher accuracy [23]. Another example of the incremental approach was in recommending videos in real time. Using both different users' actions and considering extra item factors' like video type and period, has led the authors in [24] to reach more efficient and accurate recommendations. As the incremental approach is handling the efficiency and accuracy, it also succeeded to solve scalability issues. In [25], the author proposed a buffer methodology to store the new data incrementally. So that they update their algorithm when only the buffer has appropriate rating amount not when each new rating is added.

Deep Learning Solutions

Deep learning solutions depends on merging different techniques for achieving better recommendations accuracy [26]. Smirnova and Vasile in [27] have highlighted the drawbacks of ignoring the contextual information in users' profiles like the date, time of rating transaction and user's offline time duration using Recurrent Neural Networks (RNN). Authors introduced a Contextual Recurrent Neural Networks (CRNNs), that uses the contextual information of the network in the input and output of the CRNN to form the item prediction. Finally, in [28] the authors have succeeded to enhance the classic RNN model throughout feeding the model with valuable distributed representations gathered from user ratings, user review and item properties.

Based on the findings of the above models, deep learning such as neural networks has succeeded to prove that it could improve the quality of recommendation systems. Although a lot of research papers in deep learning have tackled how to achieve better recommendation results, all these algorithms were focusing mainly on reaching more precise and relevant user-preferences recommendations rather than attacking the large-scale problems. In this light of this, this area needs to be the upcoming challenge for researchers to pay attention for and solve due to its significance.

5 Quality Challenges of Large-Scale Recommendation Systems

In this section, we have focused on the challenges of recommendation systems that affect directly the quality of the recommendation systems. The challenges are many and different however; they are generally domain independent. Thus, we took the following challenges from the quality perspective and discussed how they affect the system's accuracy and performance.

5.1 Cold Start Issue

This problem arises when a new item or a new user added in the database [29]. Since any newly added item has no ratings so it could not be recommended using a CF system. For example, MovieLens (movielens.com) cannot recommend new movies until these movies have got initial ratings. In the CB, a new user problem becomes more difficult as it is harsh to find similar users or to establish a user profile without knowing the prior user's interests [13]. Authors in [30] have succeeded in figuring out an algorithm that can recommend venues or events to users taking into consideration the new user's interest and the new user's location preference. This algorithm has shown great effectiveness and efficiency in new spatial items recommendation.

5.2 Sparse

Sparsity is one of the issues that affects the quality of recommendation systems' results due to the inconsistency that users do with rating the items. Not all users rate most of the items. Consequently, the accuracy of the resultant rating matrix is not of that good quality and it became difficult to find users with a similar rating [29]. This sparsity issue is one of the outstanding demerits of the CF approach as it leads to a hesitation in results accuracy. In [31], authors have proposed a fusing algorithm that merges both the user and item information to generate recommendations for target users. The integrated user, item and user-item information to predict the ratings of the items until they reached a solution that showed an efficiency against sparse data and produces higher recommendation quality.

5.3 Large Scale and Scalability

With the tremendous increase of data, the systems become the large scale and having millions of items. Thus, the presence of these large scales and distributed systems, recommendation techniques have been suffering from processing these amounts of data. For example, IMDB.Com recommends more than 3 billion products to more than 300 million users in 2018. In other words, the more the training data increased the poorer the quality of recommendations are predicted. Various approaches have been proposed to enhance the quality of the predicted items like clustering, reducing dimensionality, and Bayesian Network [32]. Authors in [20] have been approached the problem using CF algorithm that search in few clusters instead of searching in the whole database, by reducing dimensionality through SVD [33] and by pre-processing approaches that merges clustering and content-analysis with CF algorithms [34].

5.4 Time Threshold

Is one of the major challenges that face any recommendation system as they should ignore the old and ancient items. Ignoring this threshold threatens the quality of the suggested preferences later on. Thus, consistently new items need to be considered so as to prosper the quality of the items suggested to the users.

5.5 Latency Problem

Collaborative filtering-based systems struggle from latency problems whenever new items added frequently to the database [35]. Latency problems mean taking recommendations from already existing rated items instead of the newly added items as they are not rated yet. Which makes the results biased to the existing items only although the newly added items might be matching the user's interests. Being biased to specific items, would eventually cause eminent degradation in the efficiency and quality of the recommended results. And this quality enhancement is considered one of the known challenges for the CF. Unlike the CF, CB is more efficient towards the latency on account of overspecialization [36].

5.6 Privacy Problem

Earning a personalized service is one of the major rewards of recent technologies, especially recommendation systems. However, these systems require the collection of personal data and interests of users, which could lead to catastrophic disasters if used unethically.

The major privacy challenge that affects a recommendation system especially if it is collaborative filtering, is storing private data of users within distributed servers. Real applications such as Netflix is particularly famous with capturing highly personalized data such as location, mail addresses, and debit cards. This causes the data to be at great risk of misuse leading to poor quality of recommendation systems, thus the associated local geographical regions are greatly impacted.

According to previous incidents such as selling personal data to third parties, users became very suspicious of their commonly used systems. Therefore, recommendation systems should seek the trust and privacy of their user's data. Lastly, cryptographic algorithms were introduced to enable personalized services being delivered without the impact of other factors such as peer users [35]. Moreover, other systems [37], allow users to distribute their personal data anonymously without revealing their true identity.

6 Quality Evaluation Measures

Quality is defined as a set of rules or measurements by which the degree of excellence of certain items is measured. Recommendation algorithms or systems require critical quality evaluation to ensure the standards of the service being provided are fulfilled. Different types of metrics are adopted to evaluate the quality of such systems such as statistical, qualitative, ranking and user satisfaction metrics. The type of metric is highly dependable on the type of filtering technique used within the recommendation system and the features of the dataset that is fed into these types of systems [38].

However, deciding the correct measuring metric and utilizing it with the correct recommendation system is a very challenging process that could lead to a successful or failure status. Moreover, the objective of quality examination may differ depending on the scenario presented in the system [39]. Therefore, such evaluations are highly challenging and complex. To conclude, the classification of these metrics could be very time consuming and tiring process. The main categories of these metrics are presented well below [40, 41] as the following:

6.1 Statistical Metrics

Statistical or probabilistic metrics are extremely efficient and effective when they are utilized in assessing the reliability and validity of predictions through comparing the predicted ratings with the user's actual ratings. Problems require these methods are often considered as regression problems. The key examples of these metrics include:

- Mean Absolute Error (MAE)
- Root Mean Squared Error (RMSE)
- Logarithmic Loss (log loss)
- Cross-Entropy.

The most well-known metrics of them are Mean Absolute Error (MAE) and Root Mean Squared Error (RMSE) as represented in Eqs. (1) and (2) respectively.

$$MAE = \frac{1}{N} \sum_{u,i} |P_{u,i} - r_{u,i}| \tag{1}$$

$$RMSE = \sqrt{\frac{1}{N} \sum_{u,i} (P_{u,i} - r_{u,i})^2} \tag{2}$$

where P(u, i) is the predicted rating for user u on item i, $r_{(u,i)}$ is the actual rating and N is the total number of ratings on item set. This metric evaluates the error in predicting the predictive numeric score.

6.2 Qualitative Metrics

Qualitative metrics are popular in the recommendation systems domain and highly beneficial when the main objective is to have a model that cuts down the number of errors. Accordingly improves the quality of the recommended outputs.

In [13], these measures are heavily used in various recommendation applications. The key instances include:

- Accuracy
- Recall
- F-measure
- Kappa statistic
- Coverage.

Since items here are classified as relevant and irrelevant to the user, these metrics are the best fit for a classification problem.

6.3 Ranking Metrics

These metrics, are widely used in Recommendation Systems, are built based on the concept of how well the recommender ranks the recommended items. Accordingly improves the accuracy of the ranked output. The main examples of these measures include:

- Precision
- Recall
- Normalized Discounted
- Mean Average Precision (MAP)
- Hit Rate (HR)
- Fallout
- Area under the ROC Curve (AUC).

These metrics are concerned with evaluating the quality of a ranked list of items instead of the average quality of the raw scores produced by the recommender system. Consequently, problems being solved with these methods are categorized as a ranking problem [13].

6.4 User Satisfaction Metrics

The last category of metrics that will be discussed in this chapter is the user satisfaction metrics that involve demonstrable experiments that have taken place with users to assess their satisfaction level in recommender systems. This measure is very useful since it collects enormous, real feedbacks from users personally. However, this metric may contain biased feedbacks and lack of an objective measure for the quality assessment of recommendation systems [13].

7 Related Work

7.1 Background

Based on the facts that the most significant part of generating valid recommendations is the prediction, as any recommendation system needs to take into consideration the user's preferences [42]. The generations of recommendation approaches [43] have been varying for that reason. However; the three major approaches are Collaborative filtering (CF), content-based filtering (CBF), and knowledge-based recommendation.

As there are multiple ways for recommendations either by depending on similarity between users or similarity between items [44], these ways are recognized as Collaborative Filtering that depends on the past behavior of users having similar interests as the current user. As a user and item-based collaborative filtering has been very successful in both market and academia. As CF proved its successful intervention in the large-scale systems it also showed that after it has been extended it played an important role in the distributed systems.

In [45] CF has been extended to interact with the opportunistic network to produce a distributed recommender system based on collecting user activities then generating users' profiles for distributed users connected through a network that does not have a dedicated internet connection. Afterwards, the CF technique used to calculate the recommendations. Diversity of data in this solution has many advantages over centralized server like reducing the cost of computing data of millions of users on the central server and cost of high internet connectivity of central server—as the connection cost has been distributed among users—providing higher security for users personal information as it will not be saved on the central server and applicable in areas with limited internet connection. CF in [46, 47] mainly was based on the nearest neighbors (NN), where users with similar preferences or similar purchasing behavior [3]. That's why it relies on two different types of data: first the set of users and second the set of items. In another meaning; users rate items and receive a recommendation for different items based on other users' (that have the same preferences and similar rating criteria) ratings.

The exponential increase of web services has led the recommender systems to become the backbone of the internet-based systems such as Netflix, YouTube and Google. In general CF methods play a critical role in reaching out to newly registered users, and successfully predict and recommend their services or products to the users accurately. The newly proposed method in this paper was CF recommendation based on dimensionality reduction through SVD and clustering through k-means. The aim of the introduced CF method is to improve the performance of RS and overcome their problems of data sparsity, cold start and scalability issues. The target was met using singular value decomposition (SVD) for reducing the dimensionality of characteristics [48] because the SVD is well known for its capability to enhance the scalability of recommender systems [23, 49].

This occurs by reducing the number of features of the data group through matrix factorization methods and k-means to cluster similar users, through similar preferences in terms of ratings. Therefore, the user clustering process helps in improving the recommendation performance because the considered cluster contains many fewer users in comparison with the general population that consists of all users. Experimental results of this method proved a significant improvement in the accuracy and efficiency of the recommendation system.

The performance of the proposed method was compared with k-means and k-nearest neighbor-based recommendation systems through RMSE. As RMSE provides the distinction between the true and the predicted likelihood about the user selecting an item and of course the lower the value of RMSE, the better the performance. The proposed method has an RMSE ranging from 0.7 to 0.6 while K-means had values from 0.75 to 0.65 and finally KNN has a range of values from almost 0.9 to 0.85 which proves the effectiveness of the proposed method.

Similar researches such as [50, 51], were conducted to address the most accurate and appropriate clustering technique for recommendation systems. In other researches [52, 53] were focused on problems issued by big data, in which solutions proposed to resolve this issue by merging multiple clustering techniques together and MapReduce in the association of multiple factors that affect the quality in terms of performance of recommendation systems. Lastly, the research [54], was introducing a new method to improve the quality of recommendation systems to provide personalized services based on building trust relationships on online social media.

Content-based filtering [55, 56] which is counting on the user's current preferences to recommend on him in the future. CBF uses its two different types of data that are the user's profile, and the items' descriptions or categories, to recommend on the user based on the similarity of the item's categories' with user's preferences categories'. On the other hand, we found the knowledge-based recommendation [57, 58] approach differs a bit in the way it suggests its recommendations as it counts on rules. It has introduced the combination of the user's current preferences along with the user's constraints [59] that are implicitly mentioned in his given requirements. An example of such a constraint is topics = Java. That indicates the fact that this user is interested mainly in any java related topics [60, 61]. Finally, the authors presented a hybrid recommendation.

Data mining has shown a great intervention in the prediction field and its quality, especially in movie rating prediction. Data mining applications have solved a lot of accuracy and performance problems related to movie rating prediction. In [62], the authors have proposed a new prediction movie rating application that is mainly based on two features. These two features that are generated based on their data study are: first, the Genre that could be any of the following: action, adventure, animation, comedy, crime, drama, family, musical, horror, mystery or romance. Second, the Word groups feature that is considered as words used in movies like a bad word, terror words or drug words. They have extracted these two attributes from a movie infor mation dataset created by them. Their data mining application model was developed on the Weka [63] tool which creates a decision tree that stores the movie information then suggests the rating of the movies upon. Their proposed model has succeeded in reaching an accuracy of 80%.

7.2 Applications

Social Networks

Since social networks have been playing a great role in our lives nowadays. It has been entered the recommendation systems world aggressively and showed a great deviation in the recommended results. That's why the large-scale recommendation systems have been tackled from the social network perspective, in terms of diversity and volatility. This has forced the researchers to focus on more information about the user other than his/her ratings and preferences. Like the context (spatial, time, and social) and the context development with time.

In [17], the authors have discussed different context aware recommendation systems. Like Context-aware recommender systems (CARS), Time-aware recommender systems (TARS) and Community-aware or "Social" recommender systems. The user context can be fixed or dynamically changes over time according to [64]. In CARS, there is a pre-filtering step where only relevant and in context information is selected then ranked. And post-filtering step that re-ranks, and filters the result of traditional recommender [65]. CARS has solved the cold start, and scalability problems as it collects the information from various sources in the social network to tune the context and fill gaps [66].

Moving Ratings

Netflix was one of the pioneers to contribute to the movie rating prediction and recommendation systems industry. In October 2006, Netflix has succeeded to release a large movie rating dataset and challenged the data mining, machine learning as well as the computer science communities to develop systems that could beat the accuracy of its recommendation system in [67]. After discovering the critical relationship between what the subscribers really love to watch with the success or abandon of their service. They decided to engage their subscribers to express their opinions in the movies they watched in the form of ratings. Subsequently, they figured out their recommendation system "Cinematch" that analyzes the accumulated movie ratings collected from subscribers using a variant of Pearson's correlation to make personalized predictions to the subscribers based on their interests and preferences. "Cinematch" has improved the movie average rating by 10%.

In [68], the authors decided to take the Netflix challenge, in terms of minimizing the root mean squared error (RMSE) while predicting the ratings on their test dataset. Authors have managed to reach a RMSE score of 0.8985 which represents one of the outstanding single-method performance results according to their knowledge. In Addition to the reason behind this RMSE score was their proposed smooth and scalable method, which is based on parallel implementation of alternating-least-squares with weighted-λ-regularization (ALS-WR) that was developed based on CF algorithm. After applying their proposed solution on Netflix challenge, they succeeded to achieve performance improvement of 5.91% over Netflix's recommendation system. However, they could not win the prize as the challenge was to improve it by 10%.

Not only data mining but also machine learning participated in the movie ratings prediction system. As many prediction systems were based on multiple machine learning related algorithms, like baseline predictor, KNN, Stochastic Gradient Descent, SVD, SVD++, asymmetric SVD, Global Neighborhood model and Co-clustering by BVD (Block Value Decomposition). In [69], the authors counted on the RMSE (Root-Mean-Square-Error) as the main criteria to evaluate their performance. For Each algorithm from the above, the RMSE values were graphed versus the different k values (dimensions) of each algorithm, till the co clustering algorithm generated better performance than the other methods as it reaches the smallest RMSE at certain K value and this K value highly depends on the characteristics of the dataset. While in [70], they have proposed an approach for improving the prediction accuracy based on machine learning algorithms that reached an accuracy of 89.2%.

Going the extra mile in the rating prediction domain, we will find that merging personalization with prediction could enhance the generated recommendation systems better and better. After highlighting the inaccuracy rating predictions that still occurs with the new users like what happens in the cold start issue, the authors in [71] have introduced a personalized rating prediction method based on several extensions of the basic matrix factorization algorithm which take the user attributes into their considerations when generating the predictions. The authors have found that there is a directly proportional relation between utilizing the text-based user attributes and personalized rating predictions, which are more accurate than asking users to explicitly provide them with their preferences. They have experimented with their solution on large two datasets: MovieLens100K and IMDb1M.

Deep learning and neural networks also have shown an intervention in the movie rating prediction as well. Where in [72], the author has proposed a technique based on the artificial neural networks that attains to achieve better accuracy and sensitivity compared to the recent techniques. He has managed to achieve accuracy of 97.33%

and a sensitivity of 98.63% when applying his proposed algorithm on 150 movies collected from IMDB. As per all of the above-mentioned experiments, they have shown a great enhancement in overall prediction performance evaluation in terms of accuracy and sensitivity using different approaches as they were concerned with handling the data diversity and new item or new user issues. However, none of them have to pay much attention to the large scale and scalability challenge that has been threating the distributed contemporary recommendation systems. Our research among the studies has shed light on this gap in order to be inspected later on. We have collected many of movie ratings predictions researches in Table 3.

8 Conclusion

This chapter presents a systematic review of many quality challenges and quality evaluation factors for large scale and distributed recommendation systems. We have investigated the quality of recommendation systems from different perspectives. Also, we have shed light on the gap that can be covered in the future to enhance the quality of cold start and data sparsity problems. This study presents the challenges and the achievements that have been tackled in movie ratings recommendation systems. Moreover, showing that the most used datasets in movie rating predictions are IMDb and Netflix due to their data enrich. This study has highlighted also on the relation between the movie ratings prediction and the movie recommendation systems. Since the more accurate the movie rating predictions are, the better quality can be obtained for the recommendation systems. In future work, a model will be developed to evaluate and enhance a large-scale recommendation system's quality.

Table 3 Movie ratings prediction techniques

Authors	Objectives	Technique	Dataset
Alexander Felfernig, Michael Jeran, Gerald Ninaus, Florian Reinfrank, Stefan Reiterer, Martin Stettinger in "Basic Approaches in Recommendation Systems" 2014 [43]	• Presented a comparative analysis between el CF, CBF and Knowledge based recommendation approaches • Suggested a hybrid recommendation approach	Collaborative filtering Content based filtering Knowledge based recommendation	MovieLens dataset [73]
James Bennett, Stan Lanning in "The Netflix Prize" 2007 [67]	Proposed Cinematch movie ratings recommendation system	Collected likes/dislikes from the subscribers	Netflix dataset holding 100 million movie rating [74]
Yunhong Zhou, Dennis Wilkinson, Robert Schreiber, Rong Pan in "Large-Scale Parallel Collaborative Filtering for the Netflix Prize" 2008 [68]	Defined movie rating predictive model with performance improvement with 5.91% over Netflix's Cinematch	Alternating-least-squares with weighted-λ-regularization (ALS-WR) based on CF	Netflix dataset holding 100 million movie rating [74]
S. Kabinsingha, S. Chindasorn, C. Chantrapornchai in "A Movie Rating Approach and Application" 2012 [62]	New movie rating application based on data mining with accuracy 80%	Weka tool and data mining application using decision tree	IMDB dataset [75]
Zhouxiao Bao, Haiying Xia in "Movie Rating Estimation and Recommendation" 2012 [69]	• Evaluating the performance of several machine learning algorithms used for movie ratings prediction • Discovered that co-clustering has better performance than other method	Baseline predictor, KNN, Stochastic Gradient Descent, SVD, SVD++, asymmetric SVD, Global Neighborhood model, Co-clustering	MovieLens dataset [73]

(continued)

Table 3 (continued)

Authors	Objectives	Technique	Dataset
Lakshmi Tharun Ponnam in "Movie recommender system using item based collaborative filtering technique" 2016 [12]	• Improves prediction accuracy versus CB • Work well with data sparsity • Reduces the number of big error predictions	Item based collaborative filtering technique	Netflix dataset
Somdutta Basu in "Movie Rating Prediction System Based on Opinion Mining and Artificial Neural Networks" 2019 [72]	• Proposed new movie rating prediction system based on artificial neural networks with accuracy 97.33% and a sensitivity of 98.63%	Levenberg–Marquardt back propagation algorithm which gains steep degradation in prediction errors	IMDB dataset [75]
Tiphaine Viard, Raphaël Fournier-S'niehotta in "Movie rating prediction using content-based and link stream features" 2018 [76]	Extend the graph-based CF for correctly modelling the graph dynamics, without losing crucial information	Graph and link stream features intercorporated with content-based features paradigm	MovieLens 20M [73]

References

1. Burke, R., Felfernig, A., Goeker, M.: Recommender systems: an overview. AI Mag. **32**(3), 13–18 (2016)
2. Jannach, D., Zanker, M., Felfernig, A., Friedrich, G.: Recommender Systems—An Introduction. Cambridge University Press, Cambridge (2010)
3. Linden, G., Smith, B., York, J.: Recommendations—item-to-item collaborative filtering. IEEE Internet Comput. **7**(1), 76–80 (2003)
4. Isinkaye, F.O., Folajimi, Y.O., Ojokoh, B.A.: Recommendation systems: principles, methods and evaluation. Egypt. Inf. J. **16**(3), 261–273 (2015)
5. Adomavicius, G., Zhang, J.: Impact of data characteristics on recommender systems performance. ACM Trans. Manage. Inf. Syst. **3**(1), 3:1–3:17 (2012)
6. Stern, D.H., Herbrich, R., Graepel, T.: Matchbox: large scale online Bayesian recommendations. In: Proceedings of the 18th International Conference on World Wide Web, pp. 111–120. ACM, New York (2009)
7. Unger, M., Bar, A., Shapira, B., Rokach, L.: Towards latent context aware recommendation systems. Knowl. Based Syst. **104**, 165–178 (2016)
8. Verma, J.P., Patel, B., Patel, A.: Big data analysis: recommendation system with hadoop framework. In: Computational Intelligence & Communication Technology, pp. 92–97. IEEE (2015)
9. Arazy, O., Kumar, N., Shapira, B.: Improving social recommender systems. IT Prof. Mag. **11**(4), 31–37 (2009)
10. Bridge, D., Göker, M.H., McGinty, L., Smyth, B.: Case-based recommender systems. Knowl. Eng. Rev. **20**(03), 315–20 (2005)
11. Schafer, J.B., Frankowski, D., Herlocker, J., Sen, S.: Collaborative filtering recommender systems. In: Brusilovsky, P., Kobsa, A., Nejdl, W. (eds.) The Adaptive Web. LNCS, vol. 4321, pp. 291–324. Springer, Berlin (2007)
12. Ponnam, L.T., et al.: Movie recommender system using item based collaborative filtering technique. In: International Conference on Emerging Trends in Engineering, Technology and Science (ICETETS), vol. 1, pp. 56–60. IEEE (2016)
13. Kumar, B., Sharma, N.: Approaches, issues and challenges in recommender systems: a systematic review. Indian J. Sci. Technol. **9**(47) (2016) https://doi.org/10.17485/ijst/2016/v9i47/94892
14. Sardianos, C., Varlamis, I., Eirinaki, M.: Scaling collaborative filtering to large-scale bipartite rating graphs using Lenskit and Spark. In: BigDataService, IEEE Second International Conference, pp. 70–79 (2017)
15. Shanahan, J.G., Dai, L.: Large scale distributed data science using apache spark. In: Conference Proceedings of the 21th ACM SIGKDD International Conference on Knowledge Discovery and Data Mining, pp. 2323–2324 (2015)
16. Tao, Y.: Design of large-scale mobile advertising recommendation system. In: 2015 4th International Conference on Computer Science and Network Technology (ICCSNT), pp. 763–767. IEEE (2015)
17. Eirinaki, M., Gao, J., Varlamis, I., Tserpes, K.: Recommender systems for large-scale social networks: a review of challenges and solutions. Future Gener. Comput. Syst. **78**, 412–417 (2017)
18. Ma, H., King, I., Lyu, M.R.: Learning to recommend with explicit and implicit social relations. ACM Trans. Intell. Syst. Technol. **2**(3), Article No. 29 (2011)
19. Gurini, D.F., Gasparetti, F., Micarelli, A., Sansonetti, G.: Temporal people-to-people recommendation on social networks with sentiment-based matrix factorization. Future Gener. Comput. Syst. **78**(P1), 430–439 (2018)
20. Su, X., Khoshgoftaar, T.M.: A survey of collaborative filtering techniques. Adv. Artif. Intell. **2009**, 4 (2009)

21. Sardianos, C., Tsirakis, N., Varlamis, I.: A survey on the scalability of recommender systems for social networks. In: Social Networks Science: Design, Implementation, Security, and Challenges. Springer, Berlin, pp. 89–110 (2018)
22. Ricci, F., Rokach, L., Shapira, B.: Introduction to Recommender Systems Handbook, pp. 1–35. Springer, Berlin (2011)
23. Koren, Y., Bell, R., Volinsky, C.: Matrix factorization techniques for recommender systems. IEEE Comput. **42**(8), 30–37 (2009)
24. Huang, Y., Cui, B., Jiang, J., Hong, K., Zhang, W., Xie, Y.: Real-time video recommendation exploration. In: Conference Proceedings of the 2016 International Conference on Management of Data, pp. 35–46 (2016)
25. Liu, C.L., Wu, X.W.: Large-scale recommender system with compact latent factor model. Expert Syst. Appl. **64**, 467–475 (2016)
26. Göksedef, M., Gündüz-Ögüdücü, S.: Combination of web page recommender systems. Expert Syst. Appl. **37**(4), 2911–2922 (2010)
27. Smirnova, E., Vasile, F.: Contextual sequence modeling for recommendation with recurrent neural networks. In: 2nd Workshop on Deep Learning for Recommender Systems (DLRS 2017), pp. 2–9. ACM Press, New York (2017)
28. Zanotti, G., Horvath, M., Barbosa, L., Immedisetty, V., Gemmell, J.: Infusing collaborative recommenders with distributed representations. In: Conference Proceedings of the 1st Workshop on Deep Learning for Recommender Systems. ACM, Boston, MA, USA, pp. 35–42 (2016)
29. Ma, H., Yang, H., Lyu, M.R., King. I.: SoRec: Social Recommendation using probabilistic matrix factorization. In: Proceedings of the 17th ACM Conference on Information and Knowledge Management, pp. 931–40, Oct 2008
30. Yin, H., Sun, Y., Cui, B., Hu, Z., Chen, L.: LCARS: a location-content-aware recommender system. In: Proceedings of the 19th ACM SIGKDD International Conference on Knowledge Discovery and Data Mining, pp. 221–229. ACM (2013)
31. Niu, J., Wang, L., Liu, X., Yu, S.: FUIR: fusing user and item information to deal with data sparsity by using side information in recommendation systems. J. Network Comput. Appl. **70**, 41–50 (2016)
32. Shahabi, C., Chen, Y.S.: Web information personalization: challenges and approaches. In: Databases in Networked Information Systems, pp. 5–15. Springer, Berlin (2003)
33. Billsus, D., Pazzani, M.J.: Learning collaborative information filters. In: ICML, pp. 46–54 (1998)
34. Shahabi, C., Banaei-Kashani, F., Chen, Y.S., McLeod, D.: Yoda: an accurate and scalable web-based recommendation system. In: Cooperative Information Systems, pp. 418–432 (2001)
35. Khusro, S., Ali, Z., Ullah, I.: Recommender systems: issues, challenges, and research opportunities. In: Information Science and Applications (ICISA) 2016. LNEE, vol. 376, pp. 1179–1189. Springer, Singapore (2016)
36. Sollenborn, M., Funk, P.: Category-based filtering and user stereotype cases to reduce the latency problem in recommender systems. In: Advances in Case-Based Reasoning, pp. 395–405. Springer, Berlin (2002)
37. Polat, H., Du, W.: Privacy-preserving collaborative filtering using randomized perturbation techniques (2003)
38. Friedman, N., Geiger, D., Goldszmidt, M.: Bayesian network classifiers. Mach. Learn. **29**(2–3), 131–163 (1997)
39. Herlocker, J.L., Konstan, J.A., Terveen, L.G., Riedl, J.T.: Evaluating collaborative filtering recommender systems. ACM Trans. Inf. Syst. (TOIS) **22**(1), 5–53 (2004)
40. Véras, D., Prota, T., Bispo, A., Prudêncio, R., Ferraz, C.: A literature review of recommender systems in the television domain. Expert Syst. Appl. **42**(22), 90, 46–76 (2015)
41. Ferri, C., Hernández-Orallo, J., Modroiu, R.: An experimental comparison of performance measures for classification. Pattern Recogn. Lett. **30**(1), 27–38 (2009)

42. Herlocker, J.L., Konstan, J.A., Borchers, A., Riedl, J.: An algorithmic framework for performing collaborative filtering. In: Proceedings of the 22nd Annual International ACM SIGIR Conference on Research and Development in Information Retrieval (SIGIR'99), pp. 230–237. ACM (1999)
43. Felfernig, A., Jeran, M., Ninaus, G., Reinfrank, F., Reiterer, S., Stettinger, M.: Basic approaches in recommendation systems. In: Recommendation Systems in Software Engineering, pp. 15–37. Springer, Berlin (2014)
44. Smith, B., Linden, G.: Two decades of recommender systems at amazon.com. IEEE Internet Comput. **21**(3), 12–18 (2017)
45. Nunes Barbosa, L., Gemmell, J., Horvath, M., Heimfarth, T.: Distributed user-based collaborative filtering on an opportunistic network. In: IEEE 32nd International Conference on Advanced Information Networking and Applications (AINA), pp. 266–273, Krakow (2018)
46. Ekstrand, M.D., Riedl, J.T., Konstan, J.A.: Collaborative filtering recommender systems. Found. Trends Hum. Comput. Interact. **4**(2), 1–94 (2011)
47. Takacs, G., Pilászy, I., Németh, B., Tikk, D.: Scalable collaborative filtering approaches for large recommender systems. J. Mach. Learn. Res. **10**, 623–656 (2009)
48. Zarzour, H., Al-Sharif, Z., Al-Ayyoub, M., Jararweh, Y.: A new collaborative filtering recommendation algorithm based on dimensionality reduction and clustering techniques. In: 9th International Conference on Information and Communication Systems (ICICS), pp. 102–106, Irbid (2018)
49. Arora, S., Goel, S.: Improving the accuracy of recommender systems through annealing. In: Lecture Notes in Networks and Systems, pp. 295–304 (2017)
50. Koohi, H., Kiani, K.: User based collaborative filtering using fuzzy c-means. Measurement **91**, 134–139 (2016)
51. Kim, K., Ahn, H.: Recommender systems using cluster-indexing collaborative filtering and social data analytics. Int. J. Prod. Res. **55**(17), 5037–5049 (2017)
52. Meng, S., Dou, W., Zhang, X., Chen, J.: KASR: a keyword-aware service recommendation method on MapReduce for big data applications. IEEE Trans. Parallel Distrib. Syst. **25**(12), 3221–3231 (2014)
53. Zarzour, H., Maazouzi, F., Soltani, M., Chemam, C.: An improved collaborative filtering recommendation algorithm for big data. In: Proceedings of the 6th IFIP International Conference on Computational Intelligence and its Applications 2018, pp. 102–106. CIIA (2018)
54. Tian, H., Liang, P.: Personalized service recommendation based on trust relationship. Sci. Program. **2017**, 1–8 (2017)
55. Pazzani, M., Billsus, D.: Learning and revising user profiles: the identification of interesting web sites. In: Michalski, R.S., Wnek, J. (eds.) Machine Learning, vol. 27, pp. 313–331 (1997)
56. Pazzani, M.J., Billsus, D.: Content-based recommendation systems. In: Brusilovsky, P., Kobsa, A., Nejdl, W. (eds.) The Adaptive Web, vol. 4321, pp. 325–341. Lecture Notes in Computer Science. Springer, Berlin (2007)
57. Burke, R.: Knowledge-based recommender systems. Encycl. Libr. Inf. Syst. **69**(32), 180–200 (2000)
58. Felfernig, A., Friedrich, G., Jannach, D., Zanker, M.: An integrated environment for the development of knowledge-based recommender applications. IJEC **11**(2), 11–34 (2006)
59. Felfernig, A., Friedrich, G., Gula, B., Hitz, M., Kruggel, T., Melcher, R., Riepan, D., Strauss, S., Teppan, E., Vitouch, O.: Persuasive recommendation: exploring serial position effects in knowledge-based recommender systems. In: 2nd International Conference of Persuasive Technology 2007. LNCS, vol. 4744, pp. 283–294. Springer, Berlin (2007)
60. Mandl, M., Felfernig, A., Teppan, E., Schubert, M.: Consumer decision making in knowledge based recommendation. J. Intell. Inf. Syst. **37**(1), 1–22 (2010)
61. Felfernig, A., Shchekotykhin, K.: Debugging user interface descriptions of knowledge-based recommender applications. In: Proceedings of the 11th International Conference on Intelligent User Interfaces, IUI 2006, pp. 234–241. ACM Press, New York (2006)
62. Kabinsingha, S., Chindasorn, S., Chantrapornchai, C.: Movie rating approach and application based on data mining. IJEIT **2**(1), 77–83 (2012)

63. Weka 3: Data Mining Software in Java, University of Waikato. Available: https://www.cs.wai kato.ac.nz/ml/weka/. Last accessed on 4 Feb 2019

64. Ricci, F.: Context-aware music recommender systems: workshop keynote abstract. In: Proceedings of the 21st International Conference Companion on World Wide Web, pp. 865–866. ACM, Lyon (2012)

65. Adomavicius, G., Tuzhilin, A.: Context-aware recommender systems. In: Ricci, F., Rokach, L., Shapira, B. (eds.), Recommender Systems Handbook. Springer, Berlin, pp. 191–226 (2015)

66. Song, L., Tekin, C., Schaar, M.V.D.: Online learning in large-scale contextual recommender systems. IEEE Trans. Serv. Comput. 9(3), 433–445 (2016)

67. Bennett, J., Lanning, S.: The Netflix prize. In: Proceedings of the KDD Cup Workshop, pp. 3–6. ACM, New York (2007)

68. Zhou, Y., Wilkinson, D., Schreiber, R., Pan, R.: Large-scale parallel collaborative filtering for the Netflix prize. In: Proceedings of the 4th International Conference on Algorithmic Aspects in Information and Management 2008. LNCS, vol. 5034, pp. 337–348. Springer, Berlin (2008)

69. Bao, Z., Xia, H.: Movie Rating Estimation and Recommendation, CS229 Project, Stanford University (2012)

70. Lee, K., Park, J., Kim, I., Choi, Y.: Predicting movie success with machine learning techniques: ways to improve accuracy. Inf. Syst. Front. 20, 577–588 (2018). https://doi.org/10.1007/s10 796-016-9689-z

71. Seroussi, Y., Bohnert, F., Zukerman, I.: Personalised rating prediction for new users using latent factor models. In: De Bra, P., Gronbaek, K. (eds.) Proceedings of the 22nd ACM Conference on Hypertext and Hypermedia, pp. 47–56. Association for Computing Machinery (ACM) (2011)

72. Basu, S.: Movie rating prediction system based on opinion mining and artificial neural networks. In: Kamal, R., Henshaw, M., Nair, P. (eds.) International Conference on Advanced Computing Networking and Informatics. Advances in Intelligent Systems and Computing, vol. 870. Springer, Singapore (2019)

73. https://grouplens.org/datasets/movielens/

74. https://www.kaggle.com/netflix-inc/netflix-prizedata#movie_titles.csv

75. Internet movie database, IMDb.com, Inc. Available: https://www.imdb.com/interfaces/. Accessed on 11 Feb 2019

76. Viard, T., Fournier-S'niehotta, R.: Movie rating prediction using content-based and link stream features. Arxiv (2018). Available at: https://arxiv.org/abs/1805.02893

Predicting Student Retention Among a Homogeneous Population Using Data Mining

Ghazala Bilquise⊙, Sherief Abdallah⊙, and Thaeer Kobbaey⊙

Abstract Student retention is one the biggest challenges facing academic institutions worldwide. In this research, we present a novel data mining approach to predict retention among a homogeneous group of students with similar social and cultural background at an academic institution based in the Middle East. Several researchers have studied retention by focusing on student persistence from one term to another. Our study, on the other hand, builds a predictive model to study retention until graduation. Moreover, our research relies solely on pre-college and college performance data available in the institutional database. We use both standard as well as ensemble algorithms to predict dropouts at an early stage and apply the SMOTE balancing technique to reduce the performance bias of machine learning algorithms. Our study reveals that the Gradient Boosted Trees is a robust algorithm that predicts dropouts with an accuracy of 79.31% and AUC of 88.4% using only pre-enrollment data. The effectiveness of the algorithms further increases with the use of college performance data.

Keywords Data mining · Retention · Dropout · Attrition · Higher education

1 Introduction

In today's knowledge society, education is the key to creativity and innovation, which are essential elements of progress. Attainment of education is formalized by earning a college degree, which not only provides individuals with opportunities for pro-

G. Bilquise (✉)
Higher Colleges of Technology, Dubai, UAE
e-mail: ghazala.bilquise@hct.ac.ae

S. Abdallah
British University in Dubai, Dubai, UAE
e-mail: sherief.abdallah@buid.ac.ae

T. Kobbaey
Higher Colleges of Technology, Dubai, UAE
e-mail: thaeer.kobbaey@hct.ac.ae

© The Author(s), under exclusive license to Springer Nature Switzerland AG 2021
A. E. Hassanien et al. (eds.), *Machine Learning and Big Data Analytics Paradigms: Analysis, Applications and Challenges*, Studies in Big Data 77,
https://doi.org/10.1007/978-3-030-59338-4_13

fessional growth but also lowers unemployment rate and thereby boosts the local economy [1]. Despite the enormous socio-economic benefits of earning a college degree, nearly 30% of students leave college without earning any credential [2]. This situation is prevalent worldwide [3] with the Middle East being no exception [4].

Previous works have shown that the decision to drop out mostly occurs in the first year of college as students struggle to cope with the challenges of an academic environment and transition from high school [5]. Intervention strategies within the first year of studies can improve retention rates by up to 50% [6]. Therefore, an early identification of students at risk of premature departure enables the institution to target their resources to benefit students who need it the most.

A promising avenue to address the challenge of early dropouts is the use of data mining and machine-learning techniques. While some studies have used Educational Data Mining (EDM) techniques to study retention, this subject lacks critical investigation in the Middle East.

Our study goes beyond previous retention works using EDM in several ways. First, while other studies have differentiated students based on culture, race, ethnicity and more [7], our study is based on a homogeneous population with similar social and cultural background. The factors that influence students to drop out in other cultures do not apply to our environment. Therefore, we focus solely on performance data to predict retention.

Second, our study utilizes enrollment data and longitudinal data of two consecutive terms to predict retention until graduation, unlike other papers that focus on persistence by utilizing student data in one term only [5].

Third, we use a larger dataset than most studies and focus on student retention across multiple disciplines rather than a single discipline [8, 9]. We also apply balancing techniques to get more reliable and unbiased results.

Our research is based on a Higher Education Institution located in the Middle East that offers degree programs for nationals. Therefore, the students of the institution form a homogeneous group belonging to the same nationality, culture and heritage. The education of all students is funded either federally or through a sponsor. During the year 2011 and 2012 only 70% enrolled students completed their graduation on time.

The purpose of this study is to apply an EDM approach to predict undergraduate students at risk of dropping out without earning a degree. In addition, we seek to determine the earliest stage when an effective prediction is possible and identify the top predictive factors of retention at each stage.

2   Related Work

Several studies have undertaken the task of investigating student success in undergraduate programs by predicting student performance [10–12] or ability to progress from one term to another [5, 13]. On the other hand, only a few studies have focused on retention until graduation in higher education [7–9]. Moreover, to the best of our knowledge, no such study has been conducted till date in the Middle East.

A variety of factors have been identified by studies as predictive variables of dropouts. A study based in a Slovenia identified the type of program, full time or part time, as the major predictor of student success [14]. On the other hand, student age, credit hours and residency status are reported as top predictive factors of graduation in a US based study [15]. Ethnicity [16], transferred hours and proximity to college [17] are also revealed as crucial factors for retention. Math test scores, ethnicity, and special education needs are identified as top predictor variables of dropouts in a High School study [18].

Previous studies have focused their retention efforts in their local environment; thus making it hard to map those studies to our unique setting. The factors such as ethnicity, race, proximity, residency status and more, that influence students to drop out other cultures do not apply to the our unique setting. Moreover this data is not easily available in the institutional database. Acquiring such data would require conducting surveys which are not only time-consuming. but would also lead to reduction in the dataset size of the experiment, thus, making it counterproductive to a data mining approach.

The Decision Tree and Rule Induction algorithms have been applied to predict dropouts from the first year of High School using both institutional data and data collected from surveys [19]. Various factors such as social status, family background, psychological profile as well as academic performance were used to study retention. However, the top predictive factors for retention revealed by the study were performance score in Physics, Humanities, Math and English courses. This finding further reinforces our decision to use academic performance data rather than compromising the size of the dataset by collecting additional data via surveys.

Machine learning techniques have been used to predict attrition by investigating failure rates in cornerstone courses. An EDM approach was used to predict early failures in a first year programming course at a Brazilian University [20]. The study used weekly performance grades and demographic data. Potential dropouts were identified in an introductory engineering course using demographic, academic and engagement data [21]. In another similar study the performance in a high-impact engineering course was predicted using cumulative GPA, grades attained in pre-requisite courses as well as course work assessment scores [11].

The aforementioned studies are based on the assumption that success in fundamental courses would eventually lead to retention in the corresponding program of study. However, our research focuses on retention across several disciplines; therefore, we do not consider the outcome of a single course.

Several studies have investigated retention in a degree program, high school or online program by exploring re-enrollment of students in subsequent terms. Various machine-learning algorithms and a genetic programming algorithm were used to predict dropouts among High School at an early stage [22]. The study collected data of 419 students in progressive stages, starting with secondary school performance, pre-enrollment data and gradually adding more data on a weekly basis. A very high accuracy of 99.8% was reported with the genetic algorithm, which outperformed all other algorithms. We believe that such a high-performance figure may be a sign of overfitting, especially due to the small size of the dataset.

Machine-learning algorithms have also been used to study persistence in online studies [16, 23, 24]. Predictive modelling techniques were applied to predict the persistence of freshmen students until the next term in a US based public Univeristy [15]. None of the aforementioned studies considered graduation and measured dropouts as failure to re-enroll in the following term. In our study, we consider this as a study of persistence rather than retention. Moreover, unlike retention studies, investigation of persistence does not utilize longitudinal student data across many terms until graduation.

Student demographic data and prior performance scores have also been used frequently to predict their success in academics. Weekly performance grades and demographic data were used to predict early failures in a first year programming course at a Brazilian University [20]. Potential dropouts were identified in an introductory engineering course using demographic, academic and engagement data [21]. Similarly, the work in [11] evaluated the performance in a high-impact engineering course to predict dropouts using cumulative GPA, pre-requisite course grades and course work assessment grades.

An EDM approach was used to predict the successful graduation of first-year students in an Electrical Engineering program in a Netherlands-based University [8]. The study utilized 648 student records collected over a ten year period. The data consisted of pre-university and university performance scores and achieved a maximum accuracy of 80%.

Dropouts were investigated in a Columbian System Engineering undergraduate program using 802 student records collected from the year 2004 to 2010 [9]. The dataset of the study comprised of attributes such as admission data, course grades, and financial aid per term. The study achieved an AUC score of 94%. Although the reported performance is very good, the study is focused on a single disciplne and the dataset size is too small for the results to be generalized. Our study achieves a similar performance of 92% AUC with a much larger dataset size of 4,056 and includes several undergraduate programs.

Eight years of enrollment data comprising of over 69,000 heterogeneous student records was used to predict graduation at an American University [25]. The dataset consisted of demographic data such as ethnicity, gender and residency information. Also, external assessments such as SAT and ACT scores, academic achievement in all courses were used to predict dropouts. The study achieved an AUC score of 72.9% using the Logistic Regression algorithm.

Student retention was investigated in a System Engineering undergraduate program using 802 student records collected from the year 2004 to 2010 [9]. The dataset of the study comprised of attributes such as admission data, course grades, and financial aid per term. The study achieved an AUC score of 94%. Although the reported performance is excellent, the study is based on a single discipline and the dataset size is too small for the results to be generalized.

Factors of retention leading to graduation were investigated using Logistic Regression and Neural Networks in a US-based University using 7,293 records [7]. Data was collected for a ten year period and consisted of pre-college and first-term college data. Pre-college data included demographic information, external assessment scores, high school scores, distance from home and work status, while college data included first semester GPA and earned hours. Missing data was handled by deleting the record thus reducing the dataset size. The study achieved an AUC of 77.7%.

Similar to our study, both the aforementioned studies [7, 25] have used AUC to evaluate the performance of the predictive models. Our results surpass the results of both the studies with an AUC of 92% using Gradient Boosted Trees algorithm.

3 Methodology

The likelihood of obtaining reliable and accurate results increases with the use of a structured approach. In this study, we follow the CRISP-DM (cross-industry process for data mining) approach [26], which provides an organized framework for performing data mining tasks. We follow the five main phases of the CRISP-DM methodology. (1) **Business Understanding:** the needs of the business are assessed and the goals of data mining are determined to create a plan for our study. (2) **Data Understanding:** the data is acquired and and explored to identify quality issues. A strategy to deal with these issues is set up at in this phase. (3) **Data Preparation:** data quality issues are resolved and the data is pre-processed and prepared for generating the predictive models. (4) **Modeling:** the machine-learning algorithms are selected and applied to the pre-processed dataset to generate predictive models. (5) **Evaluation:** the results of the various algorithms are compared and findings of the experimentation are discussed.

Our study aims to answer the following three research questions:

Research Question 1: Can machine-learning algorithms effectively predict retention/dropouts in a homogeneous group of students using performance data?

To answer this research question, we generate predictive models using five standard and five ensemble classification algorithms. Due to the stochastic nature of machine-learning algorithms, predictive models cannot perform with 100% accuracy. Therefore, an effective model is one with a reasonable performance in a given domain. The reviewed literature on retention showed that predictive models have achieved accuracies ranging from 72 to 77% [7]. Therefore, we evaluate the predictions of our models against a minimum threshold requirement of 75% AUC to be labeled as effective.

Research Question 2: How early can we predict potential dropouts using machine learning? To answer this research question, we use three datasets with pre-college, college term 1 and college term 2 features. The performance of the algorithms on each dataset is compared using a pair-wise t-test to determine if the difference is significant or merely due to chance.

Research Question 3: Which attributes are the top predictors of retention? To answer this research question, we analyze the relevance of each attribute to the class label. We begin by calculating the feature weight of each attribute using information gain,

gain ratio, gini index, correlation and chi-squared statistic. We then examine the top predictors of retention identified by the Decision Tree model. We also explore the weights assigned to each attribute by the predictive models of the Support Vector Machine (SVM), Logistic Regression, Random Forest and Gradient Boosted Trees algorithms. We summarize and report the topmost predictive features for each dataset by the frequency of its appearance.

3.1 The Dataset

The enrollment dataset consists of 22,000 enrollment records for the academic year 2011–2013, inclusive, for each term (excluding the summer term since new enroll-ments do not take place in summer). The academic year 2013 is chosen as a cut off period to allow the nominal degree completion time, which is a period of 6 years.

Each enrollment record is described by 84 features, which includes student demo-graphic data, personal data, High School performance scores, IELTS performance scores, English and Math placement scores and the current term GPA. From this dataset we filter out the newly enrolled students in each term, which results in 4056 newly enrolled student records and perform the initial preparation tasks in MS Excel and MS Access as shown in Fig. 1.

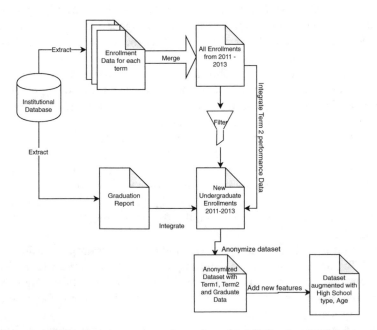

Fig. 1 Dataset extraction and preparation tasks performed in MS Excel and MS Access

In addition to the enrollment records we also extract the dataset of graduation records of all students who have graduated till date. We then integrate the graduation records with the enrollment records and generate a class label to identify new enrollments that either graduated or dropped out in their respective program of study.

Next we extract the term 2 GPA by joining enrollment records of subsequent terms. Nearly 30 attributes pertaining to student personal data are removed to anonymize the dataset. These include details such as Student ID, name, contact numbers, parent data, sponsor data, advisor details, and more.

We also impute two new features called High School Type (public or private) and Age at enrollment. High school type is generated by cross-listing the high school name attribute against a list of all high schools in the city downloaded from a public website and age is computed using the date of birth feature.

3.2   Data Preprocessing

The quality of the dataset determines the performance of the data mining algorithms. Therefore, pre-processing is a crucial step that is needed to ensure the success of the algorithms as well as the validity of the results. Three main pre-processing tasks are performed in Rapid Miner, which include handling missing values, balancing the dataset, and feature selection. Figure 2 shows the tasks performed with Rapid Miner to pre-process the data, generate, apply and evaluate the predictive models.

Missing Values: To handle missing values we use three approaches. First, any feature that has more than 20% of missing values is excluded from the study. This results in removal of around 20 attributes. Second, all other attributes, except Term 2 GPA, are imputed using the K Nearest Neighbor (k-NN) algorithm with the value of k set to

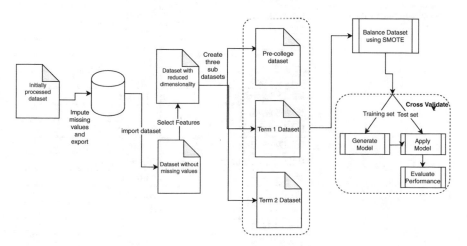

Fig. 2 Pre-processing tasks performed in Rapid Miner

3. No action is taken for Term 2 GPA since a missing value in this feature indicates that the student did not register in term 2 and dropped out in term 1. The dataset with the imputed values is exported and used in the next steps to reduce processing time for each experimentation run.

Feature Selection: The high dimensionality of a dataset can lead to many problems such as overfitting of the models thus producing to a model that cannot be generalized [21]. Therefore, we further reduce the features by removing irrelevant attributes such as enrollment status, grading system, campus codes and more and select only those features that are related to student demographic data, pre-college performance and college performance. Moreover, any feature that is used to generate new attributes, such as date of birth and high school name, is excluded from the study. The resultant dataset consists of the 20 features shown in Table 1.

Creating Three Sub-datasets: The pre-processed dataset is divided into three distinct datasets to determine at what stage we can effectively predict students who are likely to drop out. The three datasets are described below:

1. **Pre-college Dataset**: Consists of demographic data, pre-college performance data (High School, IELTS, English and Math placement test scores).
2. **College Term 1 Dataset**: Pre-college dataset + Term 1 GPA + program of study
3. **College Term 2 Dataset**: College Dataset 1 + Term 2 GPA.

Table 1 Dataset features

Type	Attribute	Data type
Demographic data	Age	Numeric
	Gender	Binary
	HS Type	Binary
	HS Stream	Nominal
Pre-college performance	HS Avg	Numeric
	HS math	Numeric
	HS english	Numeric
	HS arabic	Numeric
	IELTS band	Numeric
	IELTS listening	Numeric
	IELTS writing	Numeric
	IELTS speaking	Numeric
	IELTS reading	Numeric
	English placement	Numeric
	Math placement	Numeric
College	Term	Numeric
	Program of study	Nominal
	Term 1 GPA	Numeric
	Term 2 GPA	Numeric
Class label	Graduated	Binary

Balancing Datasets: Our dataset at this stage is unbalanced with the graduates (majority class) representing 74% of the observations while the dropouts (minority class) representing only 26% of the records. A model generated by the machine-learning algorithms using an unbalanced dataset is more likely to produce misleading results. Therefore, balancing the dataset is essential to reduce the bias caused by the majority class and improve classification performance [27, 28].

While there are many techniques of balancing a dataset, specifically, in this study we have chosen the Synthetic Minority Oversampling Technique (SMOTE) proposed by Chawla et al. [29]. The SMOTE algorithm is a popular technique used in several studies [20].This technique artificially generates new minority class observations using the k-NN algorithm until the dataset is balanced.

Training and Validation: We employ the ten-fold cross-validation with stratified sampling to split the original dataset into ten subsets while preserving the ratio of the minority and majority samples. The machine-learning algorithms are trained using nine subsets, while testing is performed using the remaining subset. This process is repeated ten times by holding out another subset and training with the remaining nine. The final performance is reported as an average of all the iterations.

The resultant dataset after the data preparation and pre-processing phase has 4,056 records and includes data of new enrollments in the academic year 2011 to 2013.

3.3 Predictive Modelling

Our study uses five standard classification algorithms, namely, Decision Tree, Naïve Bayes, Logistic Regression, SVM and Deep Learning. In addition, we also employ five ensemble algorithms Random Forest, Voting, Bagging, AdaBoost and Gradient Boosted Trees to achieve a robust prediction. The algorithms classify each instance of the testing data in into two classes Y (graduated) or N (did not graduate).

Furthermore, we use a parameter optimization operator in Rapid Miner to maximize the performance of the Decision Tree algorithm. Using multiple combinations of the parameters, the following optimal settings are determined by the optimizer and used for the rest of our experimentation:

- **Minimal gain**: 0.046,
- **Confidence**: 0.34
- **Split criteria**: Gini-index.

The aim of our research is to effectively predict dropouts using a dichotomous class label that classifies the graduates (positive class) and dropouts (negative class). Our primary interest is to accurately predict the dropouts while not misclassifying the graduates. Inaccurate classification leads to poor utilization of resources in dropout interventions for students who would have graduated, whilst also missing out on students who are actually at risk of dropping out.

To ensure that the algorithms perform effectively for our requirement, we use five evaluation metrics to measure and compare the performance of our predictive models, namely Accuracy, True Negative Rate (TNR), True Positive Rate (TPR), AUC of the Receiver Operator Characteristic (ROC) curve and F-Measure. The TNR measures the percentage of correctly classified dropouts, where as the TPR measures the percentage of accurately classified graduates. The AUC of the ROC curves focuses on correctly identifying the graduates (TPR) while lowering the misclassification of the dropouts (TNR).

Accuracy is popular metric used by several researchers [5, 8, 11, 15, 16, 27, 30]. It computes the percentage of correct classifications in a confusing matrix. However, accuracy is an unreliable metric when used with an unbalanced dataset since it can result in a performance bias. Since a balancing technique (SMOTE) is applied to our dataset, we use accuracy to measure the overall performance the models.

The True Negative Rate (TNR) measures the percentage of correctly classified dropouts, where as the True Positive Rate (TPR) measures the percentage of accurately classified graduates. The AUC of the ROC curves focuses on correctly identifying the graduates (TPR) while lowering the misclassification of the dropouts (TNR). The curve is plotted with the TPR in the y-axis and 1- TNR in the x-axis. Within the context of our study, a higher value of AUC indicates that the model has correctly classified a large number of dropouts as well as graduates.

4   Results and Discussion

In this section, we discuss our experimental findings and answer each research question that was presented in Sect. 3.

4.1   Research Questions

Question 1—Can machine-learning algorithms effectively predict retention/ dropouts in a homogeneous group of students using performance data? To answer this research question, we generate predictive models on three datasets using five standard classification algorithms as well as five ensemble algorithms. The performance of each model is evaluated to determine if the algorithm performs effectively, with an AUC of at least 75%, as described in Sect. 3.

Standard Classifiers Table 2 shows the result of the standard classifiers across all three datasets.

None of the standard classifiers are able to meet the threshold requirement of AUC 75% when the pre-college dataset is used. However, when the dataset is enhanced with college term 1 performance, all the standard algorithms perform effectively with an AUC score above the threshold requirement ranging from 77 to 84.8%. The Logistic Regression and SVM algorithms perform the best, achieving accuracies of up to 77.3%. Both algorithms predict graduates and dropouts equally well with TPR

Table 2 Standard classifier results

Model	Accuracy (%)	TNR (%)	TPR (%)	AUC (%)	F-Measure (%)
Pre-college dataset					
Decision tree	**71.59**	67.17	76.02	73.80	**72.80**
Naïve bayes	67.22	**77.10**	57.34	73.90	63.60
Logistic regression	67.92	70.22	65.63	**74.40**	67.16
SVM	67.09	72.64	61.53	73.30	65.13
Deep learning	62.80	46.40	**79.19**	70.16	68.02
College term 1 dataset					
Decision tree	75.77	70.54	**81.00**	79.40	76.96
Naïve bayes	73.97	**77.49**	70.45	81.80	73.01
Logistic regression	77.20	75.82	78.57	**84.80**	77.51
SVM	**77.31**	75.46	79.23	84.70	**77.77**
Deep learning	69.68	61.14	78.21	77.50	72.07
College term 2 dataset					
Decision tree	80.41	**88.40**	72.41	81.70	**78.70**
Naïve bayes	75.23	76.93	73.53	83.33	74.81
Logistic regression	77.42	76.51	78.34	**86.00**	77.62
SVM	77.95	76.38	79.52	85.80	78.27
Deep learning	75.20	67.01	**83.39**	82.90	77.08

Bold formatting is used to highlight the best results of the evaluation metric for each dataset

of 79.2 and TNR of 75.4%. The use of College Term 2 data further enhances the performance of the algorithms with an AUC score of above 80%. Again, the Logistic Regression and SVM algorithms perform the best with an AUC score of 86%, and overall accuracy of 77.9%.

Ensemble Classifiers Table 3 shows the result of the ensemble classifiers across all three datasets.

Overall, the ensemble algorithms perform better than the standard algorithms when the pre-college dataset is used. Except for Voting and AdaBoost all ensemble algorithms are able to meet the threshold requirement of AUC 75%.

The Gradient Boosted Trees classifier consistently performs the best across all the three datasets. It achieves an AUC score of upto 92.2% with an accuracy of 84.7%. All other classifiers have also improved in their performance with the lowest accuracy of 78.9% achieved by the Voting algorithm. AdaBoost and Bagging algorithms are the best at predicting dropouts at 88.4%.

Question 2—How early can we predict potential dropouts using machine learning? We answer this research question by examining the performance of each clas-

Table 3 Ensembler classifier results

Model	Accuracy (%)	TNR (%)	TPR (%)	AUC (%)	F-Measure (%)
Pre-college dataset					
Random forest	70.35	73.75	66.94	77.10	69.21
Gradient boosted trees	**79.31**	72.35	**86.27**	**88.40**	**80.66**
Voting	70.54	**75.00**	66.09	71.90	69.15
AdaBoost	71.23	66.74	75.72	71.20	72.47
Bagging	73.02	69.10	76.87	79.50	74.02
College term 1 dataset					
Random forest	77.79	73.00	82.57	85.00	78.84
Gradient boosted trees	**82.10**	**79.59**	**86.04**	**90.10**	**83.35**
Voting	77.26	74.93	79.59	77.70	77.77
AdaBoost	75.62	70.48	80.77	78.60	76.80
Bagging	76.69	71.63	81.75	83.40	77.79
College term 2 dataset					
Random forest	81.16	82.67	79.65	88.80	80.88
Gradient boosted trees	**84.75**	83.32	**86.17**	**92.20**	**84.97**
Voting	78.96	80.83	77.10	81.60	78.56
AdaBoost	80.21	88.40	72.02	83.90	78.44
Bagging	82.01	**88.47**	75.56	89.10	80.76

Bold formatting is used to highlight the best results of the evaluation metric for each dataset

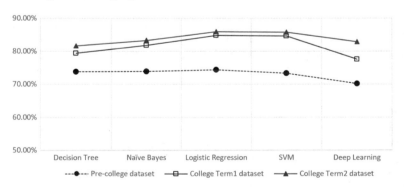

Fig. 3 Standard Classifier AUC performance

sifier across all the datasets. Figures 3 and 4 show the performance of the Standard and Ensemble Algorithms respectively.

The predictive capabilities of all the standard algorithms increases when the College Term 1 and College Term 2 datasets are used. The results indicate that although

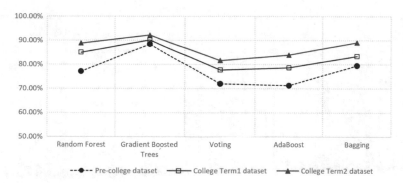

Fig. 4 Ensemble Classifier AUC performance

pre-college data can provide a good initial prediction of students likely to dropout, the Term 1 and Term 2 performance data can produce more effective and accurate predictions.

The Logistic Regression and SVM classifiers have shown the most increase in performance among the standard algorithms (up to 11% AUC), while the Decision Trees' performance has improved by 5.6% only. However, among the ensemble algorithms the increase is not very high ranging from 2 to 7%. The Gradient Boosted Trees classifier has shown very little increase proving to be a robust and reliable predictor for all the three datasets.

A pairwise t-test of significance to test the difference in performance between the Pre-college and College Term 1 dataset reveals that the increase in performance is indeed significant ($\alpha < 0.001$, for each algorithm). Although the difference in performance between the College Term 1 and College Term 2 data set is not very large, yet a pair-wise t-test shows that this increase is also significant.

Research Question 3: Which attributes are the top predictors of retention? To answer this research question, we first analyze the attributes by feature weights to study its relevance with respect to the class label. We begin by calculating the feature weight of each attribute using five feature weight algorithms, namely, information gain, gain ratio, gini index, correlation and chi-squared statistic. We rank the attributes by lowest weight to highest.

Figure 5 shows the feature ranking of all features using the five feature weight algorithms.

Term1 GPA and Term2 GPA are consistently picked as the most significant predictors by all feature weight algorithms, followed by High School average and then High School Math. Some of the least relevant attributes are Age, IELTS Speaking score, College, Placement and High School Type which are ranked the lowest by most of the feature weight algorithms.

We also use the Decision Tree algorithm for its interpretability and ability to identify the top predictor of retention. The root node of the decision tree model shows the most influential attribute in classifying dropouts. Figures 6, 7 and 8 shows

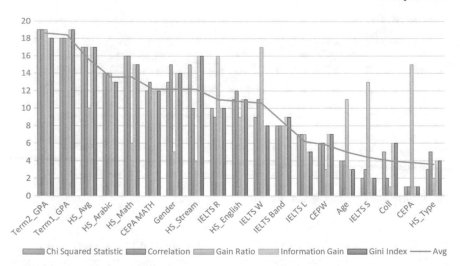

Fig. 5 Feature rankings

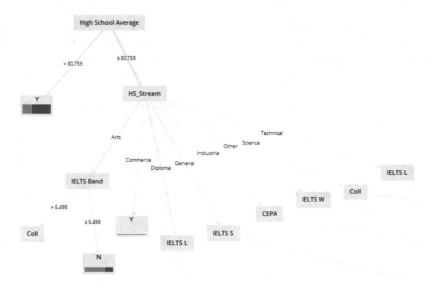

Fig. 6 Decision tree for Pre-college dataset

the Decision Tree model produced for the Pre-College, College Term 1 and College Term 2 dataset respectively.

The Decision Tree model reveals that, the High School Average, High School Stream and IELTS band are the top predictors of retention when the pre-college data set is used.

We also identify the top predictive attributes using the feature weights assigned by the SVM, Logistic regression, Gradient Boosted Trees and Random Forest algorithm

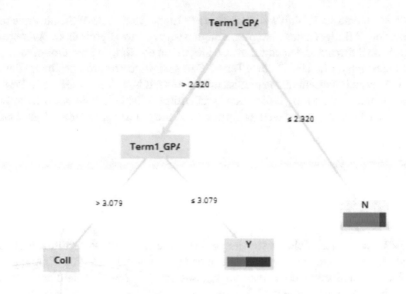

Fig. 7 Decision tree for college term 1 dataset

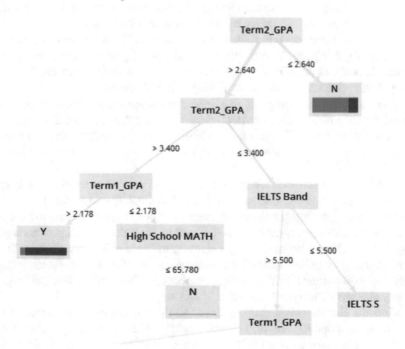

Fig. 8 Decision tree for college term 2 dataset

across each dataset. A high weight indicates a higher relevance of the attribute to the prediction. All algorithms also assign highest weight to High School Average and High School Stream and in addition also identify age as the top predictor of retention.

Interestingly, when the College Term 1 Dataset is used, the pre-enrollment features do not play an important role in predicting dropouts. It is the Term 1 GPA and Program of Study that are the top predictors of graduation. When College Term 2 dataset is used, then Term 2 GPA as well as Term 1 GPA are the top predictors of graduation.

5 Conclusion

In this chapter, we used the structured CRISP-Data Mining methodology to predict student dropout among a homogeneous population using a dataset of 4056 student records. The dataset includes student performance data prior to enrollment, in addition to the academic performance data in each term of the first-year of studies.

We evaluated ten machine learning algorithms, including standard and ensemble algorithms, for classifying a student as a successful graduate or a dropout. The ten algorithms were applied to three datasets of student performance at various stages of their academic journey form enrollment to end of the second term.

Overall, the ensemble algorithms have performed better than standard algorithms across all the three datasets, thus proving to be more reliable and better at handling misclassifications. Among these, the Gradient Boosted Trees is the most effective algorithm performing equally well on all datasets, proving to be a robust and reliable predictor. It achieves an AUC score of 88.4% for the pre-college dataset, 90.1% for College Term1 dataset and 92.2% for College Term 2 dataset.

Our research predicts dropouts at a very early stage using pre-enrollment data with an accuracy of 79.31% and AUC of 88.4% using the Gradient Boosted Trees algorithm. Our results will enable the academic institution to start remedial support at an early stage from the first term onwards by directing resources to where they are required the most.

In addition, our study also identified the top predictors of retention at each stage starting from enrollment to end of term 1. High School Average and IELTS Band were found to be the top predictors of retention when a student joins the college. This result indicates that students who did well in High School and possess a good level of English have a better chance of meeting the academic demands of college.

Interestingly, when the College Term 1 dataset is used, the pre-enrollment features do not play an important role in predicting dropouts. It is the Term 1 GPA and Program of Study that are the top predictors of graduation.

The results also show that the program of study is an important predictive factor in determining dropouts. Hence it can be said that if students do not choose their program of study wisely, it is likely they will eventually discontinue their studies. Students often choose their discipline of study based on their interest or prospective career choices without aligning it with their academic capabilities. Therefore, this often leads to academic struggle and abandonment of the studies. It is, therefore, essential for the academic institution to advise students in wisely choosing their programs of study to ensure success.

It is crucial to collect relevant data to enhance the accuracy of the predictions. While our research has achieved a good accuracy using demographic and performance data, we believe that augmenting the dataset with more detailed data (such as attendance, sponsorship status, and working status) can provide better predictions. Although such data is recorded in the system, it was unavailable at the time of this study.

Another interesting research avenue to pursue would be to expand the study to include the other campuses of the same college and other similar colleges within the region.

References

1. Tamhane, A., Ikbal, S., Sengupta, B., Duggirala, M., Appleton, J.: Predicting student risks through longitudinal analysis. In: Proceedings of the 20th ACM SIGKDD International Conference on Knowledge Discovery and Data Mining, pp. 1544–1552. ACM (2014)
2. Ma, J., Pender, M., Welch, M.: Education pays 2016: the benefits of higher education for individuals and society. Trends Higher Edu. ser, College Board (2016)
3. Levitz, R.S., Noel, L., Richter, B.J.: Strategic moves for retention success. New Dir Higher Edu. **1999**(108), 31–49 (1999)
4. Miguéis, V.L., Freitas, A., Garcia, P.J., Silva, A.: Early segmentation of students according to their academic performance: a predictive modelling approach. Decis. Support Syst. **115**, 36–51 (2018)
5. Rubiano, S.M.M., Garcia, J.A.D.: Formulation of a predictive model for academic performance based on students' academic and demographic data. In: 2015 IEEE Frontiers in Education Conference (FIE), pp. 1–7. IEEE (2015)
6. Yu, C.H., DiGangi, S., Jannasch-Pennell, A., Kaprolet, C.: A data mining approach for identifying predictors of student retention from sophomore to junior year. J. Data Sci. **8**(2), 307–325 (2010)
7. Yukselturk, E., Ozekes, S., Türel, Y.K.: Predicting dropout student: an application of data mining methods in an online education program. Eur. J. Open, Distance E-learn. **17**(1), 118–133 (2014)
8. Perez, B., Castellanos, B., Correal, D.: Applying data mining techniques to predict student dropout: a case study. In: 2018 IEEE 1st Colombian Conference on Applications in Computational Intelligence (ColCACI), pp. 1–6. IEEE (2018)
9. Jayaprakash, S.M., Moody, E.W., Lauría, E.J., Regan, J.R., Baron, J.D.: Early alert of academically at-risk students: an open source analytics initiative. J. Learn. Analytics **1**(1), 6–47 (2014)
10. Chalaris, M., Gritzalis, S., Maragoudakis, M., Sgouropoulou, C., Lykeridou, K.: Examining students graduation issues using data mining techniques-the case of tei of athens. In: AIP Conference Proceedings, vol. 1644, pp. 255–262. AIP (2015)
11. Natek, S., Zwilling, M.: Student data mining solution-knowledge management system related to higher education institutions. Expert Syst. Appl. **41**(14), 6400–6407 (2014)
12. Aguiar E, Chawla NV, Brockman J, Ambrose GA, Goodrich V (2014) Engagement versus performance: using electronic portfolios to predict first semester engineering student retention. In: Proceedings of the Fourth International Conference on Learning Analytics And Knowledge, pp. 103–112. ACM (2014)

13. Aulck, L., Velagapudi, N., Blumenstock, J., West, J.: Predicting student dropout in higher education. arXiv preprint arXiv:1606.06364 (2016)
14. Márquez-Vera, C., Cano, A., Romero, C., Noaman, A.Y.M. , Mousa Fardoun, H., Ventura, S.: Early dropout prediction using data mining: a case study with high school students. Expert Syst. **33**(1):107–124 (2016)
15. Asif, R., Merceron, A., Ali, S.A., Haider, N.G.: Analyzing undergraduate students' performance using educational data mining. Comput. Edu. **113**, 177–194 (2017)
16. Kovacic, Z.: Predicting student success by mining enrolment data (2012)
17. Bayer, J., Bydzovská, H., Géryk, J., Obsivac, T., Popelinsky, L.: Predicting drop-out from social behaviour of students. Int. Educ, Data Mining Soc. (2012)
18. Guarín, C.E.L., Guzmán, E.L., González, F.A.: A model to predict low academic performance at a specific enrollment using data mining. IEEE Revista Iberoamericana de Tecnologias del Aprendizaje **10**(3), 119–125 (2015)
19. Costa, E.B., Fonseca, B., Santana, M.A., de Araújo, F.F., Rego, J.: Evaluating the effectiveness of educational data mining techniques for early prediction of students' academic failure in introductory programming courses. Comput. Hum. Behav. **73**, 247–256 (2017)
20. Dekker, G.W., Pechenizkiy, M., Vleeshouwers, J.M.: Predicting students drop out: a case study. In: International Working Group on Educational Data Mining (2009)
21. NSCRC - National Student Clearinghouse Research Center https://nscresearchcenter.org/snapshotreport33-first-year-persistence-and-retention/. Last accessed 15 Feb 2019
22. Márquez-Vera, C., Morales, C.R., Soto, S.V.: Predicting school failure and dropout by using data mining techniques. IEEE Revista Iberoamericana de Tecnologias del Aprendizaje **8**(1), 7–14 (2013)
23. Delen, D.: Predicting student attrition with data mining methods. J. Coll. Stud. Retention: Res. Theo. Prac. **13**(1), 17–35 (2011)
24. Tinto, V.: Dropout from higher education: a theoretical synthesis of recent research. Rev. Educ. Res. **45**(1), 89–125 (1975)
25. Khaleej Times https://www.khaleejtimes.com/nation/new-ratings-system-for-uae-universities-education-quality. Last accessed 5 Feb 2019
26. Shearer, C.: The crisp-dm model: the new blueprint for data mining. J. Data Warehous. **5**(4), 13–22 (2000)
27. Huang, S., Fang, N.: Predicting student academic performance in an engineering dynamics course: a comparison of four types of predictive mathematical models. Comput. Edu. **61**, 133–145 (2013)
28. Thammasiri, D., Delen, D., Meesad, P., Kasap, N.: A critical assessment of imbalanced class distribution problem: The case of predicting freshmen student attrition. Expert Syst. Appl. **41**(2), 321–330 (2014)
29. Raju, D., Schumacker, R.: Exploring student characteristics of retention that lead to graduation in higher education using data mining models. J. Coll. Stud. Retention: Res. Theory & Pract. **16**(4), 563–591 (2015)
30. Hoffait, A.-S., Schyns, M.: Early detection of university students with potential difficulties. Decis. Support Syst. **101**, 1–11 (2017)

An Approach for Textual Based Clustering Using Word Embedding

Ehab Terra, Ammar Mohammed, and Hesham Hefny

Abstract Numerous endeavors have been made to improve the retrieval procedure in Textual Case-Based Reasoning (TCBR) utilizing clustering and feature selection strategies. SOPHisticated Information Analysis (SOPHIA) approach is one of the most successful efforts which is characterized by its ability to work without the domain of knowledge or language dependency. SOPHIA is based on the conditional probability, which facilitates an advanced Knowledge Discovery (KD) framework for case-based retrieval. SOPHIA attracts clusters by themes which contain only one word in each. However, using one word is not sufficient to construct cluster attractors because the exclusion of the other words associated with that word in the same context could not give a full picture of the theme. The main contribution of this chapter is to introduce an enhanced clustering approach called GloSOPHIA (GloVe SOPHIA) that extends SOPHIA by integrating word embedding technique to enhance KD in TCBR. A new algorithm is proposed to feed SOPHIA with similar terms vector space gained from Global Vector (GloVe) embedding technique. The proposed approach is evaluated on two different language corpora and the results are compared with SOPHIA, K-means, and Self- Organizing Map (SOM) in several evaluation criteria. The results indicate that GloSOPHIA outperforms the other clustering methods in most of the evaluation criteria.

Keywords Text clustering · Text mining · Textual case based reasoning · Word embedding · Knowledge discovery

E. Terra (✉) · A. Mohammed (✉) · H. Hefny
Department of Computer Science, Faculty of Graduate Studies for Statistical
Research (FGSSR), Cairo University, Cairo, Egypt
e-mail: ehab_terra@pg.cu.edu.eg

A. Mohammed
e-mail: ammar@cu.edu.eg

H. Hefny
e-mail: hehefny@ieee.org

© The Author(s), under exclusive license to Springer Nature Switzerland AG 2021   261
A. E. Hassanien et al. (eds.), *Machine Learning and Big Data Analytics*
Paradigms: Analysis, Applications and Challenges, Studies in Big Data 77,
https://doi.org/10.1007/978-3-030-59338-4_14

1 Introduction

Case-Based Reasoning (CBR) [1] is the process towards perceiving and tackling issues 'reasoning' in light of past encounters 'cases'. In general, an effective CBR system design is achieved by selecting the appropriate features and retrieving cases correctly [2]. CBR is used in Information Retrieval (IR) [3] and machine learning [4]. Textual Case-Based Reasoning (TCBR) is a type of CBR that is applied on unstructured text [5]. Unlike simple attribute-value or structured object types of CBR where knowledge is easily gained and represented. In TCBR, the case is more complicated, since a single case may constitutes several subject areas and complex linguistic terms. Along these lines, it is testing issue in TCBR to look for comparative cases that contains various expressions of the same topics. For the most part, the retrieval procedure in TCBR utilizing clustering to arrange cases into subsets can improve the retrieval procedure in TCBR frameworks via looking only at comparable cases. Clustering [6] is a quest for potential relations and partitioning a dataset as per these. Clustering is considered one of the most essential data mining tasks. Text clustering is necessary for many applications such as, organizing web content [7] and text mining [8, 9]. Today Natural Language Processing (NLP) methods [10] (for instance: N-gram) can be accomplished on a lot bigger datasets, and they particularly outperform the basic models [11].

SOPHisticated Information Analysis (SOPHIA) [12] approach is one of the most successful efforts in case-based clustering which is characterized by its ability to work without prior domain of knowledge or language dependency. SOPHIA is based on the entropy of conditional probabilities to intelligently extract narrow themes from the case-base, where the theme is represented by a single word that can attract cases for their clusters. Thus, the semantic distance between a case and the theme is evaluated using Jensen–Shannon divergence [13] which facilitates an advanced Knowledge Discovery (KD) framework for case-based retrieval with one-shot, rather than iterating till divergence as in some other clustering approaches. However, using one word as a theme in SOPHIA is not sufficient to construct cluster attractors. This is because of the exclusion of the other words, associated with that word-theme in the same context, could not give a full picture of the theme. For example, Poetry can be a theme and Poet would be another theme although both belongs to the same context. Thus, single word theme allows SOPHIA to have more number of clusters to incorporate the whole attractors as possible as it can. However, if we combine more similar words in a single theme, then we will be able to attract more cases by any of its words. For instance, if 'Poetry' and 'Poet' themes could be combined in a single theme, the result would be better than each of these words having its own theme, because the two topics are in fact in the same context. One successful way to measure the similarities of words within a text is the word embedding technique [14]. Basically, word embedding is a technique that maps words into real number vectors for language modeling and learning features. The whole text is represented in a D-dimensional embedding space. In 2014, Pennington et al. [15] have proposed a prosperous word embedding technique called GloVe (Global Vector). It is an unsupervised method to discover connections between words and becomes a good alternative to the external ontology such as Wikipedia [16] or WordNet [17] or the predefined classes to find these relationships. Word embedding can be used to search for similar words and create vector space of words to become themes for clustering.

To this end and to enhance the TCBR, this chapter proposes an approach to integrate Global Vector (GloVe) word embedding technique into SOPHIA clustering method to form the so-called GloVe SOPHIA (GloSOPHIA). The chapter proposes an algorithm to construct themes based on threshold similarities between words. The proposed approach is evaluated on two different languages corpora namely Arabic Watan-2004 [18–20] and English Reuters-21578 corpus [21]. The key idea of using two different language corpora is to prove that the proposed approach is language independent. The experimental results are evaluated using several clustering evaluation criteria, and is compared with other clustering methods, namely, SOPHIA, K means [22], and Self-Organizing Map (SOM).

The rest of this chapter is organized as the following. Section 2 demonstrates some research efforts related to the work of this chapter. Section 3 describes the proposed approach. Section 4 shows the evaluation criteria that are used to evaluate the performance of the proposed approach. Section 5, discusses the experimental results. Section 6 concludes the chapter.

2 Related Work

Numerous endeavors have been made to enhance the clustering of TCBR. Vernet and Golobardes [23] used clusters to represent the space of each class. Their clustering method is introduced in two levels. In the first level, clusters the training data are clustered into spheres one for each cluster. In the second level, the spheres are clustered to detect the behavior of the elements existing in the sphere. In this second level, policies are applied, Mean Sphere and (MKM) approaches. Their algorithm aims to reduce the number of features in order to reduce CPU time and to avoid noise data. However, their technique depends on the labeled data, and that is not available in every dataset. Cunninghamet, et al. [24] proposed an algorithm dependent on graph, which improves the Bag-Of-Words (BOW) by preserving word order, and based on n-grams to give a higher weight for similarity. Nonetheless, the proposed algorithm cannot discriminate between the problem and solution. Proctor et al. [25] proposed a method dependent on Information Gain (IG). The later is a statistical algorithm to identify predictive elements for classification in order to discriminate problem from solution. It is based on Shannon's theory that the value of a message increases as its likelihood of occurring decreases. The algorithm computes the IG score for all words in the corpus, then the list of score is sorted and take a subset of highly predictive words. Recio-Garcia et al. [2] proposed TCBR clustering technique based on a Lingo method [26] that is based on Singular Value Decomposition (SVD) [27]. They used Lucene to retrieve similar cases and Carrot2 to cluster the cases using the framework jcolibri. However, their previous works are not expanding well to larger dataset. Another effort which is called SOPHIA is introduced by Patterson et al. [12]. The main idea of their proposed approach is to build clusters based on the entropy of conditional probabilities to intelligently extract narrow themes from the case-base where a theme is a single word that groups semantically related cases. However, their experiment was set to a high number of clusters which in turn decreases the quality of clustering due to forming empty clusters or clusters with only few cases. Additionally, using just a single word to form a cluster theme is not sufficient (for example: Poetry can be a theme and Poet would be another theme). Fornells, et al. [28] provided a plugging called Thunder for the jcolibri framework. Thunder allows CBR experts to manage case memories organized in clusters and to incorporate a case memory organization model based on SOM [29] as a clustering technique to manage case retrieval. Guo et al. [30] integrated Growing Hierarchical Self-Organizing Map (GHSOM) [31] clustering to enhance feature selection in TCBR. They used GHSOM to solve the performance drawback occurs in SOM due to the very frequent similarity calculations that become necessary in a highly dimensional feature space that becomes a critical issue when handling large volumes of text. However, their proposed cluster technique inherits the same drawbacks of SOM in terms of random weights for the initialization and sensitivity to input order [32, 33].

3   The Proposed GloSOPHIA Approach

The primary goal of the proposed GloSOPHIA is to supply SOPHIA with word embedding. Figure 1 shows the proposed approach. The proposed approach initially begins with a preprocessing phase of the data followed by a group of steps including normalization, build word embedding vector space, construct word group vector space, narrow theme discovery, similarity knowledge discovery, and case assignment discovery steps. In the following, we describe the preprocessing phase followed by the details of those steps.

3.1   Preprocessing

Usually, preprocessing of the data is a crucial step in knowledge discovery. The final output of the data prepossessing is considered a training data set. Since the data preprocessing depends on the language of data, the way of processing might be slightly different. For example, converting a text to lowercase is applicable in English but not in Arabic and removing diacritics is applicable in Arabic but not in English. The data preprocessing phase includes the following:

- Normalize the characters to a single form آ، أ، إ → ا. (**For Arabic only**)
- Remove Arabic diacritics using "arabicStemR" package. (**For Arabic only**)
- Convert text to lowercase. (**For English only**)

- Lemmatize words and removes "pronoun, punctuation, coordinating conjunction, determiner, numeral, ad-position, particle" using the excellent package called "udpipe" (it used as Part Of Speech (POS) tagging tool) [34, 35].
- Remove stop words and noises.
- Stripe white-spaces, remove numbers and punctuation.
- Documents tokenization and convert to Document-Term matrix with term frequencies tf.
- Remove sparse terms that sparsity ratio is more than 99%.

3.2 GloSOPHIA Steps

Having described the preprocessing phase, we describe the details of the 6 steps constituting Glosophia. Figure 1 depicts steps order and the data flow between them. In step 1, case-base normalization using conditional probability is illustrated. In step 2, GloVe model is used to get the matrix of word-word co-occurrence. In step 3, the vectors is constructed of the closest words Ω that have a minimum threshold of cosine similarity. In step 4, subset of themes is selected to construct the clusters. Finally, steps 5 and 6 respectively, similarity are calculated between the clusters and all cases and the most similar clusters are assigned. It is worth emphasizing that steps from 2 to 4 do not depend on step 1.

Step 1. Normalization In this step, a case knowledge is discovered for feature (word) space [36] as step 1 in SOPHIA. Let X denotes the set of all documents in the corpus and Y the set of all words in X. The normalization process is represented by the following formula [12]:

$$P(y|x) = \frac{tf(x, y)}{\sum_{t \in Y} tf(x, t)} \tag{1}$$

Where, x is a document in X, y is a term in Y and $tf(x, y)$ is the term frequency for the term y in the document x.

Step 2. Build Word Embedding Vector Space The word embedding vector space is generated using GloVe model from the case-base. The main formula for GloVe is as following [15]:

$$J = \sum_{i,j=1}^{v} f(X_{ij})(w_i^T \tilde{w}_J + b_i + \tilde{b}_J - \log X_{ij})^2 \tag{2}$$

Where, $w \in \mathbb{R}^d$ are word vectors with transpose w^T, $\tilde{w} \in \mathbb{R}^d$ are independent context word vectors, $\mathbb{R}^d$ is real coordinate space of d dimensions, V is the size of the vocabulary, X_{ij} the frequency of term j in the context of term i, and $f(X_{ij})$ is the weighting function. b_i and $\tilde{b}_J$ represent the bias of w_i and $\tilde{w}_J$ respectively. $w_i^T \tilde{w}_J$ is calculated by Eq. 3.

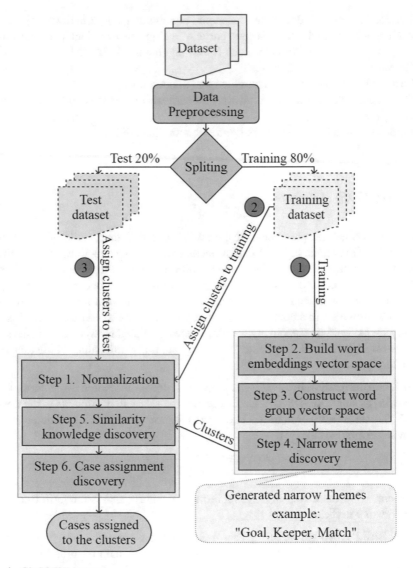

Fig. 1 GloSOPHIA steps

$$w_i^T \tilde{w}_J = \log(P_{ij}) = \log(X_{ij}) - \log(X_i) \qquad (3)$$

Where, P_{ij} is the conditional probability of the term j in the context of the term i, X_i is all the terms that appeared in context of the term i and can be calculated as follows:

$$X_i = \sum_k X_{ik} \qquad (4)$$

Step 3. Construct Word Group Vector Space In this step, we propose the algorithm 1 to construct the themes from the generated word embedding vector space in the previous step. The proposed algorithm takes a word-word matrix X generated from GloVe method with a threshold that ranges from 0 to 1 (0 means no similarity and 1 means typical word) and returns a vector space with the closest terms. X in line 1 of the algorithm is normalized using the Euclidean norm which is Square Root for Sum of Squares (SRSS) $\sqrt{x_1^2 + \cdots + x_n^2}$ to scale each element by vector-wise scaling for the whole matrix. In line 2, the algorithm calculates pairwise cosine similarity between each two words in X. Line 3 sorts in descending the similar words in each vector. Line 4 trims the vectors having only words with similarity greater than or equal the threshold t. Line 5 removes each vector Z from the vector space Ω having words count $|Z|$ less than or equal 1 1. Lines 6 to 9 demonstrate copying words from vectors that contain a mutual word(s). Finally, it should be mentioned that there will be some duplicated vectors resulted from the word copying process, these vectors will be removed so that only the unique vectors representing the output vector space (themes). The later process is described in line 10. Table 1 demonstrates a simple example of themes after narrowed by step 4 where Ψ is the narrow theme after removing themes that have less attraction and N is the number of desired clusters. The example shows 7 themes ($N = 7$) when threshold $t = 0.72$ gained from watan-2004 corpus.

Algorithm 1 Generate themes.

Input: Terms vector space from GloVe algorithm $\rightarrow X$, threshold $\rightarrow t$
Output: Closest terms Vector space which is based on a threshold $t \rightarrow \Omega$

1: Normalizing X using L2 norm
2: Calculates pairwise cosine similarities for X.
3: Sort all related terms descending by similarity value.
4: $\Omega \leftarrow$ all terms with similarity values $\geq$ threshold t.
5: Eliminates all vectors in Ω that have terms ≤ 1.
6: **for** each vector $Z \in \Omega$ vector space **do**:
7:    $\Omega(Z) \leftarrow$ all vectors that have mutual terms with Z vector.
8:    uniqueTerms $\leftarrow$ all unique terms in $\Omega(Z)$.
9:    **for** each $Z \in \Omega(Z)$ **do:** $Z \leftarrow$ uniqueTerms
10:    **end for**
11: **end for**
12: Remove duplicated vectors from Ω.

Step 4. Narrow Theme Discovery In this step, SOPHIA formula's is updated in such a way that allows the theme to be represented as a vector. The theme in GloSOPHIA is a conditional probability over a set of words that co-occur with any of the theme terms $z \in Z$. The updated equation here can be represented by:

$$P(y|Z) = \frac{\sum_{x \in X(Z)} tf(x, y)}{\sum_{x \in X(Z), t \in Y} tf(x, t)} \tag{5}$$

Table 1 Narrow themes Ψ when N = 7 and threshold = 0.72

Narrow themes	English translated
هريرة، رضي، أبي	May be pleased with, Abi, Huraira
اولمبياد، اثينا، اولمبي	Olympic, Olympiad, Olympiad
مباريات، خسر، منتخبنا	Match, lost, our team
خام، برميل، نفط، نفطي	Crude, oil, barrels
جورج، بوش، واشنطن	George, Bush, Washington
بمحافظة، ظفار، مديرية	Governorate of, Dhofar, Directorate
انتخاب، رئاسي، انتخابي	Elected, presidential, election

Where, y is the word that co-occurs with any from Z terms in the same document, $X(Z)$ is all documents that contains any word from Z terms, $tf(x, y)$ is the term frequency for the term y in document x, and Y: all terms that co-occur with any term from Z.

By applying Eq. 5, a vector of normalized values for every theme in Ω is ready to be compared with documents. Then a selection process of the most significant subset of themes in Ω is performed as a cluster attractor. The later process is performed by calculating the entropy as the following:

$$H(Y|Z) = -\sum_{y \in Y} p(y|Z) \log p(y|Z) \tag{6}$$

Where, Y is all words that co-occur with any term in the vector $Z \in \Omega$.
$Z_1 \in \Psi_i, Z_2 \in \Omega - \Psi_i \rightarrow H(Y|Z_1) \geq H(Y|Z_2)$, then:

$$\Psi = \bigcup_i^N Z_i \tag{7}$$

Where, Ψ is a set of selected narrow themes. Based on the rule that states that only themes with in-between case frequency are informative, only the terms with case frequencies range between min_cf (minimum case frequency for a single word) and max_cf (maximum case frequency for a single word) is included. Also, only themes Ω from narrow themes that have case frequencies range between theme_min_cf (minimum case frequency for the whole theme) and theme_max_cf (maximum case frequency for the whole theme) is included. It should be Noted that the length of vector space $|\Omega|$ must be greater than or equal to N in order to generate $|\Psi|$ that is equals to the number of clusters N. If $|\Omega|$ is less than N then it will fail to generate the desired number of clusters.

Step 5. Similarity Knowledge Discovery Similar to SOPHIA, Jensen- Shannon divergence (JS) [13] is used to measure the semantic similarity between document and cluster theme according to the following:

$$JS_{\{0.5,0.5\}}[p_1, p_2] = H[\overline{p}] - 0.5H[p_1] - 0.5H[p_2] \tag{8}$$

Where, $H[\overline{p}]$ is the entropy of probability distribution p and average probability distribution $= 0.5p_1 + 0.5p_2$. The lower value of JS represents a higher similarity between two probabilities.

Step 6. Case Assignment Discovery The attractor theme Z will be assigned to case when:

$$Z = \arg \min_{t \in \Omega} JS_{\{0.5,0.5\}}[p(Y|x), p(Y|t)] \tag{9}$$

Now, all cases are assigned to the clusters having the highest similarities such that every case is assigned to only one cluster. The case assignment can be also softened by adding a weight to every cluster instead of choosing the highest similar.

4 Evaluation Criteria

In order to evaluate the performance of the proposed approach, several performance measures are used like F-measure on beta = 0.4 [12], Purity [3], Normalized Mutual Information (NMI) [3], Rand Index (RI) [3], Silhouette [37], Dunn Index (DI) [38], and Connectivity [39]. These performance measures are categorized into two main groups; external and internal evaluation criteria. External criteria depend on labels of the clusters. On the other hand, internal evaluation criteria depend on the quality of the clusters and the degree of its separation. Broadly speaking, good results in internal evaluations does not mean it is effective in application [3].

4.1 External Evaluation

In the following, we explain the external criteria used in the evaluation of the proposed approach.

F-Measure: The evaluation of clustering is performed by using labels added by experts $ET(d)$ to the documents. New label $CT(d)$ is assigned by calculating the percentage of labels in the cluster that exceed a threshold β, while A, B, and C are total numbers of "true positive", "false positive" and "false negative" respectively. F-measure is calculated as following [12]:

$$A = A + |ET(d) \cap CT(d)| \tag{10}$$

$$B = B + |CT(d) \backslash ET(d)| \tag{11}$$

$$C = C + |ET(d) \backslash CT(d)| \tag{12}$$

$$F = \frac{2A}{2A + B + C} \tag{13}$$

A high F-measure is obtained by making many clusters. Simply, F measure is 1 if the number of documents = clusters number. Thus, increasing the clusters number to gain a high F measure will decrease the quality of the clustering.

Purity: Purity has a value ranging from 0 to 1 scale, where 0 indicates a bad clustering and 1 a perfect clustering. It assigns the class d for each cluster m which has the maximum number of documents to the cluster. Purity is calculated according to the following [3]:

$$Purity = \frac{1}{N} \sum \max |m \cap d| \tag{14}$$

NMI: Normalized Mutual Information performance measure is derived from maximum likelihood function [3] and is calculated according to the following:

$$NMI(M, D) = \frac{I(M, D)}{\sqrt{H(M)H(D)}} \tag{15}$$

where, M, D are class and cluster labels respectively. $I(M, D)$ is mutual information and $H(M)$, $H(D)$ is entropy of M and D respectively. In NMI, the result is not improving due to the increase in the number of clusters. That can solve the existing problem in F-measure and Purity. This is due to the fact that it is normalized by increasing the denominator with the number of clusters.

RI: Rand Index performance measure indicates how similar the clusters are to the experts classes [3] and is calculated according to the following equation:

$$RI = \frac{TP + TN}{TP + FN + FP + TN} \tag{16}$$

Where, TP, TN, FP, FN are the number of documents with the same class in the same cluster (true positive), same class in different cluster (true negative), different class in the same cluster (false positive), different class in different cluster (false negative) respectively.

4.2 Internal Evaluation

In the following, we explain the internal criteria used in the evaluation of proposed approach.

Silhouette: Silhouette is used to calculate the closeness of each point in one cluster to the points in the other clusters and to judge some attributes such as clusters geometry using a similarity distance measure;e.g., Euclidean or Manhattan distance. Silhouette performance measure has a range from $[-1$ to $1]$ [37].

For each data point i in the cluster C_i.

$$a(i) = \frac{1}{|C_i| - 1} \sum_{j \in C_i, i \neq j} d(i, j) \qquad (17)$$

Where, $d(i, j)$ is the distance between data points i and j in the cluster C_i

$$b(i) = \min_{k \neq i} \frac{1}{|C_k|} \sum_{j \in C_k} d(i, j) \qquad (18)$$

$$s(i) = \frac{b(i) - a(i)}{\max a(i), b(i)}, \text{ if } |C_i| > 1 \qquad (19)$$

If $|C_i| = 1$, then, $s(i) = 0$.

DI: Dunn Index specifies non-sparse well-separated clusters. It is defined as the ratio between the minimum inter-cluster to the maximum intra-cluster distances [38] according to the following equation.

$$DI = \frac{\min_{1 \leq i \leq j \leq n} d(i, j)}{\max_{1 \leq k \leq n} d'(k)} \qquad (20)$$

Where, $d(i, j)$ is the minimum distance between clusters i and j. $d'(k)$ is largest intra-cluster distance of cluster k. *Dunn* Index has a value between 0 and ∞. The higher the value of D, the better clustering quality.

Connectivity: Connectivity uses the k-nearest neighbors to assess connectivity degree between clusters. The connectivity value begins from 0 and has no limit. In contrast to the other measures, a minimum value of connectivity enhances the quality of the clustering [39].

5 Experimental Results

In the following subsections experimental results are discussed. In Sect. 5.1, the corpora used to evaluate the performance are discussed. In Sect. 5.2, classes distribution to clusters are shown. In Sect. 5.3, the environment setup is described. In Sect. 5.4, the performance of GloSOPHIA against several cluster methods is compared. Finally, Sect. 5.5, shows the effect of changing the threshold on clustering performance.

5.1 Datasets Description

To evaluate the performance of the proposed approach, we use two different Corpora: Watan-2004 Arabic newspaper [18–20] and Reuters-21578 English corpus [21]. We use two different languages corpora to prove language independence from the proposed approach. Watan-2004 contains 182, 240 distinct terms reduced to 2866 distinct terms after the preprocessing step. It consists of 20, 291 documents organized in 6 main topics. The corpus is divided into 6 directories named with the class names. Each directory contains file for each document. Table 2 summarizes the distributions of the topics in the corpus. The corpus is randomly partitioned by a ratio 80%, 20% for training and testing respectively.

On the other hand, the Reuters-21578 corpus contains 21, 578 documents with 37, 852 distinct terms reduced to 846 distinct terms after the preprocessing phase. This corpus has 120 overlapped labels and includes only 10, 377 non-empty documents that have at least one label. These documents are divided into 7068 and 2745 for training and testing respectively. The remaining 564 documents are marked 'NOT-USED' as according to 'lewissplit' attribute. There are 68 categories taken into consideration when the most frequented category is assigned to the document. Table 3 shows a summary of the corpus distribution on the top 10 categories ordered in descending by document frequencies.

The case frequency that is used here to include informative terms only is set from $min\_cf = 100$ to $max\_cf = 1000$, and to include only informative themes is set from $theme\_min\_cf = 500$ to $theme\_max\_cf = 5000$ as a case frequencies for the themes has to be narrowed (see step 4 in Sect. 3) and a word is represented by a vector of 75 dimensions using GloVe as described in step 2 Sect. 3. In addition, all themes which have only one word are eliminated.

Table 2 Watan-2004 dataset structure

Label name	Number of documents
Culture	2782
Economy	3468
International	2035
Local	3596
Religion	3860
Sports	4550
Total	20,291

Table 3 Reuters-21578 dataset structure of 10 categories that have highest document frequencies

Label name	Number of documents
Trade	392
Sugar	121
Ship	163
Money-fx	602
Interest	217
Grain	523
Earn	3754
Crude	475
Coffee	116
Acq	2113
Total	9671

5.2 Evaluation of Classes Distribution in Clusters

In order to evaluate GloSOPHIA, we need to distribute case-base categories into clusters. In the following, we explain the distribution of only one example for each corpus. The thresholds have been selected because they have maintained good results in all criteria, while the number of clusters N is arbitrarily chosen from the generated range.

Evaluation of Watan-2004 Corpus An example to distribute the case-base categories to clusters using GloSOPHIA when N = 7 and threshold t = 0.72 are selected.

Table 4 describes more details about the distribution of classes within clusters. To see the class distribution Figure 2 illustrates the percentage of classes in each cluster.

Evaluation of Reuters-21578 Corpus An example to distribute the case-base categories to clusters using GloSOPHIA is to set N = 10 and threshold t = 0.67 are selected. Table 5 describes more details about the distribution of classes with the highest 10 clusters. To see the class distribution, Fig. 3 illustrates the percentage of classes in each cluster.

Table 4 Watan-2004 dataset: distribution of dataset categories to clusters when N = 7 and threshold t = 0.72

Cluster label	Sports	Culture	Local	International	Economy	Religion	Total
..., Abi, Huraira		126	20		3	732	881
Olympic, ...	363	38	2				403
Match, ...	458	26	3	1			488
Crude, ...	2	34	55	10	417	9	527
George, ...	3	119	1	181	26	9	339
Governorate ...	81	148	622	1	238	8	1098
Elected, ...	3	65	16	214	9	14	321
Total	910	556	719	407	693	772	4057

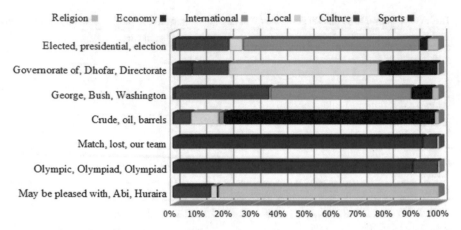

Fig. 2 Watan-2004 dataset: distribution of categories to clusters when N = 7 and threshold t = 0.72

Table 5 Reuters-21578 dataset: case-base distribution categories to clusters using GloSOPHIA when N = 10 and threshold t = 0.67

Clusters/categories	Earn	Acq	Money	Crude	Grain	Trade	Interest	Ship	Gnp	Gold
Delivery, shipment,...	0	1	0	5	101	1	0	17	0	3
Figure, show	1	0	101	2	1	21	51	0	19	1
Import, export	0	0	39	7	16	76	6	9	8	1
Merger, acquisition	11	328	0	2	0	0	0	1	0	0
Natural, gas, oil,...	2	5	1	121	0	0	0	9	1	3
Operate, operation	925	26	10	2	4	0	0	1	0	3
Output, production,...	0	0	0	8	7	0	1	0	1	0
Outstand,...	102	226	5	0	2	0	1	0	0	0
Own, subsidiary,...	4	46	3	5	0	0	4	1	0	14
Regulatory, approval,...	0	8	2	0	2	2	1	4	0	1

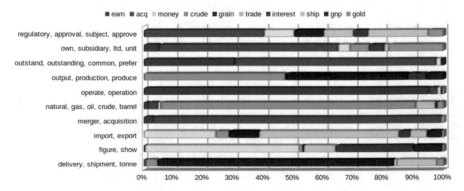

Fig. 3 Reuters dataset: distribution of categories to clusters when N = 10 and threshold t = 0.67

5.3 Experiment Environment Setup

The environment setup for the experiment is a laptop with Intel Core i5 2.5 GHz processor, 16 GB Ram. The language used for processing is R version 3.5.3 that is running on Ubuntu 18.04 operating system. The execution time of the training data (16,234 documents) takes only 5.5 s to generate narrow themes (clusters) from the training set, and the execution time of all experimental test data (4057 documents) is ranging from 4 to 7 s. It is enhanced by using a parallel c++ code depending on the number of clusters 6–24.

5.4 Evaluate Performance

In the following, we compare GloSOPHIA with the clustering methods SOPHIA, K-means, and SOM using 7 different evaluation criteria: F measure on beta = 0.4, Purity, NMI, RI, Silhouette, DI, and Connectivity described in Sect. 4. We show the performance of GloSOPHIA in the two corpora. The range of cluster numbers is set to allow testing intensively the behavior of each clustering technique. This range is constrained by clustering approaches capabilities such as K-means convergence and number of themes that can be constructed Ω.

Watan-2004 Dataset Table 6 describes the performance of GloSOPHIA when threshold = 0.72 compared to the other clustering methods. The table represents the best score of each evaluation. It shows that GloSOPHIA outperforms the other methods in all criteria except connectivity. More clearly, the superiority of the GloSOPHIA in external evaluations (F-measure, Purity, NMI, and RI) is due to the fact that it achieves a better representation of the categories. On the other hand, GloSOPHIA's superiority in internal evaluations (Silhouette, DI) is attributed to its compactness and the clusters are well-separated. SOM is the best in connectivity, due to most items are placed in the same cluster as their nearest neighbors in the data space. Figure 4 shows the performance of each clustering method. The number of clusters is set in the range between 6 and 24. It shows that the contributed approach outperforms SOPHIA in all criteria and also it has a higher performance than K-means and SOM in most of the criteria.

Reuters-21578 Dataset Table 7 describe the performance of GloSOPHIA when threshold = 0.67 compared to other clustering methods. Similar to Watan-2004, the results show that GloSOPHIA outperform the other clustering methods in all criteria except connectivity. We note that here we have the same conclusions that were discussed using watan-2004 newspaper dataset. Figure 5 shows the performance of each clustering method. The number of clusters is set in the range between 10 and 20. It also shows that the contributed approach algorithm outperforms also SOPHIA in all criteria and has a higher performance than K-means and SOM in some criteria.

Table 6 Watan-2004 dataset: best values comparison between clustering methods

Criteria	GloSOPHIA	SOPHIA	K-means	SOM
F measure	**0.8097**	0.6147	0.7275	0.7472
Purity	**0.8097**	0.6170	0.7259	0.7375
NMI	**0.5547**	0.4249	0.5153	0.4641
RI	**0.8707**	0.8288	0.8121	0.8189
Silhouette	**0.03812**	0.027366	0.009602	0.03099
DI	**0.3376**	0.2245	0.12771	0.27864
Connectivity	2752	3333	1528	**1091**

Bold represents the best results out of all methods

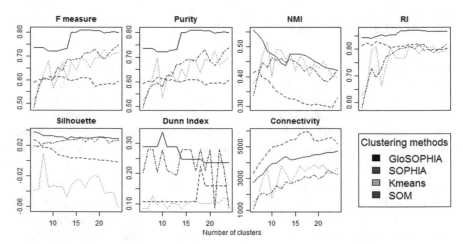

Fig. 4 Watan-2004 dataset: comparison between GloSOPHIA (on threshold = 0.72), SOPHIA, K-means, and SOM using evaluation criteria. Note that in Connectivity the lower is better

Table 7 Reuters-21578 dataset: best values comparison between clustering methods

Criteria	GloSOPHIA	SOPHIA	K-means	SOM
F measure	**0.7435**	0.6605	0.7267	0.6871
Purity	**0.7373**	0.6397	0.6718	0.6288
NMI	**0.5923**	0.385	0.4919	0.4362
RI	**0.8972**	0.788	0.7823	0.762
Silhouette	**0.10319**	0.010692	0.05013	0.04833
DI	**0.2163**	0.07832	0.1383	0.084259
Connectivity	1844	2214	1288	**968.6**

Bold represents the best results out of all methods

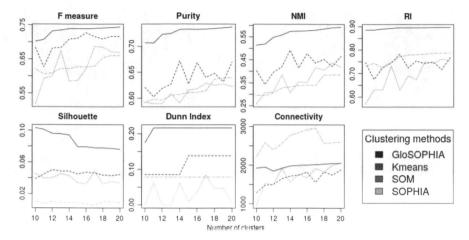

Fig. 5 Reuters-21578 dataset: comparison between GloSOPHIA (on threshold = 0.67), SOPHIA, K-means, and SOM using evaluation criteria. Note that in Connectivity the lower is better

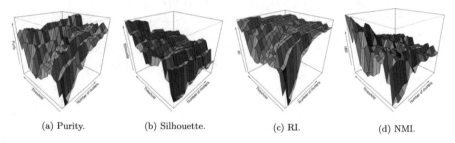

(a) Purity.   (b) Silhouette.   (c) RI.   (d) NMI.

Fig. 6 Watan-2004 dataset: the impact of the threshold on evaluations when changing the number of clusters

5.5 Investigating the Influence of Threshold on Clustering

The main goal here is to investigate the effect of changing threshold on the clustering performance. Figures 6 and 7 illustrate the evaluation of Purity, Silhouette, RI, NMI on watan-2004 and reuters-21578 respectively.

Impact on Watan-2004 Dataset: It is found that, when the range of threshold is changed between 0.67 and 0.77, the number of clusters is changed from 6 to 24.

Impact on Reuters-21578 Dataset: It is found that, when the range of threshold is changed between 0.64 and 0.74, the number of clusters is changed from 7 to 31.

It should be noted that those ranges have been chosen because if the threshold is decreased to low scores, for example 0.5, it would place a burden on the algorithm. This is caused by the construction of group with weaker relationships, which reduces the performance and quality of the algorithm. On the other hand, increasing the

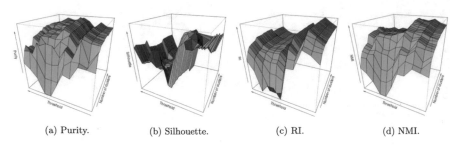

(a) Purity.   (b) Silhouette.   (c) RI.   (d) NMI.

Fig. 7 Reuters-21578 dataset: the impact of the threshold on evaluations when changing the number of clusters

threshold with very high scores, for example 0.9, will make it harder to construct a number of groups that fit the desired number of clusters.

Figures 6 and 7 describes the impact of the threshold on the purity, RI and NMI when the number of clusters is low. At the point when the number of clusters increases, the threshold effect on clustering quality gradually decreases.

Likewise, it's clear that there is a opposite connection among silhouette and the number of clusters. Hence, it is smarter to pick high threshold when the number of clusters is low. In any case, when the clusters quantity is expanded to reduce the threshold, more groups of comparable words can be created by step 3, and subsequently N is consistently $\leq |\Omega|$.

6 Conclusion

In this chapter, an enhanced clustering approach called GloSOPHIA (GloVe SOPHIA) have been introduced. The key idea of the proposed approach was to extend the textual based cluster technique SOPHisticated Information Analysis (SOPHIA) by integrating the word embedding technique Global Vector (GloVe) in order to enhance knowledge discovery in Textual Case-Based Reasoning (TCBR). A new algorithm have been introduced to generate themes that are fed to SOPHIA with similar terms vector space gained from the embedding technique. To evaluate the performance of GloSOPHIA, the approach was applied on two corpora; Watan-2004 Arabic newspaper and Reuters-21478 English corpus. Results of the proposed approach were compared with the clustering methods SOPHIA, K-means, and Self-Organizing Map (SOM). Several performance measures, including F-measure, Purity, NMI, RI, Silhouette, and DI, have been taken into consideration to evaluate the proposed approach. The results indicated that GloSOPHIA has gained a stable improvement in most of the evaluations criteria against the other clustering methods.

References

1. Aamodt, A., Plaza, E.: Case-based reasoning: foundational issues, methodological variations, and system approaches. AI Commun. **7**(1), 39–59 (1994)
2. Recio-Garcia, J.A., Diaz-Agudo, B., González-Calero, P.A.: Textual cbr in jcolibri: from retrieval to reuse. In: Proceedings of the ICCBR 2007 Workshop on Textual Case-Based Reasoning: Beyond Retrieval, pp. 217–226 (2007)
3. Manning, C., Raghavan, P., Schütze, H.: Introduction to information retrieval. Nat. Lang. Eng. **16**(1), 100–103 (2010)
4. Witten, I.H., Frank, E., Hall, M.A., Pal, C.J.: Data Mining: Practical Machine Learning Tools and Techniques, 4th edn. Morgan Kaufmann (2017)
5. Weber, R.O., Ashley, K.D., Brüninghaus, S.: Textual case-based reasoning. Knowl Eng. Rev. **20**(3), 255–260 (2005)
6. Aggarwal, C.C., Zhai, C.X.: A survey of text clustering algorithms. In: Mining text data, pp. 77–128. Springer (2012)
7. Chuanping, H.: Zheng, X., Liu, Y., Mei, L., Chen, L., Luo, X.: Semantic link network-based model for organizing multimedia big data. IEEE Trans. Emerg. Top. Comput. **2**(3), 376–387 (2014)
8. Allahyari, M., Pouriyeh, S.A., Assefi, M., Safaei, S., Trippe E.D., Gutierrez, J.B., Kochut, K.: A brief survey of text mining: classification, clustering and extraction techniques. CoRR, abs/1707.02919 (2017)
9. Silge, J., Robinson, D.: Text mining with R: a tidy approach. O'Reilly Media, Inc. (2017)
10. Hirschberg, J., Manning, C.D.: Advances in natural language processing. Science **349**(6245), 261–266 (2015)
11. Mikolov, T., Chen, K., Corrado, G., Dean, J.: Efficient estimation of word representations in vector space. arXiv preprint arXiv:1301.3781 (2013)
12. Patterson, D., Rooney, N., Galushka, M., Dobrynin, V., Smirnova, E.: Sophia-tcbr: a knowledge discovery framework for textual case-based reasoning. Knowl.-Bas. Syst. **21**(5), 404–414 (2008)
13. Lin, J.: Divergence measures based on the shannon entropy. IEEE Trans. Inf. Theo. **37**(1), 145–151 (1991)
14. Schnabel, T., Labutov, I., Mimno, D., Joachims, T.: Evaluation methods for unsupervised word embeddings. In: Proceedings of the 2015 Conference on Empirical Methods in Natural Language Processing, pp. 298–307 (2015)
15. Pennington, J., Socher, R., Manning, C.: Glove: global vectors for word representation. In: Proceedings of the 2014 Conference on Empirical Methods in Natural Language Processing (EMNLP), pp. 1532–1543 (2014)
16. Hu, X., Zhang, X., Lu, C., Park, E.K., Zhou, X.: Exploiting wikipedia as external knowledge for document clustering. In: Proceedings of the 15th ACM SIGKDD International Conference on Knowledge Discovery and Data Mining, KDD '09, pp. 389–396. ACM : New York, NY, USA (2009)
17. Wei, T., Yonghe, L., Chang, H., Zhou, Q., Bao, X.: A semantic approach for text clustering using wordnet and lexical chains. Exp. Syst. Appl. **42**(4), 2264–2275 (2015)
18. Abbas, M., Smaïli, K., Berkani, D.: Comparing TR-classifier and KNN by using reduced sizes of vocabularies. In: 3rd International Conference on Arabic Language Processing, Rabat, Morocco (2009)
19. Abbas, M.: Smaïli, K., Berkani, D.: TR-classifier and kNN evaluation for topic identification tasks. Int. J. Inf. Commun. Technol. **3**(3), 10 (2010)
20. Abbas, M., Smaili, K., Berkani, D.: Evaluation of topic identification methods on arabic corpora. J. Dig. Inf. Manage. **9**(5):8 double column (2011)
21. Lewis, D.: Reuters-21578 text categorization test collection, 1997. E-print: http://www.daviddlewis.com/resources/testcollections/reuters21578 (1997)
22. Hartigan, J.A., Wong, M.A.: Algorithm as 136: a k-means clustering algorithm. J. Roy. Stat. Soc.. Series C (Appl. Stat.), **28**(1), 100–108 (1979)

23. Vernet, D.: Golobardes, E.: An unsupervised learning approach for case-based classifier systems. Intelligence **6**, 01 (2003)
24. Cunningham, C., Weber, R., Proctor, J.M., Fowler, C., Murphy, M.: Investigating graphs in textual case-based reasoning. In: European Conference on Case-Based Reasoning, pp. 573–586. Springer, Springer Berlin Heidelberg (2004)
25. Proctor, J.M., Waldstein, I., Weber, R.: Identifying facts for tcbr. In: ICCBR Workshops, pp. 150–159 (2005)
26. Osiński, S., Stefanowski, J., Weiss, D.: Lingo: search results clustering algorithm based on singular value decomposition. In: Intelligent Information Processing and Web Mining, pp. 359–368. Springer (2004)
27. Golub, G.H., Reinsch, C.: Singular value decomposition and least squares solutions. In: Linear Algebra, pp. 134–151. Springer (1971)
28. Fornells, A., Recio-García, J.A., Díaz-Agudo, B., Golobardes, E., Fornells, E.: Integration of a methodology for cluster-based retrieval in jcolibri. In: International Conference on Case-Based Reasoning, pp. 418–433. Springer (2009)
29. Kohonen, T.: The self-organizing map. Proc. IEEE **78**(9), 1464–1480 (1990)
30. Guo, Y., Jie, H., Peng, Y.: Research of new strategies for improving cbr system. Artif. Intell. Rev. **42**(1), 1–20 (2014)
31. Chan, A., Pampalk, E.: Growing hierarchical self organising map (ghsom) toolbox: visualisations and enhancements. In: Proceedings of the 9th International Conference on Neural Information Processing, 2002. ICONIP '02., vol. 5, pp. 2537–2541 (2002)
32. Salem, M., Buehler, U.: An enhanced ghsom for ids. In: 2013 IEEE International Conference on Systems, Man, and Cybernetics, pp. 1138–1143 (2013)
33. Shi, H., Xu, H.: An enhanced ghsom for the intrusion detection. In: IET Conference Proceedings, pp. 5–5 (1) (2015)
34. Böhmová, A., Hajič, A., Hajičová, E., Hladká, B.: The prague dependency treebank. In: Treebanks, pp. 103–127. Springer, Netherlands (2003)
35. Smrž, O., Bielický, V., Kouřilová, I., Kráčmar, J., Hajič, J., Zemánek, P.: rague arabic dependency treebank: a word on the million words. In: Proceedings of the Workshop on Arabic and Local Languages (LREC 2008), pp. 16–23. European Language Resources Association, Marrakech, Morocco (2008)
36. Joachims, T.: A probabilistic analysis of the rocchio algorithm with tfidf for text categorization. In: Proceedings of the Fourteenth International Conference on Machine Learning, ICML '97, pp. 143–151. Morgan Kaufmann Publishers Inc, San Francisco, CA, USA (1997)
37. Rousseeuw, P.J.: Silhouettes: a graphical aid to the interpretation and validation of cluster analysis. J. Comput. Appl. Math. **20**, 53–65 (1987)
38. Dunn, J.C.: Well-separated clusters and optimal fuzzy partitions. J. Cybern. **4**(1), 95–104 (1974)
39. Knowles, J., Handl, J.: Exploiting the Trade-off—The Benefits of Multiple Objectives in Data Clustering, pp. 547–560. Springer, Berlin Heidelberg (2005)

A Survey on Speckle Noise Reduction for SAR Images

Ahmed S. Mashalyⓘ **and Tarek A. Mahmoud**ⓘ

Abstract Speckle noise disturbance is the most essential factor that affects the quality and the visual appearance of the synthetic aperture radar (SAR) coherent images. For remote sensing systems, the initial step always involves a suitable method to reduce the effect of speckle noise. Several non-adaptive and adaptive filters have been proposed to enhance the noisy SAR images. In this chapter, we introduce a compressive survey about speckle noise reduction in SAR images followed by two proposed non-adaptive filters. These proposed filters utilize traditional mean, median, root-mean square values, and large size filter kernels to improve the SAR image appearance while maintaining image information. The performance of the proposed filters are compared with a number of non-adaptive filters to assess their abilities to reduce speckle noise. For quantitative measurements, four metrics have been used to evaluate the performances of the proposed filters. From the experimental results, the proposed filters have achieved promising results for significantly suppressing speckle noise and preserving image information compared with other well-known filters.

Keywords Synthetic aperture radar (SAR) · Remote sensing · Non-adaptive filter · Adaptive filter · Despeckling filter · Image enhancement

1 Introduction

Synthetic Aperture Radar (SAR) is the essential system for ground mapping and remote sensing applications since its performance is independent on both weather conditions and daytime. Generally, SAR senses the interaction between the incidents radiated waves from SAR and Earth surface. However, the quality of the composed

A. S. Mashaly (✉) · T. A. Mahmoud
Military Technical College, Cairo 11766, Egypt
e-mail: mashaly@mtc.edu.eg

T. A. Mahmoud
e-mail: t.mahmoud@mtc.edu.eg

© The Author(s), under exclusive license to Springer Nature Switzerland AG 2021
A. E. Hassanien et al. (eds.), *Machine Learning and Big Data Analytics*
Paradigms: Analysis, Applications and Challenges, Studies in Big Data 77,
https://doi.org/10.1007/978-3-030-59338-4_15

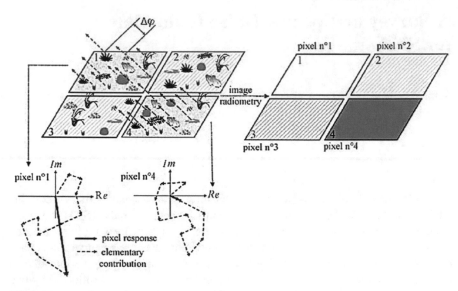

Fig. 1 Physical origin of speckle: The combination of elementary contributions within each pixel [4]

images from raw SAR data suffers from speckle noise distribution. The coherent nature of radiated wave of SAR is considered as the main source of speckle noise. For more details, let us consider a certain ground patch that has a various number of scatter elements with different level of backscatter coefficients as shown in Fig. 1. The final response of the constructed SAR image at each pixel location is the addition of amplitude and phase of each scatter element as shown in Eq. (1) [1–4].

$$Ae^{j\Phi} = \sum_{k=1}^{N} A_k e^{j\Phi_k} = X + jY \qquad (1)$$

where A … is the scatters amplitude response, Φ … is the scatters phase response, and N … is the total number of scatters for one pixel.

From Eq. (1), it is obvious that the total response of each pixel location depends on both scatter phase response and the total number of scatters. Therefore, it is accepted to get a random pixel response due to the phase differences between scatters. In fact, speckle noise can be produced due to an interference phenomenon in which the principal source of the noise is the distribution of the phase terms Φ_k. Then, the sum of Eq. (1) seems like a random walk in the complex plane, where each step of length A_k is in a completely random direction [1–4].

Finally, speckle noise is a serious problem that not only degrades the quality of the processed SAR images but also reduces the opportunity of image interpretation and analysis. Consequently, using suitable image processing method to improve SAR image quality and maintain image information is not a trivial task [1–4].

2   Speckle Noise Modelling

The speckle noise distribution could be fully estimated when [5]:

- The responses of each scatter point on the earth surface are independent.
- The amplitude A_k and the phase φ_k of Eq. (1) are independent.
- The variables A_k are distributed according to the same probability density function (PDF).
- The phases φ are uniformly distributed between $-\pi$ and π.

In terms of the central limit theorem, $Ae^{j\varphi}$ is a complex number where its real and imaginary parts, X and Y respectively, are independent random variables. These variables have Gaussian distribution with a zero mean and a standard deviation of σ [4].

For a given ground patch with a reflectivity of R, when amplitude mean $(E(A))$ increases, its standard deviation (σ_A) increases proportionally as shown in Fig. 2. This proportionality function gives SAR speckle noise the behavior of multiplicative noise [4]. Let:

$$I = R \cdot n \tag{2}$$

where n … Random variable representing speckle, R … Scatter reflectivity, and I … Echo intensity.

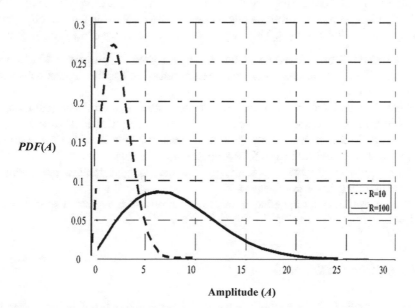

Fig. 2 Probability density function (*PDF*) of the amplitude of fully developed speckle. The mean radiometry value R is 10 or 100 [4]

Since, n and R are independent variables. Assuming that n has a unit mean, we could rewrite Eq. (2) as follows [4]:

$$E(I) = E(R \cdot n) = E(R) \cdot E(n) = R \tag{3}$$

It is obvious that reducing speckle noise by averaging nearby pixels based on the equality relationship between $E(I)$ and R, in a trade-off with spatial resolution is an efficient method [4].

3 Speckle Noise Reduction Methods

Generally, different methods and techniques are used to eliminate the effect of speckle noise. These methods could be classified as onboard methods (Multi-look methods) and offline methods [1–4].

3.1 Onboard Methods

For onboard methods, the main idea of this technique is to apply multi-look processing on a certain ground patch and the final resultant image depends on the average result of these multiple looks to eliminate the effect of speckle noise. This technique has promising results while its performance relies on several constraints to suppress speckle noise [3]. This method could be summarized as follows [6]:

- All signals, return from a ground patch, are used to process a single look image. The resultant image contains speckle noise but has the highest achievable resolution.
- Assuming that the phase parameter is not needed, multi-look technique is sufficient to reduce speckle noise effect of the final resultant SAR image.
- Using SAR digital processing, several independent images of the same ground zone could be achieved from SAR signal returns subsets.
- To create a single multi-look image, the independent images are averaged incoherently (phases not needed).
- The final processed image has a lower resolution and a neglected speckle noise level.

3.2 Offline Methods

For offline methods, these methods utilize the great development of image processing research field to find the optimum solution to reduce speckle noise. Theoretically, two types of filters are used to eliminate the speckle noise those are adaptive and non-adaptive filters. The main difference between these filters could be noticed as follows "*adaptive filters adapt their weightings and functions across the image to the speckle level, and non-adaptive filters apply the same weightings and functions uniformly across the entire image*". Moreover, such filtering also loses actual image information as well. In particular high-frequency information, the applicability of filtering and the choice of filter type involve tradeoffs [3, 4].

Intuitively, adaptive filters have better performance compared with that of the non-adaptive filter in preserving image information in high texture areas and resulting in non-blurred edges. On the other hand, adaptive filters require high computational power and their implementation is a complicated task. While, the non-adaptive filters seem to be simple and require less computational power but with some losses in image information and blurred edges [6, 7]. In this chapter, we will concern with the non-adaptive filters in reducing speckle noise effect.

4   Non-adaptive Despeckling Filters

In this section, we are going to introduce the most popular non-adaptive despeckling filters that have a wide usage for speckle noise reduction in several remote sensing and ground mapping applications.

Generally, This type of filters has some properties that can be summarized as follows [6, 7]:

1. The parameters of the whole image signal are considered.
2. Do not take into consideration the local properties of the terrain backscatter or the nature of the sensor.
3. Not appropriate for filtering of non-stationary scene signal.

Now, we are going to introduce the most popular and well-known non-adaptive filters that are used to attenuate speckle noise.

4.1   Mean Filter

In the beginning, this filter could be classified as a linear filter. The main operation and final response for the center pixel of the filter mask depends on the average value of the grey levels neighborhoods defined by the filter mask. This filter sometimes is called averaging filter or low pass filter [8].

Appling mean filter will reduce random noise that typically consists of sharp transitions in the intensities levels. However, edges (which almost always are desirable features of an image) also are characterized by sharp intensity transitions, so averaging filter has undesirable side effect that it blurs image edges [9].

The general implementation for filtering an $M * N$ image with mean filter of $m * n$ (m and n odd) is given by the expression [9]:

$$g(x, y) = \frac{\sum_{s=-a}^{a} \sum_{t=-b}^{b} w(s, t) f(x + s, y + t)}{\sum_{s=-a}^{a} \sum_{t=-b}^{b} w(s, t)} \qquad (4)$$

where $g(x, y)$ ... Filter output, $x = 0, 1, 2, ..., M - 1$, and $y = 0, 1, 2, ..., N - 1$.

This equation is evaluated for all values of the displacement variables x and y so that all elements of w visit every pixel in SAR image f. where $a = (m - 1)/2$ and $b = (n - 1)/2$.

4.2 Median Filter

On the other hand, Median filter is classified as a nonlinear order-statistic filter whose response is based on ordering (ranking) the pixels contained in the image area encompassed by the filter mask. The response of median filter for the center pixel of the filter mask depends on the median value of the grey levels neighborhoods defined by the filter mask [9].

Median filter is quite popular technique because, for certain types of random noise, it provides excellent noise reduction performance, with considerably less blurring appearance than linear mean filter of similar mask size. Median filter is particularly effective in the presence of speckle noise because it appears as white and dark dots superimposed on SAR image [9].

In order to implement median filter at certain image pixel, we first sort the greyscale values of the pixels in the neighborhood filter mask, determine their median value, and assign that value to the corresponding pixel in the filtered image [9].

4.3 Value-and-Criterion Filters

The value-and-criterion filter structure, another nonlinear filter, is a new framework for designing filters based on mathematical morphology. Whereas morphological filters use only nonlinear (minimum and maximum) operators, the value-and-criterion structure allows both linear and nonlinear operators. However, the new filtering structure retains the strong geometrical component of morphological filters. The value-and-criterion filter structure, therefore, exhibits the shape-based characteristics of mathematical morphology but adds a great deal of flexibility in the selection of filtering operations [9].

A value-and-criterion filter is a two-stage operation similar to the morphological opening and closing operators. Like opening and closing, the output of a value-and-criterion filter results from the second-stage operator that acts on the output of the first stage of the filter. Unlike the morphological operators, however, the first stage of a value-and-criterion filter consists of two separate operations, a criterion operator and a value operator. The second stage of the filter uses the results of the criterion operator to choose an output of the value operator for the final output. This new filter structure unites many different types of nonlinear filters into a single mathematical formalism [9].

The "value" function, V, and the "criterion" function, C, act on the original image $f(x,y)$ and are defined over a structuring element (filter window) FW. The "selection" operator, S, comprises the second stage of the filter and acts on the output of the criterion function. As in morphological filters, the second stage (S) is defined over the structuring element $\widehat{FW}$, a 180° rotation of FW. Let $f(x, y)$ be the input to a value-and-criterion filter, and $v(x, y)$, $c(x, y)$, and $s(x, y)$ be the output of V, C, and S respectively. The filter output $g(x, y)$ is defined by Eqs. (5), (6), and (7):

$$v(x, y) = V\{f(x, y); N\} \tag{5}$$

$$c(x, y) = C\{f(x, y); N\} \tag{6}$$

$$s(x, y) = S\left\{c(x, y); \tilde{N}\right\} \tag{7}$$

The value-and-criterion filter structure has the same comprehensive "sub-window" structure that develops naturally in morphological opening and closing. This structure is more comprehensive than those of previous sub-window based filtering schemes. The value-and-criterion filter structure with an $(n * n)$ square structuring element develops an overall filtering window of size $((2n - 1) * (2n - 1))$ that contains n^2 "sub-windows" the same size and shape as the structuring element. These sub-windows are all the possible $(n * n)$ structuring elements that fit in the overall window. Within each overall window, the structuring element that yields the "selected" criterion function value is chosen, and the value function output from that structuring element becomes the filter output for the overall window. The operation of this filter structure is very efficient, since the value and criterion function outputs for each sub-window are only computed once per image, unlike many of the earlier related filters [9].

Minimum Coefficient of Variation (MCV) Filter

Minimum Coefficient of Variant (MCV) filter is an example of the value-and-criterion filter. The MCV filter, therefore, has V = sample mean, C = sample coefficient of variation, and S = minimum of the sample coefficient of variation. At every point in an image, this filter effectively selects the $(n * n)$ sub-window with an overall window $(2n - 1 * 2n - 1)$ that has the smallest measured coefficient of variation, and outputs the mean of that sub-window [10, 11].

Mean of Least Variance (MLV) Filter

Mean of Least Variance (MLV) filter is another value-and-criterion filter that uses the sample variance as the criterion for selecting the most homogeneous neighborhood and the sample mean for computing a final output value from that neighborhood.

Since the sample variance requires the sample mean for its computation, the V and C functions are dependent in this case. Therefore, the general implementation for MLV filter will be as follows: $V =$ mean, $C =$ variance, and $S =$ minimum. The MLV filter resembles several earlier edge-preserving smoothing filters, but performs better and is more flexible and more efficient. The MLV filter smoothes homogeneous regions and enhances edges, and is therefore useful in segmentation algorithms [10, 11].

4.4 Mathematical Morphological Filters

Morphological filters are nonlinear image transformations that locally modify geometric features of images. They stem from the basic operations of a set-theoretical method for image analysis, called mathematical morphology, which was introduced by Matheron [12], Serra [13] and Maragos and Schafer [14]. The most common morphological filters that are used in speckle noise reduction are morphological opening filter as in Eq. (10). This filter can be easily expressed in terms of the value-and-criterion filter as both V, C are the minimum (erosion) operators as in Eq. (8), and S is the maximum (dilation) operator as in Eq. (9) respectively [9].

$$[f \ominus b](x, y) = \min_{(s,t) \in b} \{f(x + s, y + t)\} \tag{8}$$

$$[f \oplus b](x, y) = \max_{(s,t) \in b} \{f(x - s, y - t)\} \tag{9}$$

$$f \circ b = (f \ominus b) \oplus b \tag{10}$$

where f … input image, and b … structure element (SE).

Furthermore, morphological closing filter as in Eq. (11) can also be easily expressed in terms of the value-and-criterion filter as both V, C is the maximum (dilation) operator as in Eq. (9) and S is the minimum (erosion) operator as in Eq. (8) [9].

$$f \bullet b = (f \oplus b) \ominus b \tag{11}$$

Finally, Tophat filter is a closing-opening filter i.e. it consists of two stages where the first stage will be morphological closing (two steps dilation followed by erosion with the same structure element (SE)). The second stage will be morphological opening (two steps dilation to the output of first stage followed by erosion with the same SE of first stage) as in Eq. (12) [9].

$$Tophat = (f \bullet b) \circ b \tag{12}$$

For speckle reduction, morphological tools perform even better than classical filters in preserving edges, saving the mean value of the image and computational time according to ease of implementation [15, 16].

5 Adaptive Despeckling Filters

There are many forms of adaptive speckle filtering, including Lee filter [17], Frost filter [18], and Kuan filter [19]. They all rely upon three fundamental assumptions in their mathematical models [6, 20]:

- Based on Eq. (2), Speckle noise is proportional directly to the local greyscale level of SAR images where it is defined as a multiplicative noise.
- Both of SAR signal and speckle noise are statistically independent on each other.
- For single SAR image pixel, its sample mean and variance are equal to the sample mean and variance of the surrounding neighborhood pixels.

Now, we are going to explain some adaptive filters that are used in speckle noise reduction in SAR images.

5.1 Lee Filter

Lee filter utilizes the statistical distribution of the pixels values within the moving filter kernel to estimate the value of the pixel of interest. Lee filter based on the assumption that the mean and variance of the pixel of interest is equal to the local mean and variance of all pixels within the user-selected moving kernel. The resulting greyscale level value G for the smoothed pixel is [17]:

$$G = (I_c * W) + (I_m * (1 - W)) \tag{13}$$

$$W = 1 - \left(C_u^2/C_i^2\right) \tag{14}$$

$$C_u = \sqrt{(1/ENL)} \tag{15}$$

$$C_i = S_d/I_m \tag{16}$$

$$ENL = \mu_{image}^2/\sigma_{image}^2 \tag{17}$$

where W ... Weighting function, C_u ... Estimated noise variation coefficient, C_i... Image variation coefficient, ENL ... Effective (Equivalent) Number of Look, I_c...

Center pixel of filter window, I_m... Mean value of intensity within window, and S_d...
Standard deviation of intensity within filter window.

For noise variance estimation, *ENL* is the main factor, which determines the
required smoothing level of Lee filter for noisy SAR image. Theoretically, the true
value for the Number of Looks should be the same as the estimated *ENL*. It should be
close to the actual number of looks, but may be different if the image has undergone
resampling. The number of looks value may be experimentally adjusted where it
controls the effect of Lee filter. Conceptually, a smaller Number of Looks value
leads to more smoothing; a larger Number of Looks value preserves more image
features [17].

5.2 Lee-Sigma Filter

Lee-Sigma filter estimates the accurate value of the pixel of interest based on the
statistical distribution of the greyscale values within Lee-Sigma moving kernel. Gaussian distribution is the basic statistical distribution of Lee-Sigma filter for noisy SAR
images enhancement. Lee-Sigma assumed that 95.5% of random samples are within
a ± 2 standard deviation range (S_d) of the filter window of the center pixel $I_c(x, y)$.
A predefined range is estimated to suppress speckle noise as shown in Eq. (18). The
pixel of interest is replaced by the average value of all greyscale values within the
moving filter window that fall within the designated range [21].

$$[I_c(x, y) - 2S_d, I_c(x, y) + 2S_d] \tag{18}$$

5.3 Kuan Filter

Kuan filter transforms the multiplicative noise model into an additive noise model.
This filter is similar to Lee filter but uses a different weighting function. The resulting
greyscale level G for the smoothed pixel is [19]:

$$G = I_c * W + I_m * (1 - W) \tag{19}$$

$$W = \left(1 - C_u^2 / C_i^2\right) / \left(1 + C_u^2\right) \tag{20}$$

5.4 Frost Filter

Frost filter is able to suppress speckle noise while reserving edges and shape features. In addition, it uses an exponentially damped convolution kernel, which adapts itself to image features based on their local statistics. Frost filter differs from Kuan and Lee filters where the observed image is convolving with the impulse response of the SAR system to compute the scene reflectivity model. By minimizing the mean square error between the scene reflectivity model and the observed image, the impulse response of SAR system could be estimated, which is assumed to be an autoregressive process. For filter implementation, it defines a circularly symmetric filter with a set of weighting values M for each pixel as shown in Eq. (21) [18]:

$$M = e^{-AT} \tag{21}$$

where $A = Dampfactor * C_i^2$, $T =$ absolute value of the pixel distance from the center pixel to its neighbors in the filter window, and $Dampfactor =$ exponential damping factor.

The resulting greyscale level value G for the smoothed pixel is:

$$G = (P_1 M_1 + P_2 M_2 + \cdots + P_n M_n) / (M_1 + M_2 + \cdots + M_n) \tag{22}$$

where $P_1, \ldots P_n$ … Greyscale levels of each pixel in filter window and $M_1, \ldots M_n$ … Weights (as defined above) for each pixel.

The large value of Dampfactor leads to better preservation for sharp edges, but degrades the smoothing performance. However, the small value of Dampfactor improves the smoothing performance, but does not effectively retain sharp edges [18].

5.5 Enhanced Lee Filter

Lopes introduces an enhanced version of Lee filter. The core of this proposed filter divides SAR image into areas of three classes. The first class is the homogeneous area where speckle noise may be simply suppressed by applying a low pass (LP) filter (or equivalently, averaging, multi-look processing). The second class is the heterogeneous area in which the speckle noise is reduced while preserving image features. The last one contains isolated points. In this case, the filter should take no action to preserve the observed value. The resulting greyscale value G for the smoothed pixel is [22]:

$$G = \begin{cases} I_m & for & C_i \le C_u \\ I_m * W + I_c(1 - W) & for & C_u < C_i < C_{max} \\ I_c & for & C_i \ge C_{max} \end{cases} \tag{23}$$

$$W = e^{-(DampFactor*(C_i-C_u)/(C_{max}-C_i))} \tag{24}$$

$$C_{max} = \sqrt{1 + (2/ENL)} \tag{25}$$

5.6   Enhanced Frost Filter

The enhanced Frost filter is similar to the enhanced Lee filter where it considers three different types of image areas separately: homogeneous areas, heterogeneous area, and isolated point targets. In the heterogeneous case, the filter output is obtained by convolving the image with a circular kernel [22].

$$G = \begin{cases} I_m \ for & C_i \le C_u \\ G_f \ for \ C_u < C_i < C_{max} \\ I_c \ for & C_i \ge C_{max} \end{cases} \tag{26}$$

where G_f is the result of convolving the image with a circularly symmetric filter whose weighting values are $M = e^{-AT}$ for each pixel as shown in Eq. (21).

$$A = \left(DampFactor * (C_i - C_u)/(C_{max} - C_i) \right) \tag{27}$$

The resulting greyscale level value G_f for the smoothed pixel is:

$$G_f = (P_1 M_1 + P_2 M_2 + \cdots + P_n M_n)/(M_1 + M_2 + \cdots + M_n) \tag{28}$$

5.7   Gamma MAP Filter

Gamma MAP (Maximum a posteriori) filter is the most suitable filter for speckle noise reduction in SAR images when a priori knowledge about the scene probability density function is known. With the assumption of Gamma distributed scene, the Gamma MAP filter is derived with the following form [23]:

$$G = \begin{cases} I_m & for & C_i \leq C_u \\ G_f & for & C_u < C_i < C_{\max} \\ I_c & for & C_i \geq C_{\max} \end{cases} \tag{29}$$

$$G_f = \left(B * I_m + \sqrt{D}\right)\Big/(2A) \tag{30}$$

$$A = \left(1 + C_u^2\right)\Big/\left(C_i^2 - C_u^2\right) \tag{31}$$

$$B = A - ENL - 1 \tag{32}$$

$$D = \left(I_m^2 * B^2\right) + (4 * A * ENL * I_m * I_c) \tag{33}$$

$$C_{\max} = \sqrt{2} * C_u \tag{34}$$

6 The Proposed Non-adaptive Filters

Generally, most of non-adaptive filters performances degrade rapidly by increasing the filter mask size (MS) and the final enhanced image suffers from blur appearance and great loss in image information. The purpose of increasing MS is to speed up the process of speckle noise reduction while keeping image information. However, most of non-adaptive filters are incompatible with this idea where their performances are very limited to small MS (from (3 * 3) to (9 * 9)).

As a result, our task is to propose a non-adaptive filter that has an effective response with large filter MS. Firstly, it is important to note that achieving this goal requires using the most informative pixels within the used filter mask to estimate accurate filter response. Therefore, we will be concerned only with the pixel locations that are expected to represent the image features. In other words, not all the pixels within filter mask will be processed. We will select only the pixels in the predefined directions (Vertical, Horizontal, Right, and Left diagonals) as shown in Fig. 3.

The choice of these 4-directions is related to the case where the pixel under-test (a44) may be probably a part of an image edge. In this case, we will search for the direction that has the smallest greyscale variations. Therefore, we need to apply a suitable quantitative index to measure the variation in each direction and selecting the direction that has the smallest variation value (*the speckle existence in this direction is very low*). Sometimes, the center pixel of the filter mask may be a part of homogenous region or the whole filter mask covers a homogenous region where there is no image features. In this case, the proposed idea is still valid where the final filter response at this location will be related to the direction, which has the smallest greyscale variations. Theoretically, as the filter mask domain decreases to a significant level, the probability to estimate both accurate response value and

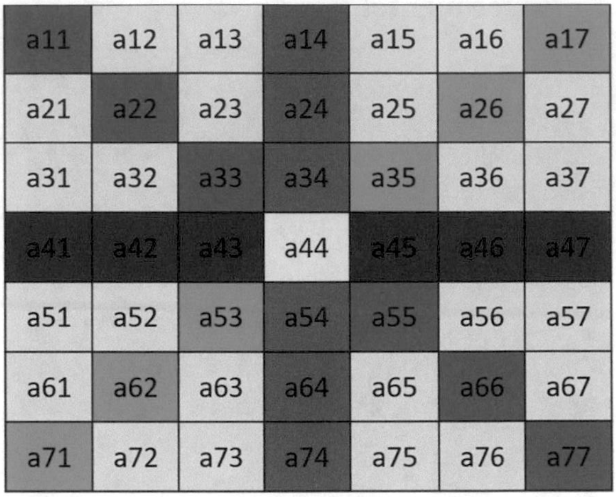

Fig. 3 Example of the
proposed filter mask (7 * 7)
with the 4-predefined
directions

a11	a12	a13	a14	a15	a16	a17
a21	a22	a23	a24	a25	a26	a27
a31	a32	a33	a34	a35	a36	a37
a41	a42	a43	a44	a45	a46	a47
a51	a52	a53	a54	a55	a56	a57
a61	a62	a63	a64	a65	a66	a67
a71	a72	a73	a74	a75	a76	a77

robust to noise increases as well. This is the key of the proposed non-adaptive filters.
Finally, the detailed steps of the proposed filters could be summarized as follows:

- For each image location covered by filter mask, assign the pixels values in the
  four predefined directions to four one-dimensional arrays.
- For each array, estimate the root-mean square (RMS) value as shown in
  Eq. (35) [24].

$$RMS = \sqrt{\frac{\sum_{i=1}^{n}(\hat{a} - a_i)^2}{n}} \tag{35}$$

where $\hat{a}$ mean value of each array, a_i greyscale levels, n number of greyscale pixels
in each array or direction.

- Select the direction that has the smallest RMS value.
- The filter response in this case could be the mean or median value of the selected
  direction.

This methodology of determining the filter response will improve the overall
performance of the traditional mean and median filter. Where using edge pixel or the
direction of minimum RMS will reduce the blurring effect at edge locations caused
by traditional Mean and Median filters when applying large filter masks. Finally,
the proposed filter could be classified as a special case of the previously mentioned
value-and-criterion filter. Where the "value" function (V) is the mean or median
value, the "criterion" function (C) is the RMS, and the "selection" operator (S) is the
minimum.

7 Experimental Results and Performance Assessment

To facilitate performance assessment of the simulated non-adaptive filters and to validate the operation of the proposed filters, we are going to introduce four quantitative evaluation indexes. The first quantitative evaluation index is called "Target to Clutter ratio" (*TCR*) index. *TCR* presents the ratio between target signal to clutter (background) signal for the filtered image (i.e. as this ratio increases, it indicates better speckle reduction level). *TCR* could be estimated as shown in Eq. (36) [16].

$$TCR = 10 \log\left(\frac{\sigma_T^2}{\sigma_C^2}\right) \tag{36}$$

where σ_T^2 target variance, and σ_C^2 clutter variance.

The second quantitative evaluation index is Normalized Mean (*NM*). This evaluation index indicates the filter ability to attenuate speckle noise while preserving image information. *NM* equals the ratio between mean value of a filtered homogenous region and the mean value of the original noisy region. The closer *NM* to one, the better filter ability to preserve image information. *NM* could be calculated as shown in Eq. (37) [16].

$$NM = \frac{\mu_{filtered}}{\mu_{original}} \tag{37}$$

where $\mu_{filtered}$ and $\mu_{original}$ are the means of the background segments of the filtered and the original image, respectively.

Standard Deviation to Mean (*STM*) index assess the ability of the filter to suppress speckle noise for a predefined homogenous patch. As *STM* decreases as speckle noise suppression gets better. *STM* could be estimated as shown in Eq. (38) [16]

$$STM = \frac{\sigma_{filtered}}{\mu_{filtered}} \tag{38}$$

where $\sigma_{filtered}$ is the standard deviation of filtered homogenous background patch.

For edge quality, Edge Index (*EI*) could be used to show the statement of image edges after despeckling operation. *EI* equals the ratio between filtered image pixels to the original image pixels. Mathematically, *EI* might be less than one that shows edge blurring or *EI* might be greater than or equals to one that indicates edge enhancement. *EI* could be calculated as shown in Eq. (39) [16].

$$EI = \frac{\sum P_f(i, j) - P_f(i - 1, j + 1)}{\sum P_O(i, j) - P_O(i - 1, j + 1)} \tag{39}$$

where $P_f(i, j)$ and $P_o(i, j)$ are the filtered and the original pixel values respectively.

Finally, the Average Processing Time (*APT*) per frame indicates the capability of the tested algorithm for real-time applications. Intuitively, there is no specific value that could be used as a reference value. Generally, such value differs from one application to another. While, the measured value gives a good indication about the tested algorithm complexity and the validity for ease implementation.

The performance of the proposed filters and the previously mentioned non-adaptive filters will be assessed based on real noisy SAR images with resolution of (2500 * 1650). In addition, all the aforementioned non-adaptive filters are implemented using LabVIEW 2018 and NI Vision Development System 2018 running on Lenovo notebook Z50-70 with Intel processor core i7 (4510u) and 8 GB RAM.

For visual inspection index, Fig. 4. introduces a sample of the resultant enhanced images of the simulated filters, the proposed mean and median filters based on RMS error value.

It is obvious that the proposed mean and median filters have promising results compared with the results of the other simulated non-adaptive filters especially, when these filters use large filter masks. We can note that MCV, MLV, and the morphological (Closing, Opening, and Tophat) filters have distorted images with large filter masks and SE. While, conventional Mean and Median filters have acceptable results with small amount of blurring at image edges. As a result, most of non-adaptive

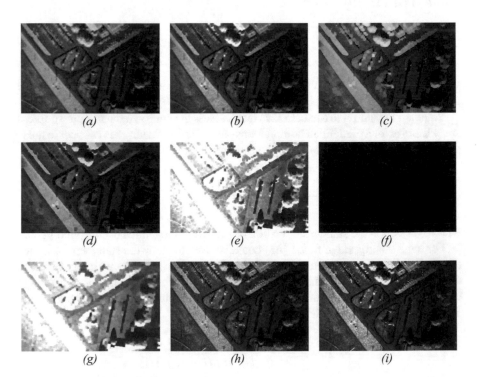

Fig. 4 Sample of enhanced SAR images using (27*27) for both filter MS and SE: **a** Mean, **b** Median, **c** MCV, **d** MLV, **e** Closing, **f** Opening, **g** Tophat, **h** Proposed Mean-RMS, **i** Proposed Median-RMS

filters performances degrade rapidly by increasing filter mask. On the other hand, the enhanced SAR images of the proposed Mean and Median filters have effectively eliminated the speckle noise while retaining image integrity.

Furthermore, Table 1 and Fig. 5 present the average processing time (*APT*) per frame for the aforementioned non-adaptive filters. It is clear that the traditional non-adaptive filters have high computational power for large filter MS. While the predefined directions of the proposed filter masks have low computational power, less complexity, and ease of implementation.

For the other quantitative evaluation indexes like *TCR*, *NM*, *STM*, and *EI*, it seems that these evaluation indexes may not introduce an accurate assessment values. Since most of non-adaptive filters (MCV, MLC, Tophat, Closing, and Opening), have distorted resultant images with large filter masks. However, these evaluation indexes still show the enhancement amount achieved by the proposed filters compared with the results of other filters.

For Target to Clutter ratio (*TCR*) index, Table 1 and Fig. 6 present the estimated TCR value for different non-adaptive filters results (Note that the original *TCR* value for noisy SAR image is 13.86). It is obvious that the proposed filters achieve the highest values with small edge blurring or distortion and with low processing time.

Table 1 Quantitative evaluation indexes results

Indexes	Tophat	Opening	Closing	MLV	MCV	Median	Mean	Proposed Median	Proposed Mean
APT (sec)	63	**30**	33	330	300	102	70	**60**	**58**
TCR	20.94	9.95	14.05	14.68	15.61	**24.42**	21.06	**24.25**	**23.15**
NM	1.237	0.087	2.397	0.953	**0.974**	0.935	1.068	**0.905**	**0.935**
STM	0.042	0.171	0.067	0.123	**0.035**	0.044	0.1	**0.16**	**0.14**
EI	0.223	0.159	0.228	0.225	0.306	0.84	**0.982**	**0.941**	**0.981**

The bold values indicate the best filter for each quantitative evaluation index

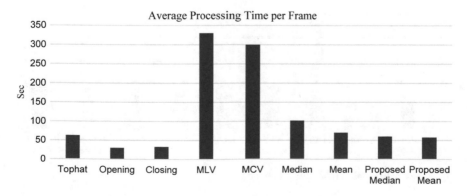

Fig. 5 Average processing time per frame for the filtering results

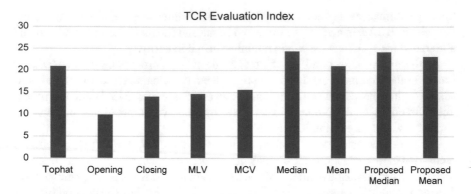

Fig. 6 Target to clutter ratio (*TCR*) evaluation index

For Normalized Mean (*NM*) index, Table 1 and Fig. 7 show the different resultant levels of *NM* index for the non-adaptive filters under test. It is clear that the proposed non-adaptive filters have the ability to preserve image information of the enhanced images compared with the original noisy image. Since the estimated *NM* values of the proposed filters are close to the value of "1".

To assess the performance of the tested filters w.r.t speckle noise cancellation, Table 1 and Fig. 8 introduce the different estimated *STM* levels for the non-adaptive filter results. It can be noticed that the proposed Mean and Median filters present moderate *STM* values (i.e. these proposed filters have the capability to attenuate speckle noise to an acceptable level while keeping image visual appearance and information). In addition, we can see that there are some non-adaptive filters, which show very small *STM* values. However, these filters have a completely distorted output images. Therefore, we cannot consider these filters values as the best *STM* results.

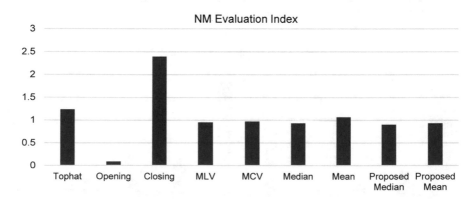

Fig. 7 Normalized mean (*NM*) evaluation index

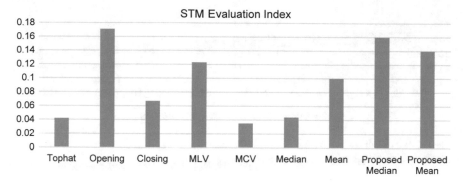

Fig. 8 Standard To Mean (*STM*) Evaluation Index

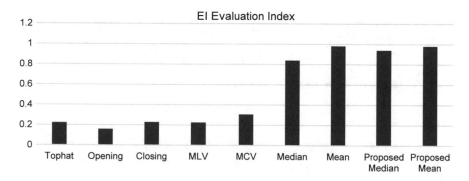

Fig. 9 Edge index (*EI*) evaluation index

Finally, for edge blurring, Table 1 and Fig. 9 introduce the evaluation index (*EI*) results of the non-adaptive filters under test. It is clear that the resultant values of *EI* are very near to "1" that means there is small edge blurring. Theoretically, these results seem to be normal since all non-adaptive filters suffer from edge blurring output after enhancement processing. Moreover, *EI* values for the proposed filter are closer to "1" compared to that of other filters. As a result, edge blurring of the proposed filters is very low compared with the other filters.

Now, we are going to study the performance of the proposed filters (Mean-RMS and Median-RMS) with different filter MS to ensure the objective and motivation of the proposed filters. Figures 10, 11, 12, 13 show the different evaluation indexes (TCR, NM, STM, and EI) with different sizes of used filter mask respectively. For TCR, it is clear that there is a value improvement with increasing filter mask size. In addition, NM evaluation index value still around "1" without any loss in image information. Furthermore, STM values decrease smoothly with increasing filter mask to confirm the objective of the proposed filters. Finally, EI values still less than "1" this means that there is small amount of edge blurring.

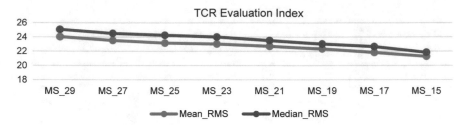

Fig. 10 TCR Evaluation index with different filter MS

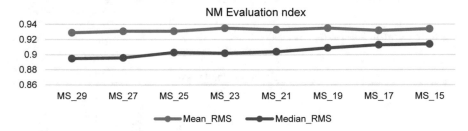

Fig. 11 NM Evaluation index with different filter MS

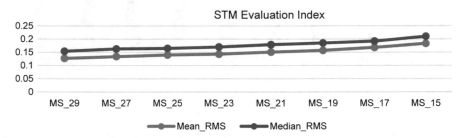

Fig. 12 STM Evaluation index with different filter MS

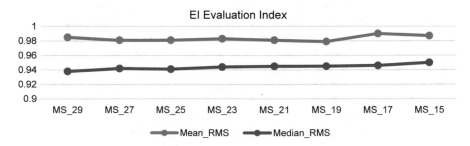

Fig. 13 EI Evaluation index with different filter MS

8 Conclusion

SAR images are useful sources of information for moisture content of the Earth surface, geometry, and roughness. SAR is an active senor that has a suitable performance for different daytime, and all-weather conditions remote sensing system, SAR images can provide information for both surface and subsurface of the Earth. Speckle noise is an additive kind of random noise, which degrades the quality of SAR images appearance. Speckle noise reduces image contrast and has a negative effect on texture-based analysis. Moreover, as speckle noise changes the spatial statistics of the images, it makes the classification process a difficult task to do. Thus, it is a vital task to reduce the effect of speckle noise. In this chapter, we introduce a comprehensive study for the different methods of speckle noise reduction (online and offline "adaptive and non-adaptive filters" processing) followed by two proposed non-adaptive filters based on RMS error value that rely on large filter mask to speedup processing time of the noisy SAR images and to enhance the signal to noise ratio. Moreover, the proposed filters aim to overcome the limitations of the traditional non-adaptive filters with large mask sizes. Different types of evaluation indexes have been used to assess the performance of the proposed filters. From the experimental results, the proposed filters achieve the best visual appearance, the highest TCR value, promising values for NM, STM, and EI values especially with large filter masks.

References

1. Sivaranjani, R., Roomi, S.M.M., Senthilarasi, M.: Speckle noise removal in SAR images using Multi-Objective PSO (MOPSO) algorithm. Appl. Soft Comput. **76**, 671–681 (2019)
2. Singh, P., Shree, R.: A new SAR image despeckling using directional smoothing filter and method noise thresholding. Eng. Sci. Technol. Int. J. **21**(4), 589–610 (2018)
3. Oliver, C., Quegan, S.: Understanding Synthetic Aperture Radar Images, 1st edn. INC, SciTech (2004)
4. Massonnet, D., Souyris, J.: Imaging with Synthetic Aperture Radar, 1st edn. EPFL Press (2008)
5. Goodman, J.W.: Some fundamental properties of speckle. J. Opt. Soc. Am. **66**, 1145–1150 (1976)
6. Tso, B., Mather, P: Classification Methods for Remotely Sensed Data, 2nd edn. CRC Press (2009)
7. Franceschetti, G., Lanari, R.: Synthetic Aperture Radar Processing. CRC Press (1999)
8. Gonzalez, R., Woods, R.: Digital Image Processing, 3rd edn. Addison-Wesley INC (2008)
9. Schulze, M., Pearce, J.: Value-and-Criterion Filters: A new filter structure based upon morphological opening and closing, Nonlinear Image Processing IV. Proc. SPIE **1902**, 106–115 (1993)
10. Schulze, M., Wu, Q.: Noise reduction in synthetic aperture radar imagery using a morphology based nonlinear filter. In: Digital Image Computing: Techniques and Applications, pp. 661–666 (1995)
11. Matheron, G.: Random Sets and Integral Geometry. Wiley, New York (1975)
12. Serra, J.: Image Analysis and Mathematical Morphology. Academic, New York (1982)
13. Maragos, P., Schafer, R.: Morphological filters-Part I: their set-theoretic analysis and relations to linear shift-invariant. IEEE Trans Acoust. Speech Signal Process. **ASSP-35**(8), 1153–1169 (1987)

14. Shih, F.: Image Processing and Mathematical Morphology Fundamentals and Applications, 1st edn. CRC Press (2009)
15. Gasull, A., Herrero, M.A. Oil spills detection in SAR images using mathematical morphology. In: Proceedings of EUSIPCO, Toulouse, France, September 3–6, 2002
16. Huang, Liu, S.: Some uncertain factor analysis and improvement in space borne synthetic aperture radar imaging. Signal Process. **87**, 3202–3217 (2007)
17. Lee, J.S.: Digital image enhancement and noise filtering by use of local statistics. IEEE Trans. Pattern Anal. Mach. Intell. **PAMI-2**, 165–186 (1980)
18. Frost, V.S., Stiles, J.A., Shanmugan, K.S., Holtzman, J.C.: A model for radar images and its application to adaptive digital filtering of multiplicative noise. IEEE Trans. Pattern Anal. Mach. Intell. **PAMI-4**, 157–166 (1982)
19. Kuan, D.T., Sawchuk, A.A., Strand, T.C., Chavell, P. Adaptive noise smoothing filter for images with signal-dependent noise. IEEE Trans. Pattern Anal. Mach. Intell. **PAMI-7**, 165–177 (1985)
20. Mashaly, A.S., AbdElkawy, E.F., Mahmoud, T.A.: Speckle noise reduction in SAR images using adaptive morphological filter. In: Proceedings of the IEEE Intelligent System Design and Application Conference (ISDA 2010), Cairo, Egypt, 29 Oct– 1 Nov 2010
21. Lee, J.S.: A simple speckle smoothing algorithm for synthetic aperture radar images. IEEE Trans. Syst. Man Cybern. **13**, 85–89 (1983)
22. Lopes, A., Touzi, R., Nezry, E.: Adaptive speckle filters and scene heterogeneity. IEEE Trans. Geosci. Remote Sens. **28**, 992–1000 (1990)
23. Kuan, D.T., Sawchuk, A.A., Strand, T.C., Chavell, P.: Adaptive restoration of images with speckle. IEEE Trans. Acoust. Speech Signal Process. **35**, 373–383 (1987)
24. GISGeography Homepage, https://gisgeography.com/root-mean-square-error-rmse-gis/. Last accessed 19 March 2019

Comparative Analysis of Different Approaches to Human Activity Recognition Based on Accelerometer Signals

Walid Gomaa

Abstract Recently, automatic human activity recognition has drawn much attention. On one hand, this is due to the rapid proliferation and cost degradation of a wide variety of sensing hardware. On the other hand there are urgent growing and pressing demands from many application domains such as: in-home health monitoring especially for the elderly, smart cities, safe driving by monitoring and predicting driver's behavior, healthcare applications, entertainment, assessment of therapy, performance evaluation in sports, etc. In this paper we focus on activities of daily living (ADL), which are routine activities that people tend to do every day without needing assistance. We have used a public dataset of acceleration data collected with a wrist-worn accelerometer for 14 different ADL activities. Our objective is to perform an extensive comparative study of the predictive power of several paradigms to model and classify ADL activities. To the best of our knowledge, almost all techniques for activity recognition are based on methods from the machine learning literature (particularly, supervised learning). Our comparative study widens the scope of techniques that can be used for automatic analysis of human activities and provides a valuation of the relative effectiveness and efficiency of a potentially myriad pool of techniques. We apply two different paradigms for the analysis and classification of daily living activities: (1) techniques based on supervised machine learning and (2) techniques based on estimating the empirical distribution of the time series data and use metric-theoretic techniques to estimate the dissimilarity between two distributions. We used several evaluation metrics including confusion matrices, overall accuracy, sensitivity and specificity for each activity, and relative computational performance. In each approach we used some of the well-known techniques in our experimentation and analysis. For example, in supervised learning we applied both support vector machines and random forests. One of the main conclusions from our analysis is that the simplest techniques, for example, using empirical discrete distri-

W. Gomaa (✉)
Cyber-Physical Systems Lab, Egypt-Japan University of Science and Technology, Alexandria, Egypt
e-mail: walid.gomaa@ejust.edu.eg

Faculty of Engineering, Alexandria University, Alexandria, Egypt

© The Author(s), under exclusive license to Springer Nature Switzerland AG 2021 303
A. E. Hassanien et al. (eds.), *Machine Learning and Big Data Analytics Paradigms: Analysis, Applications and Challenges*, Studies in Big Data 77, https://doi.org/10.1007/978-3-030-59338-4_16

butions as models, have almost the best performance in terms of both accuracy and computational efficiency.

Keywords Activities of daily living · Human activity recognition · Supervised learning · Dissimilarity measure · Empirical probability distribution · Time series · Accelerometer · Wearable devices

1 Introduction

Automatic recognition of human activities has become a very substantial research topic. A wide range of computer applications depend mainly on *human activity recognition (HAR)* in their work such as monitoring patients and elderly people, surveillance systems, robots learning and cooperating, and military applications. The idea of an automatic HAR system depends on collecting measurements from some appropriate sensors which are affected by selected human motion attributes. Then, depending on these measurements, some features are extracted to be used in the process of training activity models, which in turn will be used to recognize these activities later.

Based on the data acquisition paradigm, HAR systems can be divided into two categories: surrounding fixed-sensor systems and wearable mobile-sensor systems. In the first category, the required data are collected from distributed sensors attached to fixed locations in the activity environment (where users activities are monitored) such as surveillance cameras, microphones, motion sensors, etc. Alternatively, the sensors are attached to interactive objects in order to detect the type of interaction with them, such as a motion sensor attached to the cupboard doors or microwave ovens (to detect opening and/or closing), or on water tap to feel turning it on or off, and so on. Although this method can detect complex actions efficiently, it has many limitations due to its fixed nature.

In the wearable based systems, the measurements are taken from mobile sensors mounted to human body parts like wrists, legs, waist, and chest. Many sensors can be used to measure selected attributes of human body motion, such as accelerometers, gyroscopes, compasses, and GPSs. Or to measure phenomena around the user, such as barometers, magnetometers, light sensors, temperature sensors, humidity sensors, microphones, and mobile cameras. On the contrary of the fixed-sensor based systems, wearables are able to measure data from the user everywhere, while sleeping, working, or even traveling anywhere since it is not bounded by a specific place where the sensors are installed. Also, it is very easy to concentrate on directly measuring data of particular body parts efficiently without a lot of preprocessing that are needed, for example, in fixed depth cameras. However, carrying and installing a lot of sensors, wires, and power supplies mounted to the user may be uncomfortable and annoying. A comprehensive review on the use of wearables in the medical sector can be found in [1].

During the recent few years, there has been a tremendous evolution in the manufacturing of mobile devices. Particularly mobile phones, tablet PCs, and smart watches. All such devices contain various types of sensors, such as accelerometers, barometers, gyroscopes, microphones, GPS, etc. The evolution of the Internet of Things and ubiquitous sensing have encouraged mobile device manufacturers to provide more types of sensors and improve the accuracy and efficiency of the existing ones. Smart phones have also become more and more popular. Recent statistics show that the total number of smart phone subscribers reached 3.9 billion in 2016 and is expected to reach 6.8 billion by 2022 [2]. Other statistics show that in some countries, the percentage of smart phone subscribers reaches 88% of the population [3]. Therefore, the disadvantages of wearable mobile-sensors of being intrusive, uncomfortable, and annoying have vanished to a great extent, making this method of on-board sensing from smart devices very suitable for HAR data acquisition.

Many kinds of attributes can be measured using wearable sensors [4]. These include: (i) environmental data such as barometric pressure, temperature, humidity, light intensity, magnetic fields, and noise level, (ii) vital health signs such as heart rate, body temperature, blood pressure, and respiratory rate, (iii) location data which are typically identified by longitudes and latitudes using GPS sensors, and (iv) body limbs motion such as acceleration and rotation of body parts like arms, legs, and torso using accelerometers and gyroscopes. For the purpose of activity recognition, the latter type of attributes have proven to represent human motion accurately, where some methods depending only on acceleration measurements have achieved very high accuracy [5, 6]. However, it is very difficult to extract the pattern of motion from acceleration raw data directly because of high frequency oscillation and noise. So, feature extraction techniques should be applied on the raw data before using it. Many types of features can be extracted from acceleration data: (1) time-domain features like mean and variance [7], (2) frequency-domain features like Fourier transform, discrete cosine transform, and wavelet transform, and (3) applying dimensionality reduction techniques like PCA and LDA [5].

An essential function of HAR systems is to provide a general model for each activity. The activity models are mostly generated based on the features extracted from training data using supervised machine learning techniques. Thus, the role of machine learning techniques in HAR systems is to build general models to describe each activity and use these models to detect or classify the target activities later. Many classification techniques are used such as support vector machine (SVM), random forests, C4.5 decision trees, hidden Markov models (HMM), k-nearest neighbor (KNN), Bayesian networks, and artificial neural networks (ANN) [8, 9].

Among all wearable sensors, accelerometers are considered as the most used sensors in HAR systems [8]. Being small-sized, inexpensive, and embedded in most of smart mobile devices has encouraged many researchers to use acceleration in their work. Compared to cameras, accelerometers are more suitable for HAR systems, it is very difficult to fix a camera to monitor a user everywhere; also mounting the camera to the user's body is very annoying and uncomfortable. From the privacy point of view, it is not acceptable nor convenient by many people to be monitored all the time. As well, the videos or images collected using cameras are very sensitive to many environmental conditions like lighting and surrounding barriers. However, accelerometers can be easily mounted to users or embedded into many devices such as smart phones and/or smart watches which are naturally carried by many users everywhere most of the time. Also acceleration data preserve user privacy and are not affected by any outside conditions.

In this paper we perform a comparative study of the predictive power of several paradigms to model and classify activities of daily living (ADL). ADLs are routine activities that people tend to do every day without needing assistance. We have used a public dataset of accelerometer data collected with a wrist-worn accelerometer [10, 11]. The data are streamed while performing 14 activities using several human subjects and with varying number of samples for the different activities. Table 1 provides a summary of the monitored activities along with each activity sample size. Each experiment consists of three time-series data for each of the tri-axial accelerometer directions. To the best of our knowledge, almost all techniques for activity recognition are based on methods from the machine learning literature (particularly, supervised learning). Our comparative study in this paper widens the scope of techniques that can be used for the automatic analysis of human activities and provides a valuation of the relative effectiveness and efficiency of a potentially myriad pool of techniques. We apply two different paradigms for the analysis and classification of daily living activities. The first approach is based on traditional supervised machine learning and the second approach is based on estimating the empirical distribution of the time series data and use measure-theoretic techniques to estimate the dissimilarity between two distributions. In each approach we used some of the well-known techniques in our experimentation and analysis. For example, in supervised learning we applied both support vector machines and random forests.

As mentioned above the rationale is to compare the effectiveness of several different paradigms in the analysis of sensory data in order to deduce human activities. Hence, in order to focus exclusively on that goal we use as little sensory information as possible, which is the tri-axial accelerometer. Such study can provide insights into the proper tools to analyze sensory data for daily human activities and human activities in general.

The paper is organized as follows. Section 1 is an introduction. The bulk of the paper is given in Sect. 2, it describes the experimentation and the associated analysis. It is divided into subsections as follows. Section 2.1 describes the data set used in our experiments. In Sect. 2.2 we describe the general setup of all the experiments along with our evaluation criteria. Supervised machine learning experiments are presented in Sect. 2.3. Probabilistic and measure theoretic techniques are presented in Sect. 2.4. Finally, Sect. 3 concludes the paper with discussions and ideas for future research.

2 Experimentation and Analysis

This is the main section of the paper. We first describe the acceleration data that we use for our experiments and the consequential analysis. This is followed by four subsections each describe the approach used for activity recognition, the results of the test experiments, and our conclusion about that approach. As we described above

1. Hardcore supervised machine learning techniques. We experimented with support vector machines and random forests.
2. Probabilistic and measure-theoretic approach.
3. Statistical analysis using goodness-of-fit tests.
4. Time-series analysis using distance metrics between time signals.

The first three approaches ignore the temporal nature of the data and just treat each sample as just unordered collection of data points.

2.1 Data Description

We have obtained our labeled accelerometer activity data from the UCI machine learning public repository. The dataset is: 'Dataset for ADL Recognition with Wrist-worn Accelerometer Data Set' located at the url https://archive.ics.uci.edu/ml/datasets/Dataset+for+ADL+Recognition+with+Wrist-worn+Accelerometer#. The data is collected by [12]. A detailed description of the data along with a detection system can be found in [10, 11].

The data are collected using a wrist-worn accelerometer with sampling rate 32 Hz. The measurement range is $[-1.5, 1.5]$ (it is in g-force units). One data sample is a tri-axial acceleration: x-axis pointing toward the hand, y-axis pointing toward the left, and z-axis perpendicular to the plane of the hand. Acceleration data are collected for 14 activities with varying number of samples for each. Table 1 lists the activities along with the number of collected samples per each.

Table 1 List of activities and associated number of samples

No.	Activity	Number of samples	No.	Activity	Number of samples
1	Brush teeth	12	8	Pour water	100
2	Climb stairs	102	9	Eat meat	5
3	Comb hair	31	10	Walk	100
4	Drink glass	100	11	Liedown bed	28
5	Getup bed	101	12	Standup chair	102
6	Sitdown chair	100	13	Descend stairs	42
7	Use telephone	13	14	Eat soup	3

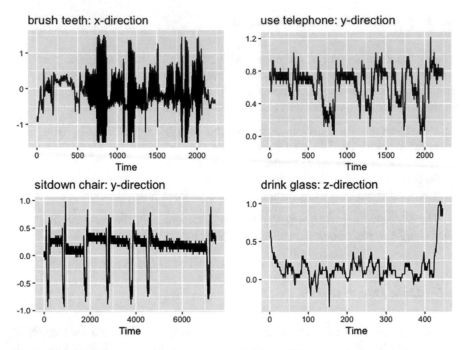

Fig. 1 Snapshots of some signals from various activities in different accelerator directions

Figure 1 shows snapshots of the collected data signals for various activities in different accelerator directions. The results obtained in all experiments are essentially the confusion matrices and accuracies averaged over a number of experiments (15) of the same settings.

2.2 Evaluation

We here describe the general process of modeling and testing. For each of the forthcoming techniques we use the following general steps for evaluation:

1. We form several recognition problems starting from the simplest, with only two activities, to the most complex, using the whole set of 14 activities (except for the machine learning techniques where we end up with only 13 activities due to lack of enough data for training for the 'eat soup' activity).
2. The order in which the activities are added to form more complex recognition problems follow that of Table 1. Note that this choice of ordering is completely arbitrary.
3. A model is constructed for each activity using part of the data set, 65% of the collected time-series samples. Such samples are selected at random.
4. The remaining 35% of the available data are used for testing.

5. The modeling-testing duality is repeated 15 times with random selection of the data samples used for modeling/testing. The results are averaged over the 15 experiments.
6. The evaluation metrics are the following: the accuracy of each model is computed, the evolution of accuracy with increasing model complexity (increasing the number of activities) is plotted, the sensitivity and specificity for each model and each activity are computed, and the confusion matrix is computed. In some cases we illustrate the confusion matrix for the complete set of activities.

As a reminder we give the definitions of the basic performance indices: sensitivity, specificity, and accuracy (let TP and TN denote true positives and true negatives respectively, and FP and FN denote false positives and false negatives respectively):

$$
sensitivity = \frac{TP}{TP + FN}
$$
$$
specificity = \frac{TN}{TN + FP}
$$
$$
accuracy = \frac{TP + TN}{TP + TN + FP + FN} \tag{1}
$$

For multiclass classification the dichotomy for any specific class A is that: the sample is either classified as belonging to the A (the positive case) or classified as belonging to any one of the remaining classes, hence not belonging to A (the negative case). An important factor that should generally be taken into consideration in evaluation is the amount of computational resources required to execute the given method. Such resources are essentially in terms of time, either asymptotic and/or physical, and amount of storage. In the current paper we do not focus on such metric, though we make a relative comparison among the different approaches. The experiments are performed on a 2.5 GHz Intel Core i7, with 16 GB 1600 MHz DDR3. We have not tried any kind of high performance computing whether CPU or GPU parallel computation.

2.3 Supervised Machine Learning

Our first paradigm is based on applying supervised learning using typical techniques from the machine learning literature. Here we use support vector machines SVM and random forests RF. The 'eat soup' activity is excluded from our analysis since it has very few samples, as shown in Table 1. Given a tri-axial time series sample s of an activity A, the features are extracted from s as follows.

1. For each of the tri-axial accelerometer directions (x, y, and z), we build a Gaussian mixture model (GMM) consisting of 2 components (this number has empirically given the best results).

2. The GMM models are built using Expectation Maximization (EM), where the means and variances are initialized by running k-means clustering over the data.
3. The maximum number of iterations for EM is taken to be 1500.
4. The feature space for sample s consists then of the output of the three generative GM models (over the three tri-axial directions). That is, for each direction we have the responsibilities (ratios of the two components), the means of the two components, and the variances of the two components. Then, overall for sample s we have a total of 18 features.
5. For SVM we have used radial basis function kernel $k(\mathbf{x}, \mathbf{x}') = e^{-\frac{\|\mathbf{x}-\mathbf{x}'\|^2}{2\sigma^2}}$.

In all the techniques and experiments we have employed repeated cross-validation: 5-fold cross validation and 5 repetitions of cross validations. In some experiments, median filtering, with window of width 15, has been applied to reduce the noise on the data as a preprocessing step. As mentioned above the tri-axial signals are treated individually as independent three univariate signals. We call such case 'disjoint tri-axial GM features'. Figure 2 shows the accuracy against number of activities for both SVM and RF with and without filtering. As can be seen filtering does not have much effect on the performance. This can be attributed to the fact that most of the activities

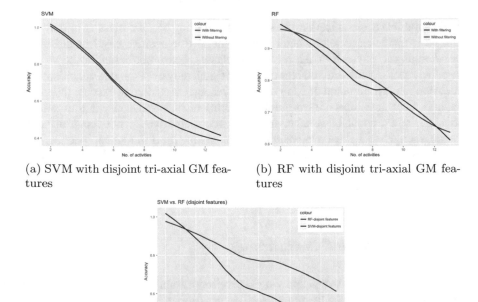

(a) SVM with disjoint tri-axial GM features

(b) RF with disjoint tri-axial GM features

(c) SVM vs. RF on disjoint tri-axial GM features (data not filtered)

Fig. 2 SVM and RF applied to both filtered and raw data (disjoint tri-axial GM features)

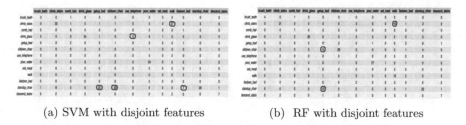

(a) SVM with disjoint features   (b) RF with disjoint features

Fig. 3 Confusion matrix for SVM (left) and RF (right) over disjoint features

signals satisfy the following properties: (1) most of the frequency components in the data are actually low frequency and (2) the data is not much noisy, this can be visualized from some snapshots of the given time series in Fig. 1. We have tried different window sizes and different kinds of filtering with almost same performance or even worse in some cases. So from now on we will work directly on the collected raw data. It is also apparent from the figure that when the number of activities exceeds 8, RF significantly outperforms SVM by about 20%. This in part is due to the fact that RF is an ensemble based method that is more robust and less prune to over fitting than SVM. A random forest is implicitly a form of *distributed representation* of the underlying data and hence, in some sense, can be considered as a form of deep learning [13]. Another reason for the apparent superiority of RF over SVM in HAR is that, though multiclass classification comes natural with random forests, support vector machines are designed mainly for binary classification and each method for generalizing to the multiclass case suffers from inherent problems.

Figure 3 shows the confusion matrices for both SVM and RF over disjoint features. The confusion matrix of the SVM indicates a large degree of misclassification between 'standup chair' and each one of the activities: 'sitdown chair', 'getup bed', and 'liedown bed'. This is indicated by the very low sensitivity (0, 0.15, 0 respectively) of these latter classes (though they have $\simeq$ 1 specificity). Most of the positives of 'standup chair' are false positives and should belong to the three latter classes. So this would lessen the specificity of 'standup chiar' as well as lessening the sensitivity of the other three activities. There is also a large degree of confusion between the activities 'climb stairs' and 'walk', and between the activities 'drink glass' and 'use telephone' (sensitivity of 0 and specificity of 1 for the first activity).

So, all the problematic activities 'sitdown chair', 'getup bed', 'liedown bed', 'walk', and 'use telephone' have common features: very low sensitivity, almost 0, and specificity of 1. This cannot be attributed to data imbalance for most of these classes have the largest sample sizes. Thus, these results indicate that the class distributions for these activities are completely overlapped (masked) by the counterpart distributions (with which they exchange the class assignment). So SVM does not perform very good, at least under this particular choice of features. We have already tried to increase the number of GM components, however, this actually led to worse results. As RF is more robust and is naturally providing distributed representation, it has much more capability of separating the activities distributions over the selected

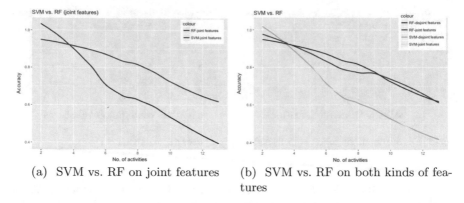

(a) SVM vs. RF on joint features   (b) SVM vs. RF on both kinds of features

Fig. 4 Comparison of accuracy between SVM and RF over joint features

feature space, making it more robust and transparent to the particular hand-engineered feature space. As can be seen from the confusion matrix in Fig. 3b, even the worst cases (circled in blue) are not that bad compared to the SVM case. By measuring the sensitivity and specificity for each activity in the RF case we have found the following: (1) sensitivity in general is above 0.7 for most of the classes and above 0.5 for almost all classes excluding 'liedown bed' (and 'eat meat' which has very few samples) and (2) specificity for all classes is above 0.9. Hence, the misclassifications are not concentrated on specific classes, somehow the false positives are evenly distributed among all activities.

The next couple of experiments are similar to the previous ones except that the constructed GM models in the new experiments are tri-variate for the tri-axial accelerometer directions. So any given sample is treated as a tri-variate signal. We performed our experiments with the following partameters: the number of Gaussian components is 2 and the maximum number of iterations for the EM algorithm is 500 or EM terminates if the change is less than 10^{-2}. We call such a setting 'joint accelerometer features'. The feature space consists of the output of this generative GM model: the responsibilities and the mean vectors, so the dimension of the feature space is 8. Figure 4a shows a comparison of accuracy between SVM and RF when using joint accelerometer features. It can be seen that the results are almost the same as in the disjoint features case (see Fig. 2c). Figure 4 shows an overall comparison in accuracy between SVM and RF when using both disjoint and joint accelerometer features. It shows that the performance is almost the same as in the disjoint features case. So disjoint feature space can perform well (as compared with the joint case), in addition it is computationally more efficient. So from now on for all the subsequent experimentation and analysis we will treat each individual sample as composed of three independent uni-variate signals as opposed to one tri-variate signal.

One advantage of the supervised learning approach is that most of the computational processing is done in the offline phase (the training phase). Hence, its online prediction performance can be considered efficient from the computational perspective.

2.4 Probabilistic and Measure Theoretic

In this paradigm we have used a variety of tools from the theory of probability and measure theory. Here again we deal with the statistical properties of the raw signal without taking time into consideration. In all of the following experiments we take all 14 activities into consideration (in the previous section we excluded 'eat soup' due to lack of enough samples for training). In this approach we rely on probabilistic modeling of the signals and measuring the distance among these probability distributions in order to classify the given activity signal.

2.4.1 Histogram Based Measures

This technique can be summarized as follows.

1. For each sample s of an activity A, we build a histogram for each of the tri-axial accelerometer directions. So we have three histograms: x-histogram, y-histogram, and z-histogram. The parameters of the histogram are as follows:

   (a) Lower bound of interval: -1.5 (lower bound of the accelerometer measurements).
   (b) Upper bound of interval: 1.5 (upper bound of the accelerometer measurements).
   (c) Step size: 0.1 (we have tried finer resolution, but the results were the same).

2. For each activity A, some of its collected samples s are randomly selected to act as a model of the underlying activity. We call these samples 'support samples'. They act as temples for the activity they are drawn from. The remaining samples are used for testing. In experimentation we use 65% of the samples as 'support samples'.

Given a test sample s, to classify to which activity s belongs we perform the following procedure:

1. For each activity A, find an overall distance between the sample s and the activity A as follows:

   (a) For each sample s' of the 'support samples' of A find the distance between s and s'; this is done as follows:
       (1) Compute the distance between each corresponding pairs of histograms $h_s = (h_x, h_y, h_z)$ and $h_{s'} = (h_{x'}, h_{y'}, h_{z'})$ using the Manhattan distance.

For example, the distance between the histograms corresponding to the x-direction is computed by:

$$d_M(h_x, h_{x'}) = \sum_i |h_x(i) - h_{x'}(i)| \tag{2}$$

where i runs over the bins of the histogram.

(2) Combine the three tri-axial scores by taking their average: $d(s, s') = \frac{d(h_x, h_{x'}) + d(h_y, h_{y'}) + d(h_z, h_{z'})}{3}$.

(b) Now the distance between sample s and activity A is taken to be the minimum among all computed distances with support samples:

$$d(s, A) = \min_{s' \in Supp(A)} d(s, s') \tag{3}$$

2. Now assign s to the activity A such that $d(s, A)$ is minimum:

$$s \in A \iff A = \underset{A'}{argmin}\, d(s, A') \tag{4}$$

Ties are broken arbitrarily.

Note that the above procedure is equivalent to k-NN with $k = 1$ using Manhattan distance. Figure 5a shows the accuracy curve for this technique. It shows a very good performance, especially when compared with the above machine learning techniques. For most of the activities (10), the classification accuracy is above 90%. And the worst accuracy among all the 14 activities is about 80%. This technique is very simple and computationally efficient. The training phase comprises of just computing the histograms of the template samples and the predictive phase comprises of computing the Manhattan distance between the given test sample and all the templates. The number of templates is relatively small and as well as the number of bins in the histogram, hence, computing such distance function is rather efficient. Figure 7a shows the confusion matrix for the 14 activities. It can easily be seen that the performance loss is due mainly to the confusion between two activities: 'getup bed' and 'sitdown chair' where most of their misclassifications (false positives) are attributed to the activity 'standup chair'.

2.4.2 Kernel Density Estimate

Here we use an approach that is also based on estimating a density function for the given time series. However, there are two main differences. The first is the use of kernel density estimates instead of building histograms to approximate the empirical density of the data samples. The second difference is the use of Kullback–Leibler KL-divergence as a dissimilarity measure instead of the Manhattan distance. The KL-divergence is not a proper distance metric since it is asymmetric with respect to

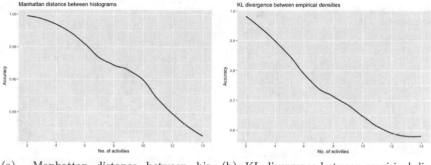

(a) Manhattan distance between histograms

(b) KL-divergence between empirical distributions

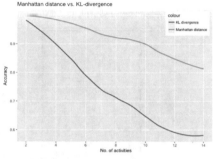

(c) Manhattan distance vs. KL-divergence

Fig. 5 Estimating densities and measuring dissimilarity among densities

its two arguments, so we compute both $KL \parallel pq$ and $KL \parallel qp$ and take the average to be the final distance between p and q. The KL-divergence can be computing by the following formula, assuming continuous random variables as is the case for accelerometer data:

$$KL \parallel pq = \int_{-\infty}^{\infty} p(x) \ln \frac{p(x)}{q(x)} dx \qquad (5)$$

Kernel density estimation is a *non-parametric method* for estimating the probability density function of a continuous random variable (here the acceleration). It is non-parametric since it does not presuppose any specific underlying model or distribution for the given data (see Sect. 2.5.1 in [14]). At each data point, a kernel function (we assumed Gaussian for our analysis here) is created with the data point at its center. The density function is then estimated by adding all of these kernel functions and dividing by the size of the data.

Figure 5b shows the performance of the KL-divergence for predicting the given ADL activities; followed by Fig. 5c for comparing the performance of the two techniques: Manhattan distance between histograms and KL-divergence between kernel

Fig. 6 Sample histograms
and kernel density
estimations

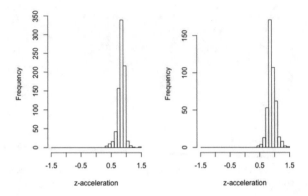

(a) Histograms for two samples of activity 'pour water' (z-direction)

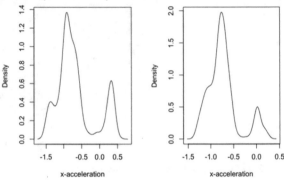

(b) Kernel density estimates for two samples of activity 'descend stairs' (x-direction)

density estimations. As is evident, the Manhattan distance achieves about 20% performance improvement over the KL-divergence. Also the computational complexity for building the histograms and computing the corresponding Manhattan distance outperform that required for building the kernel estimation and computing the corresponding KL-divergence. Figure 6 shows the histograms and kernel densities for some samples from two activities in order to give a sense of how time series data can be transformed into random variables with empirical distributions.

By studying the sensitivity and specificity of each activity we have found a general remark: the specificity is very high for all activities in both the Manhattan distance and KL-divergence cases. This means that for all activities the true negatives are high and the false positives are low which implies that the misclassifications in any activity are roughly distributed among many classes (of course, up to a factor that takes into consideration the distribution of sample data across all activities).

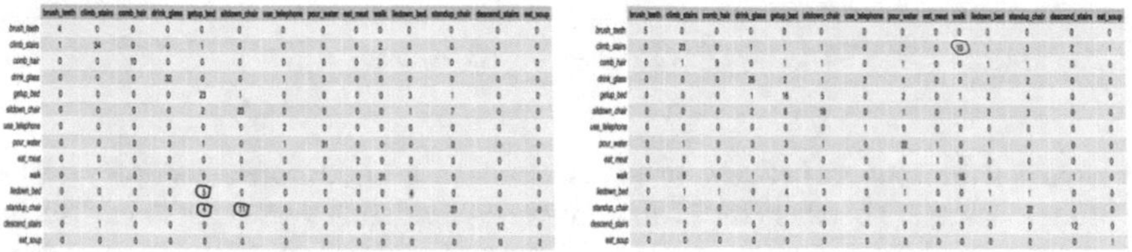

(a) Confusion matrix using Manhattan distance between histograms

(b) Confusion matrix using KL-divergence between empirical distributions

Fig. 7 Confusion matrices for both Manhattan distance and KL-divergence

2.4.3 Estimations Using Discrete Distributions

Here we consider the points of a time series as generated from a discrete distribution. Hence, the empirical distribution of the time series is estimated using the discrete distribution that corresponds directly to the observed data. We then apply several distance measures among the discrete discrete distributions in order to measure the dissimilarity among different time series samples and according decide on activity membership. An important thing to notice is that the resolution of the accelerometer data is fine enough to provide a comparable discrete distributions among the time series samples. We mean by resolution both the sampling rate relative to the accelerometer range ($[-1.5, 1.5]$), the precision of measurements as real numbers, and the length of each sampled time series. In the following we briefly describe four distance measures that we have used to measure dissimilarity among discrete probability distributions.

The first is the *Hellinger distance* which is defined in terms of the Hellinger integral and can be computed for discrete distributions as follows [15]. Given two discrete distributions $P = (p_1, \ldots, p_n)$ and $Q = (q_1, \ldots, q_n)$, then

$$d_H(P, Q) = \frac{1}{\sqrt{2}} \sqrt{\sum_{i=1}^{n} \left(\sqrt{p_i} - \sqrt{q_i}\right)^2} \tag{6}$$

Hellinger distance is a proper metric and its normalized by the factor $1/\sqrt{2}$ such that it is upper bounded by 1. The second distance metric is the *total variation distance*. Given two probability measures P and Q on the same σ-algebra $(\Omega, \mathcal{F})$, the total variation distance between P and Q can be computed as follows:

$$d_T(P, Q) = \sup\{|P(A) - Q(A)| : A \in \mathcal{F}\} \tag{7}$$

So it is largest difference between the two measures for any given event in the σ-algebra. In the assumed discrete case, as also the support is finite, this can be reduced to the following formula:

$$d_T(P, Q) = \frac{1}{2} \sum_{i=1}^{n} |p_i - q_i| \tag{8}$$

Note that this reduces to the Manhattan distance, Eq. (2), however, we apply it here in order to compare between the histogram-based modeling versus the finer discrete distribution modeling of the data. The third distance metric is the Cramér-von Mises distance, which computes the distance between the empirical cumulative distribution functions that are estimated from the given data [16]. This distance can be computed as follows:

$$d_C^2(P, Q) = \int (P(\{y \in [-1.5, 1.5]: y \le x\})$$
$$- Q(\{y \in [-1.5, 1.5]: y \le x\}))^2 \mu(dx) \tag{9}$$

where μ is the integration measure, which in our experiments taken to be the probability measure Q. The last distance metric that we adopt in our experimentation is the Kolmogorov distance. It is inspired by the Kolmogorov-Smirnov goodness-of-fit test [16]. It computes the supremum distance between the two given probability distributions evaluated over the sets in the σ-algebra defined over $[-1.5, 1.5]$:

$$d_K(P, Q) = \sup\{|P(\{y \in [-1.5, 1.5]: y \le x\})$$
$$- Q(\{y \in [-1.5, 1.5]: y \le x\})| : x \in [-1.5, 1, 5]\} \tag{10}$$

The offline modeling phase for the above techniques consists of the easy, and computationally efficient, task of estimating the discrete probability distributions of the given time series samples. The testing phase consists essentially of computing the distance, using one of the four distance metrics, between the test sample and all of the support samples of the 14 activities. For each time series sample we have three estimated empirical distributions for the tri-axial accelerometer directions. When a new test sample s is classified (to predict to which of the 14 activities this sample belongs), the following is done:

1. For each activity A, we find the distance between s and every member s' of the supporting samples of A. The test sample s has three empirical distributions associated with it for the tri-axial directions. Each such distribution is compared against the corresponding distribution in the supporting sample s'. So we have three scores. These scores are combined together by taking their summation.

$$d(s, s') = d_X(s_x, s_x') + d_X(s_y, s_y') + d_X(s_z, s_z') \tag{11}$$

where d_X is either one of the four distance metrics described above.

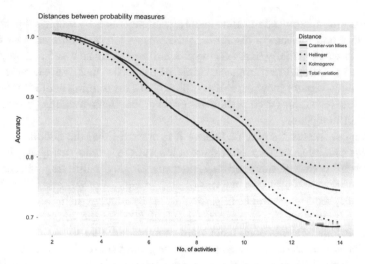

Fig. 8 Comparing the accuracy using different distance metrics between discrete empirical distributions

2. Given that we have computed the distance scores of the given test sample s against all the supporting samples of activity A. We need to choose a single score representing the distance between the test sample s and activity A. We choose the minimum score:

$$d(s, A) = \min_{s' \in Supp(A)} d_X(s, s') \qquad (12)$$

3. The decision is then taken by assigning s to the activity A' with minimum distance: $A' = argmin_A d(s, A)$.

Notice that this procedure is equivalent to a k-NN classifier with $k = 1$. Figure 8 illustrates a comparison in prediction accuracy over the 14 activities using the four distance measures. From this figure it is clear that working with probability mass functions, as in Hellinger and total variation, outperforms working with cumulative distribution functions, as in Cramer-von Mises and Kolmogorov. Also working with a L_2 norm style as in the Hellinger case outperforms working with an L_1 norm style as in the total variation case. Of course, the former is smoother and more uniform in accounting for distances among the various parts of the two input signals. The best performance is achieved using the Hellinger distance, using the whole set of 14 activities its accuracy reaches about 80%.

3 Discussion, Conclusion, and Future Work

We have focused in this paper on the modeling and analysis of activities of daily living. We have used a public data set of acceleration data in order to focus on the underlying analytical tool. We have experimented with techniques from two general paradigms: supervised machine learning and modeling by empirical probability distributions and measuring dissimilarity between such distributions. Our evaluation criteria are essentially based on the predictive performance and coarsely on the relative computational performance. Some of the techniques, for example, supervised learning, require more extensive offline phase for training and/or building the model. Other techniques such as those based on probabilistic modeling require light-weighted offline phase.

Looking at the results obtained above it is apparent that the simplest modeling and analysis methods perform almost the best results. Modeling using empirical discrete probability distributions and analysis using simple dissimilarity measures have shown to be the most effective in terms of both predictive power and computational performance; see Figs. 5 and 8. In particular, the Manhattan distance and Hellinger distance each gives predictive performance about 80% for the whole set of 14 activities. The offline phase of such techniques requires building the empirical discrete distribution which needs only one pass over the given data set $\mathcal{O}(N)$, where N is the size of the data set. The testing phase only requires computing a simple distance function between the given sample and the set of supporting samples of the considered activities. So depending on the number of bins of the empirical discrete distribution we can balance between the predictive performance and the computational efficiency of the underlying technique.

Supervised learning consists of a huge compendium of tools, methods, techniques, and technologies for feature selection, extraction, model building, and prediction. In this paper we have very lightly touched on such paradigm of analysis. We have tried two popular methods: random forest and support vector machines. The predictive performance of RF is rather good. Random forests can be though of as a form of *deep learning* [13]. Two factors affect the predictive performance of our experimentation with SVM and RF: the size of the data set and feature extraction and selection. Several activities have very small data sets such as 'eat meat', 'brush teeth', and 'use telephone'. Several ways can be used to remedy this without collecting more data. One way works as follows. Typically a sample time signal consists of a repetitive train of patterns each of which represents a complete cycle of the same activity. We can use techniques from time series analysis to extract each such pattern. Hence, one sample signal can give rise to several samples leading to an increase of the size of the data set of the given activity. As for feature selection/extraction a Gaussian mixture model is built for each sample and the parameters of the generative model are used as features for the training procedure. Deeper consideration can be taken into account in the selection of features, again using techniques from time series analysis. For example, we can compute the Fourier transform and take the first few components as features. Also we can compute the autocorrelation function up to certain gap and use the values as features, etc. Two disadvantages of Gaussian mixture models are the relatively poor computational efficiency and the high possibility of the occurrence of degeneracy (collapse of one of the components) leading to failure of building the model. The latter can be alleviated by using Bayesian approach instead of expectation maximization based on the likelihood function, however, this comes at the cost of steep increase of the required computational resources. As for the learning technique itself there are a myriad literature of learning algorithms that we can choose from. Most notably the different variations of deep neural networks such as recurrent neural networks, particularly, long short term memory, deep belief networks, and convolution neural networks. However, such application dictates the availability of enough data samples (time signals) in order to reliably train the network. One of the advantages of using deep neural networks is that there is an upsurge in the realization of the trained network in hardware such as FPGA, which can be embedded on mobile devices such as smart phones. This gives the ability for real-time accurate prediction of activities on commodity mobile devices. Of course, this is in addition to the possible realization on embedded GPU on smart phones [17].

A general observation is that the specificity (true negative rate) for individual activities is good for almost all techniques and almost all activities. This means that the misclassifications are in some sense evenly distributed among the activities. So any degradation in accuracy is essentially due to low sensitivity. From the confusion matrices we can generally see that there are some activities with high sensitivity across almost all the techniques such as 'brush teeth' and 'comb hair'. On the contrary there are other activities that generally have low sensitivity among most of the techniques such as 'walk', 'sitdown chair', and 'standup chair'. Another observation is that degradation in accuracy starts to occur when the number of activities exceeds 6 or 7, after which the task of activity classification becomes more challenging.

One possible way to increase the predictive accuracy is to merge different methods together. For example, we can take the majority vote of the predictions of several techniques, though this may not work with large number of activities. Another possibility is to take the predictions of the different models as inputs to a new metamodel, say a logistic regression learning algorithm, that fuses these different opinions into one final, more reliable, estimate. We also need to study the effect of the size of the set of 'support samples' on the predictive accuracy of the different techniques. In our experimentation we have unified that size to be 65% of the given data set.

References

1. Patel, S., Park, H., Bonato, P., Chan, L., Rodgers, M.: A review of wearable sensors and systems with application in rehabilitation. J. Neuro Eng. Rehabil. **9**(1), 21 (2012)
2. Ericsson Mobility Report on The Pulse of The Networked Society. Technical report, Ericsson, 11 (2016)
3. Poushter, J.: Smartphone ownership and internet usage continues to climb in emerging economies (2016)
4. Lara, O.D., Labrador, M.A.: A survey on human activity recognition using wearable sensors. IEEE Commun. Surv. Tutor. **15**(3), 1192–1209 (2013)
5. Bruno, B., Mastrogiovanni, F., Sgorbissa, A., Vernazza, T., Zaccaria, R.: Analysis of human behavior recognition algorithms based on acceleration data. In: 2013 IEEE International Conference on Robotics and Automation (ICRA), pp. 1602–1607. IEEE (2013)
6. Khan, A.M., Lee, Y.-K., Lee, S.Y., Kim, T.-S.: A triaxial accelerometer-based physical-activity recognition via augmented-signal features and a hierarchical recognizer. IEEE Trans Inf Technol. Biomed. **14**(5), 1166–1172 (2010)

7. Kilinc, O., Dalzell, A., Uluturk, I., Uysal, I.: Inertia based recognition of daily activities with anns and spectrotemporal features. In: 2015 IEEE 14th International Conference on Machine Learning and Applications (ICMLA), pp. 733–738. IEEE (2015)
8. Cornacchia, M., Ozcan, K., Zheng, Y., Velipasalar, S.: A survey on activity detection and classification using wearable sensors. IEEE Sens. J. **17**(2), 386–403 (2017)
9. Ye, L., Ferdinando, H., Seppänen, T., Huuki, T., Alasaarela, E.: An instance-based physical violence detection algorithm for school bullying prevention. In: 2015 International Wireless Communications and Mobile Computing Conference (IWCMC), pp. 1384–1388. IEEE (2015)
10. Bruno, B., Mastrogiovanni, F., Sgorbissa, A., Vernazza, T., Zaccaria, R.: Human motion modelling and recognition: a computational approach. In: CASE, pp. 156–161. IEEE (2012)
11. Bruno, B., Mastrogiovanni, F., Sgorbissa, A., Vernazza, T., Zaccaria, R.: Analysis of human behavior recognition algorithms based on acceleration data. In: ICRA, pp. 1602–1607. IEEE (2013)
12. Bruno, B., Mastrogiovanni, F., Sgorbissa, A.: Laboratorium—Laboratory for Ambient Intelligence and Mobile Robotics, DIBRIS, University of Genova, via Opera Pia 13, 16145, Genova, Italia (IT), Version 1 released on 11th, 2014
13. Bengio, Y.: Learning Deep Architectures for AI (foundations and trends(r) in Machine Learning). Now Publishers Inc, 10 2009
14. Bishop, C.: Pattern Recognition and Machine Learning. Springer (2006)
15. Miescke, F.L.-J.: Statistical decision theory: estimation, testing, and selection. In: F. Liese (ed.) Springer Series in Statistics (2008-06-11). Springer (1616)
16. Rieder, H.: Robust Asymptotic Statistics: Volume I (Springer Series in Statistics). Springer, softcover reprint of the original 1st ed. 1994 edition, 7 (2012)
17. Nurvitadhi, E., Venkatesh, G., Sim, J., Marr, D., Huang, R., Ong Gee Hock, J., Liew, Y.T., Srivatsan, K., Moss, D., Subhaschandra, S., Boudoukh, G.: Can fpgas beat gpus in accelerating next-generation deep neural networks? In: Proceedings of the 2017 ACM/SIGDA International Symposium on Field-Programmable Gate Arrays, FPGA '17, pp. 5–14. ACM. New York, NY, USA (2017)

Deep Learning Technology for Big Data Analytics

Soil Morphology Based on Deep Learning, Polynomial Learning and Gabor Teager-Kaiser Energy Operators

Kamel H. Rahouma and Rabab Hamed M. Aly

Abstract Soil Morphology is considered the main observable characteristics of the different soil horizons. It helps farmers to determine what kind of soil they can use for their different plants. The observable characteristics include soil structure, color, distribution of roots and pores. The main concept of this chapter is to classify the different soils based on their morphology. Furthermore, the chapter contains a comparison between polynomial neural network and deep learning for soil classification. The chapter introduces a background about the different methods of feature extraction including the Gabor wavelet transform, Teager-Kaiser operator, deep learning, and polynomial neural networks. The chapter, also, includes two goals. The first goal is to improve the extraction of soil features based on Gabor wavelet transform but followed by the Teager-Kaiser Operator. The second goal is to classify the types of different morphological soil based on two methods: deep learning and polynomial neural network. We achieved accuracy limits of (95–100%) for the polynomial and deep learning classification achieved accuracy up to 95% but the deep learning is more accurate and very powerful. Finally, we compare our work results with the previous work and research. Results show an accuracy range of (98–100%) for our work compared with (95.1–98.8%) for the previous algorithms based on PNN. Furthermore, the accuracy of using DNN in this chapter comparing with pervious works achieved a good accuracy rather than the others.

Keywords Soil detection · Gabor wavelet · Deep learning · Polynomial neural network (PNN) · Teager-kaiser

K. H. Rahouma (✉)
Electrical Engineering Department, Faculty of Engineering, Minia University, Minia, Egypt
e-mail: kamel_rahouma@yahoo.com

R. H. M. Aly
The Higher Institute for Management Technology and Information, Minia, Egypt

© The Author(s), under exclusive license to Springer Nature Switzerland AG 2021   325
A. E. Hassanien et al. (eds.), *Machine Learning and Big Data Analytics Paradigms: Analysis, Applications and Challenges*, Studies in Big Data 77,
https://doi.org/10.1007/978-3-030-59338-4_17

1 Introduction

Recently, deep learning, machine learning and computer vision play an important role in environment image analysis, especially in the detection of features processing [2]. Furthermore, the effective role of computer vision and image processing is to classify the type of different soils. The researchers try to make the classification very easier using the modern research technologies. Researchers try to improve the methods of extraction or classification. On the other hand, some researches have introduced some other methods for the feature detection of soils or diseases of leaves of trees. In the following, we will show some of the recent researches and how the authors try to improve methods to classify and detect the medical images and environments images such as soils [2, 16, 18].

Furthermore, deep learning is considered one of the important methods in different applications such as agricultures, medical and communications. Some researchers focused on prediction methods for image analysis for medical applications such as [10, 24]. Tekin and Akbas [24] described how to use the Adaptive Neuro Fuzzy Inference System (ANFIS) to predict soil features and comparing to another type of soil. They tried groutability of granular soils with a piece of cement. It is one of the papers which helped us in the work of this chapter. We will introduce some of another previous researches in the next section. The remainder of this chapter is organized as follows: Sect. 2 concerns with a background and a literature review of previous work of prediction and classification techniques. Section 3 illustrates the main methodology and it gives the main algorithm used in this chapter. Section 4 discusses the results of the applied technique and compares it with the previous works. Section 5 presents a brief conclusion of our work and suggests some future work that can be accomplished.

2 A Literature Review

There are a lot of techniques that achieved accurate results in detection and classification. In the following, we will introduce the important cases of the previous researches about detection methods. Actually, the hybrid technique plays an important role in detection processing. Sweilam et al. [23] used a hybrid method based on some of information and support vector machine (SVM) to classify the types of tumors. The accuracy of this method is 90.3% for some specified types of cancer. On another hand, Cheng and Han [4] introduced a survey about object detections for optical remote and how to apply machine learning for this detection. Further, Ford and Land [8] applied a new model based on latent support vector machine as a model for cancer prognosis. The operations of this model are based on microarray operations and some of the gene expression and they improved the algorithm and technique of this model. The results showed that the increase of quality of curve receiver operating characteristic (ROC) when replacing least regression to SVM.

Some authors invested new approaches of cancer features in soil detection and classification. Khare et al. applied ANFIS (Neuro fuzzy) as a classification method. They used fuzzy system to detect the severity of the lung nodules depending on IF-Then rules method. They also applied 150 images in computer aided diagnosis (CAD) system. They achieved (sensitivity of 97.27%, specificity of 95% with accuracy of 96.66%). Potter and Weigand [19] introduced a study about the image analysis of soil crusts to get the properties of surface heating. Actually, the study approved moderate skewness toward negative tails and the other results can help to improve future mapping for any place having biocrust surfaces such as in the Mojave Desert.

Recently, some authors presented a survey about the most recent image segmentation processing especially for medical images [6]. They improved the classification techniques to increase the classification rate accuracy by 99.9%. Dallali et al. [6] introduced a new classification algorithm based on fuzzy clustering method and improved the performance by neural networks to classify heart rate (HR) and RR intervals of the ECG signal and they called this method fuzzy clustering method neural network (FCMNN). Nabizadeh and Kubat [15] applied Gabor wavelet to extract features of MRI images. They also compared results with statistical features methods. The comparing technique for the feature's extraction method was based on some classifiers. Authors have evaluated the methods and Gabor wavelet achieved 95.3% based on some of the classification techniques.

Recently, some authors introduced deep learning to improve the methods of classifications and predictions based on using huge datasets. The deep learning helps to use any big size of datasets and images in classification and prediction and how can achieve it in a little time [21] authors introduced deep learning for classification and the detection of foliar esca symptoms during summer. They applied deep learning by integrated the deep learning base network within a detection network (RetinaNet). That allows performing detection in practical time. They used six frames per second on GPU and achieved an accuracy of 90% Hu et al. [9], authors introduced three algorithms for predictions based on modern machine learning techniques. They approved that the three methods considered as extend for machine learning. The methods included particle swarm optimization (PSO) with support vector regression (SVR), back-propagation neural network (BPNN), and extreme learning machine (ELM). The study applied in real-world data for forecasting the surface settlement in cities of China. The PSO with SVR achieved the highest accuracy in prediction other than the other methods. On another hand, Singh et al. [22], authors applied more than prediction methods in artificial intelligent such as Gaussian process (GP), random forest (RF) and multi-linear regression (MLR)-based models. They achieved that SVM is the best method other than the others methods. Furthermore, authors introduced 95 datasets, a total of 66 data sets were selected for preparing different algorithms and 29 for testing model. They measured outputs soil parameters such as percentage of sand (S), percentage of fly ash (Fa), specific gravity (G), time (T) and head (H) and also coefficient of permeability (k). They also proved that time and water head are the most effective parameters for the estimation of permeability of the soil. Torabi et al. [25] applied SVM with Cluster-Based Approach (CBA) and ANN to predict the solar radiations. The CBA-ANN-SVM approach is validated other than the ANN or SVM. They achieved an acceptable percentage error. Vargas et al. [26] achieved better performance when applying deep learning (DNN) other than RNN. The results showed that DNN better on catching semantic from texts and RNN is better on catching the context information and modeling complex temporal characteristics for stock market forecasting. Cramer [5] applied 7 methods of machine learning.

For rainfall prediction. The methods are: the current state-of-the-art (Markov chain extended with rainfall prediction) and six other popular machine learning algorithms, namely: Genetic Programming, Support Vector Regression, Radial Basis Neural Networks, M5 Rules, M5 Model trees, and k-Nearest. They compared between the performances of these methods. They also run tests using rainfall time data sets for 42 cities.

Eslami et al. [7] introduced study using deep learning for forecast ozone concentrations over a specified area. They used the method to predict the hourly ozone concentration. The study achieved that CNN can predict the next 24 h of ozone concentrations and they identified several limitations for deep learning model for real time forecasting.

Some authors introduced new classification methods for environment images. Wang [27] applied deep learning for hyper spectral remote sensing images. They applied classification methods to achieve multi features learning and the accuracy is 99.7%. Furthermore, Perez et al. [17] introduced deep learning classification for soil related to illegal tunnel activities. They proposed a new method in handling imbalance learning. The result showed that the method improved the performance significant of soil detection. Boudraa and Salzenstein [3] proposed review about Teager Kaiser Operator (TKO) for image enhancement and improved the enhanced technique by following it by an energy operator of TKO. In the next section, based on the previous researches, we will introduce the Gabor wavelet transform followed by TKO in feature extraction of soil image datasets [2] and after that we will classify the result based on polynomial neural network (PNN). Furthermore, we will introduce how to use DNN as classification techniques for soil Morphology system.

3   Methodologies and Algorithms

The Method of this chapter consist of two main parts:

(1) First part: how to apply Gabor wavelet followed by Teager Kaiser to extract features
(2) Second part: Classification process and it is consisting of two methods:

    (a) Applying polynomial neural network (PNN)
    (b) Applying deep learning DNN.

3.1 Features extractions using Gabor Wavelet Followed by Teager-Kaiser operators:

(a) First part is for enhancement and features extraction based on Gabor wavelet and Teager Kaiser.
(b) Second part is for classification based on the polynomial neural network (PNN) or DNN. We, also, will compare the system methods with the other previous work of the soil classification and detection.

In this chapter, we used soil datasets taken from a set of online recorded images from different places and also based on the database images which have used by Bhattacharya and Solomatine [2].

(a) The first part: in this part, the practical work consists of two main steps:

- The enhancement process of the soil datasets (see Fig. 1).

  - Find limits to contrast stretch an image (to increase the contrast of image).
  - Convert image to gray level.

- Gabor wavelet features extraction followed by Teager Kaiser Operator:

Gabor wavelet is considered one of the most practical methods to extract the optimal features from images after the enhancement process. Bhattacharya and Solomatine [2] extracted features based on boundary energy. Furthermore Boudraa and Salzenstein [3], approved that 2D Teager Kaiser Operator reflects better local activity than the amplitude of classical detection operators. The quadratic filter also is used to enhance the high frequency and combined with image gray values to estimate the edge strength value and all of that is used in the enhancement process.

The main function of this method is to generate energy pixels based on 2D Teager Kaiser Operator (TKO). The Teager Kaiser Energy operator is defined by the energy of the signal x(t) as follows [3, 20]:

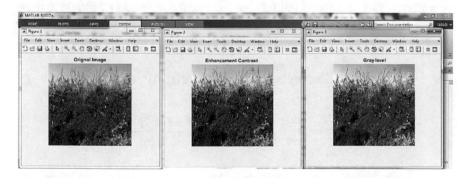

Fig. 1 Enhancement process strategy

$$\Psi_c[x(t)] = \left[x(t)\right]^2 - x(t)\,x(t) \tag{1}$$

where x(t) is the signal, x(t) is the first derivative and x(t) is the second derivative.

Actually, we applied Teager Kaiser Energy operator in the discrete as follows Eq. (2):

$$\Psi_d[x(t)] = x_n^2 - x_{n+1}x_{n-1} \tag{2}$$

On the other hand, the Gabor wavelet equations which we applied in our system are based on the Gabor transform employed in one dimension as a Gaussian window shape as shown in Eq. (3) but in the two dimension case, it provides the spectral energy density concentrated around a given position and frequency in a certain direction [3].

$$g_{\alpha\mathcal{E}}(x) = \sqrt{\frac{\alpha}{\pi e^{-\alpha x^2}}} e^{-i\mathcal{E}x} \tag{3}$$

where $\mathcal{E}, x \in R, \alpha \in R^+$ and $\alpha = (2\sigma^2)^{-1}$, σ^2 is the variance and $\mathcal{E}$ is the frequency.

Then, the mother wavelet of the Gabor wavelet as follows:

$$g_{\alpha,\mathcal{E},a,b}(x) = |a|^{-0.5} g_{\alpha,\mathcal{E}}\left(\frac{x-b}{a}\right) \tag{4}$$

where $a \in R^+$(scale), and $b \in R$(shift).

Note that, the Gabor wavelet doesn't form orthonormal bases. Actually, the Gabor wavelet can detect edge corner and blob of an image [3]. We applied blob detection and used the main function of energy operator of Teager Kaiser in the calculation of Gabor energy. Gabor wavelet followed by Teager Kaiser achieved a set of features from soils datasets images which helped us to classify the types of soil. Actually, we extracted 98 as a general but the optimal calculated features from 98 features are Correlation, Energy and Kurtosis based on Eqs. (5–7) [19]. We will discuss the details in Sect. 4.

$$Energy = \sum_{i=1}^{m}\sum_{j=1}^{n}(GLCM(i, j))^2 \tag{5}$$

$$Correlation = \sum_{i=1}^{m}\sum_{j=1}^{n}\frac{\{ij\}GLCM(i, j) - \{\mu_x\mu_y\}}{\sigma_x\sigma_y} \tag{6}$$

where i, j index instant $\mu_x\mu_y$ and $\sigma_x\sigma_y$ are the mean and standard deviations of probability matrix Gray level coherence Matrix (GLCM) along row wise x and column wise y.

$$\text{Kurtosis} = \frac{1}{\sigma^4} \sum_{i=0}^{m-1} (i - \mu)^4 x(i) - 3 \tag{7}$$

In the following section, we introduce the classification method to classify the types of the soils based on the previous features.

3.2 Classification Process

The classification process in this chapter applying two methods. One of these methods is PNN and the second method is DNN. In the following, we will explain the both of them.

3.2.1 Polynomial Neural Network Classification (PNN)

There are many types of ANN based on a mathematical classification equation such as PNN (Polynomial neural network) which will discuss in this section. On the other hand, ANN is used in classification in data mining and also to predict future data. GMDH is a multilayer network which used quadratic neurons offering an effective solution to modeling non-linear systems. The PNN is one of the most popular types of neural networks based on polynomial equation. It is used for classification and regression. It is more practical and accurate in prediction of behavior of the system model [20]. A class of polynomials (linear, modified quadratic, cubic, etc.) is utilized. We can obtain the best description of the class by choosing the most significant input variables and polynomial according to the number of nodes and layers. Ivakhnenko used a polynomial (Ivakhnenko Polynomial) with the grouping method of data handling (GMDH) to obtain a more complex PNN. Layers connections were simplified and an automatic algorithms was developed to design and adjust the structure of PNN neuron (see Fig. 2).

To obtain the nonlinear characteristic relationship between the inputs and outputs of the PNN a structure of a multilayer network of second order polynomials is used. Each quadratic neuron has two inputs (x_1, x_2) and the output is calculated as described

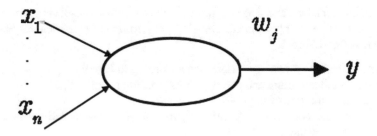

Fig. 2 The neuron inputs and output

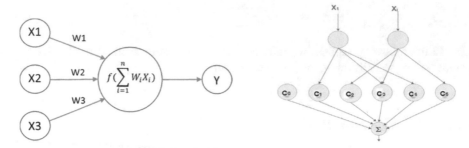

Fig.3 The polynomial network structures

in Eq. (8) where "Fig. 3"shows the structure of PNN.

$$g = w_0 + w_1 x_1 + w_2 x_2 + w_3 x_1 x_2 + w_4 x_1^2 + w_5 x_2^2 \tag{8}$$

where $w_i; i = 0, \ldots, 5$ are weights of the quadratic neuron to be learnt. The main equations for PNN structure which is the basic of GMDH-PNN are:

$$(X_i, y_i) = x_{1i}, x_{2i}, \ldots \ldots, x_{Ni}, y_i \tag{9}$$

where X_i, y_i are data variables and $i = 1; 2; 3; \ldots; n$.
The input and output relationship of PNN-structure of "Fig. 3" is:

$$Y = F(x_1, x_2, \ldots x_N) \tag{10}$$

The estimated output is:

$$\begin{aligned}
\grave{y} = \grave{f}(x_1, x_2, \ldots x_N) &= c_0 + \sum_i c_i x_i \\
&+ \sum_i \sum_j c_{ij} x_i x_j \\
&+ \sum_i \sum_j \sum_k c_i c_j c_k x_i x_j x_k + \cdots
\end{aligned} \tag{11}$$

The PNN is the best and fastest solution in classification data techniques of data.
The main steps to apply the GMDH-PNN algorithm for classification based on polynomial Eq. (8) are:

(a)   Determine the system input variables according to Eq. (9).
(b)   Formulate the training and testing data according to Eq. (10, 11.
(c)   Select the structure of the PNN.
(d)   Estimate the coefficients of the polynomial of nodes to estimate the error between y_i, $\grave{y}_i$ then:

$$E = \frac{1}{n_{tr}} \sum_i^{n_{tr}} (y_i - \hat{y}_i)^2 \tag{12}$$

where n_{tr} is the number of training data subsets, i is the node number, k is the data number, n is the number of the selected input variables, m is the maximum order and n_0 is the number of estimated coefficients [20]. By using the training data, the output is given by a linear equation as:

$$Y = X_i C_i \tag{13}$$

$$C_i = (X_i^T X_i)^{-1} X_i^T Y \tag{14}$$

where $Y = [y_1, y_2, y_3, \ldots, y_{ntr}]^T$, $X_i = [X_{1i}, X_{2i}, X_{3i}, \ldots, X_{ntri}]^T$,

$$X_{ki}^T = [X_{ki1}, X_{ki2}, X_{kin}, \ldots, X_{ki1}^m, X_{ki2}^m, \ldots, X_{kin}^m]^T \text{ and}$$

$C_i = [C_{0i}, C_{1i}, \ldots, C_{n'i}]^T$, and after that, check the stopping criterion.

(e) Determine the new input variables for the next layer.

Note That: The database which have applied in this chapter for classification process based on the online datasets which applied in [2]. Furthermore, the Fig. 10 shows the general flow chart of the system.

The general algorithm of the detection and the classification PNN for soil datasets

Start

(1) Enter number of images n, Maximum Number of Neurons (Nu) = 50 and the Maximum of Layer (NL) = 10, Alpha (AL) = 0.6,The train ratio (TR) = 0.7.
(2) Enter n datasets of soil images.
(3) Loop i = 1: n

    (a) Enhance the soil image using Low pass filter
    (b) Apply the gray level of the image.

(4) Extract the image features (nf = 98 features) based on Gabor wavelet and TKO energy operator.
(5) Calculate Correlation, Energy, and Kurtosis from discussed Eqs. (5–7).

Start a loop

Use 80% of the datasets for training using PNN structure to obtain the system coefficient.

Use the trained system to estimate the classification the rest of 20% of datasets.

End Loop

(6) Compute the error of the accuracy of features as follows:
    Err = (actual value − Estimated value)/actual value * 100%

Table 1 Layer architecture of CNN

Type	Kernel size	Filters	Activations
Convolution	3 × 3	64	ReLU
Max-Pooling	2 × 2	–	–
Convolution	3 × 3	128	ReLU
Convolution	3 × 3	256	RelU
Max-Pooling	2 × 2	–	–
Convolution	3 × 3	512	ReLU
Convolution	3 × 3	512	RelU
Full-Connected	–	100	ReLU
Full-Connected	–	1	Linear

(7) Calculate the accuracy $= 100 - $ Err.
(8) Print the results.

End

3.2.2 Deep Learning Classification (DNN)

Actually, deep learning has many characteristics that make it more powerful than classical machine learning techniques. Deep learning is performing by a convolutional multilayer neural network, which has many hidden layers and free parameters. Unlike the commonly used multilayer neural network, each soil input image passes through basics steps; convolution layers with filters (kernels), pooling layers, fully connected (FC) layers and SoftMax function.

Convolution Neural Network (CNN)

CNN includes asset of making decisions on hyperparameter (Number of layers and rate of learning) [27].

The Network architecture of CNN consist of sequence of layers to build the neural network (see Table 1) (see Figs. 4 and 5). There are four main operations in CNN (see Table 1).

- Convolution
- Non-Linearity (ReLU)
- Pooling or Sub Sampling
- Classification (Fully Connected Layer)

These operations are the basic building blocks of every Convolutional Neural Network.

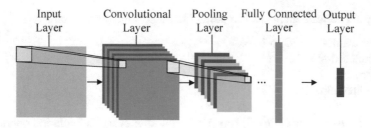

Fig. 4 General architecture of CNN

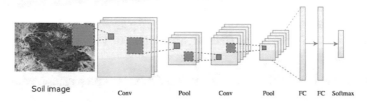

Fig. 5 Architecture of CNN for Soil classification

Note that:

(a) The size of the feature map is controlled by three parameters:

- **Depth:** is the number of filters used for the convolution operation.
- **Stride:** is the number of pixels by which filter matrix over the input matrix. Stride specifies how much we move the convolution filter at each step.
- **Padding:** is used for input matrix with zeros around the border, matrix. We use the padding to maintain the same dimensionality.

On another hand, **total** $= \sum \frac{1}{2}$ (target probability $-$ output probability)^2 and Computation of output of neurons that are connected to local regions in the input. This may result in volume such as [32 × 32 × 16] for 16 filters.

There are some others important construction parameters such as:

Pooling: After a convolution operation we usually perform pooling to reduce the dimensionality. This enables us to reduce the number of parameters, which both shortens the training time and combats overfitting.

Pooling layers: is a down sample each feature map independently, reducing the height and width, keeping the depth intact. The most common type of pooling is max pooling which just takes the max value in the pooling window. Contrary to the convolution operation, pooling has no parameters. It slides a window over its input, and simply takes the max value in the window. Similar to a convolution, we specify the window size and stride. The simple definition is a down sampling operation along the spatial dimensions (width, height), resulting in volume such as [16 × 16 × 16] which reduces the dimensionality of each feature map but retains the most important information.

Hyperparameters: there are 4 important hyperparameters to decide on:

- Filter count from 32 to 1024.
- Default stride = 1.
- Filter size.
- Usual padding.

Rectified Linear Unit (ReLU) Layer: is used after every Convolution operation. It is applied per pixel and replaces all negative pixel values in the feature map by zero. This leaves the size of the volume unchanged ([32 × 32 × 16]). ReLU is a non-linear operation.

Fully Connected: After the convolution and pooling layers, we add a couple of fully connected layers to wrap up the CNN architecture. The fully connected layers then act as a classifier.

Training: CNN is trained the same way like ANN, backpropagation with gradient descent.

In this chapter, we perform the convolution operation by sliding this filter over the input. If we use a 32 × 32 × 3 image, so we have to use a filter of size 5 × 5 × 3 (note that the depth of the convolution filter matches the depth of the image, both being 3). When the filter is at a particular location it covers a small volume of the input, and we perform the convolution operation. The only difference is this time we do the sum of matrix multiply in 3D instead of 2D, but the result is still a scalar. We slide the filter over the input and perform the convolution at every location aggregating the result in a feature map. This feature map is of size 32 × 32 × 1.

On another hand the structure of the deep learning code introduced into 4 methods for classifications:

(1) Conv2D: to create convolution layers.
(2) MaxPooling2D: to create a maxpooling layer.
(3) Flatten.
(4) Dropout: is used to prevent overfitting.

A ConvNet is able to successfully capture the Spatial and Temporal dependencies in an image through the application of relevant filters. The architecture performs a better fitting to the image dataset due to the reduction in the number of parameters involved and reusability of weights. In other words, the network can be trained to understand the sophistication of the image better.

We use the DNN based on basic structure of CNN classifications and that achieved very strong classifications with a higher accuracy. The ConvNet depends on reducing the images into a form which is easier to process, without losing features which are critical for getting a good prediction. That is very important in learning features and scalable to massive datasets (see example of CNN Fig. 6) [13].

We use the DNN based on basic structure of CNN classifications and that achieved very strong classifications with a higher accuracy.

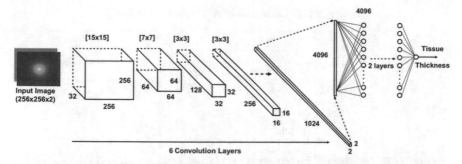

Fig. 6 Example of proposed CNN architecture and it consists of six convolution layers and three fully connected layers for regression

4   Results and Discussions

In this chapter, we applied our system based on image processing toolbox of MATLAB2017a.

Actually, the chapter operated two cases: The first case uses the Gabor Wavelet for features extraction and multi support vector machine (MSVM) for classification and the Gabor Wavelet followed by Teager Kaiser Energy Operator for features extraction and the PNN for classification. The second case uses DNN for classification after using Gabor Wavelet followed by Teager Kaiser Energy Operator for filteration to enhance performances.

4.1   Results for GW-TK and PNN Classifications

Actually, the accuracy of this case consists of the three optimal features (Correlation, Energy, and Kurtosis) is from 95 to 98%. The three optimal features came from 98 features which are extracted from soil images-based GW-TK operators as discussed in pervious section. In Figs. 7 and 8 show the accuracy of the energy feature for all datasets and also Tables 2 and 3.

The datasets of this work consist of 6 types of soils and each type has 20 different images. Actually, the CPU time approved the operations of GW-Tk-PNN in 13 s for each image in datasets so, applying Gabor wavelet followed by Teager-Kaiser Energy operator improved the accuracy of the extraction which is found to be 98.8% or higher. We compared between our results and the results from previous work [2] and show that in Fig. 9. Actually, the most of pervious work based on SVM or MSVM so, we tried to improve the accuracy as discussed before. The applied techniques of this chapter can be utilized in the detection and diagnosis of plants' problems, defects, and diseases. A future work may be done to study images of the plant's leaves, roots, and stalks and then extract their features and classify their problems, defects and diseases (Fig. 10).

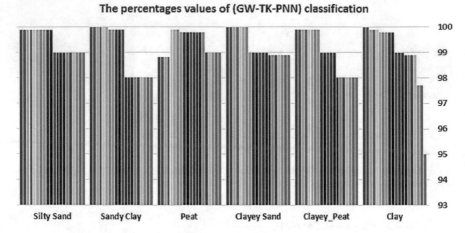

Fig. 7 The result of GW-multi-SVM of the previous research (energy)

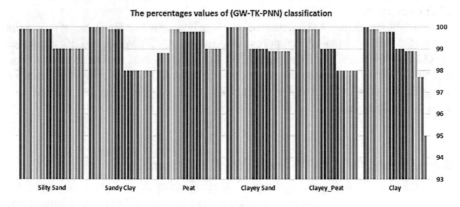

Fig. 8 The result of applying-Gabor wavelet—TK-PNN (energy)

4.2 Results of DNN Classification Method

Actually, in this part to improve accuracy, we applied the DNN in a set of tests. The tests approved a group of objects (see Table 4). The first test with 12 objects approved 81% accuracy as shown in Figs. 11, 12 and 13. The first test is a simple test contains 12 objects and 600 iterations with 100 epochs as shown in Fig. 11.

The main concept of this part is the classification of the Soil images based on DNN classification method. The soil datasets contain 3 different types of categories. In DNN method, we applied a maximum iteration from 500 to 600 iteration. We applied our system on laptop with CPU (core i3 2.5GH and 6 GB Ram) and using MATLAB 2019a.

Table 2 The percentages values of GW-MSVM classification (energy feature)

Datasets	The percentages values of classification (GW-MSVM) result %																			
Clay	95.1	95.1	95.1	95.1	95.1	96	96	96	96	96	96.6	96.7	96.7	97	97	97.1	97.1	97.1	97.1	98
Clayey_peat	95.1	95.1	96.7	96.7	96	96.77	96.7	96.7	96.7	96	96.6	96.77	96.77	98	98	98	98	98	98	98
Clayey sand	96	96.7	96.7	96.7	96.7	96.7	96.7	98	98	98.1	98.1	98.1	98	98	98	98	98.7	98.7	98.7	98.7
Peat	96.7	96.7	96.7	96.7	96.7	96.7	96.7	96.7	96.7	98.3	98.3	98.3	98.3	98.3	98.8	98.8	98.8	98.8	98.8	98.8
Sandy clay	96	96.8	96.77	96.77	96.77	96.77	96.77	98.2	98.2	98.2	98.2	98.2	98.37	98.37	98.37	98.4	98.4	98.4	98.4	98.4
Silty sand	96.77	96.77	96.77	96.77	96.77	98	98	98.3	98.3	98.3	98.2	98.2	98.2	98.2	98.3	98.3	98.3	98.3	98.3	98.3

Table 3 The percentages values of GW-TK-PNN classification (Energy Feature)

Datasets	The percentages values of Classification (GW-TK-PNN) result %																
Clay	95	97.7	97.7	98.9	98.9	99	99	99.8	99.8	99.8	99.8	99.8	99.9	99.9	99.9	100	100
Clayey_peat	98	98	98	98	98	99	99	99	99	99.9	99.9	99.9	99.9	99.9	99.9	99.9	99.9
Clayey sand	98.9	98.9	98.9	98.9	98.9	99	99	99	99	100	100	100	100	100	100	100	100
Peat	99	99	99	99	99.8	99.8	99.8	99.8	99.8	99.9	99.9	99.9	99.9	98.8	98.8	98.8	98.8
Sandy clay	98	98	98	98	98	98	99.9	99.9	99.9	99.9	100	100	100	100	100	100	100
Silty sand	99	99	99	99	99	99	99.9	99.9	99.9	99.9	99.9	99.9	99.9	99.9	99.9	99.9	99.9

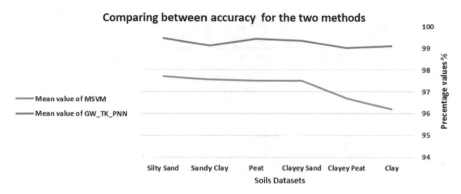

Fig. 9 The comparing between our method with pervious work

In the following comparing between our method in classification using deep learning and the accuracy with the others pervious works.

As shown as in Table 5, The pervious works of the other authors achieved accuracy from 90 to 92% based on DNN but in our work, we tried to improve the accuracy and also by using CPU.

5 Conclusions

This chapter aimed to extract features based on Gabor wavelet followed by Teager Kaiser Energy operator and to apply the polynomial learning technique to classify the different types of the soil datasets images. Furthermore, we applied another method to classify the different types of soil but based on deep learning. We obtained our results and compared them with the results of previous research. The previous algorithms achieved accuracy limits of (95.1–98.8%) while our results achieved accuracy limits of (95–100%) in GW-TK-PNN. On another hand, the previous algorithms achieved accuracy limits of (90–92%) while our results achieved accuracy limits of (up to 95%) in DNN. However, the applied techniques can be utilized in the analysis, feature extraction, and classification of the plants' problems, defects, and diseases.

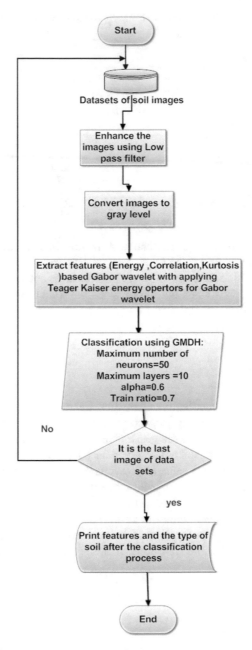

Fig. 10 The general flow chart of the GW-TK-PNN system

Table 4 The details of the results for each different category

Objects categories	Number of images	Iterations	Time	Accuracy (%)
12	176	600	37.33 min	81.38
12	704	600	150 min	81.38
113	10,000	500	332.38 h	Up to 95.3

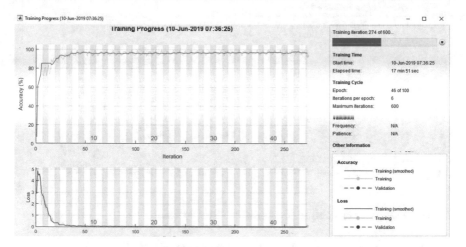

Fig. 11 The First part of the training progress for 12 objects

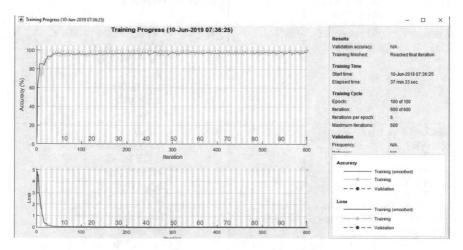

Fig. 12 The Finial part of the training progress for 12 objects categories

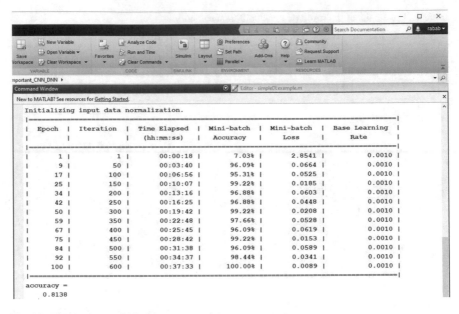

Fig. 13 The accuracy of 12 objects categories

Table 5 The comparing of our work with the pervious works

Paper	Number of data sets images	Method	Epochs	Processor	Accuracy (%)
Manit et al. [13]	900 categories (4096 image)	DNN	200	NVIDIA GTX 980 GPU and an Intel® CoreTM i5 CPU (3.40 GHz), running Ubuntu 14.04	90.4
Mukherjee et al. [14]	10,000 images	CNN	500	Intel i7 processor with 32 GB RAM and with two linked GPU of Titan	90.58
Ayan and Ünver [1]	10,000 images	CNN-Keras	25	Two intel processors E5-2640 Xenon with 64 GB RAM and NVIDIGPU card	82

(continued)

Table 5 (continued)

Paper	Number of data sets images	Method	Epochs	Processor	Accuracy (%)
Kwasigroch et al. [11]	Over 10,000 images	VGG19, Residual Networks (ResNet) and CNN-SVM	Up to 100	GeForce GTX 980Ti GPU with 6 GB memory, Intel Core i7-4930 K processor and 16 GB RAM memory	90–95
Maia et al. [12]	200 images	VGG19 + LR	Up to 100	Non limited	92.5
This work	500 images	GW-TK-PNN	50	CPU (core i3 2.5GH and 6 GB Ram)	98–100
This work	Up to 10,000	CNN classifier	100	CPU (core i3 2.5GH and 6 GB Ram)	81–95

References

1. Ayan, E., Ünver, H.M.: Data augmentation importance for classification of skin lesions via deep learning. In: *2018 Electric Electronics, Computer Science, Biomedical Engineerings' Meeting (EBBT)*, pp. 1–4. IEEE (2018)
2. Bhattacharya, B., Solomatine, D.P.: Machine learning in soil classification. Neur. Netw. **19**(2), 186–195 (2006)
3. Boudraa, A.O., Salzenstein, F.: Teager–Kaiser energy methods for signal and image analysis: a review. Digit. Signal Process. **78**, 338–375 (2018)
4. Cheng, G., Han, J.: A survey on object detection in optical remote sensing images. ISPRS J. Photogramm. Remote Sens. **117**, 11–28 (2016)
5. Cramer, S., et al.: An extensive evaluation of seven machine learning methods for rainfall prediction in weather derivatives. Expert Syst. Appl. **85**, 169–181 (2017)
6. Dallali, A., Kachouri, A., Samet, M.: Fuzzy C-means clustering neural network, WT, and HRV for classification of cardiac arrhythmia. ARPN J Eng Appl Sci **6**(10), 112–118 (2011)
7. Eslami, E., et al.: A real-time hourly ozone prediction system using deep convolutional neural network. arXiv preprint arXiv:1901.11079 (2019)
8. Ford, W., Land, W.: A latent space support vector machine (LSSVM) model for cancer prognosis. Procedia Comput. Sci. **36**, 470–475 (2014)
9. Hu, M., et al.: modern machine learning techniques for univariate tunnel settlement forecasting: a comparative study. In: Mathematical Problems in Engineering. Hindawi (2019)
10. Jahmunah, V., et al.: Computer-aided diagnosis of congestive heart failure using ECG signals—a review. Physica Med. **62**, 95–104 (2019)
11. Kwasigroch, A., Mikołajczyk, A., Grochowski, M.: 'Deep neural networks approach to skin lesions classification—a comparative analysis. In: 2017 22nd International Conference on Methods and Models in Automation and Robotics (MMAR), pp. 1069–1074. IEEE (2017)
12. Maia, L.B., et al.: Evaluation of melanoma diagnosis using deep features. In: 2018 25th International Conference on Systems, Signals and Image Processing (IWSSIP), pp. 1–4. IEEE (2018)

13. Manit, J., Schweikard, A., Ernst, F.: Deep convolutional neural network approach for forehead tissue thickness estimation. Curr. Direct. Biomed. Eng. **3**(2), 103–107 (2017)
14. Mukherjee, S., Adhikari, A., Roy, M.: Malignant melanoma classification using cross-platform dataset with deep learning CNN architecture BT. In: Recent Trends in Signal and Image Processing, pp. 31–41. Springer Singapore, Singapore
15. Nabizadeh, N., Kubat, M.: Brain tumors detection and segmentation in MR images: Gabor wavelet vs. statistical features. Comput. Electr. Eng. **45**, 286–301 (2015)
16. Odgers, N.P., McBratney, A.B.: Soil material classes. In: Pedometrics, pp. 223–264. Springer (2018)
17. Perez, D., et al.: Deep learning for effective detection of excavated soil related to illegal tunnel activities. In: 2017 IEEE 8th Annual Ubiquitous Computing, Electronics and Mobile Communication Conference (UEMCON), pp. 626–632. IEEE (2017)
18. Pham, B.T., et al.: Hybrid integration of Multilayer Perceptron Neural Networks and machine learning ensembles for landslide susceptibility assessment at Himalayan area (India) using GIS. CATENA **149**, 52–63 (2017)
19. Potter, C., Weigand, J.: Imaging analysis of biological soil crusts to understand surface heating properties in the Mojave Desert of California. CATENA **170**, 1–9 (2018)
20. Rahouma, K.H., et al.: Analysis of electrocardiogram for heart performance diagnosis based on wavelet transform and prediction of future complications. Egypt. Comput. Sci. J., 41 (2017)
21. Rançon, F., et al.: Comparison of SIFT encoded and deep learning features for the classification and detection of Esca disease in bordeaux vineyards. Remote Sens. **11**(1), 1 (2019)
22. Singh, B., et al.: Estimation of permeability of soil using easy measured soil parameters: assessing the artificial intelligence-based models. ISH J. Hydraul. Eng, 1–11 (2019)
23. Sweilam, N.H., Tharwat, A.A., Moniem, N.K.A.: Support vector machine for diagnosis cancer disease: a comparative study. Egypt. Inf. J. **11**(2), 81–92 (2010)
24. Tekin, E., Akbas, S.O.: Predicting groutability of granular soils using adaptive neuro-fuzzy inference system. In: Neural Computing and Applications, pp. 1–11. Springer (2017)
25. Torabi, M., et al.: A hybrid machine learning approach for daily prediction of solar radiation. In: International Conference on Global Research and Education, pp. 266–274. Springer (2018)
26. Vargas, M.R., De Lima, B.S.L.P., Evsukoff, A.G.: Deep learning for stock market prediction from financial news articles. In: 2017 IEEE International Conference on Computational Intelligence and Virtual Environments for Measurement Systems and Applications (CIVEMSA), pp. 60–65. IEEE (2017)
27. Wang, L., et al.: Spectral–spatial multi-feature-based deep learning for hyperspectral remote sensing image classification. Soft Comput. **21**(1), 213–221 (2017)

Deep Layer Convolutional Neural Network (CNN) Architecture for Breast Cancer Classification Using Histopathological Images

Zanariah Zainudin, Siti Mariyam Shamsuddin, and Shafaatunnur Hasan

Abstract In recent years, there are various improvements in computational image processing methods to assist pathologists in detecting cancer cells. Consequently, deep learning algorithm known as Convolutional Neural Network (CNN) has now become a popular method in the application image detection and analysis using histopathology image (images of tissues and cells). This study presents the histopathology image related to breast cancer cells detection (mitosis and non-mitosis). Mitosis is an important parameter for the prognosis/diagnosis of breast cancer. However, mitosis detection in histopathology image is a challenging problem that needs a deeper investigation. This is because mitosis consists of small objects with a variety of shapes, and is easily confused with some other objects or artefacts present in the image. Hence, this study proposed four types of deep layer CNN architecture which are called 6-layer CNN, 13-layer CNN, 17-layer CNN and 19-layer CNN, respectively in detecting breast cancer cells using histopathology image. The aim of this study is to detect the breast cancer cell which is called mitosis from histopathology image using suitable layer in deep layer CNN with the highest accuracy and True Positive Rate (TPR), and the lowest False Positive Rate (FPR) and loss performances. The result shows a promising performance for deep layer CNN architecture of 19-layer CNN is suitable for this MITOS-ATYPHIA and AMIDA13 dataset.

Keywords Breast cancer image classification · Deep learning; histopathology image · Convolutional neural network (CNN)

Z. Zainudin (✉) · S. M. Shamsuddin · S. Hasan
School of Computing, Faculty of Engineering, Universiti Teknologi Malaysia, 81310 Skudai, Johor, Malaysia
e-mail: zanariah86@gmail.com

S. M. Shamsuddin
e-mail: sitimariyams@gmail.com

S. Hasan
e-mail: shafaatunnur@gmail.my

© The Author(s), under exclusive license to Springer Nature Switzerland AG 2021   347
A. E. Hassanien et al. (eds.), *Machine Learning and Big Data Analytics
Paradigms: Analysis, Applications and Challenges*, Studies in Big Data 77,
https://doi.org/10.1007/978-3-030-59338-4_18

1 Introduction to Breast Cancer and Histopathology Image

Cancer has become the most popular health problem that can be dangerous to human life especially for women. This is supported by the statistics from Europe and United State health reports that 30% women from a total of 852,630 was estimated to be suffering from breast cancer. Another supporting statistics is based on The National Cancer Institute in 2014, which reported on breast cancer based on the population statistics in the US population. It shows that the number of cancer patients kept increasing year by year. Based on this report, the study conducted by the National Cancer Registry (NCR), US women are more exposed to breast cancer where in 1999–2013, some 231,840 new estimated cases related to breast cancer were reported to NCR in 2014 and the number kept increasing from year to year. The number of new cases of female breast cancer increased to 852,630 in 2017 [1, 2]. Moreover, it is also stated that the age-adjustment for female breast cancer amongst women was lower compared to several age gaps such as stated and illustrated below (Fig. 2) 20 years (0.00%), 20–34 year (1.8%), 35–44 years (9.1%), 45–54 years (21.6%), 55–64 year (25.6%), 65–74 years (21.9%), 75–84 years (14.2%) and 84 year and above (5.7%). It can be summarized that, Fig. 1 provides comparisons for age–adjustment for female breast cancer per 100,000 women from 2008 until 2012 for all women in United State (US) [1, 2].

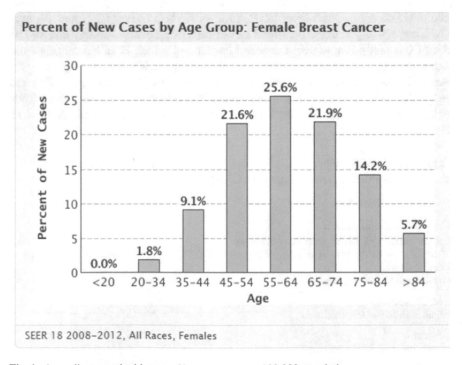

Fig. 1 Age-adjustment incidences of breast cancer per 100,000 population

Table 1 Estimating new cancer cases in the US in 2018 and 2017	Female new cases	(878,980 person) (%)	(852,630 person) (%)
	Breast	30	30
	Lung and bronchus	13	13
	Colon and rectum	8	7
	Uterine corpus	7	7
	Non-Hodgkin lymphoma	4	4

*Supported by Europe and United States health reports

Another supported by the statistic from Europe and United States health reports in year 2018 and year 2017. In year 2017, Table 1 show that 30% women's from 852,630 women's has been estimated getting breast cancer. Table 1 also shows that year 2018, it stated that breast cancer is the number one based on the population statistics in Europe and United States population. Based on this report, the study conducted by National Cancer Registry (NCR), US women are more exposed on breast cancer where the number of the new cases are slightly increases, where 878,980 new estimated cases involved in breast cancer are reported to NCR for year 2018 and the number keep increasing in last two years. The number of new cases of female breast cancer was increased to 852,630 in 2017 [1, 2].

This situation can be reduced if there are improvements in early detection and treatment [3, 4]. In order to detect breast cancer, various clinical tests have been used in hospitals such as ultrasound scan, mammogram, and histopathology. Normally, many patients use mammogram to detect possibility of having cancerous cells. However, these tests have several disadvantages where younger women may get non-accurate results as their breasts give clearer picture compared to older women. As an alternative test, younger women are advised to use histopathology image for more accurate result.

Usually biopsy test is a supportive test after the first test is conducted. Using histopathology image is the most efficient test to detect cancerous cells [4, 5]. Although histopathology image has been proven to be the best way to classify the mitosis and nuclei for breast cancer, getting True Positive Rate (TPR) result from the image however requires the skills and experience of a pathology expert [6, 7]. Breast cancer is the most fearful disease to all middle age women. It also major cause of death to all women [3]. Breast cancer is a very big problem to all developing countries.

As shown in Fig. 2a, the normal female breast is made up lobules where it produce milk for newborn, duct where tiny tubes which bring the milk to the nipple and stroma where contain some fat tissue and connective tissues such as blood and lymphatics vessels. Mostly the breast cancers cases started in the cells called the duct, lobules and other tissues as shows in Fig. 2b. The nodes of Lymph are small, look like a bean where it connected by the lymphatic vessels where it keep all the immune system

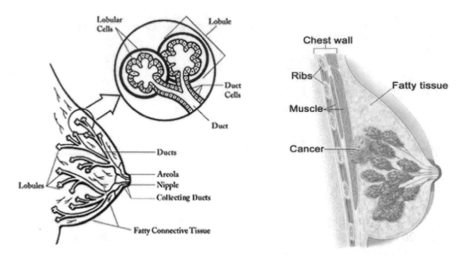

Fig. 2 **a** Anatomy of normal breast. **b** Anatomy of breast cancer

cell. Lymphatic vessel are very small where it carry some liquid away from the breast. The Lymph contains the tissues, waste, immune systems cell where the breast cancer can entering the system and begin to growth inside the lymph nodes [3, 8].

Histopathology image is an image which the image are been examined under microscope with higher resolution and magnifier. The systematic biopsy test procedure been illustrated in Fig. 3.

First off all, the tissues are be taken using needle biopsy into small cut. After that, the smaller tissues will be cut into smaller slices and put inside cassette and be put some chemical. Then, pathologist will be putting the specimen inside glass

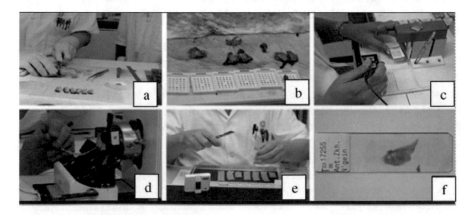

Fig. 3 **a** Tissues will be cut into small pieces. **b, c** The smaller pieces are put into the cassettes and put some chemical named paraffin. **d** The smaller pieces are made from paraffin and microtome. **e** The smaller pieces called section are put on glass slides. **f** Histopathology slide are created

for staining process. The main goal of this research is to classify the mitosis or non-mitosis in histopathology and help pathology to classify the histopathology image clearly. The problems that have arisen in histopathology image are:

1. Normalisation of the image into clear image
2. Identification of the mitosis inside the image
3. Performance analysis in classified mitosis or non mitosis using deep learning algorithm.

Even though there are many types of histopathology image in this research area, but the focus for this research are histological image from images from breast cancer only. In general, there are four main procedures to get histopathology image, which are:

1. Squash procedure
2. Smears procedure
3. Whole-mounts procedure
4. Sections procedure.

A smear procedure contain cell in fluid such as blood. Meanwhile squash procedure will take information such as chromosome in DNA. Whole-mounts procedure will get the entire specimen into microscopic slide to get information otherwise the cell has cancerous cell or not. Usually the tissues be cut into tiny part and will undergo staining process. This procedure are the most easiest to help pathology in classified cancerous or non-cancerous cells into tissues compared to sections. Sections procedure is where the tissues are been cut into tiny slices from the tissues where it already been frozen inside freezing chamber.

The specimen will undergo a few steps after that and be put a chemical agents when it will convert the tissues became solid material so it will be easy to cut into small part. After that, the specimen will put away to stops all the cells and enzymes activity and harden the specimen. It been suggested that all the procedure has been quickly apply to the tissues after the separation of the cancerous cell from body.

The chemical call formalin is making the specimen turn into colourless. Figure 4 shows the example of the histopathology image when the chemical called formalin been putted inside the specimen.

The next step in getting histopathology image is by staining process where the colourless image be converted into purple or pink image. This process called Hematoxylin & Eosin (H & E) where it can see the structural of the cell image more clearly. But, after H & E stain process are not enough to classify the cancerous cell or not. Therefore using ImmuHistoChemical (IHC) chemical can overcome this problem where it provides more additional information about histopathology image in Fig. 5.

After all the staining process, the cell will covered with glass slide and achieve better visual quality for microscopic examination. The slides are then sent to pathologist who will examines the slide under microscope and will classify the slide contain the cancer cell. Early detection will help the patient get a better attention to cure cancer. Figure 6 shows some example on mitotic figure where the cancerous cells are marked with green arrows.

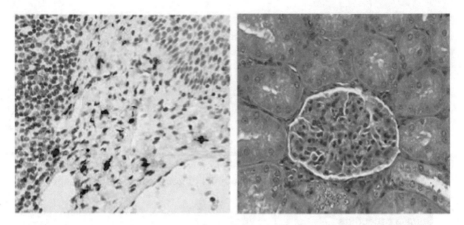

Fig. 4 Example of histopathology image after put chemical named formalin

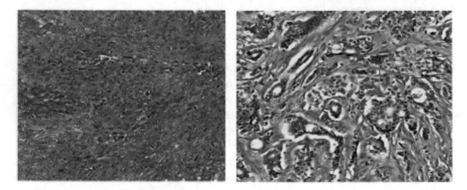

Fig. 5 Example of histopathology image

2 Literature Review on Deep Learning

Nowadays, there are popular technique, which called deep learning. Deep learning and unsupervised feature learning have shown great promise applications. State-of-the-art performance has been reported in several domain recognition [9–11] visual object recognition [11] to text processing [12–17].

Recently, using Deep Learning algorithm is better compared to other machine learning algorithm due to the speed up the computational time in large dataset and misclassification of the mitosis with non-mitosis [18–20]. This is caused by the complicated structure of the histopathology image and also the difficulty in the detection of the mitosis part due to its similarity [3, 20].

Using deep learning can skip the feature extraction using standard machine learning for image classification. Other than that, it has also been observed that increasing the scale of deep learning, with respect to the number of training examples,

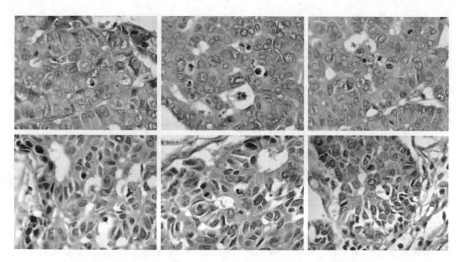

Fig. 6 Examples of mitosis where the mitosis are marked with green arrow

the number of model parameters, or both, can drastically improve ultimate classification accuracy [11, 21]. These results have led to a surge of interest in scaling up the training and inference algorithms used for these model and in improving applicable optimization procedures [16, 21].

Based on previous researcher as stated by Khuzi et al. and Singh et al., the misclassification of the mitosis and non-mitosis are based on many criteria such as image quality is not very good, physician eye, poor segmentation problem in image and many other causes where it can lead to incorrect misclassification, which is also called False Positive Rate (FPR) [6, 7]. The main contribution of this studies is to find the suitable layer for Deep Learning algorithm for histopathology image dataset.

Previous research shows that the False Positive Rate (FPR) results on diagnosis of mitosis and non-mitosis using the normal deep learning algorithm were quite high [8, 22–25]. Ciresan et al. [20], in their paper entitled "Mitosis detection in breast cancer histology images with deep neural networks" using Deep Neural Network and MITOS dataset stated that their result of using Deep Neural Network is that the precision is 0.88, re-call is 0.70 and F1-Score is 0.782 which better than the algorithm. Meanwhile, Su et al. [26], said that using deep learning algorithm are improving the result on detection of the breast cancer cell. They used image of size 1000×1000 for the experimental result and they enable to shorten the computational time. Most of the traditional algorithm are not good in term of computational time and the performance of the breast cancer detection.

Spanhol et al. [27], also discussed on DeCAF features where they compare three type of layer Deep Neural Network and reusing the CAFFE architecture and parameters of a pre-trained neural network. Wahlstr mentioned that using deep learning algorithm successfully decreased the error for his testing model for detecting Sensor Network Data [28]. The result for the joint training error is 0.0011 and separate training error is 0.0011. His also said that, for the future work that he will used the sampling-based approaches to get better result. Feng et al. stated that Deep Learning model is the best algorithm compare other machine learning classifiers [29].

Su et al., has done experimental using deep convolutional neural network and the experiments execution time is 2.3 s using the High Performance Computers (HPC) [26]. Meanwhile, Adriana Romero et al., also stated that using trained model network are very effective and suggested that using the spatial information to get better result, combine the output feature and max pooling steps and adding the new layer to deep learning to improve kappa agreement score. Chan et al. [30] used Deep Learning called PCANet and the experiments results shows that for pattern recognition 99.58% for Extended Yale B Dataset, and 86.49% accuracy using 60% Extended Yale B Dataset for training. On the other side, Couprie et al., also tested using Deep Learning for Stanford Background Dataset and getting some over segmentation for some images [31]. Sukre et al., also using Deep Learning Restricted Boltzmann Machine and has better performance measure [32]. Wahab et al. also mentioned that using Deep Learning Convolutional Neural Network getting better training for the image dataset [33]. Some researchers also stated that using complex image can effects the performance measure of classification.

Based on previous researchers, it can be concluded that using Deep Learning algorithm can improve performance detecting the cancer cell. It also support the evidence where the misclassification of the images are based on many criteria such as quality of images, lacking of physician eye, over or poor results using histopathology image and many other cause where it can lead to incorrect misclassification. Figure 2 shows the Deep Layer CNN framework for breast cancer image detection in the next subsequent section.

3 The Proposed Deep Layer CNN Architecture for Breast Cancer Image Classification

Various applications of classifiers especially in medical image has great accuracy using Deep Learning classifier. The Deep Learning can be trained with many training images inside the algorithm. The advantage of the Deep Learning is it can divide each image into sub regions. For an example, 4096×3 sub regions can be extracted from a Region of Interest (ROI) with an image size of $64 \times 64 \times 3$ for MITOS-ATYPHIA dataset and 6400×3 sub regions can be extracted from a Region of Interest (ROI) with an image size of $80 \times 80 \times 3$ for AMIDA13. As a result, the training of Deep Learning is done on sub regions basis; not on the whole region basis as shown in Fig. 7.

Deep Learning with huge number of samples from the training sample dataset can allow some enrichment process which will help the training to avoid over fitting problem of an Artificial Neural Network (ANN) [33–35]. Deep learning algorithm which is called Convolutional Neural Network (CNN) is used in this studies and we proposed suitable layer architecture for CNN for this histopathology image. The

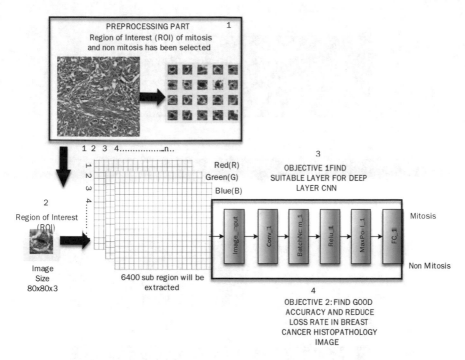

Fig. 7 Deep Layer CNN framework for breast cancer image classification

deep learning framework for breast cancer image detection (mitosis and non-mitosis) is shown in Fig. 2 from pre-processing part until classification part.

For data preparation, the sample of mitosis images were collected from opened source database which is called MITOS ATYPHIA and AMIDA13 (See Fig. 9). The dataset was divided into two datasets; 70% for training dataset and 30% for testing dataset. This was to ensure the learning generalization had been achieved by feeding the classifier to a new datasets.

In this study, four types of deep layer CNN architecture have been proposed for breast cancer image classification, namely, 6-layer CNN, 13-layer CNN, 17-layer CNN and 19-layer CNN. The proposed deep layer of CNN architecture is shown in Fig. 8a–d.

4 Parameter Setup and Performance Measure for Breast Cancer Image Classification

Parameter setting The performance measures for previous researchers on image cancer classification were typically measured using the Receiver Operating Characteristic (ROC) curve analysis. It had four types of performance measures, and the

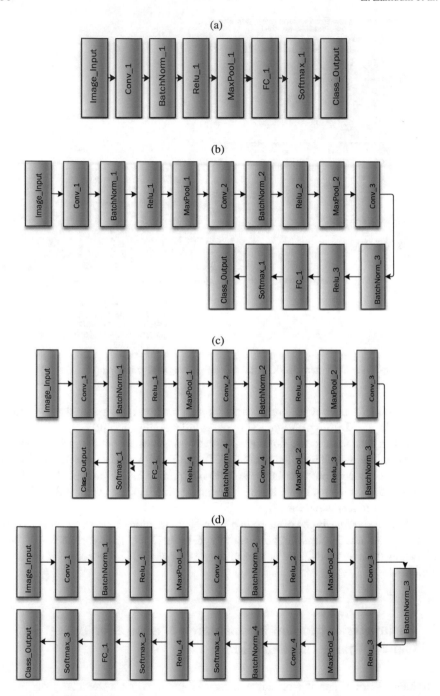

Fig. 8 **a** The proposed Deep Layer CNN Architecture of 6-layer CNN. **b** The proposed Deep Layer CNN Architecture of 13-layer CNN. **c** The proposed Deep Layer CNN Architecture of 17-layer CNN. **d** The proposed Deep Layer CNN Architecture of 19-layer CNN

Table 2 Performance Measure Test (TP, FP, TN and FN) are represented for this study

Performance measurement					
Actual versus predicted	Predicted Yes	Parameter	Predicted No	Parameter	Total parameter
Actual Yes	True Positive (TP)	TP	False-Negative (FN)	FN	TP + FN
Actual No	False Positive (FP)	FP	True Negative (TN)	TN	FP + TN
Total parameter		TP + FP		FN + TN	TP + FP + FN + TN

tests were called true positive rate (TPR), false positive rate (FPR), true negative rate (TNR) and false negative rate (FNR). In this case, the sensitivity and specificity tests had also been used to get the accuracy from the histopathology image with mitosis being the True Positive Rate (TPR) class, and non-mitosis samples being the True Negative Rate (TNR) class [36, 37]. Hence, the performance measure for the breast cancer image classification is described in Table 2.

Where the elaboration of all tests for the performance measurement are listed below:

- **True Positive Rate/Recall/Sensitivity (TP/(TP + FN))** is the number of correct predictions that an instance is positive,
- **False Positive Rate (FP/(TP + FP))** is the number of incorrect predictions that an instance is positive,
- **Accuracy ((TP + TN)/(TP + FN + FP + TN))** is the number of correct classifications, and
- **Misclassification Rate/Error Rate/Loss ((FP + FN)/(TP + FN + FP + TN)** is the number of wrong classifications.

The experimental results will be discussed in the next section.

5   Experimental Results

The experimental results for breast cancer image classification using the proposed deep Convolutional Neural Network (CNN) layer are reported in Tables 3, 4 and 5 using MITOS-ATYPHIA dataset, accordingly. The Graphical Processing Unit (GPU) device NVDIA GeForce 940MX has been implemented in this study for better speed up (Table 6).

In overall performance, the proposed deep layer CNN architecture with 19-layer CNN gave better result compared to 6-layer CNN, 13-layer CNN and 17-layer CNN is suitable for MITOS-ATYPHIA dataset. The results from Tables 2, 3, 4 and 5 shows that deeper layer of CNN architecture produced better result than the least layer. For example, the highest accuracy of 84.49% and True Positive Rate (TPR)

Table 3 Experimental result using 6-layer CNN using MITOS-ATYPHIA dataset

Experiments	Accuracy (%)	Loss (%)	True positive rate (%)	False positive rate (%)
1	68.06	31.94	75.00	38.89
2	77.78	22.22	80.56	25.00
3	62.50	37.50	66.67	41.97
4	50.00	50.00	100	100
5	66.67	33.33	58.33	25.00
6	61.11	38.89	63.89	41.67
7	59.72	40.68	66.67	47.22
8	65.28	34.72	58.33	27.78
9	73.61	26.39	66.67	19.44
10	55.56	44.44	55.56	44.44
Average	64.03	35.97	69.17	41.44

Table 4 Experimental Result using 13-layer CNN using MITOS-ATYPHIA dataset

Experiments	Accuracy (%)	Loss (%)	True Positive Rate (%)	False Positive Rate (%)
1	75.00	25.00	69.44	19.44
2	76.39	23.61	72.22	19.44
3	72.22	27.78	69.44	25.00
4	76.39	23.61	75.00	22.22
5	68.06	31.94	66.67	30.56
6	68.06	31.94	66.67	30.56
7	76.39	23.61	80.56	27.78
8	72.22	27.78	80.56	36.11
9	65.28	34.72	58.33	27.78
10	73.61	26.39	75.00	27.78
Average	72.36	27.64	71.39	26.67

of 80.55%; while the least loss of 15.51% and False Positive Rate (FPR) of 11.66% performances were produced by 19-layer CNN compared to 17-layer CNN, 13-layer CNN and 6-layer CNN. Furthermore, the performance comparison of accuracy, loss, true positive rate and false positive rate using the proposed deep layer CNN architecture is illustrated in Fig. 9, accordingly.

The experimental results for breast cancer image classification using the proposed deep Convolutional Neural Network (CNN) layer are reported in Tables 7, 8, 9 and 10 using AMIDA13 dataset, accordingly.

Based on Tables 7, 8, 9 and 10, the proposed 19-layer CNN and 17-layer CNN produced the best accuracy compared to 6-layer CNN and 13-layer CNN with the most of the accuracy results from 10 experiments were maintained above 90%. While Fig. 9 shows that the proposed 19-layer CNN is the best model for MITOS-ATYPHIA

Table 5 Experimental Result using 17-layer CNN using MITOS-ATYPHIA dataset

Experiments	Accuracy (%)	Loss (%)	True positive rate (%)	False positive rate (%)
1	79.17	20.83	86.11	27.79
2	83.33	16.67	86.11	19.44
3	86.61	13.39	83.33	11.11
4	81.94	18.06	83.33	19.44
5	84.72	15.28	83.33	13.89
6	86.11	13.89	80.56	8.33
7	81.94	18.06	86.11	22.22
8	81.94	18.06	77.78	13.89
9	83.33	16.67	83.33	16.67
10	80.56	19.44	83.33	22.22
Average	82.96	17.04	83.33	17.50

Table 6 Experimental result using 19-layer CNN using MITOS-ATYPHIA dataset

Experiments	Accuracy (%)	Loss (%)	True positive rate (%)	False positive rate (%)
1	81.94	18.06	77.78	13.89
2	86.11	13.89	80.56	8.33
3	86.61	13.39	83.33	11.11
4	86.11	13.89	80.56	8.33
5	81.94	18.06	77.78	13.89
6	86.11	13.89	80.56	8.33
7	81.94	18.06	86.11	22.22
8	81.94	18.06	77.78	13.89
9	86.11	13.89	80.56	8.33
10	86.11	13.89	80.56	8.33
Average	84.49	15.51	80.55	11.66

dataset as it produced the lowest loss range between 3 and 9%. In conclusion, the aim of this study has been achieved where deep layer CNN architecture was capable on detecting the breast cancer (which is known as mitosis and non-mitosis) with feasible results of accuracy, loss, TPR and FPR. Furthermore, the performance comparison of accuracy, loss, true positive rate and false positive rate using the proposed deep layer CNN architecture for AMIDA13 dataset is illustrated in Fig. 10, accordingly.

In conclusion for all performance, the proposed deep layer CNN architecture with 19-layer CNN gave better result compared to 6-layer CNN, 13-layer CNN and 17-layer CNN is suitable for MITOS-ATYPHIA dataset and. The results from Tables 7, 8, 9 and 10 shows that deeper layer of CNN architecture produced better result than the least layer for AMIDA13 Dataset.

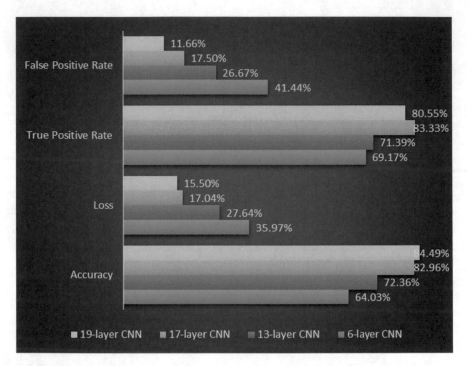

Fig. 9 Comparison performance measurement for MITOS-ATYPHIA dataset

Table 7 Experimental result using 6-layer CNN using AMIDA13 dataset

Experiments	Accuracy (%)	Loss (%)	True positive rate (%)	False positive rate (%)
1	92.16	7.84	95.03	4.97
2	93.14	6.86	93.79	6.21
3	92.16	7.84	87.58	12.42
4	88.24	11.76	95.03	4.97
5	92.40	7.60	95.03	4.97
6	90.44	9.56	95.03	4.97
7	90.44	9.56	95.03	4.97
8	92.16	7.84	95.03	4.97
9	88.24	11.76	95.03	4.97
10	90.44	9.56	95.03	4.97
Average	90.98	9.02	94.16	5.84

Table 8 Experimental result using 13-layer CNN using AMIDA13 dataset

Experiments	Accuracy (%)	Loss (%)	True positive rate (%)	False positive rate (%)
1	92.16	7.84	87.58	12.42
2	97.06	2.94	95.65	4.35
3	96.81	3.19	97.52	2.48
4	92.16	7.84	87.58	12.42
5	97.79	2.21	94.41	5.59
6	96.81	3.19	97.52	2.48
7	92.16	7.84	87.58	12.42
8	97.06	2.94	95.65	4.35
9	92.16	7.84	87.58	12.42
10	90.44	9.56	95.03	4.97
Average	94.46	5.54	92.61	7.39

Table 9 Experimental result using 17-layer CNN using AMIDA13 dataset

Experiments	Accuracy (%)	Loss (%)	True positive rate (%)	False positive rate (%)
1	92.16	7.84	87.58	12.42
2	90.44	9.56	95.03	4.97
3	98.28	1.72	96.27	3.73
4	96.81	3.19	97.52	1.86
5	90.44	9.56	95.03	4.97
6	92.16	7.84	87.58	12.42
7	96.81	3.19	97.52	1.86
8	98.53	1.47	98.14	1.86
9	90.44	9.56	95.03	4.97
10	92.16	7.84	87.58	12.42
Average	93.82	6.18	93.73	6.15

6 Conclusion

In this study, the proposed deep layer CNN architecture which is called 6-layer CNN, 13-layer CNN, 17-layer CNN and 19-layer CNN has been successfully implemented in detecting the mitosis and non-mitosis breast cancer histopathology images using two types of dataset which is MITOS-ATYPHIA and AMIDA13. Hence, we can conclude that, the deeper the CNN layer, the better performance result for breast cancer image classification. This is because more training layers had been fed into the proposed model and the learning multiple level of abstraction yielded better generalization of high level features.

Table 10 Experimental result using 19-layer CNN using AMIDA13 dataset

Experiments	Accuracy (%)	Loss (%)	True positive rate (%)	False positive rate (%)
1	92.16	7.84	87.58	12.42
2	98.53	1.47	97.52	2.48
3	99.26	0.74	98.76	1.24
4	92.16	7.84	87.58	12.42
5	99.26	0.74	99.38	0.62
6	96.81	3.19	97.52	1.86
7	96.81	3.19	97.52	1.86
8	99.02	0.98	98.14	1.86
9	92.16	7.84	87.58	12.42
10	98.28	1.72	96.89	3.11
Average	96.45	3.56	94.85	5.03

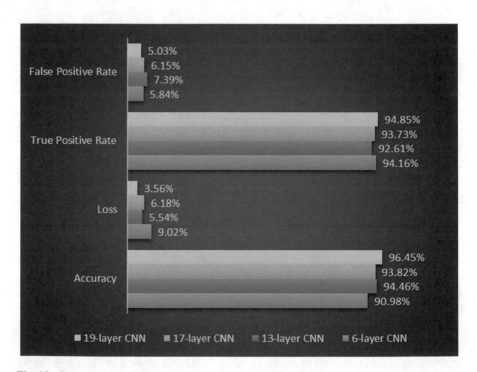

Fig. 10 Comparison performance measurement for AMIDA13 dataset

Acknowledgements This work is supported by Ministry of Education (MOE), Malaysia, Universiti Teknologi Malaysia (UTM), Malaysia and ASEAN-Indian Research Grant. This book chapter is financially supported by MYBRAIN, University Grant No. 17H62, 03G91, and 04G48. The authors would like to express their deepest gratitude to all researchers for their support in providing the datasets to ensure the success of this research, as well as School of Computing, Faculty of Engineering for their continuous support in making this research possible.

References

1. Howlader, N., Noone, A.M., Krapcho, M., Garshell, J., Miller, D., Altekruse, S.F., Kosary, C.L., Yu, M., Ruhl, J., Tatalovich, Z., Mariotto, A., Lewis, D.R., Chen, H.S., Feuer, E.J.: Cancer Statistics Review 1975–2012: Introduction, pp. 1–101 (2015)
2. Siegel, R., Naishadham, D., Jemal, A., Ma, J., Zou, Z., Jemal, A.: Cancer statistics, 2014. CA. Cancer J. Clin. **64**(1), 9–29 (2014)
3. Veta, M., et al.: Assessment of algorithms for mitosis detection in breast cancer histopathology images. Med. Image Anal. **20**(1), 237–248 (2015)
4. Zhang, S., Grave, E., Sklar, E., Elhadad, N.. Longitudinal analysis of discussion topics in an online breast cancer community using convolutional neural networks. J. Biomed. Inform. **69**, 1–9 (2017)
5. Dalle, J., Leow, W.K., Racoceanu, D., Tutac, A.E., Putti, T.C.: Automatic Breast Cancer Grading of Histopathological Images, no. 2 (2000)
6. Mohd. Khuzi, A., Besar, R., Wan Zaki, W.M.D., Ahmad, N.N.: Identification of masses in digital mammogram using gray level co-occurrence matrices. Biomed. Imag. Interv. J. **5** (2009)
7. Singh, S., Gupta, P., Sharma, M.: Breast cancer detection and classification of histopathological images. Int. J. Eng. **3**(5), 4228–4232 (2010)
8. Veta, M., van Diest, P.J., Kornegoor, R., Huisman, A., Viergever, M.A., Pluim, J.P.W.: Automatic nuclei segmentation in H&E stained breast cancer histopathology images. PLoS One **8** (2013)
9. Suzuki, K., Doi, K.: How can a massive training artificial neural network (MTANN) be trained with a small number of cases in the distinction between nodules and vessels in thoracic CT?1. Acad. Radiol. **12**(10), 1333–1341 (2005)
10. Dahl, G.E., Yu, D., Deng, L., Acero, A.: Context-dependent pre-trained deep neural networks for large-vocabulary speech recognition. IEEE Trans. Audio Speech Lang. Process. **20**(1), 30–42 (2012)
11. Ciresan, D.C., Meier, U., Gambardella, L.M., Schmidhuber, J.: Deep big simple neural nets excel on handwritten digit recognition. Neural Comput. **22**(12), 1–14 (2010)
12. Collobert, R., Weston, J.: A unified architecture for natural language processing: deep neural networks with multitask learning. In: Proceedings of the 25th International Conference on Machine Learning, pp. 160–167 (2008)
13. Jain, N.L., Friedman, C.: Identification of findings suspicious for breast cancer based on natural language processing of mammogram reports. Proc. AMIA Annu. Fall Symp., 829–833 (1997)
14. González, F.A., et al.: Automatic annotation of histopathological images using a latent topic model based on non-negative matrix factorization. J. Pathol. Inform. **2**, S4 (2011)
15. Hinton, G., Bengio, Y., Lecun, Y.: Deep learning tutorial. Nips (2015)
16. Martens, J.: Deep learning via Hessian-free optimization. In: 27th International Conference on Machine Learning, vol. 951, pp. 735–742 (2010)
17. Krizhevsky, A., Hinton, G.E. Imagenet, pp. 1–9
18. Roux, L., et al.: Mitosis detection in breast cancer histological images An ICPR 2012 contest. J. Pathol. Inform. **4**(1), 8 (2013)
19. Veta, M., Pluim, J.P.W., van Diest, P.J., Viergever, A.A.: Breast Cancer Histopathology Image Analysis: A Review, vol. 61 (2014)
20. Ciresan, D.C., Giusti, A., Gambardella, L.M., Schmidhuber, J.J.: Mitosis detection in breast cancer histology images with deep neural networks. In: Lect. Notes Comput. Sci. (including Subser. Lect. Notes Artif. Intell. Lect. Notes Bioinformatics), vol. 8150 LNCS, no. PART 2, pp. 411–418 (2013)

21. Coates, A., Arbor, A., Ng, A.Y.: An analysis of single-layer networks in unsupervised feature learning. Aistats **2011**, 215–223 (2011)
22. Lim, G.C.C., Halimah, Y.: Cancer incidence in Peninsular Malaysia 2003–2005. Natl. Cancer Regist. (2008)
23. Khan, A.M., Rajpoot, N., Treanor, D., Magee, D.: A nonlinear mapping approach to stain normalization in digital histopathology images using image-specific color deconvolution. IEEE Trans. Biomed. Eng. **61**(6), 1729–1738 (2014)
24. Kothari, S., Phan, J., Wang, M.: Eliminating tissue-fold artifacts in histopathological whole-slide images for improved image-based prediction of cancer grade. J. Pathol. Inform. **4**, 22 (2013)
25. Shen, W., et al.: Multi-crop Convolutional Neural Networks for lung nodule malignancy suspiciousness classification. Pattern Recognit. **61**, 663–673 (2017)
26. Su, H., Liu, F., Xie, Y., Xing, F., Meyyappan, S., Yang, L.: Region segmentation in histopathological breast cancer images using deep convolutional neural network. In: 2015 IEEE 12th International Symposium Biomedical Imaging, pp. 55–58 (2015)
27. Spanhol, F.A., Oliveira, L.S., Petitjean, C., Heutte, L.: Breast cancer histopathological image classification using Convolutional Neural Networks. In: 2016 International Joint Conference on Neural Networks, pp. 2560–2567 (2016)
28. Wahlstr, N.: Learning deep dynamical models from image pixels (2016)
29. Feng, Y., Zhang, L., Yi, Z.: Breast cancer cell nuclei classification in histopathology images using deep neural networks. Int. J. Comput. Assist. Radiol. Surg. **13**(2), 179–191 (2018)
30. Chan, T., Jia, K., Gao, S., Lu, J., Zeng, Z., Ma, Y.: PCANet: a simple deep learning baseline for image classification? arXiv Prepr., pp. 1–15 (2014)
31. Couprie, C., Najman, L., Lecun, Y.: Learning hierarchical features for scene labeling. Pattern Anal. Mach. Intell. IEEE Trans. **35**(8), 1915–1929 (2013)
32. Sukre, K.M., Rizvi, I.A., Kadam, M.M.: Deep Learning Method for Satellite Image Classification : A Literature Review, vol. 31, no. 31, pp. 111–116 (2015)
33. Wahab, N., Khan, A., Lee, Y.S.: Two-phase deep convolutional neural network for reducing class skewness in histopathological images based breast cancer detection. Comput. Biol. Med. **85**(April), 86–97 (2017)
34. Kotzias, D.: From group to individual labels using deep features (2015)
35. Wu, S., Zhong, S., Liu, y.: Deep residual learning for image steganalysis. Multimed. Tools Appl., 1–17 (2017)
36. Demir, C., Yener, B.: Automated cancer diagnosis based on histopathological images: a systematic survey. Department of Computer Science at Rensselaer Polytechnic Institute, Troy, NY, USA, Technical Report, vol. TR-05–09, pp. 1–16 (2005)
37. Bhattacharjee, S., Mukherjee, J., Nag, S., Maitra, I.K., Bandyopadhyay, S.K.: Review on histopathological slide analysis using digital microscopy. Int. J. Adv. Sci. Technol. **62**, 65–96 (2014)

A Survey on Deep Learning for Time-Series Forecasting

Amal Mahmoud and Ammar Mohammed

Abstract Deep learning, one of the most remarkable techniques of machine learning, has been a major success in many fields, including image processing, speech recognition, and text understanding. It Is powerful engines capable of learning arbitrary mapping functions, not require a scaled or stationary time series as input, support multivariate inputs, and support multi-step outputs. All of these features together make deep learning useful tools when dealing with more complex time series prediction problems involving large amounts of data, and multiple variables with complex relationships. This paper provides an overview of the most common Deep Learning types for time series forecasting, Explain the relationships between deep learning models and classical approaches to time series forecasting. A brief background of the particular challenges presents in time-series data and the most common deep learning techniques that are often used for time series forecasting is provided. Previous studies that applied deep learning to time series are reviewed.

Keywords Deep learning · Time series

List of Acronyms

ANN	Artificial Neural Network
AR	Auto Regression
ARMA	Auto-Regressive Moving Average
ARIMA	Auto Regressive Integrated Moving Average
CNN	Convolutional Neural Network
DBN	Deep Be-Lief Networks
DL	Deep learning

A. Mahmoud (✉) · A. Mohammed
Faculty of Graduate Studies for Statistical Research, Cairo University, Cairo, Egypt
e-mail: amal.mahmoud@pg.cu.edu.eg

A. Mohammed
e-mail: Ammar@cu.edu.eg

© The Author(s), under exclusive license to Springer Nature Switzerland AG 2021   365
A. E. Hassanien et al. (eds.), *Machine Learning and Big Data Analytics*
Paradigms: Analysis, Applications and Challenges, Studies in Big Data 77,
https://doi.org/10.1007/978-3-030-59338-4_19

GRU   Gated Recurrent Unit
LSTM   Long Short Term Memory
MA   Moving Average
MAE   Mean Absolute Error
MFE   Mean Forecast Error
ML   Machine Learning
MLP   Multi-Layer Perception
MMSE   Minimum Mean Square Error
MPE   The Mean Percentage Error
RBM   Restricted Boltzmann Machine
RMSE   The Root Mean Squared Error
SAE   Stacked-Autoencoders
SARIMA   Seasonal Autoregressive Integrated Moving Average Models

1 Introduction

In recent years, deep neural networks have attracted the undivided attention of both academics and industry. In research funding and volume, many factors are contributing to this rapid rise. Some of the following are briefed [1]:

- A surge in the availability of large training data sets with high-quality labels.
- Specialty software platforms such as PyTorch [2], Tensorflow [3], Caffe [4] Chainer [5], Keras [6], BigDL [7] etc.
  Better techniques of regularisation introduced avoid overfitting. Techniques such as batch normalization, dropout, data augmentation, and early stopping are very effective in preventing overfitting.

In the computational intelligence and machine learning literature, numerous forecasting models for time series data were proposed. Multilayer perceptron models or feedforward networks, recurrent neural networks, support vector machines, kernel methods, ensemble methods, and other graphical models are all examples of methods that have recently gained popularity [8].

The classical methods of time series such as Autoregressive Integrated Moving Average (ARIMA) have been used for many decades in predicting the future, especially in the field of industry, for example, but not exclusively manufacturing supply chain, and finance. But the classical methods are limited and not able to deal with many real-world problems, for example, It only focuses on complete data and cannot handle missing or corrupt data, Focus on linear relationships, Deal with fixed temporal dependence which means that the relationship between the number of lag observations presented as input and observations at different times must be diagnosed and determined, Focus on univariate data While in the real - world a lot of problems that have multiple input variables, and Focus on one-step forecasts While real-world problems require forecasts with a long time horizon [9, 10].

Deep learning techniques have an effective and important role in solving time series forecasting problems, and this is reflected in their ability to handle multiple input variables, support multivariate inputs, complex nonlinear relationships, and may not require a scaled or stationary time series as input [11, 12].

This paper's main research question is to study and compare the efficacy of time series models to make predictions of real data.

The rest of this paper is structured as follows: Sect. 1.1 explains Time-series characteristics and classical model. Section 1.2 introduces the basic types of Artificial Neural Network (ANN) modules that are often used to build deep neural structures. Section 2 describes how the present paper relates to other works in the literature. Finally, Sect. 3 concludes the paper.

1.1 Time Series Definition

Time series can be defined as anything that is observed sequentially over time. For example, Sales trends, Stock market prices, weather forecasts, etc. The nature of these observations can be numbers, labels, colors, and much more. Moreover, the times at which the observations were taken can be regularly or irregularly spaced. On the other hand, time can be discrete or continuous [13].

Prediction of the time series methods is based on the idea that historical data includes intrinsic patterns that convey useful information for the future description of the investigated phenomenon. Usually, these patterns are not trivial to identify, and their discovery is one of the primary goals of time series processing: the circumstances that the patterns found will repeat and what types of changes they may suffer over time [14].

Time series represent as (y_t) where t means the time at which the observation was taken. Mostly $t \in \mathbb{Z}$, where $\mathbb{Z} = (0, \pm1, \pm2, ...)$ is the set of positive and negative integer values [15]. However, only a finite stretch of data is available. In such situations, time series can write as $(y_1, y_2, y_3, ...)$. A time series (y_t) corresponds to a stochastic process, which in turn is composed of random variables observed across time.

Time series have many characteristics that distinguish them from other types of data. Firstly, the time-series data contain high dimensionality and much noise. Dimensionality reduction, wavelet analysis or filtering are Signal processing techniques that can be applied to eliminate some of the noise and reduce the dimensionality. The next characteristic of time-series data, it is not certain that the available information is enough to understand the process. Finally, Time series data is high-dimensional, complex, and has unique characteristics that make time-series challenging to model and analyze. Time series data representation is important for extract relevant information and reduces dimensionality. The success of any Application depends on the right representation.

1.1.1 Components of a Time Series

Time series are generally affected by four basic components, and these components are (Trend, Seasonal, Cyclical, and Irregular) components.

When a time series exhibits an upward or downward movement in the long run, it is known that this time series has a general trend [14]. In general, the Trend is a Long-term increase or decrease in the data over time. For example, Population growth over the years can be seen as an upward trend [16].

When a series was affected by seasonal factors in this case, the seasonality pattern exists, such as Quarterly, Yearly, monthly, weekly, and daily data Seasonality always knows it is fixed and a known period, sometimes known as periodic time series. For example, trips are increasing in the Hajj and Umrah seasons as well, the sales of ice cream increase in the summer.

The cyclic occurs when data rises and falls Which means it is not a fixed period. The duration of the Cycle is over a long period, which may be two years or more. For example, five years of economic growth, followed by two years of economic recession, followed by seven years of economic growth followed by one year of economic recession. The irregular component (sometimes also known as the residual) refers to the variation that exists because of unpredictable factors and also does not repeat in specific patterns, for Example, Variations caused by an incident like an earthquake, floods, etc.

1.1.2 Types of Time Series Decomposition

Three different types of models are generally used for a time series viz. Multiplicative and Additive models and The Mixed decomposition model [17, 18].

Multiplicative Model

$$Y_t = T_t \times S_t \times C_t \times I_t \tag{1}$$

Additive Model

$$X_t = T_t + S_t + C_t + I_t \tag{2}$$

Y_t Original data at time t
T_t Trend value at time t
S_t Seasonal fluctuation at time t,
C_t Cyclical fluctuation at time t,
I_t Irregular variation at time t

1.1.3 Procedure for Decomposition

In the 1920s, the classical method of time series decomposition originated and was commonly used until the 1950s. These classic methods are still being used in time series decomposition methods [19].

- Moving Averages

Moving Averages method is used to smooth out the short-term fluctuations and identify the long-term trend in the original series (i.e., to create a smooth series to reduce the random fluctuations in the original series and estimates the trend-cycle component).

A moving average of order can be defined by [15]:

$$\hat{T}_t = \frac{1}{m} \sum_{j=-k}^{k} yt + j \tag{3}$$

where m = 2 k + 1. That is, the estimate of the cycle-trend at time t is known by calculating the average values of the time series within periods. There is a possibility that the observations that occur at close times are near in terms of value.

- Exponential Smoothing

For in fixed $\alpha \in [0, 1]$, the one-sided moving average $\hat{m}_t, t = 1, ..., $ n, Define by the following equation [20]:

$$\hat{m}_t = \alpha X_t + (1 - \alpha)\hat{m}_{t-1}, t = 2....n \tag{4}$$

And

$$\hat{m}_1 = X_1 \tag{5}$$

1.1.4 Time Series Classical Models

Machine learning and deep learning techniques can achieve impressive results in challenging time series forecasting problems. However, there are many classical methods such as SARIMA, and exponential smoothing readily achieves impressive results in time series. These methods must be understood and evaluated before exploring more advanced methods.

- Autoregressive Moving Average (ARMA) Models

An ARMA (p, q) model is a combination of AR(p) and MA(q) models; it is suitable for univariate time series modeling [21]. (AR) A model that uses the dependent relationship between an observation and some number of lagged observations. AR can be defined, followed by [22].

$$y_t = f(y_{t-1}, y_{t-2}, y_{t-3} \ldots)$$
$$y_t = \beta_0 + \beta_1 y_{t-1} + \beta_2 y_{t-2} + \beta_3 y_{t-3} \ldots \tag{6}$$

Y_t depends only on past values $y_{t-1}, y_{t-2}, y_{t-3}$ etc.

(MA) A model that uses the dependency between an observation and a residual error from a moving average model applied to lagged observations. MA can define, followed by [22] [23].

$$y_t = f(\varepsilon_t, \varepsilon_{t-1}, \varepsilon_{t-2}, \varepsilon_{t-3} \ldots)$$
$$y_t = \beta + \varepsilon_t + \theta_1 \varepsilon_{t-1} + \theta_2 \varepsilon_{t-2} + \theta_3 \varepsilon_{t-3} + \cdots \tag{7}$$

Y_t depends only on the random error term.

Mathematically an ARMA(p, q) model is represented as [22].

$$y_t = c + \varepsilon_t + \sum_{i=1}^{p} \varphi_i y_{t-i} + \sum_{j=1}^{q} \theta_j \varepsilon_{t-j} \tag{8}$$

p, q refers to p autoregressive and q moving average terms.

- Autoregressive Integrated Moving Average (ARIMA) Models

Autoregressive Integrated Moving Average (ARIMA) is the most popular and frequently model used in time series models [21, 24]. The assumption of this model is based on the fact that the time series is linear and follows a known statistical distribution. It is a combination of AR(p), I(d) Integrated the use of differencing of raw observations (e.g., subtracting an observation from observation at the previous time step) to make the time series stationary and MA(q).

In ARIMA models a non-stationary time series is made stationary by applying finite differencing of the data points [25].

The mathematical formulation of the ARIMA(p, d, q) model using lag polynomials defined below [22, 26].

$$\varphi(L)1 - L^d y_t = \theta(L)\varepsilon_t, \ i.e.$$
$$\left(1 - \sum_{i-1}^{p} \varphi_i L^i\right)(1 - L)^d y_t = \left(1 + \sum_{j=1}^{q} \theta_j L^j\right) \tag{9}$$

A nonlinear variant called the nonlinear autoregressive-moving-average (NARMA) model can be used to capture nonlinear structure with ARMA models. The generalized autoregressive conditional heteroscedasticity (GARCH) model addresses time series data volatility and has common structural features of the ARMA model.

- Seasonal Autoregressive Integrated Moving Average (SARIMA) Models

The ARIMA model is for non-stationary and non-seasonal data. Box and Jenkins have generalized this model to deal with seasonality [21]. In the SARIMA model, seasonal differencing of appropriate order is used to eliminate non-stationarity from the series.

A first-order seasonal difference is a difference between an observation and the corresponding observation from the previous year) $Z_t = Y_t - Y_t - S($, for example, monthly time series $s = 12$ [16]. This model is generally termed as the SARIMA $(p, d, q) \times (P, D, Q)$ s model. SARIMA mathematical formulation defined by [27]:

$$\Phi_P(L^s)\varphi_p(L)(1 - L)^d(1 - L)^D y_t = \Theta_Q(L^s)\theta_q(L)\varepsilon_t,$$
$$i.e.\ \Phi_P(L^s)\varphi_p(L)z_t = \Theta_Q(L^s)\theta_q(L)\varepsilon_t. \tag{10}$$

Z_t Is the seasonally differenced series

1.2 Artificial Neural Network

ANN are computational models inspired by the information processing performed by the human brain [28]. Schmidhuber [29] define neural network (ANN) as "A standard neural network consists of many simple, connected processors called neurons, each producing a sequence of real-valued activations." input neurons (xi), that get activated from the environment, and other neurons, "get activated through weighted (wi) connections from previously active neurons" Such as hidden or output neuron. The output of the neural network control by the assignment of weights for the connections, the process of tuning weights to achieve certain output is termed learning. Each layer transforms, often nonlinearly, the aggregate activation of the previous layer and propagates that to further layers. The reason for the widespread use of neural networks in business-related problems is that, first, neural networks are highly suitable for handling incomplete, missing or noisy data; second, neural networks do not require prior assumptions about data distribution; third, they are capable of mapping complex and approximate continuous function [30].

The following equation explains that "the neuron sums up the inputs multiplied by the weights using. "

$$A = \sum xiwi + bq \tag{11}$$

A   is the net sum.
b   is the threshold.

The output applied to a function called activation function F(A).

Weights related to Connections between two neurons correspond to how strong between them. Typically, the computation done by a neuron is divided into two steps: the aggregation function and the activation function. Applying the aggregation function mostly corresponds to calculating the sum of the inputs obtained by the neurons via all its incoming connections. The activation function receives the resulting values. The common choices for network architectures are the sigmoid and the hyperbolic-tangent. Rectified linear units using a ramp function have become increasingly popular [12, 31].

Rectified Linear: is the most common activation function due to its simplicity and good results. A subset of neurons fire at the same time; this makes the network sparser and improving efficiency.

Sigmoid: a very popular activation function. It squeezes the output between 0 and 1

$$\sigma(x) = \frac{1}{1 + e^{-x}} \tag{12}$$

Tanh: is similar to the sigmoid function, only that it outputs values from -1 to 1.

Hyperbolic tangent tanh

$$(x) = \frac{2}{1 + e^{-x}} \tag{13}$$

Ramp function

$$R(x) = \max(0; x) \tag{14}$$

Softmax: similar to the sigmoid function. It outputs continuous values from 0 to 1 and is often used at the output layer as a classifier because it outputs the probabilities distributed over the number of classes.

After reviewing an overview of classical methods dealing with time series and Artificial neural networks, we can summarize the disadvantages and advantages of each of them in Table 1 support multivariate inputs, complex nonlinear relationships, and may not require a scaled or stationary time series as input.

Since deep learning was presented in Science magazine in 2006, In the machine learning community, it has become an extremely hot topic of research. Deep ANNs are most commonly referred to as deep learning (DL) or DNNs [33], allowing computational models that are composed of multiple processing layers to learn complex data representations using multiple levels of abstraction. One of the DL main advantages is that in some cases, the model itself performs the feature extraction step. DNN's are simply an ANN with multiple hidden layers between the input and output layers and can be supervised, partially supervised, or even unsupervised. Over the past few years, several deep learning models have been developed. The most typical models of deep learning are convolutional neural network (CNN), recurrent neural network

Table 1 Summary of qualitative comparison for ANN and Classical methods for time series forecasting technique

Model	Advantage	Disadvantage
ANN	– Ability to map input and output relationships accurately [32] – Good performance for nonlinear time series [32] – More general and flexible – Ability to handle multiple input variables – Support multivariate inputs – May not require a scaled or stationary time series as input	– It depends on the weight initialization [32] – Hardware dependence – Overfitting and hard to generalize – Local minima problem
Classical Method	– Uses historical data lag and shift – A moving average and regression model, "improve accuracy" [32]	– Limited and not able to deal with many real world problems – Cannot handle missing or corrupt data – Not suitable for long-term Forecasting

(RNN), stacked auto-encoder (SAE), and deep belief network (DBN) [29, 33]. In the following parts, we review the four typical deep learning models.

1.2.1 Convolutional Neural Network

A convolutional neural network (CNN), inspired by visual cortex studies in mammals, is a type of feed-forward neural network. For video and image processing, CNNs were the most effective. In a variety of restricted tasks, for example, handwriting recognition, identifying objects in photos, tracking movements in videos, etc., they have been shown to approach or even exceed human-level accuracy [34]. Motivated by the success of these (CNN) architectures in these different fields, researchers have begun to adopt them for the analysis of time series [12].

A hidden layer (also called a convolutional layer in this case) consists of several groups of neurons. The weights of all the neurons in a group are shared. Generally, each group consists of as many neurons as necessary to cover the entire image. Thus, it is as if each group of neurons in the hidden layer calculated a convolution of the image with their weights, resulting in a processed version. Figure 1 shows the structure of a Convolutional Neural Network (CNN).

The CNN consists of several types of layers that are stacked on top of each other. There is no way to stack the various layers that is up to the designer. Using object classification is a very intuitive example that goes through the basics of CNNs, but it can be used on other types of data, such as text or sound [35].

Various types of layers will be described in the following sub-sections, input layer, convolution layer, pooling layer, fully connected layer, and batch normalization [36].

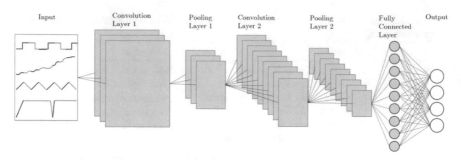

Fig. 1 Structure of a Convolutional Neural Network (CNN)

Input Layer: The input layer stores the raw input data. It is a three-dimensional input consisting of the image's width and height, and the color channels represent the depth, typically three for RGB.

Convolutional Layer: CNN's key layer is the convolution layer. It uses filters or kernels that are smaller than the input image. Convolution is done with a part of the input and the kernel, and this is done in a sliding window (Window sliding is the strategy of taking the previous time steps to predict the subsequent time step), which ultimately covers the entire image input. This process's output is called a feature map or activation map. The region of the input, the feature map, is called the receptive field. The result of each filter is a feature map. For each filter, the activation map is stacked to output a 3-dimensional tensor. They learn to recognize edges and patterns as the filters are trained, and they can recognize more advanced shapes deeper in the network. The ReLU activation function is very commonly used in CNNs. This layer takes all the negative inputs and sets them to zero. There are no hyperparameters in the ReLU layer, i.e., parameters selected by the designer.

Pooling Layer: The pooling layer reduces the size of the data. Max pooling is the most common version that outputs the maximum value of the given window size and ignores the rest. Over the entire input, it does this operation. The designer chooses the step [37, 38].

Fully Connected Layer: Typically, we have one or more fully connected layers at the CNN output or classification. For the different classes, the classifier outputs probabilities.

Batch Normalization: Before entering the data into the NN, it is common to normalize the data. Batch normalization normalizes the mean output activation of the layer close to 0 and its standard deviation close to 1. Usually, this method is used to speed up CNNs training.

The result of convolution on an input time series X (one filter) can be considered as another univariate time series C that has undergone a filtering process. Applying multiple filters on a time series will result in multivariate time series with dimensions equal to the number of filters used. An intuition behind applying multiple filters to an input time series would be to learn multiple discriminatory features that are useful for the task of classification [39].

Convolution can be seen throughout the time series as applying and sliding a filter. Unlike images, instead of two dimensions (width and height), the filters display only one dimension (time). The filter can also be viewed as a time series generic non-linear transformation.

1.2.2 Recurrent Neural Network (RNN)

RNN will be first introduced in this section, and then the LSTM and GRU will be discussed.

For all sequential forms of data such as video frames, text, music, etc., recurrent neural networks (RNN) can be used [40]. RNNs are suitable for multivariate time series data, capable of capturing temporary dependencies over variable periods [41–44]. RNNs have been used in many time series applications, including speech recognition [45], electricity load forecasting [46], and air pollution [47]. RNN feed-forward networks input some value to the network and return some value based on that input and the network parameters. RNN has an internal state which is fed back to the input [48]. It uses the current information on the input and the prediction of the last input. Feed-forward networks have a fixed size input and output. Figure 2 shows the structure of the typical RNN model.

Generally, when there are loops on the network, it is called an RNN. The Back-propagation algorithm can be adapted to train a recurrent network by unfolding the network through time and restricting some of the connections to keep the same weights at all times [50]. RNN can encode the prior information in the current hidden layer's learning process so that data from the time series can be learned effectively [51].

In contrast to feed-forward neural networks, RNNs use their internal "memory" state to process sequences of input [42]. RNNs have a problem with "vanishing

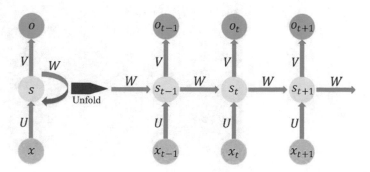

Fig. 2 Illustration of RNN structure [49]

gradient descent" [52], and this can occur when the gradient of the activation function becomes very small. When back-propagating through the network, the gradient becomes smaller and smaller, further back in the network, and this makes it hard to model long dependencies. The opposite of the problem of the "vanishing gradient" is the exploding gradient problem where the gradient becomes large.

1.2.3   Long Short-Term Memory (LSTM)

LSTM has been shown to be efficient in dealing with various issues such as speech and handwriting recognition as it has the ability to learn from inputs even when there are long delays between important events. The LSTM units consist of cell and three gates; the forgotten, the input, and the output gates, Fig. 3 shows an LSTM units architecture. The cell is the LSTM memory that is used over arbitrary time intervals to remember values.

LSTM's "gate" is a special network structure with a vector input and a range of outputs from 0 to 1. No information is allowed to pass if the output value is 0. When the output value is 1, it is allowed to pass all information.

The output of the forgotten gate f_t is calculated as follows, where W* and U* respectively are weight matrices whose parameters are to be learned during model training:

$$f_t = \text{sigmoid}\left(W_t^* x_t + U_t^* h_{t-1} + b_t\right) \tag{21}$$

The model can learn by taking into account the previous output to accumulate certain information j_t from the current time step. Similar to the forget gate, the update

Fig. 3 Illustration of LSTM [53]

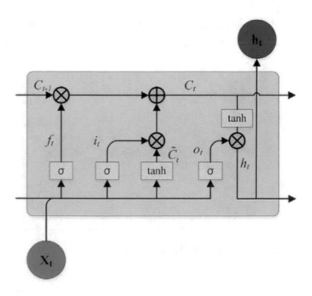

gate that it then decides which information will be added to the cell state from the current time step.

$$i_t = \text{sigmoid}(W_i x_e + u_i h_{t-1} + b_i) \tag{22}$$

$$j_t = \tanh(W_j x_t + U_j h_{t-1} + b_j) \tag{23}$$

$$c_t = f_t \odot c_{t-1} + i_t \odot j_t \tag{24}$$

Finally, the updated state c_t can be calculated from the previous state c_{t-1}, the forget gate output f_t, and the update gate output as shown in Eq. 24.

Here, the operation $\odot$ denotes the element-wise vector product. The output of the cell h_i at the current time step is subsequently calculated with the updated cell states c_t:

$$o_t = \text{sigmoid}(W_o x_t + U_o h_{t-1} + b_o) \tag{25}$$

$$h_t = 0, \odot \tanh(c_t) \tag{26}$$

In many recent publications, LSTMs have proven to perform well and are fairly easy to train. Therefore, LSTMs have become the baseline architecture for tasks where it is necessary to process sequential data with temporal information. However, this architecture has many extensions, as the purpose of the individual components is disputed, and therefore, more optimal architectures may exist [54].

LSTMs became well known; researchers proposed various variants of LSTMs that, under certain circumstances, proved to be highly effective. To enable the model to capture more structure, researchers have added new gates to the original LSTM design. Others did the opposite and merged various gates to create a simplified LSTM variant called a Gated Recurrent Unit (GRU) [55] that has fewer parameters than LSTMs but has similar results in practice. In contrast to conventional recurrent neural networks, long-term memory (LSTM) and Gated Recurrent Unit (GRU) are more suitable for classifying, processing forecasting time series with arbitrary time steps from experience.

1.2.4 Gated Recurrent Unit (GRU)

Gated Recurrent Unit is very similar to long-term memory; it proposed by Cho et al. [55]. The main difference is the way gates are applied. There are only two gates instead of three for GRU, which is the reset and update gate. The reset gate determines how the new input and previous memory can be combined and controls the extent to which previous memory is neglected. The smaller value of the reset gate is the more previous memory will be neglected. The update gate determines how much of the previous memory will be kept. The larger value of the update gate is the more previous information will be brought. Set all gates to 1 and update gates to 0, then return to the recurrent neural network model. LSTM and GRU have few more differences. First, LSTM can control memory exposure while

GRU exposes the uncontrollable memory. Then, as in LSTM, GRU does not have the output gate. The input gate and forget gate in LSTM are replaced by the update gate, so the reset gate is applied directly to the previous hidden state. Because GRU has fewer parameters than LSTM, GRU is going to be faster than LSTM and require fewer data. If the data scale is large, however, LSTM could lead to a better Result [56].

1.2.5   Deep Autoencoders (AEs)

A deep AE is an unsupervised learning neural network trained to reproduce its input to its output. An AE has a hidden layer h, which defines a code used to represent the input [33]. There are two parts of an AE neural network: the encoder function $h = f(x)$ and the decoder function, which reproduce the input $r = g(h)$. The encoder gets the input and converts it into an abstraction, commonly referred to as a code. Subsequently, to reconstruct the original input, the decoder acquires the constructed code that was originally produced to represent the input. Hence, weights in encoding and decoding are forced to be the transposed of each other. The training process in AEs with minimum reconstruction error should be performed [57].

1.2.6   Restricted Boltzmann Machines (RBM)

Restricted Boltzmann Machine (RBM) [58] can be interpreted as a stochastic neural network. It is one of the popular deep learning frameworks because of its ability to learn both supervised and unsupervised distribution of input probability. Paul Smolensky first introduced it in 1986 with the name Harmonium [59].

However, it became popular by Hinton in 2002 [60]. After that, it has a wide range of applications in various tasks such as dimensionality reduction [61], prediction problems [62], learning representation [63].

An RBM is a completely undirected model that has no link between any two nodes in the same layer. RBM model containing two layers, visible and hidden layers. The known input is held by the visible layer, whereas the hidden layer consists of multiple layers that include the latent variables. RBMs hierarchically understand data features, and the features captured in the initial layer are used in the following layer as latent variables. Temporal RBM [64] and conditional RBM [65] Proposed and used for modeling multivariate time series data. The ability of a single RBM to represent features is limited. However, by stacking two or more RBMs to form a DBN, RBM can be applied substantially. This process is discussed in the section below.

1.2.7 Deep Belief Networks (DBNs)

DBNs are generative methods [66]. A DBN consists of stacked RBMs performing greedy layer-wise training in an unsupervised environment to achieve robust performance. Training in a DBN is performed layer by layer, each of which is executed as an RBM trained on top of the formerly trained layer (DBNs are a set of RBM layers that are used in the pre-training phase and then become a feed-forward network for weight fine-tuning with contrastive convergence) [67]. In the pre-training phase, the initial features are trained by a greedy layer-wise unsupervised method, whereas in the fine-tuning phase a softmax layer is applied to the top layer to fine-tune the features of the labeled samples [68].

DBNs are unsupervised learning methods for significant feature representation trained iteratively with unlabeled data. In most applications, this approach to deep architecture pre-training led to the state of performance in the discriminative model, such as recognizing detecting pedestrians, handwritten digits, time series prediction etc. Even if the number of data labeled is limited

2 Deep Learning-Based Time Series Previous Work

Recently, deep learning has been applied to the time-series data that have been a popular subject of interest in other fields such as solar power forecasting [69–71], weather forecasting [72], electricity load forecasting [73–79], electricity Price forecasting [80, 81] and financial prediction [82–84]. In several time-series studies, models such as Deep Be-Lief Networks (DBN), Autoencoders, Stacked-Autoencoders (SAE), and Long Short-Term Memory (LSTM) were used instead of simple ANN. Many references in the literature suggest the use of Deep Learning, or a variation of them, as a powerful tool for forecasting time series. The most important works are explained in detail below. A random walk is followed by FOREX and the stock market, and all profits made by chance. The ability of NN to adapt to any non-linear data set without any static assumption and prior knowledge of the data set can be attributed [85]. Deep Learning have both fundamental and technical analysis data, which is the two most widely used techniques for financial time series forecasting to trained and build deep learning models [8].

Hossain et al. [86] created a hybrid model to predict the S&P 500 stock price by combining the LSTM and Gated Recurrent Unit (GRU). The model was compared with various architectural layers against standalone models such as LSTM and GRU. The hybrid model outperformed all other algorithms.

Siami-Namini et al. [87] compared the accuracy of ARIMA and LSTM when it was forecasting time series data as representative techniques. These two techniques have been implemented and applied to a set of financial data, and the results have shown that LSTM is superior to ARIMA.

Fisher and Krauss [88] adapted LSTM for stock prediction and compared its performance to memory-free algorithms such as random forest, logistic regression classifier, and deep neural network. LSTM performed better than other algorithms, but random forests surpassed LSTM during the 2008 financial crisis. Overall, they have demonstrated successfully that an LSTM network can efficiently extract meaningful information from noisy data from the financial time series.

Wei Bao et al. [49] presented a novel deep learning framework in which stacked autoencoders, long-short term transforms and wavelet (WT) are used together to predict stock price. There are three stages in this deep learning framework. First, WT decomposes the stock price time series to eliminate noise. Then, SAEs are applied for stock price prediction to generate deep high-level features. Finally, high-level denoising features are fed into long short-term memory to predict the next day's closing price.

Santos and Zadrozny [89] Presented the use of a simple LSTM neural network with embedded character level for stock market forecasting using as predictors only financial news. The results suggested that the use of embedded character levels is promising and competitive with more complex models using, in addition to news articles, technical indicators and methods for event extraction.

Wang et al. [90] proposed a novel approach for forecasting stock prices using a wavelet de-noising-based backpropagation neural network. They show significantly improved performance compared to a standard multi-layer perceptron by using monthly closing price data from the Shanghai Composite Index over a period from 1993 to 2009. They attribute the improved performance to de-noising and preprocessing using wavelet transform (allows us to analyze non-stationary signals to discover local details [91]).

Chen et al. [92] predicted China stock returns using the LSTM network. With past China Stock Market data, two-dimensional data points were formed with 30-day-long sequences with ten features. Data points were assigned as class labels for 3-day earning rates. Compared to the method of random prediction, the LSTM model improved the performance of the stock return classification.

The input data must be transformed or adapted for CNN to apply CNN to the financial market [93]. By using a sliding window, Gudelek el at [94] used images generated by taking stock time series data snapshots and then fed into 2D-CNN to perform daily predictions and trend classification (whether down or up). On 17 exchange-traded fund data set, the model was able to obtain 72% accuracy.

Türkmen and Cemgil [95] used SAE, a deep learning structure that is unsupervised, to find the direction of share prices traded on the US Nasdaq Stock Exchange. As an input to the deep learning model, technical indicators calculated from the price data were given. Model performance was evaluated with accuracy, and F-Measure metrics and the proposed model delivered the best performance using the SVM method.

Ding et al. [96] designed a model for stock market prediction driven by events. First, events are extracted from financial news and represented by word embedding as dense vectors. They trained a deep CNN to model on stock price events both short-term and long-term influences. Their proposed model in S&P 500 index prediction and individual stock prediction gave better performance than SVM.

Chen et al. [97] Predicted on the stock market using GRU, which also achieved good predictive performance with small error.

Lago et al. [81], A new modeling framework is proposed to forecast electricity prices. Although many predictive models have already been proposed for this task, the field of deep learning algorithms. Three of the four proposed DL forecasters, i.e., the Deep Neural Network (DNN) model, the Long-Short-Term Memory (LSTM) model, and the Gated Recurrent Unit (GRU) model, are shown to have statistically significantly better predictive accuracy than any other model.

Nichiforov et al. [98] presented the application of sequence models, implemented through the Recurrent Neural Network with LSTM, for the forecasting of electrical-energy loads in large commercial buildings. Results showed good performance for modeling and forecasting accuracy, a value of about 50 LSTM units in the hidden layer was found to yield the best estimates for the underlying time series based on the number of layers selected. There was a strong tendency for the network to overfit input data and perform poorly on test samples. The results were evaluated from a reference benchmarking dataset on selected building-energy traces over a year, and the relative MAPE metric for all investigated buildings was found to be between 0.5 and 1%.

Zahid et al. [99], Two models are proposed: one for predicting electricity charges and the other for predicting electricity prices. The primary objective is to increase predictive accuracy using these models. Using GS and cross-validation, data pre-processing (feature selection and extraction) and classifier parameter tuning are done for better accuracy. The performance metrics simulation results and values show that the models proposed are more accurate than the benchmark schemes. A hybrid feature selection technique is used in these models to select the best features from the dataset. RFE is used to eliminate data redundancy and decrease dimensionality. Enhanced classifiers (ECNN and ESVR) are used to predict electricity prices and loads. It also reduced the problem of overfitting and minimized the time for execution and computation.

Shi et al. [100] presented a comprehensive, in-deep learning assessment on load forecasting across power systems at different levels. The assessment is demonstrated in two extreme cases: (1) regional aggregated demand with New England data on electricity load, (2) disaggregated household demand with examples from Ireland of 100 individual households The proposed deep model improved the predictive accuracy in terms of MAPE by 23% at aggregated level and RMSE by 5% at disaggregated level compared to the shallow neural network.

Hernández et al. [101] proposed an Auto-encoder and Multilayer perceptron-based deep learning model to predict the daily accumulated rainfall. The authors used Auto-encoder to select features and MLP to predict. The results showed a better performance of their proposed model than other approaches. To make predictions, they used the deep feedforward neural network.

Zhao et al. [102] proposed a DL ensemble approach called SDAE-B for "crude oil price forecasting" due to the complex relationship between crude oil prices and different factors. The SDAE Deep Learning Model is used to learn useful data representations and generate predictions. As a powerful ensemble method, bagging combines the strength of multiple SDAEs and thus creates a better performance ensemble model. In the empirical study, the proposed SDAE-B surpasses benchmark models such as econometric models (i.e., RW and MRS), shallow architecture machine learning models (i.e., FNN, FNN-B, SVR, and SVR-B) and it is base model SDAE. Furthermore, statistical tests confirmed the superiority of the proposed model, indicating that the proposed approach can be used as a promising tool for the price of crude oil.

Ni et al. [103] to predict the power output of the near-shore WEC, a Convolutional Neural Network with multiple inputs has been developed. To convert time-series data into image data, the proposed CNN applied 1D to 2D data conversion. The result shows that MCNN provides much better prediction results compared with others.

Fu et al. [104] used (LSTM) and (GRU) methods to predict short-term traffic flow and experiments demonstrate that (RNN) based deep learning methods such as LSTM and GRU perform better than (ARIMA) model and GRU have a little better performance than LSTM. Meanwhile, on 84% of the total time series, GRU has better performance than LSTM.

For the prediction of travel time, Duan et al. [105] explored a deep learning model, the LSTM neural network model. Using Highways England's travel time data, Duan et al., built 66 series of LSTM neural networks for 66 links in the data set. Use the trained model to predict multi-step travel times for each link in the test set. Evaluation results show that the prediction error for the 1-step ahead travel time is relatively small, while the MRE median for the 66 links is 7.0% on the test set.

Du et al. [106] proposed an adaptive multimodal deep learning model HMDLF for short-term traffic flow forecasting. The model integrated both one-dimensional CNN and GRU as one basic module to capture correlation features between local trends and long dependencies of single-modality traffic data. The experimental results show that the proposed multimodal deep learning model is capable of responding with satisfying accuracy and effectiveness to complex nonlinear urban traffic flow forecasting.

Alhassan et al. [107], Temporary predictive deep neural network models were investigated for the diagnosis of T2DM. The proposed models (LSTM and GRU) can achieve very high accuracy with as short as three sequences using clinical time-stamped data and without intensive feature engineering. The models were trained and tested using unique and large datasets (KAIMRCD) with different input sizes. The results were compared using the same dataset to common baseline classifiers (LR, SVM, and MLP). LSTM and GRU models outperformed the baseline classifiers and achieved 97.3% accuracy, and it is a very important finding as it would reduce the time and associated costs required for further testing and early diagnosis.

Choi et al. [108] proposed Doctor AI system, a model, based on RNN, can learn from a large number of longitudinal patient records efficient patient representation and predict future patient events. They tested Doctor AI on a large real-world EHR dataset, achieving 79.58% recall and outperforming many baselines significantly. A medical expert's empirical analysis confirmed that Doctor AI not only mimics human doctors' predictive power but also provides clinically meaningful diagnostic results. One limitation of Doctor AI is that incorrect predictions can sometimes be more important in medical practice than correct predictions because they can degrade the health of patients.

Lipton et al. [109] proposed a model to apply LSTM to a clinical dataset. Using 13 laboratory test results, the authors used LSTM on a Children's Intensive Care Unit (ICU) dataset to predict multiple disease diagnosis (such as asthma, hypertension, and anemia). The LSTM model was designed for the competitive accuracy classification of 128 diseases. The reason for using LSTM is that their ability to memorize sequential events could improve modeling of the different time delays between the onset of clinical emergencies.

Lv et al. [110] propose a deep learning approach with an SAE model for traffic flow prediction. They applied the greedy layerwise unsupervised learning algorithm to pre-train the deep network, and then they did the process of fine-tuning to update the parameters of the model to improve the performance of predictions. Results indicate that the method proposed is superior to competing methods.

Yang et al. [111] use convolutional neural networks for recognition of human activity (HAR). Their methodology, which capitalizes on the fact that a combination of unsupervised learning and supervised classification of features can increase the discriminative power of features.

Mehdiyev et al. [112] proposed a new multi-stage approach to deep learning for multivariate time series classification issues. They used the stacked LSTM autoencoders after extracted the features from the time series data in an unsupervised manner. The objective of the case study is to predict post-processing activities depending on the detected steel surface defects using the time series data obtained from the sensors installed in various positions of the steel casting process facility and the steel's chemical properties.

A summary of the studies reviewed with highlighted the deep learning architecture applied and the Time Series domain considered in Table 2.

3 Conclusion

In this paper, Literature on the definitions of time series, its classification, especially with deep neural networks, has been reviewed. Time Series Classical methods and ANN methods are explained. These methods are then compared in terms of their advantages and disadvantages. Some important performance measures to evaluate the accuracy of forecasting models have been explained. A comprehensive review of the potential uses of Deep Learning methods in Time series forecasting in many different domains such as electricity load forecasting, electricity price forecasting, solar power forecasting, and financial prediction is provided. Some of these methods have treated the input as static data, but the most successful are those that modified

Table 2 Selected studies focusing on deep learning for time series

Author	Focus Area	Model	Dataset	Data type	Forecasting period	Performance
Hossain, Mohammad Asiful et al. [86]	Stock Market	LSTM GRU GRU-LSTM (Hybrid)	Yahoo Finance API[a]	Daily	1950–2016	MAE: LSTM: 0.051 GRU: 0.041 GRU-LSTM(Hybrid): 0.023
Siami-Namini et al. [87]	Stock Market	ARIMA LSTM	8372 observation	Monthly	1985–2018	RMSE: ARIMA: 511:481 LSTM: 64:213
Fisher and Krauss [88]	Stock Market	LSTM	500 index constituents from Thomson Reuters	Daily	1989–2015	Accuracy: 50%
Wei Bao, et al. [49]	Stock Market	WSAEs-LSTM (Hybrid)	six stock indices	Daily	2008–2016	MAPE: 0.019
Gudelek et al. [94]	Stock Market	CNN	Google finance, 17 different most commonly used ETFs	Daily	2000–2017	Accuracy: 72%
Santos and Zadrozny [89]	Stock Market	LSTM	260,853 Documents	Daily	2006–2013	Accuracy:≈65%
Di persio and Honchar [113]	Stock market	MLP CNN RNN Wavelet-CNN	16,706	Daily	2006–2016	Accuracy: MLP: 81% CNN: 81% RNN: 79% Wavelet-CNN: 83%
Nichiforov et al. [98]	Electrical-energy load forecasting	LSTM	8.670 data points	Hourly	Unclear	CV (RMSE) Min: 0.92 Max: 1.78

(continued)

Table 2 (continued)

Author	Focus Area	Model	Dataset	Data type	Forecasting period	Performance
Zahid, Maheen et al. [99]	Electricity prices forecasting	ECNN	9314 Records	Hourly	November 2016–January 2017	MAE: 1.38 MAPE: 32.08 MSE: 3.22 RMSE: 1.79
Lago J, et al. [81]	Electricity prices forecasting	DNN GRU LSTM MLP SOM-SVR SVR-ARIMA CNN	43,536 data points	Daily	2014–2016	MAPE (%) DNN:12.34 GRU: 13.04 LSTM: 13.06 MLP: 13.27 SOM-SVR: 13.36 SVR-ARIMA: 13.39 CNN: 13.91
Ugurlu et al. [79]	Electricity prices forecasting	LSTM GRU	Turkish day-ahead market	Hourly	2013–2016	MAE: LSTM: 5.47 GRU: 5.36
Ni et al. [103]	Energy	CNN	100,352 records	Daily	February–April 2017	RMSE: 3.11 MAE: 1.92
Dedinec, Aleksandra [114]	Energy (electricity)	DBN	Macedonian hourly electricity consumption data	Hourly	2008–2014	MAPE: 8.6%
Hernández et al. [101]	Energy (Rainfall)	Autoencoder -MLP MLP	4216 samples	Daily	2002–2013	MSE: Autoencoder and MLP: 40.11 MLP: 42.34

(continued)

Table 2 (continued)

Author	Focus Area	Model	Dataset	Data type	Forecasting period	Performance
Du et al. [106]	Traffic Flow Prediction	RNN CNN LSTM GRU CNN-LSTM CNN-GRU	37,564 records	Monthly	2013–2014	RMSE: RNN: 12.18 CNN: 9.34 LSTM: 11.14 GRU: 11.15 CNN-LSTM: 9.75 CNN-GRU: 9.09
Ke et al. [115]	Passenger demand forecasting	ARIMA ANN LSTM	1,000,000 order	Hourly	2015–2016	MAE: ARIMA: 0.0178 ANN: 0.0194 LSTM: 0.0181
Zhao et al. [116]	Traffic forecast	ARIMA SVM RBF SAE RNN LSTM	data are collected from over 500 observation stations with a frequency of 5 min. 25.11 million validated records	Daily	January 2015–June 2015	MRE (%): ARIMA: 8.25 SVM: 10.63 RBF: 6.98 SAE: 6.32 RNN: 6.35 LSTM: 6.05
Duan et al. [105]	Travel time prediction	LSTM	66 link on highway M25	Unclear	2013	MRE: 7.0%
Fu et al. [104]	Traffic flow prediction	ARIMA LSTM GRU	15,000 sensors	Unclear	Unclear	MSA: ARIMA: 841.0065 LSTM: 710.0502 GRU: 668.9304

(continued)

Table 2 (continued)

Author	Focus Area	Model	Dataset	Data type	Forecasting period	Performance
Lv et al. [110]	Traffic flow prediction	SAE	Traffic data are collected every 30 s from over 15 000 individual detectors	Weekly	first three months of the year 2013	Accuracy: 93% MAE: 43.1
Alhassan et al. [107]	Medical	MLP LSTM GRU	(KAIMRCD)14,609 patient 41 million time-stamped	Unclear	2010–2015	Accuracy: MLP: 95.09% LSTM: 95.08% GRU: 97.3%
Choi et al. [108]	Medical (Diagnosis, prescription, and timing)	RNN	263,706 patient	Unclear	Unclear	Recall: 79%
Lipton et al. [109]	Medical	LSTM	10,401 PICU episodes	Hourly	Unclear	Accuracy: 0.856
Yang, Jianbo et al. [111]	Human Activity Recognition	DBN CNN	136,869	Daily	Unclear	Accuracy: DBN: 82% CNN: 96%
Mehdiyev et al. [112]	Industry	LSTM	3 Terabytes from 89 sensors	Unclear	Unclear	Accuracy: 87.49%

[a]https://finance.yahoo.com

the models of deep learning to better handle data from time series. Deep Learning Techniques provide better representation and classification on time series problems compared to classical methods when configured and trained properly.

References

1. Sengupta, S., et al.: A Review of deep learning with special emphasis on architectures. Applications and Recent Trends. arXiv preprint arXiv:1905.13294 (2019)
2. Paszke, A., et al.: Automatic differentiation in pytorch (2017)
3. Abadi, M., et al.: TensorFlow: large-scale machine learning on heterogeneous systems, 2015. Software available from tensorflow.org, vol. 1, no. 2 (2015)
4. Jia, Y., et al.: Caffe: convolutional architecture for fast feature embedding. In: Proceedings of the 22nd ACM International Conference on Multimedia, pp. 675–678. ACM (2014)
5. Tokui, S., Oono, K., Hido, S., Clayton, J.: Chainer: a next-generation open source framework for deep learning. In: Proceedings of Workshop on Machine Learning Systems (LearningSys) in the Twenty-Ninth Annual Conference on Neural Information Processing Systems (NIPS), vol. 5, pp. 1–6 (2015)
6. e. a. Chollet, F.: Keras. https://github.com/fchollet/keras (2015)
7. Dai, J., et al.: BigDL: a distributed deep learning framework for big data. arXiv preprint arXiv: 1804.05839 (2018)
8. Cavalcante, R.C., Brasileiro, R.C., Souza, V.L., Nobrega, J.P., Oliveira, A.L.: Computational intelligence and financial markets: A survey and future directions. Expert Syst. Appl. **55**, 194–211 (2016)
9. Dorffner, G.: Neural networks for time series processing. In: Neural Network World. Citeseer (1996)
10. I. Sutskever, O. Vinyals, and Q. V. Le, "Sequence to sequence learning with neural networks," in *Advances in neural information processing systems*, (2014), pp. 3104–3112
11. Tkáč, M., Verner, R.: Artificial neural networks in business: two decades of research. Appl. Soft Comput. **38**, 788–804 (2016)
12. J. C. B. Gamboa, Deep learning for time-series analysis. arXiv preprint arXiv:1701.01887 (2017)
13. Palma, W.: Time Series Analysis. Wiley (2016)
14. Boshnakov, G.N.: Introduction to Time Series Analysis and Forecasting, Wiley Series in Probability and Statistics (Montgomery, D.C., Jennings, C.L., Kulahci, M. (eds.)). Wiley, Hoboken, NJ, USA (2015). Total number of pages: 672 Hardcover: ISBN: 978-1-118-74511-3, ebook: ISBN: 978-1-118-74515-1, etext: ISBN: 978-1-118-74495-6, J. Time Ser. Anal. **37**(6), 864 (2016)
15. Fuller, W.A.: Introduction to Statistical Time Series. Wiley (2009)
16. Adhikari, R., Agrawal, R.K.: An introductory study on time series modeling and forecasting. arXiv preprint arXiv:1302.6613 (2013)
17. Otoo, H., Takyi Appiah, S., Wiah, E.: regression and time series analysis of loan default. Minescho Cooperative Credit Union, Tarkwa (2015)
18. Dagum, E.B., Bianconcini, S.: Seasonal Adjustment Methods and Real Time Trend-Cycle Estimation. Springer (2016)
19. Hyndman, R.J., Athanasopoulos, G.: Forecasting: Principles and Practice. OTexts (2018)
20. Brockwell, P.J., Davis, R.A., Calder, M.V.: Introduction to Time Series and Forecasting. Springer (2002)
21. Box, G.E., Jenkins, G.M., Reinsel, G.C., Ljung, G.M.: Time Series Analysis: Forecasting and Control. Wiley (2015)
22. Hipel, H.W., McLeod, A.I.: Time Series Modelling of Water Resources and Environmental Systems. Elsevier (1994)

23. Cochrane, J.H.: Time Series for Macroeconomics and Finance. University of Chicago, Manuscript (2005)
24. Zhang, G.P.: A neural network ensemble method with jittered training data for time series forecasting. Inf. Sci. **177**(23), 5329–5346 (2007)
25. Hamzaçebi, C.: Improving artificial neural networks' performance in seasonal time series forecasting. Inf. Sci. **178**(23), 4550–4559 (2008)
26. Chatfield, C.: Time series forecasting with neural networks. In: Neural Networks for Signal Processing VIII. Proceedings of the 1998 IEEE Signal Processing Society Workshop (Cat. No. 98TH8378), pp. 419–427. IEEE (1998)
27. Kihoro, J., Otieno, R., Wafula, C.: Seasonal time series forecasting: A comparative study of ARIMA and ANN models (2004)
28. Haykin, S.S.: Neural Networks and Learning Machines. Pearson Education, Upper Saddle River (2009)
29. Schmidhuber, J.: Deep learning in neural networks: an overview. Neur. Netw. **61**, 85–117 (2015)
30. Vellido, A., Lisboa, P.J., Vaughan, J.: Neural networks in business: a survey of applications (1992–1998). Expert Syst. Appl. **17**(1), 51–70 (1999)
31. Rojas, R.: Neural Networks: A Systematic Introduction. Springer Science & Business Media (2013)
32. Deb, C., Zhang, F., Yang, J., Lee, S.E., Shah, K.W.: A review on time series forecasting techniques for building energy consumption. Renew. Sustain. Energy Rev. **74**, 902–924 (2017)
33. Bengio, Y., Goodfellow, I.J., Courville, A.: Deep learning. Nature **521**(7553), 436–444 (2015)
34. LeCun, Y., Bengio, Y., Hinton, G.: Deep learning. Nature 521(7553), 436 (2015)
35. Chen, Z., Yi, D.: The game imitation: deep supervised convolutional networks for quick video game AI. arXiv preprint arXiv:1702.05663 (2017)
36. Øyen, S.: Forecasting Multivariate Time Series Data Using Neural Networks. NTNU (2018)
37. Ciresan, D.C., Meier, U., Masci, J., Gambardella, L.M., Schmidhuber, J.: Flexible, high performance convolutional neural networks for image classification. In: Twenty-Second International Joint Conference on Artificial Intelligence (2011)
38. Scherer, D., Müller, A., Behnke, S.: Evaluation of pooling operations in convolutional architectures for object recognition. In: International Conference on Artificial Neural Networks, pp. 92–101. Springer (2010)
39. Fawaz, H.I., Forestier, G., Weber, J., Idoumghar, L., Muller, P.-A.: Deep learning for time series classification: a review. Data Mining and Knowledge Discovery, pp. 1–47 (2019)
40. Graves, A., Mohamed, A.-r., Hinton, G.: Speech recognition with deep recurrent neural networks. In: 2013 IEEE International Conference on Acoustics, Speech and Signal Processing, pp. 6645–6649. IEEE (2013)
41. Palangi, H., et al.: Deep sentence embedding using long short-term memory networks: Analysis and application to information retrieval. IEEE/ACM Trans. Audio Speech Lang. Process. (TASLP) **24**(4), 694–707 (2016)
42. Che, Z., Purushotham, S., Cho, K., Sontag, D., Liu, Y.: Recurrent neural networks for multivariate time series with missing values. Sci. Rep. **8**(1), 6085 (2018)
43. Palangi, H., Ward, R., Deng, L.: Distributed compressive sensing: a deep learning approach. IEEE Trans. Signal Process. **64**(17), 4504–4518 (2016)
44. Graves, A.: Sequence transduction with recurrent neural networks. arXiv preprint arXiv:1211.3711 (2012)
45. Graves, A.: Generating sequences with recurrent neural networks. arXiv preprint arXiv:1308.0850 (2013)
46. Walid, A.: Recurrent neural network for forecasting time series with long memory pattern. J. Phys.: Conf. Ser. **824**(1), 012038 (2017)
47. Gómez, P., Nebot, A., Ribeiro, S., Alquézar, R., Mugica, F., Wotawa, F.: Local maximum ozone concentration prediction using soft computing methodologies. Syst. Anal. Model. Simul. **43**(8), 1011–1031 (2003)

48. Pascanu, R., Gulcehre, C., Cho, K., Bengio, Y.: How to construct deep recurrent neural networks. arXiv preprint arXiv:1312.6026 (2013)
49. Bao, W., Yue, J., Rao, Y.: A deep learning framework for financial time series using stacked autoencoders and long-short term memory. PLoS ONE **12**(7), e0180944 (2017)
50. Rumelhart, D.E., Hinton, G.E., Williams, R.J.: Learning representations by back-propagating errors. Cogn. Model. **5**(3), 1 (1988)
51. Xu, L., Li, C., Xie, X., Zhang, G.: Long-short-term memory network based hybrid model for short-term electrical load forecasting. Information **9**(7), 165 (2018)
52. Bengio, Y., Simard, P., Frasconi, P.: Learning long-term dependencies with gradient descent is difficult. IEEE Trans. Neural Networks **5**(2), 157–166 (1994)
53. Li, X., et al.: Long short-term memory neural network for air pollutant concentration predictions: Method development and evaluation. Environ. Pollut. **231**, 997–1004 (2017)
54. Jozefowicz, R., Zaremba, W., Sutskever, I.: An empirical exploration of recurrent network architectures. In: International Conference on Machine Learning, pp. 2342–2350 (2015)
55. Cho, K., et al.: Learning phrase representations using RNN encoder-decoder for statistical machine translation. arXiv preprint arXiv:1406.1078 (2014)
56. Y. Song, "Stock trend prediction: Based on machine learning methods," UCLA, (2018)
57. Mohammadi, M., Al-Fuqaha, A., Sorour, S., Guizani, M.: Deep learning for IoT big data and streaming analytics: a survey. IEEE Commun. Surv. Tutor. **20**(4), 2923–2960 (2018)
58. Fischer, A., Igel, C.: An introduction to restricted Boltzmann machines. In: iberoamerican Congress on Pattern Recognition, pp. 14–36. Springer (2012)
59. Smolensky, P.: Information processing in dynamical systems: Foundations of harmony theory. Colorado Univ at Boulder Dept of Computer Science (1986)
60. Hinton, G.E.: Training products of experts by minimizing contrastive divergence. Neural Comput. **14**(8), 1771–1800 (2002)
61. Hinton,G.E., Salakhutdinov, R.R.: Reducing the dimensionality of data with neural networks. Science 313(5786), 504–507 (2006)
62. Larochelle, H., Bengio, Y.: Classification using discriminative restricted Boltzmann machines. In: Proceedings of the 25th International Conference on Machine Learning, pp. 536–543. ACM (2008)
63. Coates, A., Ng, A., Lee, H.: An analysis of single-layer networks in unsupervised feature learning. In: Proceedings of the Fourteenth International Conference on Artificial Intelligence and Statistics, pp. 215–223 (2011)
64. Sutskever, I., Hinton, G.: Learning multilevel distributed representations for high-dimensional sequences. In: Artificial Intelligence and Statistics, pp. 548–555 (2007)
65. Taylor, G.W., Hinton, G.E., Roweis, S.T.: Modeling human motion using binary latent variables. In: Advances in Neural Information Processing systems, pp. 1345–1352 (2007)
66. Hinton, G.E., Osindero, S., Teh, Y.-W.: A fast learning algorithm for deep belief nets. Neural Comput. **18**(7), 1527–1554 (2006)
67. Nweke, H.F., Teh, Y.W., Al-Garadi, M.A., Alo, U.R.: Deep learning algorithms for human activity recognition using mobile and wearable sensor networks: state of the art and research challenges. Expert Syst. Appl. **105**, 233–261 (2018)
68. Zhang, Q., Yang, L.T., Chen, Z., Li, P.: A survey on deep learning for big data. Information Fusion **42**, 146–157 (2018)
69. Gensler, A., Henze J., Sick, B., Raabe, N.: Deep Learning for solar power forecasting— an approach using AutoEncoder and LSTM neural networks. In: 2016 IEEE International Conference on Systems, Man, and Cybernetics (SMC), pp. 002858–002865. IEEE (2016)
70. Ahmad, A., Anderson, T., Lie, T.: Hourly global solar irradiation forecasting for New Zealand. Sol. Energy **122**, 1398–1408 (2015)
71. Sharma, V., Yang, D., Walsh, W., Reindl, T.: Short term solar irradiance forecasting using a mixed wavelet neural network. Renew. Energy **90**, 481–492 (2016)
72. Grover, A., Kapoor, A., Horvitz, E.: A deep hybrid model for weather forecasting. In: Proceedings of the 21st ACM SIGKDD International Conference on Knowledge Discovery and Data Mining, pp. 379–386. ACM (2015)

73. Marino, D.L., Amarasinghe, K., Manic, M.: Building energy load forecasting using deep neural networks. In: IECON 2016-42nd Annual Conference of the IEEE Industrial Electronics Society, pp. 7046–7051. IEEE (2016)
74. Ryu, S., Noh, J., Kim, H.: Deep neural network based demand side short term load forecasting. Energies **10**(1), 3 (2016)
75. Tong, C., Li, J., Lang, C., Kong, F., Niu, J., Rodrigues, J.J.: An efficient deep model for day-ahead electricity load forecasting with stacked denoising auto-encoders. J. Parallel Distrib. Comput. **117**, 267–273 (2018)
76. Lu, S., et al.: Electric load data characterising and forecasting based on trend index and auto-encoders. J. Eng. **2018**(17), 1915–1921 (2018)
77. Shi, H., Xu, M., Li, R.: Deep learning for household load forecasting—a novel pooling deep RNN. IEEE Trans. Smart Grid **9**(5), 5271–5280 (2017)
78. Bouktif, S., Fiaz, A., Ouni, A., Serhani, M.: Optimal deep learning LSTM model for electric load forecasting using feature selection and genetic algorithm: comparison with machine learning approaches. Energies **11**(7), 1636 (2018)
79. Ugurlu, U., Oksuz, I., Tas, O.: Electricity price forecasting using recurrent neural networks. Energies **11**(5), 1255 (2018)
80. Kuo, P.-H., Huang, C.-J.: An electricity price forecasting model by hybrid structured deep neural networks. Sustainability **10**(4), 1280 (2018)
81. Lago, J., De Ridder, F., De Schutter, B.: Forecasting spot electricity prices: deep learning approaches and empirical comparison of traditional algorithms. Appl. Energy **221**, 386–405 (2018)
82. Kim, K.-J., Ahn, H.: Simultaneous optimization of artificial neural networks for financial forecasting. Appl. Intell. **36**(4), 887–898 (2012)
83. Adebiyi, A.A., Adewumi, A.O., Ayo, C.K.: Comparison of ARIMA and artificial neural networks models for stock price prediction. J. Appl. Math. **2014** (2014)
84. Göçken, M., Özçalıcı, M., Boru, A., Dosdoğru, A.T.: Integrating metaheuristics and artificial neural networks for improved stock price prediction. Expert Syst. Appl. **44**, 320–331 (2016)
85. Lu, C.-J., Lee, T.-S., Chiu, C.-C.: Financial time series forecasting using independent component analysis and support vector regression. Decis. Support Syst. **47**(2), 115–125 (2009)
86. Hossain, M.A., Karim, R., Thulasiram, R., Bruce, N.D., Wang, Y.: Hybrid deep learning model for stock price prediction. In: 2018 IEEE Symposium Series on Computational Intelligence (SSCI), pp. 1837–1844. IEEE (2018)
87. Siami-Namini, S., Namin, A.S.: Forecasting economics and financial time series: Arima vs. LSTM. arXiv preprint arXiv:1803.06386 (2018)
88. Fischer, T., Krauss, C.: Deep learning with long short-term memory networks for financial market predictions. Eur. J. Oper. Res. **270**(2), 654–669 (2018)
89. dos Santos Pinheiro, L., Dras, M.: Stock market prediction with deep learning: a character-based neural language model for event-based trading. In: Proceedings of the Australasian Language Technology Association Workshop 2017, pp. 6–15 (2017)
90. Wang, J.-Z., Wang, J.-J., Zhang, Z.-G., Guo, S.-P.: Forecasting stock indices with back propagation neural network. Expert Syst. Appl. **38**(11), 14346–14355 (2011)
91. Rafiei, M., Niknam, T., Khooban, M.-H.: Probabilistic forecasting of hourly electricity price by generalization of ELM for usage in improved wavelet neural network. IEEE Trans. Industr. Inf. **13**(1), 71–79 (2016)
92. Chen, K., Zhou, Y., Dai, F.: A LSTM-based method for stock returns prediction: a case study of China stock market. In: 2015 IEEE International Conference on Big Data (Big Data), pp. 2823–2824. IEEE (2015)
93. Sengupta, S., et al.: A review of deep learning with special emphasis on architectures, applications and recent trends. Networks, 21 (2006)
94. Gudelek, M.U., Boluk, S.A, Ozbayoglu, A.M.: A deep learning based stock trading model with 2-D CNN trend detection. In 2017 IEEE Symposium Series on Computational Intelligence (SSCI), pp. 1–8. IEEE (2017)

95. Türkmen, A.C., Cemgil, A.T.: An application of deep learning for trade signal prediction in financial markets. In: 2015 23rd Signal Processing and Communications Applications Conference (SIU), pp. 2521–2524. IEEE (2015)
96. Ding, X., Zhang, Y., Liu, T., Duan, J.: Deep learning for event-driven stock prediction. In: Twenty-Fourth International Joint Conference on Artificial Intelligence (2015)
97. Chen, W., Zhang, Y., Yeo, C.K., Lau, C.T., Lee, B.S.: Stock market prediction using neural network through news on online social networks. In: 2017 International Smart Cities Conference (ISC2), pp. 1–6. IEEE (2017)
98. Nichiforov, C., Stamatescu, G., Stamatescu, I., Făgărăşan, I.: Evaluation of sequence-learning models for large-commercial-building load forecasting. Information 10(6), 189 (2019)
99. Zahid, M., et al.: Electricity price and load forecasting using enhanced convolutional neural network and enhanced support vector regression in smart grids. Electronics 8(2), 122 (2019)
100. Shi, H., Xu, M., Ma, Q., Zhang, C., Li, R., Li, F.: A whole system assessment of novel deep learning approach on short-term load forecasting. Energy Procedia 142, 2791–2796 (2017)
101. Hernández, E., Sanchez-Anguix, V., Julian, V., Palanca, J., Duque, N.: Rainfall prediction: a deep learning approach. In: International Conference on Hybrid Artificial Intelligence Systems, pp. 151–162. Springer (2016)
102. Zhao, Y., Li, J., Yu, L.: A deep learning ensemble approach for crude oil price forecasting. Energy Econ. 66, 9–16 (2017)
103. Ni, C., Ma, X.: Prediction of wave power generation using a convolutional neural network with multiple inputs. Energies 11(8), 2097 (2018)
104. Fu, R., Zhang, Z., Li, L.: Using LSTM and GRU neural network methods for traffic flow prediction. In: 2016 31st Youth Academic Annual Conference of Chinese Association of Automation (YAC), pp. 324–328. IEEE (2016)
105. Duan, Y., Lv, Y., Wang, F.-Y.: Travel time prediction with LSTM neural network. In: 2016 IEEE 19th International Conference on Intelligent Transportation Systems (ITSC), pp. 1053–1058. IEEE (2016)
106. Du, S., Li, T., Gong, X., Yu, Z., Huang, Y., Horng, S.-J.: A hybrid method for traffic flow forecasting using multimodal deep learning. arXiv preprint arXiv:1803.02099 (2018)
107. Alhassan, Z., McGough, R.S., Alshammari, R., Daghstani, T., Budgen, D., Al Moubayed, N.: Type-2 diabetes mellitus diagnosis from time series clinical data using deep learning models. In: International Conference on Artificial Neural Networks, pp. 468–478. Springer (2018)
108. Choi, E., Bahadori, M.T., Schuetz, A., Stewart, W.F., Sun, J.: Doctor AI: predicting clinical events via recurrent neural networks. In: Machine Learning for Healthcare Conference, pp. 301–318 (2016)
109. Lipton, Z.C., Kale, D.C., Elkan, C., Wetzel, R.: Learning to diagnose with LSTM recurrent neural networks. arXiv preprint arXiv:1511.03677 (2015)
110. Lv, Y., Duan, Y., Kang, W., Li, Z., Wang, F.-Y.: Traffic flow prediction with big data: a deep learning approach. IEEE Trans. Intell. Transp. Syst. 16(2), 865–873 (2014)
111. Yang, J., Nguyen, M.N., San, P.P., Li, X.L., Krishnaswamy, S.: Deep convolutional neural networks on multichannel time series for human activity recognition. In: Twenty-Fourth International Joint Conference on Artificial Intelligence (2015)
112. Mehdiyev, N., Lahann, J., Emrich, A., Enke, D., Fettke, P., Loos, P.: Time series classification using deep learning for process planning: a case from the process industry. Procedia Comput. Sci. 114, 242–249 (2017)
113. Di Persio, L., Honchar, O.: Artificial neural networks approach to the forecast of stock market price movements. Int. J. Econ. Manag. Syst. 1 (2016)
114. Dedinec, A., Filiposka, S., Dedinec, A., Kocarev, L.: Deep belief network based electricity load forecasting: an analysis of Macedonian case. Energy 115, 1688–1700 (2016)
115. Ke, J., Zheng, H., Yang, H., Chen, X.M.: Short-term forecasting of passenger demand under on-demand ride services: a spatio-temporal deep learning approach. Transp. Res. Part C: Emerg. Technol. 85, 591–608 (2017)
116. Zhao, Z., Chen, W., Wu, X., Chen, P.C., Liu, J.: LSTM network: a deep learning approach for short-term traffic forecast. IET Intell. Transp. Syst. 11(2), 68–75 (2017)

Deep Learning for Taxonomic Classification of Biological Bacterial Sequences

Marwah A. Helaly, Sherine Rady, and Mostafa M. Aref

Abstract Biological sequence classification is a key task in Bioinformatics. For research labs today, the classification of unknown biological sequences is essential for facilitating the identification, grouping and study of organisms and their evolution. This work focuses on the task of taxonomic classification of bacterial species into their hierarchical taxonomic ranks. Barcode sequences of the 16S rRNA dataset—which are known for their relatively short sequence lengths and highly discriminative characteristics—are used for classification. Several sequence representations and CNN architecture combinations are considered, each tested with the aim of learning and finding the best approaches for efficient and effective taxonomic classification. Sequence representations include k-mer based representations, integer-encoding, one-hot encoding and the usage of embedding layers in the CNN. Experimental results and comparisons have shown that representations which hold some sequential information about a sequence perform much better than a raw representation. A maximum accuracy of 91.7% was achieved with a deeper CNN when the employed sequence representation was more representative of the sequence. However with less representative representations a wide and shallow network was able to efficiently extract information and provide a reasonable accuracy of 90.6%.

Keywords DNA · RNA · Biological sequences · Deep learning · Classification · Convolutional neural networks · Feature representation

M. A. Helaly · S. Rady (✉) · M. M. Aref
Faculty of Computer and Information Sciences, Ain Shams University, Cairo 11566, Egypt
e-mail: srady@cis.asu.edu.eg

M. A. Helaly
e-mail: marwah.ahmad.helaly@cis.asu.edu.eg

M. M. Aref
e-mail: mostafa.m.aref@gmail.com

393

1   Introduction

Cells are the basic building blocks of all living things. There are two types of cells, namely: prokaryotic and eukaryotic cells. Prokaryotic cells are small and structurally simple cells that do not contain a nucleus. They are found in prokaryotes such as bacteria and archaea, in which cells are prokaryotic cells. Eukaryotic cells are much more complex cells that contain a membrane-bound nucleus. They are found in eukaryotes, which are found in more complex organisms such as animals, plants and fungi.

A biological sequence is a single, continuous molecule of nucleic acid or protein. Deoxyribo-nucleic Acid (DNA) and Ribo-nucleic Acid (RNA) are the two main types of nucleic acids that exist in living organisms. DNA contains the genetic code that is considered as the instruction book for creating both RNA and proteins and with which they are responsible to pass on to later generations [1]. DNA uses its genetic information directly to create the intermediate molecule called RNA, which is then transformed to proteins.

DNA and RNA are polymer molecules composed of a very long series of smaller monomer molecules called the *nucleotides*. A nucleotide consists of three compo-nents: (1) a sugar molecule, (2) a phosphate group, and (3) a nitrogen-containing base. There are four types of nucleotides, all identical to each other except for the base. The four nucleotide bases are, namely: adenine (A), guanine (G)—also called the *purines*—cytosine (C), and thymine (T)—also called the *pyrimidines*. Thymine (T) is replaced with a uracil (U) base when a DNA sequence is converted to RNA. At times it is not certain which of the four basic nucleotide bases is present at a certain point in the sequence, hence *ambiguity codes* (or *ambiguity characters*) were proposed. There are 11 ambiguity characters: each represent a certain subset of nucleotide bases that could possibly exist at a certain position. For example, the ambiguous character *B* means that a single base in a sequence could be one of the nucleotide bases C or G or T.

DNA sequences have one main important function—which is to encode informa-tion, while there are several types of RNA sequences each responsible for a different function. A single DNA sequence is split into many contiguous segments called *genes*. Each gene contains the information needed to build either a RNA molecule or a protein [2].

Barcode (i.e. *marker*) gene sequences are short specific regions of biological sequences that represent genetic markers in species. They are unique in their high mutation rates that easily result in highly discriminative characteristics, which make them favorable for the classification and discovery of species. The main target of barcoding is to find certain regions of a sequence that have minimal intra-variation within a species and maximum inter-variation between it and other species. Differ-ent species usually have certain standard and agreed-upon genes that are used for barcoding [3]. For example, in eukaryotes, the mitochondrial gene Cytochrome C Oxidase subunit 1 (commonly abbreviated the *COX1*, *CO1* or *COI* gene, or *MT-CO1* exclusively in humans) is commonly used for barcoding [4, 5], while in prokaryotes, the 16S rRNA barcode gene is commonly used. Common sources of barcode sequences are the Barcode of Life Data system (BOLD) Systems and the RDP II repository [6].

Recent large biological projects such as the Human Genome project [7], the 100,000 Genomes Project [8] and others, have made an incredible amount of biological data available. As a result, *Bioinformatics* emerged as the science needed to analyze and manage this data. Bioinformatics is defined as the research, development and application of computational tools to capture and interpret biological data [9]. Current problems in bioinformatics include DNA and protein sequence classification, gene finding, gene-function prediction, protein structure prediction, protein structure-function relationship prediction and sequence alignment [10, 11].

DNA sequence classification is a key task in Bioinformatics. There are various ways sequences may be classified according to the area of research. This includes identifying species at different taxonomic levels, recognizing promoters, enhancers and splice site recognition. *Biological systematics* is a field in biology concerned with the identification and classification of organisms and the depiction of their evolutionary relationships accurately via taxonomies that organize organisms into hierarchical categories. In a taxonomy, in general similar *species* are grouped together under the same *genera*, similar genera into *families*, continuing similarly into *orders*, *classes*, *phyla*, *kingdoms* until finally reaching the *domain* [12].

The task of DNA sequence classification is a very common bioinformatics task that has been studied using various Machine Learning (ML) methods. Two main things are important for obtaining a proper performance from a ML model, namely: a good input data representation and a suitable machine learning technique that is capable of generalization. Deep Learning (DL) is an area of ML that has boomed in recent years. It provided promising performance in many areas of research, including bioinformatics. There are many types of DL architectures, but in general they are all computational neural networks that all consist of many multiple non-linear information processing layers. These layers learn hierarchical representations of complex data representations with multiple increasing levels of abstraction. DL models require large memory and powerful computation resources due to its need for a large amount of training data and representation sizes. Due to the large availability and need of biological sequences, DL is therefore very fitting for biological sequence classification due to biological sequences and their representations typically very large in size.

A promising key characteristic of DL methods is that they are *representation-learning methods*, which allows the computational model to automatically extract useful features from raw input data without direct human input. This replaces the traditional non-trivial *feature engineering* process on a raw input data set, which is one of the most time-consuming phases of the ML process. Feature engineering is used to extract useful, strong and distinctive features to discriminate one input sample from another and then feed to a model. Because it requires very little hand engineering, it is anticipated that DL will provide many more improvements and success in the near future [13, 14].

Convolutional Neural Networks, commonly called *CNNs* or *ConvNets* is a popular type of DL models that were inspired by the visual cortex of the brain [15, 16]. Probably due to this fact lies the reason that CNNs have had enormous progress with image-related tasks, such as image recognition. They consist of two general phases: (1) a feature extraction phase, which commonly include convolutional and pooling layers, and (2) a classification phase, which includes fully-connected nonlinear classification layers that take the features extracted from the first phases and classifies them, working as a normal Neural Network (NN) classifier. *Convolutional* layers consist of a number of filters, each with a certain size and stride that pass over the input and produce an output image, called a *feature map*. *Pooling* layers have several types, the most common of which are *max-pooling* layers. These layers reduce the size of input images and assist in extracting more abstract and general features from images, causing the features to be invariant to the slight spatial differences in different inputs.

Since biological sequences are normally acquired in a textual form, they must be represented using some meaningful representation technique that is capable of capturing salient and discriminative features from the sequences to effectively differentiate between them. Although this is not always the case, but the sequence representation used may result in a representation that is very large in size, that in turn will increase computational and time costs. Other biological challenges may include that sometimes only small parts of sequences are available due to their direct collection from the environment, which is known as *metagenomic studies*. Such a problem may affect training of a model negatively if counted on. The problem of taxonomic sequence classification has recently had very few works, while other problem areas in bioinformatics are usually more commonly researched [17].

This book chapter classifies the 16S rRNA gene barcode dataset for three different bacterial phyla, on five different taxonomic ranks. Many experiments were performed, experimenting with various combinations of the most common feature representations methods and different CNN architectures, some of which are related works on this same task. This work aims to highlight the benefits and drawbacks of every feature representation method and CNN model design as a step to ultimately reach the goal of finding the best methods that would produce the best performance for DNA sequence classification.

The book chapter is partitioned as follows: Sect. 2 reviews some of the related DL works in DNA sequence classification. The various representation methods and the experimented CNN models are explained in detail in Sect. 3, which also highlights which representations were used with which CNNs. Section 4 presents the experimental environment, the results and discussions. Finally, the conclusions are deduced in Sect. 5.

2 Related Works

Several works compared traditional ML and DL methods for the various tasks of DNA sequence classification, in which DL methods proved to be more effective. One of the common successfully used traditional ML methods was Support Vector Machines (SVM) [18, 19].

Taxonomic species classification as recently been frequently studied. Being a bacterial barcode dataset, 16S rRNA datasets have been quite popular for this task. In [20], a 6-layer CNN was employed on such a dataset for sequence classification at five different taxonomic ranks. A k-mer spectral representation of the sequences was used, where subsequences of a variable size k-bases from the original sequence were extracted, captured by a sliding window of size k and a certain stride. In the meantime the observed k-mers of each specific sequence are counted into frequency vectors. The CNN was trained once and tested on both the full-sequence and fragment sequences (commonly needed for metagenomic studies). Compared to five other traditional classifiers, the CNN outperformed all the others concluding that the CNN had better generalization capabilities. The same CNN was used with a frequency chaos game representation (FCGR) sequence representation in [21]. In regards to the value of k, it was concluded that increasing its value will only slightly increase the accuracy but will greatly increase the computational cost. The CNN outperformed a SVM in both tests; slightly for full-length sequences and greatly for sequence fragments.

Also using a 16S rRNA dataset, [22] compared a Long Short-Term Memory (LSTM) Recurrent NN and CNN, where the CNN was similar to that of [20, 21]. Sparse one-hot encodings of sequences were converted to dense continuous input vectors by means of an embedding layer. *Multi-task learning* improved the generalization of the LSTM for all tasks. The CNN provided superior performance for the first four and more general taxonomic ranks, while the LSTM performed better for the last and most fine-grained one.

In [23], the authors used a single dataset partitioned into two classes of Hepatitis-B virus (HBV). A CNN model was proposed for the classification between the classes. Representing sequences using one-hot vectors, the effect of using different lengths of sequences (500–2000 bp) on the proposed model was studied, with accuracies in the range between 93.3 and 95.6%, increasing as the length increased. Comparisons of the proposed CNN model to 6 other traditional classifiers were performed, with the CNN presenting the highest accuracy of 96.83% while the other classifiers ranged from 67.4 to 71.6%. This proved that a CNN has the capability of extracting useful, representative and reliable features that can be used for classifying sequences into their correct categories.

To test the generalization of their proposed classification methods, [24] applied a CNN on 12 different datasets (10 concerning histone-wrapping, 1 splice-junctions and 1 E.Coli promoters) and compared its performance to a SVM by other authors. Dataset sizes ranged from approximately 100 to 37,000 samples. A combination of k-mer extraction, region grouping and a one-hot vector representation of the sequences resulted in an 2D numerical input to the network [25]. It was shown that CNNs provided significant improvement for all datasets, ranging from 1 to 6% in improved accuracy.

In [26] two different datasets were used and were represented using a normal one-hot vector representation. Five different CNN architectures—differing in the types and numbers of layers—were trained on a permissive enhancer dataset to learn the general features of enhancers. It was concluded that max-pooling layers and using batch normalization increases the performance and that deeper architectures decrease the performance of the model. The best CNN out of the five produced an AUROC of 0.916, a AUPRC of 0.917. Compared to methods of another work with a gapped k-mer representation SVM (gkm-SWM) [27], their permissive deep learning models were always superior in performance. *Transfer learning* was then used to fine-tune the model on specific cell types, where the training was continued on 9 enhancer datasets. The model produced a better accuracy for all cell-types than the gkm-SVM, with an increase of an average of 7% in AUROC and AUPRC. CNNs proved to be powerful for learning deep features from raw sequences and for classification.

Two 16S rRNA gene datasets were used with a CNN and a Deep Belief Network (DBN) in [17] for the task of metagenomic taxonomic classification. The two datasets originated from 1000 reference sequences from the RDP database [6] which were further used with two different Next Generation Sequencing (NGS) technique simulations: Whole Genome Shotgun (WGS) and Amplicon (AMP), to create two larger datasets. This resulted in datasets of 28,224 sequences with WGS and 28,000 with AMP. A k-mer representation of sequences was used with discrete values of k in the range [3, 7] inclusively. The proposed methods were compared to the Naïve Bayes RDP classifier. The CNNs and DBNs were independently trained and tested with ten-fold cross validation for each type of dataset and k-values. The largest value of k always performed better but increased the computation time incredibly. The DBN had a more stable increase in performance than the CNN. At the genus level both networks gave a very similar performance, with a small win of accuracy for the DBN on the AMP dataset and a larger win for the CNN on the SG dataset. The accuracy of the AMP dataset topped that of the WGS dataset. Both the proposed CNN and DBN outperformed the RDP, especially on the AMP dataset. The AMP dataset gave approximately 91% accuracy for both networks at the most fine-grained classification task as opposed to the 83% given by the RDP, while the WGS dataset gave 85% with the CNN and 80% with the RDP.

From the related work, it can be concluded that it is still worth experimenting with the task of DNA sequence classification in order to obtain efficient DL models. Tuning the best combination of layers in a CNN that is capable of extracting the most representative features and classifying them desirably—while taking memory and computation resources into consideration—is highly required. Specifically for biological sequences, the optimal goal in selecting a representation method to capture short and long-term interactions between the sequence regions to efficiently discriminate between different types of sequences. Moreover, using datasets in which the sequences are best representatives of their kinds is very important, which can handled by using accurate barcode sequences or by alternatively longer sequences.

3   DL Methods

This work performs various combinations of sequence representation techniques and CNN architectures, some of which are have been previously proposed by Rizzo et al. [20, 21], Lo Bosco and Di Gangi [22] and others. Figure 1 is a block diagram that demonstrates the considered DL approach and sequence representations.

3.1   Pre-processing Input

The bacterial DNA sequences are parsed from a FASTA file to extract required information, then the dataset is shuffled—a very important step. Uncertain ambiguous characters are not considered and were removed from all sequences in all the experiments. Some representations do not require all the sequences to be the same length, therefore shorter sequences are zero-padded to unify the lengths of all the sequences—only when needed.

3.2   Sequence Representation

Biological sequences are generally available in a textual sequence of characters. To be used with ML methods, they must be represented by some numerical form. Common sequence representations include k-mer based representations, image representations, one-hot encodings and sometimes combinations of several together.

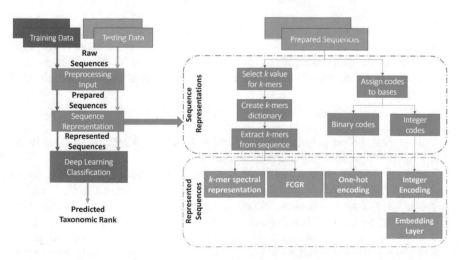

Fig. 1 The proposed DL DNA sequence classification process

In this work two k-mer-based representations are selected, as well as the common integer and one-hot encoding sequence representations, as shown in Fig. 1:

1. One-hot encoding: A common representation used in many bioinformatics research works, one-hot encoding is considered a *raw* representation of sequences [17]. Setting ambiguous characters aside, a 4-digit code is assigned to each nucleotide base where a 1 is placed in a unique position. The alphabet also includes an extra character Z to pad shorter sequences, hence the code is as follows:

$$Z : [0, 0, 0, 0]$$
$$A : [1, 0, 0, 0]$$
$$G : [0, 1, 0, 0]$$
$$C : [0, 0, 1, 0]$$
$$T : [0, 0, 0, 1]$$

When representing a sequence, each base is represented with its one-hot vector code. The sequence of representing vectors are then flattened into one vector to be presented to 1-dimensional CNNs. Ambiguous characters are not considered in this work. However, in the case of their consideration, the vector becomes a *probability vector*, where the vector elements will instead represent the probability that each of the possible bases that an ambiguous character symbolizes will occur. For example, the ambiguity character W symbolizes a A or T base, and the probability that each will occur is 0.5, so the digitization of this base is [0.5, 0, 0, 0.5]. It is important to note that this is a very sparse representation. The final size of this representation depends on the sequence size:

$$size = 4 \times Sequence\ Size$$

2. Character-level integer encoding: Similar to one-hot encoding, integer encoding also creates an alphabet, that consists of a unique integer value in the range [1, 4] for each of the four basic nucleotides bases (A, G, C and T), as well as an extra character Z (with code 0) to pad shorter sequences, resulting in a dictionary of five codes:
   $$Z = 0,\ A = 1,\ G = 2,\ C = 3,\ T = 4$$

   This representation may be extended and passed to an Embedding layer in the CNN, which converts it into a denser representation.

3. k-mer spectral representation: k-mers are similar to the idea of n-grams, where k is a variable value. If only the four basic nucleotide bases (A, G, C and T) are considered and all the possible k-length sequence combinations of these four bases are found, then this will result in the creation of a dictionary of 4^k possible k-length sequences. A sliding window of size k is then passed over the sequence to capture subsequent regions of length k-bases. The window is slid with a certain stride or step (commonly 1), while keeping count of the number of occurrences

of each k-mer in the given sequence. This is commonly known as a *frequency vector* or a *bag-of-words* representation. Three different values of $k = 4$, 5 and 6 are investigated. Recent works found that the values of k in the range [3, 7] are generally balanced in terms of computational complexity and information content [17]. The size of the resulting k-mer representation depends on the value of k and is independent of the sequence size, hence no sequences are zero-padded for this representation:

$$size = 4^k$$

4. Frequency Chaos Game Representation (FCGR): A CGR uses k-mers to produce a fractal-like image. Since there are four basic nucleotide bases, the 2D-matrix representation is hence split into 4 quadrants, where each quadrant represents one of the four bases. This partitioning continues recursively k times. After extracting normal k-mers with the selected size k and their frequencies from the sequence, the *FCGR* matrix is then filled with the k-mer *frequencies* and placed in their respective k-mer position in the 2D-matrix. The matrix is often normalized. We also experiment with the three different values of $k = 4$, 5 and 6. The final FCGR matrix has the size:

$$size = \sqrt{4^k} \times \sqrt{4^k}$$

3.3 Deep Learning Classification

CNNs are feedforward NNs, in which signals pass only in a forward direction through the network: from the input layer to the output layer. CNNs, like any NN, can be built in various ways with many different factors that may be taken into consideration. In this work we experiment with different architectures of a CNN, with the ultimate aim of being guided to the best factors that will result in the best performance for the given task. Table 1 presents the CNN architectures, while Table 2 summarizes the combinations of sequence representations and CNN architectures investigated in this work. The CNNs have a minimum of 7 layers and a maximum of 9 layers and are described as follows:

1. *CNN1-1D* starts with two convolutional layers with filter sizes of 5 and number of filters 10 and 20, respectively, each followed by maxpooling layers with sizes and strides of 2. The network is closed with two dense fully-connected layers. Their sizes are 500 neurons and the number of categories in the taxonomic rank considered, respectively.
2. *CNN2-2D*: Since CNNs are known to work relatively well with image-based tasks, some authors have proposed representing sequences in a 2D and sometimes a 3D representation. As proposed in [21], a Frequency Chaos Game Representation (FCGR) sequence representation is used with a CNN model to receive the 2-dimensional image input with the usage of 2-dimensional convolutional layers with 5×5 filters and 2×2 maxpooling layers.

Table 1 Configurations of the considered CNN architectures

Layers	Layer sizes					
	CNN1-1D	CNN2-2D	CNN3-Embd	CNN4-Deep	CNN5-Wide	CNN6-WideDeep
Embedding	–	–	10×1	–	–	–
Conv1	$10 \times (5 \times 1)$	$10 \times (5 \times 5)$	$10 \times (5 \times 1)$	$10 \times (5 \times 1)$	$32 \times (5 \times 1)$	$32 \times (5 \times 1)$
Conv2	–	–	–	$10 \times (5 \times 1)$	–	$32 \times (5 \times 1)$
Maxpool1	(1×2)	(2×2)	(1×5)	(1×2)	(1×2)	(1×2)
Conv3	$20 \times (5 \times 1)$	$20 \times (5 \times 5)$	$20 \times (5 \times 1)$	$20 \times (5 \times 1)$	$64 \times (5 \times 1)$	$64 \times (5 \times 1)$
Conv4	–	–	–	$20 \times (5 \times 1)$	–	$64 \times (5 \times 1)$
Maxpool2	(1×2)	(2×2)	(1×5)	(1×2)	(1×2)	(1×2)
Flatten	✓					
Dense1	500					
Dense2	NumCategories					

The notation used for the description of layers is $a \times (b)$, where a represents the number of filters and b represents the sizes of filters or windows. Dense layers are characterized by the number of neurons in the layers

3. *CNN3-Embd* receives a simple representation of the input sequence and uses an embedding layer at the beginning of the network for a denser sequence representation, followed by a typical CNN. This architecture is similar to *CNN1*, with a difference of an embedding layer before the network and larger maxpooling layers of size and stride 5. It was experimented with an Integer encoding representation only, as Keras's Embedding layers require input as positive integer indexes [28]. These indexes can be thought of as the indexes of the ones in a one-hot representation of a sequence. This configuration was proposed by [22].

4. *CNN4-Deep*, inspired from the CNNs in [26], experiments with deeper networks with four convolutional layers and two max-pooling layers. The authors found that max-pooling layers increased the performance, hence they were considered in *CNN4-Deep*'s architecture. They also found that even deeper networks decreased the performance, for which further study was needed to explain.

5. All the previous networks use a rather small number of filters in each convolutional layer. *CNN5-Wide* features a shallow but wider network with more filters, to experiment the effect of wider networks on bacterial taxonomic classification.

6. *CNN6-WideDeep* is a combination of the two previous CNNs—*CNN4-Deep* and *CNN5-Wide*, where capturing the advantages of each network is observed.

Table 2 Valid combinations of CNN architectures versus sequence representations

	Main characteristics	k-mers	FCGR	One-hot encoding	Integer encoding
CNN1-1D	1D shallow	✓	×	✓	✓
CNN2-2D	2D shallow	×	✓	×	×
CNN3-Embd	Embedding	×	×	×	✓
CNN4-Deep	1D deep	✓	×	✓	✓
CNN5-Wide	1D wide	✓	×	✓	✓
CNN6-WideDeep	1D wide & deep	✓	×	✓	✓

Symbol meanings: ✓: performed, ×: not allowed

4 Experiments

4.1 Dataset Description

The *16S rRNA barcode dataset* used in this work was originally collected from RDP II repository [20] from the largest 3 bacterial phyla: Actinobacteria, Firmicutes and Proteobacteria, with 1000 sequences from each.

Each sequence in the dataset is categorized into 5 hierarchical taxonomic ranks. The total number of categories increase greatly as the taxonomic rank in consideration becomes more fine-grained; 3, 6, 23, 65 and 393 categories from phylum to genus rank respectively, as shown in Table 3 [20]. It is important to note that given the large number of categories for the *genus* rank, this makes the *genus* classification task the most difficult to learn out of all the ranks. These categories (per rank) are what will be learned by the CNNs. As the taxonomic ranks become more fine-grained, the distribution of the sequences from the original phyla become more unbalanced. Specifically, Actinobacteria reaches a total 79 *genera* categories, while Firmicutes reaches 110 and Proteobacteria reaches 204.

Table 3 Distribution of categories for the different taxonomic ranks onto the general bacterial phyla in the 16S rRNA dataset

Phyla	Ranks				
	Phylum	Class	Order	Family	Genus
Actinobacteria	1	2	5	12	79
Firmicutes	1	2	5	19	110
Proteobacteria	1	2	13	34	204
Total num. of categories	3	6	23	65	393

4.2 Experimental Setup

Experiments were performed on an Intel i7 7th generation processor and an AMD Radeon R7 M440 GPU, all implemented with the Keras Python library with the PlaidML backend. PlaidML is a DL framework written by Vertex. AI, which was acquired by Intel in 2018. It is one of the few libraries that support AMD GPU computing. All CNN models were trained and tested using 10-fold cross validation. The following model configurations were set: a fixed number of epochs per fold (200), a fixed batch-size (64), a categorical cross-entropy loss function and the AdaDelta optimizer, where it was found that this combination produced the best results. For each experiment, a number of performance measures were calculated: the *average* (1) precision, (2) recall, (3) F1-score, (4) accuracy, (5) AUPRC, (6) AUROC and (7) the *average* 10-fold execution run time across all ranks (in minutes).

4.3 Experimental Results and Discussion

Table 4 presents the performance results for *CNN1-1D*, which is a simple CNN with 2 convolutional and 2 maxpooling layers that accepts 1-dimensional input. All of the three spectral k-mer representations, one-hot encoding and integer encoding are investigated. It may be said that at the most general level—the *Phylum* rank—all of the representations may have produced very similar performance results. However, all these results degrade quickly with one-hot encoding and integer encoding as we move down to the more specific ranks. The spectral k-mer representation remains very stable over the different values of k, where the accuracy slightly increases with the increase of k. However, the execution time increases exponentially with the increase of k.

Table 5 shows the only CNN experiment that utilizes 2-dimensional convolutional layers; the *CNN2-2D*. Hence, the only fitting representation is the 2D *FCGR* sequence representation that is also experimented at the three different values of $k = 4$, 5 and 6. The performance is slightly less than that of *CNN1-1D* over all the ranks and k-values, all still increasing slightly as k increases. The execution time also increases exponentially with the increase in k but at a much lower rate, where all the execution times were fairly less than that of CNN1-1D.

In Table 6, the performance measures for *CNN3-Embd* are presented. This CNN employs a Keras Embedding layer, built to accept only an integer encoding representation, as explained in Sect. 3.2. As is the general goal with any type of embedding, the Embedding layer creates a denser representation of the input integer-encoding. At all the taxonomic ranks the performance is fairly similar to the previous two CNNs (with their k-mer based representations), but heavily degrades at the last most fine-grained *Genus* rank.

Deeper and wider CNNs are experimented in Tables 7 and 8. Similar to the *CNN1-1D* experiments, both the one-hot and integer encoding perform fairly well for the

Table 4 Performance measures for all the experiments of *CNN1-1D* with each of the valid sequence representation methods

		Rank	Precision	Recall	F1-score	Accuracy	AUPRC	AUROC	Execution time
k	4	Phylum	1.0	1.0	1.0	1.0	1.0	1.0	**35.2**
		Class	0.997	0.999	0.998	0.999	0.999	0.999	
		Order	0.986	0.987	0.986	0.987	0.987	0.993	
		Family	0.977	0.981	0.977	0.981	0.981	0.99	
		Genus	0.877	0.901	0.885	0.901	0.901	0.95	
	5	Phylum	1.0	1.0	1.0	1.0	1.0	1.0	49.47
		Class	0.997	0.999	0.998	0.999	0.999	0.999	
		Order	0.99	0.991	0.99	0.991	0.991	0.995	
		Family	0.984	0.987	0.984	0.987	0.987	0.993	
		Genus	0.892	0.913	0.899	0.913	0.913	0.956	
	6	Phylum	1.0	1.0	1.0	1.0	1.0	1.0	137.517
		Class	0.997	0.999	0.998	0.999	0.999	0.999	
		Order	0.991	0.992	0.991	0.992	0.992	0.996	
		Family	0.985	0.987	0.985	0.987	0.987	0.993	
		Genus	0.9	0.916	0.904	0.916	0.916	0.958	
One-hot encoding		Phylum	0.963	0.963	0.963	0.963	0.969	0.972	183.93
		Class	0.957	0.959	0.958	0.959	0.963	0.976	
		Order	0.768	0.771	0.76	0.771	0.776	0.88	
		Family	0.617	0.618	0.599	0.618	0.621	0.806	
		Genus	0.434	0.46	0.43	0.46	0.461	0.729	
Integer-encoding		Phylum	0.978	0.978	0.978	0.978	0.981	0.983	71.8
		Class	0.968	0.97	0.969	0.97	0.973	0.982	
		Order	0.855	0.856	0.848	0.856	0.859	0.925	
		Family	0.706	0.704	0.69	0.704	0.707	0.85	
		Genus	0.511	0.527	0.501	0.527	0.528	0.763	

first two (most general) ranks and heavily degrade for the rest of the more specific ranks. Hence, the spectral k-mer representation still tops for both the *CNN4-Deep* and *CNN5-Wide*. In comparison to each other, the performances are very close, but the wider network *CNN5-Wide* slightly tops the deeper network *CNN4-Deep*. Compared to *CNN1-1D*, *CNN5-Wide* also slightly tops. The execution time in both *CNN4-Deep* and *CNN5-Wide* increases exponentially with the size of k and is extremely high with one-hot encoding and average with integer encoding. However, the execution time in *CNN5-Wide* always exceeds that of *CNN4-Deep*, due to the increase of the number of parameters in the network with the widening of the layers.

CNN6-WideDeep was proposed to observe the performance of a CNN that combines the advantages of both a wide and deep network. Shown in Table 9, the performance of the *CNN6-WideDeep* did not improve over using a wide network or a deep network alone. *CNN5-Wide* still had higher or the same accuracies for $k = 4$ and 5 on the most fine-grained task, and *CNN4-Deep* still performed better on $k = 6$ on the

Table 5 Performance measures for all the experiments of *CNN2-2D* with each of the valid sequence representation methods

k	Rank	Precision	Recall	F1-score	Accuracy	AUPRC	AUROC	Execution time
4	Phylum	0.999	0.999	0.999	0.999	0.999	0.999	25.8
	Class	0.997	0.998	0.997	0.998	0.998	0.999	
	Order	0.976	0.974	0.974	0.974	0.975	0.986	
	Family	0.935	0.931	0.928	0.931	0.932	0.965	
	Genus	0.724	0.741	0.721	0.741	0.741	0.87	
5	Phylum	0.999	0.999	0.999	0.999	0.999	0.999	37.6
	Class	0.995	0.997	0.996	0.997	0.997	0.998	
	Order	0.978	0.978	0.977	0.978	0.978	0.988	
	Family	0.958	0.956	0.954	0.956	0.956	0.978	
	Genus	0.826	0.84	0.824	0.84	0.84	0.92	
6	Phylum	0.999	0.999	0.999	0.999	0.999	0.999	94.9
	Class	0.996	0.998	0.997	0.998	0.998	0.999	
	Order	0.983	0.984	0.983	0.984	0.985	0.992	
	Family	0.974	0.975	0.973	0.975	0.976	0.987	
	Genus	0.864	0.88	0.867	0.88	0.88	0.94	

Table 6 Performance measures for all the experiments of *CNN3-Embd* with each of the valid sequence representation methods

	Rank	Precision	Recall	F1-score	Accuracy	AUPRC	AUROC	Execution time
Integer encoding	Phylum	0.996	0.996	0.996	0.996	0.997	0.997	147.95
	Class	0.991	0.993	0.992	0.993	0.994	0.996	
	Order	0.958	0.958	0.955	0.958	0.959	0.978	
	Family	0.917	0.921	0.912	0.921	0.921	0.96	
	Genus	0.721	0.76	0.728	0.76	0.76	0.88	

most fine grained task. It is valuable to note that the difference in performance is not highly significant. However, on the more raw representations—one-hot and integer encoding—the *CNN6-WideDeep* did slightly improve the performance than most of the other CNNs.

Figure 2 visualizes all the average execution times (over all the taxonomic ranks) for each CNN model and sequence representation method. *CNN3-Embd* does not use any k-mer based representation nor one-hot encoding, so it is not present in neither Figure 2a nor in the one-hot encoding columns in Figure 2b. Figure 2a shows that the k-mer representations on all the given CNN models increase exponentially in execution time as k-increases. One-hot encoding produces the highest execution

Table 7 Performance measures for all the experiments of *CNN4-Deep* with each of the valid sequence representation methods

		Rank	Precision	Recall	F1-score	Accuracy	AUPRC	AUROC	Execution time
k	4	Phylum	0.999	0.999	0.999	0.999	0.999	1.0	37.18
		Class	0.996	0.998	0.997	0.998	0.998	0.999	
		Order	0.984	0.985	0.984	0.985	0.986	0.992	
		Family	0.976	0.978	0.976	0.978	0.978	0.989	
		Genus	0.862	0.889	0.871	0.889	0.889	0.944	
	5	Phylum	1.0	1.0	1.0	1.0	1.0	1.0	77.1
		Class	0.997	0.999	0.998	0.999	0.999	0.999	
		Order	0.992	0.993	0.992	0.993	0.993	0.996	
		Family	0.981	0.983	0.98	0.983	0.983	0.991	
		Genus	0.885	0.909	0.897	0.909	0.909	0.954	
	6	Phylum	1.0	1.0	1.0	1.0	1.0	1.0	221.37
		Class	0.997	0.999	0.998	0.999	0.999	0.999	
		Order	0.991	0.993	0.992	0.993	0.993	0.996	
		Family	0.984	0.987	0.984	0.987	0.987	0.993	
		Genus	0.901	0.917	0.906	0.917	0.917	0.958	
One-hot encoding		Phylum	0.969	0.969	0.969	0.969	0.974	0.976	315.63
		Class	0.965	0.967	0.966	0.967	0.97	0.98	
		Order	0.792	0.796	0.785	0.796	0.8	0.893	
		Family	0.643	0.65	0.628	0.65	0.652	0.822	
		Genus	0.467	0.491	0.459	0.491	0.492	0.745	
Integer-encoding		Phylum	0.977	0.977	0.977	0.977	0.981	0.983	105.18
		Class	0.974	0.976	0.975	0.976	0.978	0.986	
		Order	0.863	0.864	0.856	0.864	0.867	0.929	
		Family	0.732	0.737	0.72	0.737	0.739	0.866	
		Genus	0.519	0.544	0.513	0.544	0.545	0.772	

times over all of the experiments, with no real gain in performance. Integer encoding produces reasonable execution times, but also with no real gain in performance.

Figure 3 presents the accuracies for each of the considered taxonomic ranks (phylum, class, order, family and genus) for each sequence representation method and CNN model combination. One-hot encoding representation provides the least accuracy with every CNN model. It is closely followed by integer encoding, with an exception of the usage of an Embedding layer in *CNN3-Embd*, with which it performs with higher accuracies than when using integer encoding alone. However, its accuracies are still less than that of the networks with the k-mer-based representations on the fine-grained tasks. The k-mer-based representations tops the other representations over all the experimented CNN models. All values of k (4, 5 and 6) give almost

Table 8 Performance measures for all the experiments of *CNN5-Wide* with each of the valid sequence representation methods

		Rank	Precision	Recall	F1-score	Accuracy	AUPRC	AUROC	Execution time
k	4	Phylum	1.0	1.0	1.0	1.0	1.0	1.0	39.56
		Class	0.997	0.998	0.998	0.998	0.998	0.999	
		Order	0.987	0.988	0.987	0.988	0.989	0.994	
		Family	0.982	0.983	0.981	0.983	0.983	0.992	
		Genus	0.887	0.906	0.893	0.906	0.906	0.953	
	5	Phylum	1.0	1.0	1.0	1.0	1.0	1.0	99.55
		Class	0.997	0.999	0.998	0.999	0.999	0.999	
		Order	0.99	0.991	0.99	0.991	0.992	0.995	
		Family	0.982	0.986	0.983	0.986	0.986	0.993	
		Genus	0.896	0.914	0.901	0.914	0.914	0.957	
	6	Phylum	1.0	1.0	1.0	1.0	1.0	1.0	320.91
		Class	0.997	0.999	0.998	0.999	0.999	0.999	
		Order	0.992	0.993	0.992	0.993	0.993	0.996	
		Family	0.985	0.987	0.985	0.987	0.987	0.993	
		Genus	0.896	0.914	0.902	0.914	0.914	0.957	
One-hot encoding		Phylum	0.962	0.962	0.962	0.962	0.968	0.971	472.76
		Class	0.957	0.958	0.957	0.958	0.962	0.975	
		Order	0.773	0.778	0.766	0.778	0.783	0.884	
		Family	0.62	0.621	0.6	0.621	0.624	0.807	
		Genus	0.445	0.466	0.436	0.466	0.467	0.732	
Integer-encoding		Phylum	0.977	0.977	0.977	0.977	0.981	0.983	139.95
		Class	0.97	0.973	0.971	0.973	0.975	0.984	
		Order	0.851	0.85	0.843	0.85	0.854	0.922	
		Family	0.708	0.709	0.692	0.709	0.711	0.852	
		Genus	0.516	0.53	0.503	0.53	0.531	0.765	

identical accuracies for the phylum, class and order ranks, which are ranks with a smaller number of categories. As we move on to family and genus ranks, the three values of k start to distance from each other in terms of accuracy, where the highest experimented value of $k = 6$ always produces the highest accuracy but also the highest execution time. Even though one-hot encoding is especially very common in recent works for the representation of DNA sequences for the general task of DNA sequence classification—it does not perform well with any of its valid CNNs, resulting in the least of performance measures and the highest execution times in comparison to all the performed experiments. The lower performance is probably due to the fact the one-hot encoding is a very sparse *raw* representation of a sequence, with no information on how sequence base patterns occur nor where the bases are positioned. The k-mer based representations are more stable and perform much better, having an execution time generally between that of one-hot encoding and integer encoding. These results conformed with the results obtained in a previous study [29]—where the best

Table 9 Performance measures for all the experiments of *CNN6-WideDeep* with each of the valid sequence representation methods

		Rank	Precision	Recall	F1-score	Accuracy	AUPRC	AUROC	Execution time
k	4	Phylum	1.0	1.0	1.0	1.0	1.0	1.0	78.56
		Class	0.996	0.998	0.997	0.998	0.998	0.999	
		Order	0.988	0.989	0.988	0.989	0.989	0.994	
		Family	0.976	0.979	0.976	0.979	0.979	0.989	
		Genus	0.883	0.905	0.889	0.905	0.905	0.952	
	5	Phylum	1.0	1.0	1.0	1.0	1.0	1.0	182.8
		Class	0.997	0.999	0.998	0.999	0.999	0.999	
		Order	0.99	0.992	0.991	0.992	0.992	0.996	
		Family	0.984	0.987	0.984	0.987	0.987	0.993	
		Genus	0.896	0.914	0.902	0.914	0.914	0.957	
	6	Phylum	1.0	1.0	1.0	1.0	1.0	1.0	589.9
		Class	0.998	0.999	0.998	0.999	0.999	0.999	
		Order	0.992	0.992	0.992	0.992	0.992	0.996	
		Family	0.984	0.987	0.985	0.987	0.987	0.993	
		Genus	0.899	0.915	0.904	0.915	0.915	0.958	
One-hot encoding		Phylum	0.977	0.977	0.977	0.977	0.98	0.983	923.5
		Class	0.969	0.972	0.97	0.972	0.974	0.983	
		Order	0.804	0.8135	0.805	0.814	0.818	0.903	
		Family	0.676	0.678	0.659	0.678	0.68	0.837	
		Genus	0.459	0.482	0.451	0.482	0.482	0.74	
Integer-encoding		Phylum	0.988	0.988	0.988	0.988	0.99	0.991	255.1
		Class	0.983	0.986	0.984	0.986	0.987	0.991	
		Order	0.873	0.873	0.867	0.873	0.875	0.933	
		Family	0.726	0.733	0.714	0.733	0.735	0.864	
		Genus	0.508	0.536	0.506	0.536	0.537	0.768	

accuracies were still obtained through the *k*-mer based representations. The values of *k* used are directly proportional to its accuracy and execution time, increasing as *k* increases. However, the improvement in accuracy is not high enough to overlook the exponential increase in execution time. Hence a value of *k* = 5 still provides a good balance in terms of performance and execution time in all experiments. The performance always decreases as the ranks become more specific—where a more *specific* rank means that it has more classification categories. This may be because the dataset size needs to be larger to capture more variations and features, especially in the more fine-grained ranks. The dataset is also imbalanced at the last three taxonomic ranks which may have caused biasing to certain categories during training. The Actinobacteria phylum at the last three ranks generally had the least number of categories, with Firmicutes following with a small increase in the number of categories and lastly Proteobacteria with the most number of categories—almost double or more of that in the previous two phyla.

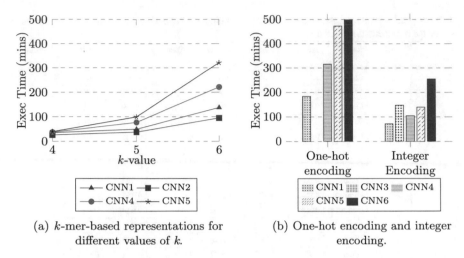

(a) k-mer-based representations for different values of k.

(b) One-hot encoding and integer encoding.

Fig. 2 The average execution time over all ranks (in minutes) for the different CNN architectures that accept the usage of the considered representations

It is hence deduced that using a sequence representation method that holds some sequential information about a sequence is preferred over the usage of a raw sequence representation. Employing 2-dimensional inputs and convolutional layers is individual in terms of accelerating the model and execution time reduction, but it did not excel in terms of performance. Wider networks are generally more effective in performance improvement, while deeper networks seem to need more information in order to provide a high performance. Given these observations, it might prove useful to experiment with wider and deeper CNNs with 2D convolutional layers and a more representative sequence representation.

5   Conclusions

This book chapter presents a set of experiments on various combinations of common sequence representation methods and CNN architectures for the task of DNA taxonomic classification on the bacterial 16S rRNA barcode gene dataset. Different types, sizes and numbers of layers were investigated for the CNN architectures. For the sequences representations; k-mer spectral representation, Frequency Chaos Game Representation (FCGR), one-hot encoding and integer encoding are studied. The experimental results have shown that representations that hold information about neighboring base sequences (i.e. k-mer-based representations) performed better than those considering raw representations. A value of $k = 5$ was generally found to be a fair balance between performance and execution time. On the same sequence representations, it has been shown that a wider CNN with more filters per layer performed

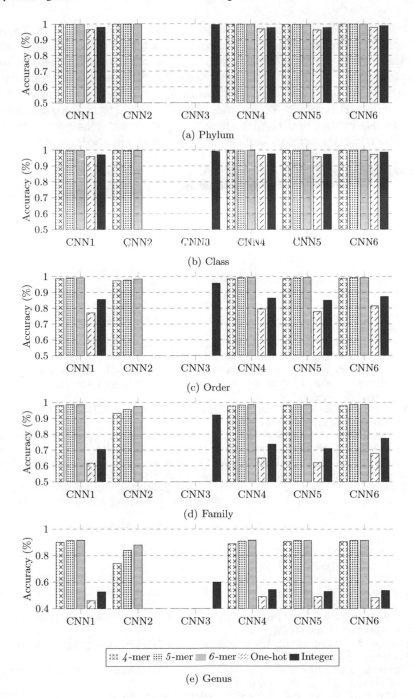

Fig. 3 The accuracy for each of the taxonomic ranks, considering all the considered CNNs and suited representations

the best, with a 91.4% accuracy. The second best CNN is a narrow and shallow networks with an accuracy of 91.3%. A deeper network slightly outperformed both of the previous networks by a 0.1–0.3% accuracy. This indicates that adding more layers to a network does not entirely mean increasing the classification accuracy, but possibly a larger dataset and more samples can contribute to accuracy improvement. Future study will target improving the accuracy of a CNN at the more fine-grained ranks and possibly increasing the samples by data augmentation techniques. Additional tests for the generalization capabilities of the networks will also be conducted.

Acknowledgements The authors would like to thank Massimo La Rosa for his assistance with the dataset organization on which we continued this study.

References

1. Brandenberg, O., et al.: Introduction to Molecular Biology and Genetic Engineering (2011)
2. Setubal, J., Meidanis, J.: Introduction To Computational Molecular Biology (1997)
3. Jalali, S.K., Ojha, R., Venkatesan, T.: DNA barcoding for identification of agriculturally important insects. In: New Horizons in Insect Science: Towards Sustainable Pest Management. Springer, pp. 13–23 (2015)
4. Paul, D.N., Hebert, S.R., deWaard, J.R.: Barcoding animal life: cytochrome c oxidase subunit 1 divergences among closely related species (2003)
5. National Center for Biotechnoloy Information (NCBI): MT-CO1 mitochondrially encoded cytochrome c oxidase I [Homo sapiens (human)]. Accessed on 25 Apr 2019. https://www.ncbi.nlm.nih.gov/gene/4512
6. Michigan State University Center for Microbial Ecology. Ribosomal Database Project (RDP). Accessed on 18 June 2019. https://rdp.cme.msu.edu/
7. National Human Genome Research Institute. The Human Genome Project (HGP). Accessed on 17 June 2019. https://www.genome.gov/human-genome-project
8. Genomics England. The 100,000 Genomes Project. Accessed on 17June 2019. https://www.genomicsengland.co.uk/about-genomics-england/the-100000-genomes-project/
9. Huerta, M., et al.: Nih Working Definition of Bioinformatics and Computational Biology (2000)
10. Libbrecht, M.W., Noble, W.S.: Machine learning applications in genetics and genomics. Nat. Rev. Genet. **16**(6), 321–332 (2015)
11. Min, S., Lee, B., Yoon, S.: Deep learning in bioinformatics. Briefings Bioinform. **18**(5), 851–869 (2017)
12. Reece, J.B., et al.: Biology: Concepts & Connections, 7th edn. Pearson Benjamin Cummings, San Francisco, California (2012)
13. LeCun, Y, Bengio, Y., Hinton, G.: Deep learning. In: Nature 521.7553, p. 436 (2015)
14. Najafabadi, M.M., et al.: Deep learning applications and challenges in big data analytics. J. Big Data **2**(1), 1 (2015)
15. Hubel, D.H., Wiesel, T.N.: Receptive fields of single neurones in the cat's striate cortex. J. Phys. **148**(3), 574–591 (1959)
16. Buduma, N., Locascio, N.: Fundamentals of Deep Learning: Designing Next-generation Machine Intelligence Algorithms. O'Reilly Media, Inc. (2017)
17. Fiannaca, A., et al.: Deep learning models for bacteria taxonomic classification of metagenomic data. BMC Bioinform. **19**(7), 198 (2018)
18. Kristensen, T., Guillaume, F.: Different regimes for classification of DNA sequences. In: IEEE 7th International Conference on Cybernetics 20 Marwah A. Helaly, Sherine Rady, and Mostafa M. Aref and Intelligent Systems and IEEE Conference on Robotics, Automation and Mechatronics, pp. 114–119. IEEE (2015)

19. Alhersh, T., et al.: Species identification using part of DNA sequence: evidence from machine learning algorithms. In: Proceedings of the 9th EAI International Conference on Bio-inspired Information and Communications Technologies, pp. 490–494. ICST (2016)
20. Rizzo, R., et al.: A deep learning approach to dna sequence classification. In: International Meeting on Computational Intelligence Methods for Bioinformatics and Biostatistics, pp. 129–140. Springer (2015)
21. Rizzo, R., et al.: Classification experiments of DNA sequences by using a deep neural network and chaos game representation, pp. 222–228 (2016)
22. Lo Bosco, G., Di Gangi, M.A. (2017) Deep learning architectures for DNA sequence classification, pp. 162–171 (2017)
23. Kassim, N.A., Abdullah, A.: Classification of DNA sequences using convolutional neural network approach. Innovations Comput. Technol. Appl. 2, (2017)
24. Nguyen, N.G., et al.: DNA sequence classification by convolutional neural network. J. Biomed. Sci. Eng. 9, 280–286 (2016)
25. Yin, B., et al.: An image representation based convolutional network for DNA classification. arXiv preprint arXiv:1806.04931 (2018)
26. Min, X., et al.: DeepEnhancer: predicting enhancers by convolutional neural networks. In: IEEE International Conference on Bioinformatics and Biomedicine, pp. 637–644. IEEE (2016)
27. Ghandi, M., et al.: Enhanced regulatory sequence prediction using gapped k-mer features. PLoS Comput. Biol. 10(7) (2014)
28. Keras: The python deep learning library. Keras Documentation—Embedding Layers. Accessed on 5 Sep 2019. https://keras.io/layers/embeddings/
29. Helaly, M.A., Rady, S., Aref, M.M.: Convolutional neural networks for biological sequence taxonomic classification: a comparitive study. In: Accepted for Publication in the International Conference on Advanced Intelligent Systems and Informatics (2019)

Particle Swarm Optimization and Grey Wolf Optimizer to Solve Continuous *p*-Median Location Problems

Hassan Mohamed Rabie

Abstract The continuous *p*-median location problem is to locate *p* facilities in the Euclidean plane in such a way that the sum of distances between each demand point and its nearest median/facility is minimized. In this chapter, the continuous *p*-median problem is studied, and a proposed Grey Wolf Optimizer (GWO) algorithm, which has not previously been applied to solve this problem, is presented and compared to a proposed Particle Swarm Optimization (PSO) algorithm. As an experimental evidence for the NFL theorem, the experimental results showed that the no algorithm can outperformed the other in all cases, however the proposed PSO has better performance in most of the cases. The experimental results show that the two proposed algorithms have better performance than other PSO methods in the literature.

Keywords *p*-median · Particle swarm · Grey wolf · Location problem · NFL theorem

1 Introduction

The center location problem answers the question of where to locate a facility or a service, and it is considered a critical element in strategic planning for a wide range in the public and private sectors [1]. The term "location problem" refers to the modeling, formulation, and solution of a class of problems that can best be described as locating facilities in some given spaces [2]. Center location problems most commonly also arise in emergency service location, where the concern for saving human life is far more important than any transportation costs that may be incurred in providing that service. Quick delivery of an emergency service is significantly more important in optimally placing, for example, ambulances and police patrol units, than the cost of delivering that service [3]. The common denominator in all these circumstances is that there is a time delay between the call for service and the actual time of

H. M. Rabie (✉)
Institute of National Planning, Cairo, Egypt
e-mail: Hassan.rabie@inp.edu.eg

© The Author(s), under exclusive license to Springer Nature Switzerland AG 2021   415
A. E. Hassanien et al. (eds.), *Machine Learning and Big Data Analytics*
Paradigms: Analysis, Applications and Challenges, Studies in Big Data 77,
https://doi.org/10.1007/978-3-030-59338-4_21

beginning to provide that service that is a direct consequence of the time spent during transportation. All other factors being constant, it makes sense to model such circumstances so that the maximum distance traversed during transportation is as small as possible [3].

Location problems are a major area of Operations Research and Management Science (OR&MS), called Location Science [4]. Facility location, location science and location models are terms that can be used instead. Center location problem is a branch of Operations Research related to locating or positioning at least a facility among a set of known demand points in order to optimize at least one objective function such as (cost, revenue, profit, service, travel distance, waiting time, coverage and market shares) [4].

Decision makers have to select locations according to their preferring criteria, actually, finding optimal facility locations is thus a difficult task [2]. Location theory is considered an important research area of environment engineering, regional and city planning, management and transportation science and other fields with a lot of applications. Many application areas including public and private facilities, business areas, and military environment can be seen in the related literature. From an application point of view there is no limitation for location science [4]. Center location models are an important topic in logistic management [5]. Locating facilities, in optimal locations and assigning the customers to them in the best possible manner, not only improves flow of material and services offered by the facility to customers, but also utilizes the facilities in an optimal manner, thereby reducing a need for multiple duplicating or redundant facilities [5]. Location of a facility and allocating customers to that facility determines the distribution pattern and associated characteristics, such as time, efficiency, and cost. The importance of location research results from several factors [6]:

1. Location decisions are frequently made at all levels of human organization from individuals and households to firms, government agencies and even international agencies.
2. Location decisions are often strategic in nature. That is, they involve large sums of capital resources and their economic effects are long term. In the private sector they have a major influence on the ability of a firm to compete in the market-place. While, for public sector, they have a significant influence at the efficiency by which governments provide public services and the ability to attract households and economic activities.
3. They frequently impose economic externalities. Such externalities include pollution, congestion, and economic development, among others.
4. Location models are often extremely difficult to solve, at least optimally, since most of the basic models are computationally intractable for large-scale problem. In fact, the computational complexity of the location models is a major reason that the widespread interest in formulating, developing and implementing such models did not occur until the spread of computers.

Center location problems could be defined as attempting to solve the following concern: given a space and a set of known n demand points, we need to determine a number of additional points (p) so as to optimize a function of the distance between new and existing demand points [3]. Therefore, the goal of the continuous p-median location problem is to locate p medians/facilities within a set of n demand points or customer locations with ($n > p$) in such a way that the sum of distances between each demand point and its nearest median/facility is minimized. The p-median location problems are intended to find the median points among the demand points, so that the sum of costs can be minimized through this target function [2]. These kinds of location problems include the establishment of the public services including schools, hospitals, firefighting, Ambulance, technical audit stations of cars, and etc. The objective function in the median problems is of the minisum type [2].

For the continuous p-median location problems; the demand points and medians/facilities are located in some subset of the d-dimensional real space R^d. Continuous location problems will have continuous variables associated with them, indicating the coordinates of the facilities that are to be located.

Location science is an important component of the strategy of private and public organizations, because the establishment of new facilities requires time and cost which should be managed due to the objectives of the organization [7]. The continuous p-median location problem is NP-hard problem [8], therefore, it is unlikely and difficult to obtain an optimal solution through polynomial time-bounded algorithms [9], and specially, when the size is relatively large.

Continuous location models assume that facilities/medians can be located anywhere in the plane. With a single facility to be located, this problem can be solved numerically very effectively. When two or more facilities ($p > 1$) are to be located, we need to simultaneously decide how to allocate demand to the facilities and where they should be. This problem is considerably more difficult and multiple optima are likely to exist [10].

The continuous p-median location problem, is a well-studied model in location theory, with the objective to generate optimal locations in a continuous plane, R^2, for a given number of demand points in order to minimize a sum of distances to a given set of customers at known point locations [11]. In many cases, the distance between demand and service points is Euclidean, but other distance functions, such as the Manhattan norm have also been employed [12].

Thus, we wish to locate p points/medians $X_j = (x_j, y_j), j = 1, ..., p$ in the plane R^2 in order to service a set of n demand points/customers at known locations (a_i, b_i), $i = 1, ..., n$. The *basic* version of continuous p-median location problem with Euclidean distance may be rewritten in the following equivalent form [12]:

$$\min_{X_j} \text{sum}_i \; w_i \min_j \left[(a_i - x_j)^2 + (b_i - y_j)^2 \right]^{1/2}, \quad j = 1, ..., p. \tag{1}$$

where

- n: the number of demand points (fixed points or customers).
- p: the number of medians/facilities to be located.
- (a_i, b_i): the location of demand point i ($i = 1, ..., n$).

- $w_i > 0$: the weight of demand point i $(i = 1, ..., n)$. in this chapter $w_i = 1$.

$X = (X_1, ..., X_p)$: the decision variables vector related to these p facility locations with $X_j = (x_j, y_j)$ representing the location of the new facility j with $X_j \in R^2$; $j = 1$, ..., p.

Location problem is the oldest topic of Operations Research since most location problems are interesting mathematical problems and were of interest to mathematicians for many years. According to [13]; Pierre de Fermat, Evangelistic Torricelli, and Battista Cavallieri; each one of them independently proposed the basic Euclidean median location problem in the seventeenth century. They approached the problem as a mathematical puzzle [13]. Their location model was locating a point on a Cartesian plane that minimizes the sum of distances to a predefined set of points. The basic location problem (finding a point that minimizes the weighted sum of distances from a given set of points) was posed by the famous French mathematician Pierre de Fermat in the early 1600s. Fermat posed the question of finding a point such that the sum of the distances to three given points is minimum. Sylvester posed the question of the smallest circle enclosing a set of points which is the one-center problem and proposed a solution approach [13].

Since Weber's book in 1909 on location theory [14], there have been many researchers, such as Hakimi [15, 16] and Cooper [17, 18], who studied location problems. Many extensive studies of location problems were developed after Cooper's research [14]. We should refer to survey papers in this area such as Melo et al. [19] and ReVelle et al. [20]. A survey on exact and heuristic algorithms can be found in Farahani [2], who reviewed different solution techniques for p-median problems and Mladenovic et al. provided a review on the p-median problem focusing on metaheuristic methods [21].

Based on Weber's book, the general location model can be extended into various location models. In the 1960s, five different problem definitions emerged in the location literature: the p-median problem [15–18], the p-center problem [16], the simple plant location problem, the plant layout problem, and the vertex cover problem [22]. The p-median location problem is one of the most widely applied models [14].

According to [23], the first well-known heuristic algorithm to solve p-median location problem was developed by Cooper [18], and according to [11] the early attempts to solve this problem by exact methods was through proposing a branch and bound algorithm proposed by Kuenne and Soland in [24] and Ostresh in [25], however, their methods were capable of solving very small instances. In [26], Rosing proposed improvements to the branch and bound algorithm but also still with small instances. p-median location problem has been widely addressed in the literature.

Brimberg et al. [11], presented a new local search approach for solving continuous p-median location problems. In [27], Drezner et al. proposed four heuristic procedures to solve p-median problem: two versions of variable neighborhood search, a genetic algorithm, and a combination of both. In [28], Drezner et al. proposed two new approaches to solve the p-median problem; the first is a variable neighborhood search and the second one is a concentric search. All these researches conducted an extensive empirical experiment on four well-known datasets. These datasets include the well-known the 50 customer problem, the 287 customer ambulance problem, and the 654-and the 1060-customer problems listed in the traveling salesman problems (TSP).

In some applications, p-median problems may involve a large number of demand points and potential medians/facilities. These problems arise, for example, in urban or regional areas where the demand points are individual private residences. In [29], the authors mentioned that it may be impossible and time consuming to solve large-scale location problems of demand points and to simplify the problem. They suggested reducing the number of demand points from n to m points so that the approximated problem can be solved within a reasonable amount of computing time; it is quite common to aggregate demand points when solving large-scale location problems.

Irawan et al. [30], proposed a multiphase approach that incorporates demand points aggregation, variable neighborhood search (VNS) and an exact method for the solution of large-scale p-median problems. The method consists of four phases. In the first phase several aggregated problems are solved to generate potential locations which are then used to solve a reduced problem in Phase 2 using VNS or an exact method. The new solution is then fed into an iterative learning process which tackles the aggregated problem; Phase 3. Phase 4 is a post optimization phase applied to the original (disaggregated) problem. The method is tested on different type of datasets such as TSP datasets.

Despite the advancement in the methods solving p-median problems, large scale problems found in the literature are still unsolvable by exact methods [11], and, aggregation methods may introduce errors in the data as well as the model output, thus resulting in less accurate results [11]. Optimal solutions may not be found for relatively large problems, and hence meta-heuristic algorithms are usually considered to be the best way for solving such problems [30]. Furthermore, the newer algorithms tend to be highly sensitive to the starting solution, and so, require state-of-the-art heuristics to obtain the best initial solution possible. Thus, advances in heuristic approaches are continually sought [11]. p-median location problems are nonconvex and nonlinear, and known algorithms cannot solve large scale problems optimally [14]. In addition, Cooper [17] proved that p-median objective function is neither concave nor convex and may contain several local minima. Hence, the multisource Weber problem falls in the realm of global optimization problems. The continuous location problem is the oldest topic of Operations Research, however, most of the methods developed for solving the continuous Euclidean problem are geometrical in nature, which involves complex, time consuming search methods for finding the smallest enclosing circle. And, few researches solved the large-scale continuous Euclidean absolute p-median location problems.

Because PSO has good performance in solving large scale continuous problems [14, 31], in this chapter, PSO algorithm is proposed to solve the continuous p-median location problem. However, and due to the No Free Lunch (NFL) theorem which logically proved that there is no meta-heuristic best suited for solving all optimization problems, i.e. a meta-heuristic algorithm may show promising results on a problem, but the same algorithm may show poor performance on another one [32]. Obviously, NFL makes this field of study highly active which results in enhancing and comparing current algorithms. This motivates us to compare the performance of PSO versus one of the recent meta-heuristic named GWO for solving p-median location problem.

The focus of this chapter is to compare solving the continuous p-median location models using two different meta-heuristic models. The p-median problem is categorized as NP-hard, however, there are many different algorithms and approaches for solving the p-median problem. In the following, some related works are mentioned.

The chapter is organized as follows. In section two, the proposed GWO algorithm and PSO algorithm are described. Section three presents the computational results of the two proposed algorithms, including a comparison with some PSO algorithms in the literature. The last section, the findings are summarized.

2   GWO and PSO for Solving p-median Location Problem

A swarm is a group of homogenous agents, which interact locally among themselves, and their environment, to allow for global behavior to emerge [33]. Recently, swarm-based algorithms have sparked as a family of intelligent approach, nature-inspired, and population-based algorithms that are able to solve complex optimization problems with low cost, fast, and robust solutions [33]. The swarm intelligence approach is a relatively new artificial approach that is used to model the collective behavior of social swarms in nature, such as bird flocks, ant colonies, and honey bees; Although these agents are relatively interacting in a simple way, this interaction create new patterns to cooperatively achieve tasks [33, 34].

The behaviors of bird flocks, ant colonies, honey bees, etc., were the field of study during earlier days [35]. Biologists and natural scientists have been studying the behavior of social insects and birds because of the efficiency of these natural swarm systems and later on 80s, computer scientists in the field of artificial life proposed the scientific insights of these natural swarm systems to the field of Artificial Intelligence [33]. Such collective motion of insects and birds is known to as "swarm behavior." Recently the interest of engineers is increasing rapidly since the resulting swarm intelligence is applicable in optimization problems in various fields like telecommunication systems, robotics, electrical power systems, consumer appliances, traffic patterns in transportation systems, military applications, and many more [35].

Swarm intelligence is the collective intelligence of groups of simple autonomous agents. The agent is a subsystem that interacts with its environment, which probably consists of other agents, but acts relatively independently from all other agents. There is no global plan or leader to control the entire group of agents [35]. In swarm intelligence, agents in the swarm or a social group are coordinating to achieve a specific goal by their behavior. This kind of collective intelligence arises from large groups of relatively simple agents. The actions of the agents are governed by simple local rules. The intelligent agent group achieves the goal through interactions of the entire group. A type of "self-organization" emerges from the collection of actions of the group [35]. For example, the movement of a bird in a flock, the bird adjusts its movements such that it coordinates with the movements of its neighboring flock mates. The bird tries to move along with its flock maintaining its movement along with the others and moves in such a way to avoid collisions among them. There is no leader to assign the movements therefore the birds try to coordinate and move among themselves. Any birds can fly in the front, center, and back of the swarm. Swarm behavior helps birds take advantage of several things including protection from predators, and searching for food [35].

Metaheuristic intelligent approaches are high level strategies for exploring search spaces by using different strategies like Simulated Annealing, Tabu Search, Genetic Algorithms and Evolutionary Algorithms, Ant Colony, Particle Swarm Optimization and recently Grey Wolf Optimizer. Metaheuristic techniques considered to be more efficient search approaches for solving hard problem [32]. Recent researches implemented metaheuristic techniques in order to overcome the complexity of problems.

Two meta-heuristic algorithms were proposed to solve the continuous *p*-median location problems; Particle Swam Optimization (PSO) and Grey Wolf Optimizer (GWO). Meta-heuristic algorithms have become remarkably common, that's mainly because [32]:

1. Simplicity: meta-heuristic algorithms are very simple. They have been inspired by very simple ideas; therefore, the developers can learn metaheuristics quickly and apply them to solve many real-life applications.
2. Flexibility which refers to the applicability of meta-heuristics algorithms to solve many type of problems without any changes in the structure of the algorithm.
3. Most of meta-heuristics algorithms are derivation-free mechanisms, which make them highly suitable for real problems with hard or unknown derivative information.
4. Meta-heuristics algorithms have the ability to avoid local optima compared to traditional techniques, because of the stochastic nature of meta-heuristics algorithms.

2.1 Grey Wolf Optimizer

Grey Wolf Optimizer (GWO) was introduced in 2014 by Mirjalili et al. [32] as a new meta-heuristic optimization algorithm; by 2019, GWO was found to be the most cited advances in engineering software articles [36]. GWO mimics the leadership hierarchy of wolves in the wild i.e. mimics the social behavior of wolves during the hunting process. This section briefly introduces the mathematical model and the inspiration behind the Grey Wolf Optimizer.

GWO algorithm is inspired from the leadership hierarchy and hunting strategy of grey wolves. Grey wolves are one of the most common predators, and they are on at the top of the food chain. They mostly prefer to live in a group ranging from 5 to 12 on average, with very strict social dominant hierarchy, As Fig. 1 shows [32]. Social

Fig. 1 Hierarchy of grey
wolf (dominance increases
from down top) [32]

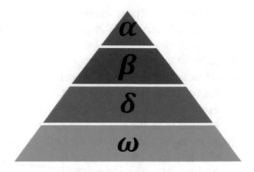

hierarchy considered the main feature of the wolves' pack. When hunting a prey, the pack can be categorized into four types:

1. The first type consists from the leader wolf, who called alpha (α). Alpha can be male or female, and it is responsible for making decisions about everything concerning the group/pack such as hunting, sleeping place, time to sleep, etc.… The alpha's decisions are dictated to the pack, and these orders must followed by the pack [32].
2. The second type named beta (β). Beta is at the second line of command. Beta is liable to deliver the orders and messages of alpha (α) wolf to the other wolves in the pack. Beta helps the alpha in decision-making and any other pack activities. The beta is probably the best candidate to be the next alpha in case of alpha died or becomes very old. The beta gives orders to the lower-level wolves. The beta shows the respect to the alpha and give him the advice. The beta reinforces the alpha's commands throughout the pack and gives feedback to the alpha [32].
3. Omega (ω) is on the bottom of the line of command. Omega acts as a scapegoat, they always have to submit to all the other dominant wolves and are the last wolves that are allowed to eat [32].
4. If a wolf is not an alpha, or beta, or omega, then the wolf called delta (δ). Deltas have to submit to alphas and betas, but they dominate the omega. This type includes the caretakers, hunters and sentinels are included. Caretaker wolves takes care of wounded wolves in the pack [32].

GWO Mathematical Model

The social hierarchy

By analyzing the social hierarchy of wolves and in order to mathematically model this social hierarchy of wolf pack, Mirjalili et al. [32] represented the hunting technique of the wolves such that; the best/fittest solution is considered as alpha (α), while the second best solution is considered as beta (β), and the third best solutions is called delta (δ). The other remaining candidate solutions are named omega (ω). In the GWO; alpha (α), Beta (β), and Delta (δ) are guided the searching process (hunting) and the set of (ω) are followers [37].

Encircling prey

Grey wolves encircling the prey during the hunting, therefore, the encircling strategy can be mathematically modelled by proposing the following equations [32]:

$$\vec{D} = \left| \vec{C} . \vec{X_p}(t) - \vec{X}(t) \right| \tag{2}$$

$$\vec{X}(t+1) = \vec{X_p}(t) - \vec{H} . \vec{D} \tag{3}$$

$$\vec{H} = 2\vec{h} . \vec{r_1} - \vec{h} \tag{4}$$

$$\vec{C} = 2 . \vec{r_2} \tag{5}$$

where t indicates the current iteration number, $\vec{H}$ and $\vec{C}$ are two coefficient vectors, $\vec{X_p}$ represents the position vector of the prey, and $\vec{X}$ represents the position vector of a wolf. r_1, r_2 are two uniformly random vectors in the range [0, 1], and $\vec{h}$ component is linearly decreased from 2 to 0 over the course of iterations, expressed as $a = 2 - 2.\left(\frac{t}{\text{max #iterations}}\right)$.

Hunting

After the wolves recognize the location of a prey and encircle them. The hunt is guided usually by alpha wolf (α). Beta (β) and Delta (δ) should participate in the hunting. However, during the hunting within the abstract search space, the location of the optimum (prey) is unknown. Therefore, to mathematically model the hunting behavior of grey wolves, we will assume that alpha (α), beta (β), and delta (δ) have the better knowledge about the potential location of prey (optimum solution). The alpha guided and lead the hunting with the participation of beta and delta. Hunting of the grey wolves can be mathematically modelled [32] as following.

$$\vec{D_\alpha} = \left| \vec{C_1}.\vec{X_\alpha} - \vec{X} \right|, \vec{D_\beta} = \left| \vec{C_2}.\vec{X_\beta} - \vec{X} \right|, \vec{D_\delta} = \left| \vec{C_3}.\vec{X_\delta} - \vec{X} \right| \tag{6}$$

$$\vec{X}_1 = \vec{X_\alpha} - \vec{H_1}.\vec{D_\alpha}, \vec{X}_2 = \vec{X_\beta} - \vec{H_2}.\vec{D_\beta}, \vec{X}_3 = \vec{X_\delta} - \vec{H_3}.\vec{D_\delta} \tag{7}$$

$$\vec{X}(t+1) = \frac{\vec{X}_1 + \vec{X}_2 + \vec{X}_3}{3} \tag{8}$$

where $X_\alpha, X_\beta, X_\delta$ are the positions of α, β and δ solutions (wolves) with the help of Eq. (2). Alpha, beta, and delta estimate the position of the prey (optimum solution), and other wolves updates their positions randomly around the prey as shown in Fig. 2.

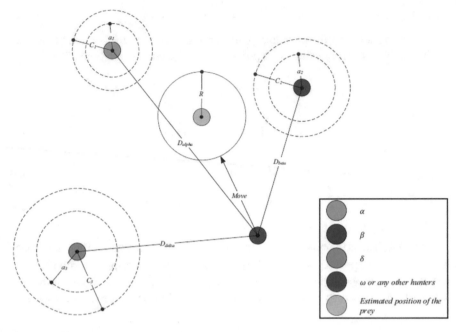

Fig. 2 Position updating in GWO [32]

Attacking prey

When the prey stops moving, the grey wolves finish the hunting; this can mathematically model by decreasing the value of $\vec{h}$.

Implementation of GWO on p-median location problem

For the continuous p-median location each candidate solution consists from p (x, y) pairs, which representing the sites of medians/facilities to be located, and p represents the number of medians/facilities. The candidate solution (wolf) can be represented as follows:

candidate solution: $[(x_1, y_1), (x_2, y_2), (x_3, y_3), \ldots, (x_j, y_j), \ldots, (x_p, y_p)]$,

where the coordinate (x_j, y_j) denotes the location of the jth facility, $j = 1, \ldots, p$.

The initial population is usually constructed by choosing the candidate solution randomly between the lower and upper bound/limit of the coordinates of the customer locations (demand points). The initial population can be represented as follows [23]:

$$\left[(x_{11}, y_{11}), \quad (x_{12}, y_{12}), \quad (x_{13}, y_{13}), \ldots, (x_{1j}, y_{1j}), \ldots, (x_{1p}, y_{1p})\right] \rightarrow f_1$$
$$\left[(x_{21}, y_{21}), \quad (x_{22}, y_{22}), \quad (x_{23}, y_{23}), \ldots, (x_{2j}, y_{1j}), \ldots, (x_{2p}, y_{2p})\right] \rightarrow f_2$$
$$\vdots$$
$$\left[(x_{k1}, y_{11}), \quad (x_{k2}, y_{12}), \quad (x_{k3}, y_{k3}), \ldots, (x_{kj}, y_{kj}), \ldots, (x_{kp}, y_{kp})\right] \rightarrow f_k$$
$$\vdots$$
$$\left[(x_{K1}, y_{K1}), \quad (x_{K2}, y_{K2}), \quad (x_{K3}, y_{K3}), \ldots, (x_{Kj}, y_{Kj}), \ldots, (x_{Kp}, y_{Kp})\right] \rightarrow f_K$$

where (x_{kj}, y_{kj}) are the (x, y) coordinates of the jth facility, $i = 1, ..., p$, of the kth candidate solution, $x_{kj} \in [x_{\min}, x_{\max}]$ and $y_{kj} \in [y_{\min}, y_{\max}]$, f_k is the fitness function (objective function value) of the kth candidate solution, and K is the population size.

In practices, customers location (demand points) are clustered, therefore, generating the median points randomly without any knowledge about the customer clusters may result in poor initial locations; since initial solutions could be located in large empty regions or far away from most of demand points (customers) [23]. To overcome this problem, k-means cluster algorithm [38] was used to find demand point (customer) clusters where $(k = p)$, and then the initial solutions generated within each cluster for each p. The using of k-means algorithm will produce, with high probability, initial solutions within demand point clusters. The proposed basic steps of the Grey Wolf algorithm to solve the continuous p-median location problem is given in Fig. 3.

2.2 Particle Swarm Optimization

PSO was developed by Kennedy and Eberhart in 1995 [39] as an evolutionary approach, and has become one of the most widely used swarm intelligence-based algorithms due to its simplicity and flexibility [14]. As shown in Fig. 4, bird flocking

---

Step 1. Read the set of demand points (A , B).
Step 2. Let p (number-of-medians), ss (population-size), *iterations* (number-of-iterations).
Step 3. Generate the candidate solutions $X_{ss \times p}$ (wolves) randomly by using k-Means.
Step 4. Evaluate the fitness/objective function, using Equation (1).
Step 5. Update the position of the candidate solutions (wolves).
  for z <= iterations,
− h=2- z *(2/*iterations*)
  for t <= ss;
− Identify; the fittest/best search wolf (X_α), the second best (X_β), and the third best (X_δ).
− Generate r_1 and r_2; two random numbers in range [0,1]
  Calculate H_1=2*h*r_1-h, **and** C_1=2*r_2 and Calculate $D_\alpha = (C_1*X_\alpha - X_t)$ **and** X_1= $(X_\alpha - H_1*D_\alpha)$
− Generate r_1 and r_2; two random numbers in range [0,1]
  Calculate H_2=2*h*r_1-h, **and** C_2=2*r_2 and Calculate $D_\beta = (C_2*X_\beta - X_t)$ **and** X_2= $(X_\beta - H_2*D_\beta)$
− Generate r_1 and r_2; two random numbers in range [0,1]
  Calculate H_3=2*h*r_1-h, **and** C_3=2*r_2 and Calculate $D_\delta = (C_3*X_\delta - X_t)$ **and** X_3= $(X_\delta - H_3*D_\delta)$
− X_{t+1}=$(X_1+X_2+X_3)$/3
− if $f(X_{t+1}) \leq f(X_\alpha)$; $f(X_\alpha)$=$f(X_{t+1})$; $X_\alpha = X_{t+1}$; endif
− if $f(X_{t+1}) > f(X_\alpha)$ **and** $f(X_{t+1}) < f(X_\beta)$; $f(X_\beta)$=$f(X_{t+1})$; $X_\beta = X_{t+1}$; endif
− if $f(X_{t+1}) > f(X_\alpha)$ **and** $f(X_{t+1}) > f(X_\beta)$ **and** $f(X_{t+1}) < f(X_\delta)$; $f(X_\delta)$=$f(X_{t+1})$; X_δ=X_{t+1}; endif
  end for
  end for
Step 6. Report X_α and $f(X_\alpha)$

Fig. 3 The basic steps of the proposed GWO for the p-median location problem

Fig. 4 The flocking birds [33]

can be defined as the social collective motion behavior of a large number of interacting birds with a common group objective, this bird flocking can be simulated
through simple three flocking rules [33]:

1. Flock centering: flock members attempt to stay close to nearby flock mates
   by flying in a direction that keeps them closer to the centroid of the nearby
   flockmates,
2. Collision avoidance: flock members avoid collisions with nearby flockmates
   based on their relative position, and
3. Velocity matching: flock members attempt to match velocity with nearby
   flockmates.

PSO was inspired from the social behavior of birds flocking. The algorithm
employs multiple particles that chase the position of the best particle and their own
best positions obtained so far. In other words, a particle is moved considering its
own best solution as well as the best solution the swarm has obtained [33]. PSO was
originally used to solve non-linear continuous optimization problems, and recently
has been used in many practical, real-life application problems [40].

PSO success in sharing the experience of each particle with part or the whole
swarm, leading the overall swarm motion towards the most promising areas detected
so far in the search space. Therefore, the moving particles, at each iteration, evaluate
their current position with respect to the problem's fitness function to be optimized,
and compare the current fitness of themselves to their historically best positions,

as well as to the other individuals of the swarm. Then, each particle updates that experience (if the current position is better than its historically best one), and adjusts its velocity to imitate the swarm's global best particle (or, its local superior neighbor, i.e., the one within its neighborhood whose current position represents a better solution than the particle's current one) by moving closer towards it. Before the end of each iteration of PSO, the index of the swarm's global best particle (or, the local best particle in the neighborhood) is updated if the most recent update of the position of any particle in the entire swarm (or, within a predetermined neighborhood topology) happened to be better than the current position of the swarm's global best particle (or, the local best particle in the neighborhood) [33].

PSO is one of the most common population-based search algorithms. The swarm consists from a number of particles that frequently move in the search space [14]. An objective function f must be defined to compare candidate solutions fitness within the search space. Each particle p in the population has two state variables.

1. Current position $\overrightarrow{x(t)}$
2. Current velocity $\overrightarrow{v(t)}$.

The PSO algorithm starts with initiating the particles of the swarm randomly. The particles are uniformly distributed in decision space. The velocity is also assigned randomly and uniformly. After the initialization, the iterative process optimization starts, such that, each particle's position and velocity is updated according to the following equations:

$$Vel_t^{z+1} = wt_z Vel_t^z + c_1 r_1 \left(Pbest_t^z - X_t^z \right) + c_2 r_2 \left(Gbest - X_t^z \right) \quad (9)$$

$$X_t^{z+1} = X_t^z + Vel_t^{z+1} \quad (10)$$

where:

- The inertia weight wt, determines the influence of the previous velocity $\overrightarrow{v(t)}$ and controls the particle's ability to explore the search space, and also affects the speed at which particles converge or de-converge [14]. The inertia weight provides a balance between the global and local search abilities. Therefore, a linearly decreasing inertia weight, wt, multiplied to that previous velocity term, as shown in Eq. (11). Intuitively, the linearly decreasing inertia weight is initially set to a high value wt_{max}, from 0.9 to 1.2 [33], inertia weight factor which is reduced dynamically to decrease the search area in a gradual fashion. The variable wt_z is updated as [41]:

$$wt_z = (wt_{max} - wt_{min}) * \frac{(k_{max} - z)}{k_{max}} + wt_{min}, \quad (11)$$

Step 1. Read the set of demand points (A , B).

Step 2. Define p (number-of-medians), ss (population-size), *iterations* (number-of-iterations), wt_{max} (the maximum value of inertia weight), wt_{min} (the minimum value of inertia weight), c_1 (cognitive parameter), and c_2 (social parameter).

Step 3. Generate the candidate solutions $X_{ss \times p}$ (wolves) randomly by using k-Means clustering algorithm. Set $Vel_{ss \times p} = 0$ to initiate the velocity of the swarm.

Step 4. Evaluate the objective function for $X_{ss \times p}$ using Equation (1)

Step 5. Let $Pbest = X$, with $f(Pbest) = f(X)$.

Step 6. Update the swarm fitness/position and velocity

        for $z <=$ iterations,

        for $t <= ss$,

-     Let $Gbest = Pbest$ (arg(min($f(Pbest)$)))
-     Generate r_1 and r_2; two random numbers in range [0,1]
-     Calculate inertia weight (wt_t) using Equation (11)
-     Calculate the velocity (Vel_t^{z+1}) using Equation (9)
-     Calculate the new position (X_t^{z+1}) using Equation (10)
-     Evaluate $f(X_t^{z+1})$ of the new position using Equation (1)
-     if $f(X_t^{z+1}) < f(Pbest_t^z)$; then $Pbest_t^z = X_t^{z+1}$; $f(Pbest_t^z) = f(X_t^{z+1})$; endif

        end for

        end for

Step 7.   Report $Gbest$ and $f(Gbest)$

Fig. 5 The basic steps of the proposed PSO for the p-median location problem

- Vel_t^z represents the velocity of particle t at iteration z.
- $Pbest_t^z$, the best previous position of particle t in iteration z.
- $Gbest$, the best previous among all particles.
- r_1 and r_2 are two random numbers between 0 and 1. c_1 and c_2 are two acceleration constants. These random numbers are set at each calculation of a velocity so that particles may vary the influence between different sources of information. These parameters determine whether a particle moves toward its previous best location *Pbest* or the global best location *Gbest*. They also called the acceleration constants or the learning factors, were initially used in the original version of PSO [35].
- X_t^z represents the position of particle t at iteration z.

Implementation of PSO on *p*-median location problem

    This section describes the implementation of PSO algorithm for solving continuous *p*-median location. As mentioned in the previous section, the candidate solutions are generated by using k-means. The basic steps of the proposed PSO for the continuous *p*-median location problem is presented in Fig. 5.

3   Computational Results

In this section, the computational results for the proposed GWO and PSO algorithms are reported. To investigate the performance of the two proposed algorithms to find good feasible solutions, we considered the following well-known three continuous

Table 1 Parameters setting for the two proposed algorithms

Parameters		PSO	GWO
Population size		100	100
Max. number of iterations		100	100
Inertia weight	wt_{min}	0.2	
	wt_{max}	1.2	
Learning coefficients	c_1	1.3	
	c_2	1.3	

p-median location problems: the 50-customer problem, the 654-customer problem and the 1060-customer problem, they are listed in the TSP library [42]. According to [43], the first problem has solved optimally by Brimberg et al. in [43], the second and the third have not been solved by deterministic methods, and only the best known solutions exist in the literature by Drezner et al. in [27, 43].

The two proposed algorithms coded in MATLAB on and compiled Windows 10 running on PC, Intel Core i7@2.2 GHz with 8 GB RAM. Due to the random nature of PSO and GWO; we conducted extensive experiments by performing 30 runs for each case (p), with total of 1920 experiments.

The two proposed algorithms have a similar execution condition; the population size is 100 and the maximum number of iterations is 100. The number of facilities p vary from 2 to 25 for the 50-customers problem and from 2 to 10 for the second, and from 5 to 150 for the 1060-customers.

The parameters of the PSO algorithm have been chosen after experimenting with several possible values by using trial and error approach. Thus, the combination that gave better results on average were listed in Table 1.

The numerical results of the two proposed algorithms experiments are summarized in Table 2 for the first problem, Table 3 for the second problem and Table 4 for the third problem. The first column of the tables shows the number of medians/facilities (p) to be located in each problem; the second column lists the optimal or the best-known solution. Other columns are expressed as a percent deviation from the optimal/best-known solutions; so that **Best** is the best of the objective function and **Avg.** is the average of the objective function found from 30 runs of the algorithm for each case. The deviation is computed as follows:

$$\text{deviation} = \frac{F_{best} - F^*}{F^*} * 100\%$$

where F_{best} is the value of the objective function found by the two proposed algorithms methods and F^* refers to the optimal or the best-known value of the objective function found in the literature [23]. Some general conclusions could be inferred from the numerical results listed in Tables 2, 3, and 4:

1. The proposed PSO algorithm found the optimal/best-known solution for all $p \leq$ 15 (the number of medians less than 15) for the three problems.

Table 2 Summary results over 30 runs for the 50-customer problem

P	Optimal	Proposed GWO		Proposed PSO	
		Best (%)	Avg (%)	Best (%)	Avg (%)
2	135.52	0.00	0.04	0.00	0.00
3	105.21	0.02	0.28	0.00	0.00
4	84.15	0.00	0.09	0.00	0.01
5	72.24	0.00	0.73	0.00	0.01
6	60.97	0.01	1.85	0.00	0.00
7	54.50	0.00	1.28	0.00	0.00
8	49.94	0.00	2.12	0.00	0.66
9	45.69	0.01	1.89	0.00	1.90
10	41.69	0.02	2.96	0.00	3.82
11	38.02	0.01	3.43	0.00	4.86
12	35.06	0.35	2.84	0.00	0.62
13	32.31	0.02	3.48	0.00	6.46
14	29.66	1.54	4.92	0.00	2.68
15	27.63	0.25	4.61	0.00	5.55
20	19.36	1.81	6.84	4.24	4.24
25	13.30	0.02	9.15	5.67	9.97
Average		0.25	2.91	0.62	2.55

Table 3 Summary results over 30 runs for the 654-customer problem

P	Best-known	Proposed GWO		Proposed PSO	
		Best (%)	Avg (%)	Best (%)	Avg (%)
2	815,313.30	0.00	0.01	0.00	0.00
3	551,062.88	0.00	0.13	0.00	0.20
4	288,190.99	0.00	0.09	0.00	0.00
5	209,068.79	0.00	0.23	0.00	0.00
6	180,488.21	0.00	1.61	0.00	0.50
7	163,704.17	0.00	0.89	0.00	0.08
8	147,050.79	0.00	0.99	0.00	0.73
9	130,936.12	0.00	1.66	0.00	0.43
10	115,339.03	0.06	3.52	0.00	0.73
Average		0.01	1.01	0.00	0.30

Table 4 Summary results over 30 runs for the 1060-customer problem

P	Best-known	Proposed GWO		Proposed PSO	
		Best (%)	Avg (%)	Best (%)	Avg (%)
5	1,851,879.90	0.00	0.02	0.00	0.02
10	1,249,564.80	0.02	0.21	0.00	0.24
15	980,132.10	0.06	1.04	0.00	0.68
20	828,802.00	0.30	1.24	0.01	1.36
50	453,164.00	1.78	3.29	1.04	2.22
75	340,242.00	2.55	3.61	1.42	2.51
150	212,926.00	4.03	4.30	2.48	3.34
Average		1.25	1.96	0.71	1.48

2. The particle swarm optimization algorithm can be effectively used to obtain good solutions. The results obtained by the proposed PSO are more effective than the proposed GWO algorithm.
3. The Comparison between the two proposed meta-heuristics algorithms suggests that no one algorithm is best in all cases. This could be considered as an experimental evidence for No Free Lunch (NFL) theorem.
4. By comparing the average results in Tables 2, 3, and 4, it can be seen that, the proposed PSO has better performance than the proposed GWO for the first and second problem. As for the average of the best results the proposed PSO outperform the proposed GWO.

The performance of the two proposed algorithms were compared to five different PSO and local search methods developed by Birto et al. [44]. These methods are explained in details in [44] and they named:

1. StPSO: Standard PSO,
2. NPSO: New proposed PSO,
3. StPSO + LS: Standard PSO with local search,
4. NPSO + LS: New PSO with local search, and
5. MS + LS: Multi-start algorithm with local Search.

These methods have been used to solve 654-customer problem. Table 5 shows the deviation of the five methods from the best-known solution versus the deviation of the two proposed algorithms. The numerical results of the experiments shows that the two proposed algorithms have better performance than the five methods proposed in [44].

Figure 6 shows the convergence performance of the two proposed algorithms for the value for $p = 5$ as an illustrative example, the figure contains three charts for the three problems. Each chart shows the convergence curve—Average of the runs—to the optimal/Best-known solution; where the horizontal axis represents the number of iterations and the vertical axis represents the objective function value. Figure 6 shows that the proposed PSO converges much faster than the proposed GWO for the

Table 5 PSO and local search methods Versus the two proposed methods

P	StPSO (%)	NPSO (%)	StPSO + LS (%)	NPSO + LS (%)	MS + LS (%)	GWO (%)	PSO (%)
2	0.09	0.04	0.00	0.00	0.00	0.00	0.00
3	3.79	1.18	0.00	0.00	0.00	0.00	0.00
4	31.91	20.19	0.00	0.00	0.00	0.00	0.00
5	50.53	36.39	0.01	0.00	0.01	0.00	0.00
6	69.19	57.50	0.07	0.04	0.03	0.00	0.00
7	64.89	43.11	0.24	0.15	0.18	0.00	0.00
8	63.04	48.43	0.85	0.40	0.74	0.00	0.00
9	79.86	81.38	0.60	0.83	1.31	0.00	0.00
10	104.09	99.24	2.09	1.48	1.25	0.06	0.00
Avg	51.93	43.05	0.43	0.32	0.39	0.01	0.00

first and second problem, however the proposed GWO converges faster in the third problem.

4 Conclusions

In this chapter, the continuous p-median location problem was studied. There are many algorithms have been proposed for solving this problem. However, Grey Wolf Optimizer (GWO) has not been implemented yet. A proposed GWO algorithm was presented and compared with the proposed PSO algorithm. As an experimental evidence for the NFL theorem; the experimental results showed that the no algorithm can outperformed the other in all cases, however the proposed PSO has better performance in most of the cases. Therefore, for future studies, it will be helpful to investigate hybrid algorithms to solve p-median location problem.

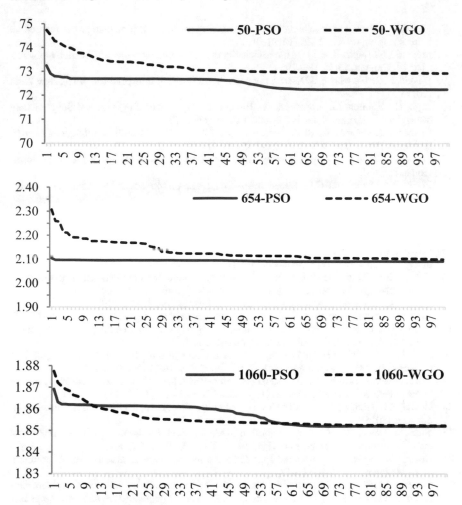

Fig. 6 The Proposed PSO&GWO Convergence curves for $p = 5$

References

1. Christofides, N.: Graph Theory, an Algorithmic Approach. Academic Press, New York (1975)
2. Farahani, R.Z., Hekmatfar, M.: Facility Location Concepts, Models Algorithms and Case Studies. Springer, Berlin Heidelberg (2009)
3. Eiselt, H.A., Marianov, V.: Foundations of Location Analysis (International Series in Operations Research & Management Science). Springer (2011).
4. Farahani, R.Z., SteadieSeifi, M., Asgari, N.: Multiple criteria facility location problems: a survey. Appl. Math. Model. **34**, 1689–1709 (2010)
5. Sule, D.R.: Logistics of Facility Location and Allocation. Marcel Dekker, Inc. (2001)
6. Drezner, Z., Hamacher, H.W.: Facility Location Applications and Theory. Springer, Berlin (2004)

7. Panteli, A., Boutsinas, B., Giannikos, I.: On solving the multiple p-median problem based on biclustering. Oper. Res. 1–25 (2019)
8. Megiddo, N., Supowit, K.J.: On the complexity of some common geometric location problems. SIAM J. Comput. 13(1), 182–196 (1984)
9. Daskin, M.S.: Network and Discrete Location: Models, Algorithms, 2nd edn Wiley, New York. (2013)
10. Zijm, H., Klumpp, M., Regattieri, A., Heragu, S.: Operations, Logistics and Supply Chain Management (Lecture Notes in Logistics). Springer (2019).
11. Brimberg, J., Drezner, Z., Mladenovic, N., Salhi, S.: A new local search for continuous location problems. Eur. J. Oper. Res. 232, 256–265 (2014)
12. Love, R.F., Morris, J.G., Wesolowsky, G.O.: Facilities Location: Models & Methods. North-Holland (1988).
13. Eiselt, H.A., Sandbiom, C.L.: Decision Analysis, Location Models, and Scheduling Problems. Springer, Berlin Heidelberg GmbH (2004)
14. Ghaderi, A., Jabalameli, M.S., Barzinpour, F., Rahmaniani, R.J.N., Economics, S.: An efficient hybrid particle swarm optimization algorithm for solving the uncapacitated continuous location-allocation problem 12(3), 421–439 (2012).
15. Hakimi, S.L.: Optimum locations of switching centers and the absolute centers and medians of a graph. Oper. Res. 12(3), 450–459 (1964)
16. Hakimi, S.L.: Optimum distribution of switching centers in a communication network and some related graph theoretic problems. Oper. Res. 13(3), 462–475 (1965)
17. Cooper, L.: Location-allocation problems. Oper. Res. 11(3), 331–343 (1963)
18. Cooper, L.: Heuristic methods for location-allocation problems. SIAM Rev. 6(1), 37–53 (1964)
19. Melo, M.T., Nickel, S., Saldanha-Da-Gama, F.: Facility location and supply chain management–a review. Eur. J. Oper. Res. 196(2), 401–412 (2009)
20. ReVelle, C.S., Eiselt, H.A., Daskin, M.S.: A bibliography for some fundamental problem categories in discrete location science. Eur. J. Oper. Res. 184, 817–848 (2008)
21. Mladenovic, N., Brimberg, J., Hansen, P., Moreno-Perez, J.A.: The p-median problem: a survey of metaheuristic approaches. Eur. J. Oper. Res. 179, 927–939 (2007)
22. Church R.L., Murray, A.: Location Covering Models: History, Applications and Advancements. Springer (2018)
23. Salhi, S., Gamal, M.: A genetic algorithm based approach for the uncapacitated continuous location–allocation problem. Ann. Oper. Res. 123(1–4), 203–222 (2003)
24. Kuenne, R., Soland, R.: Exact and approximate solutions to the multisource Weber problem. Math. Program. 3, 193–209 (1972)
25. Ostresh, J.L.M.: Multi-exact solutions to the M-center location-allocation problem. In: Rushton, G., Goodchild M.F., Ostresh L.M. Jr. (eds.) Computer Programs for Location-Allocation Problems. Monograph No. 6, University of Iowa, IA (1973).
26. Rosing, K.E.: An optimal method for solving the (generalized) multi-Weber problem. Eur. J. Oper. Res. 58, 414–426 (1992)
27. Drezner, Z., Brimberg, J., Mladenović, N., Salhi, S.: New heuristic algorithms for solving the planar p-median problem. Comput. Oper. Res. (2014)
28. Drezner, Z., Brimberg, J., Mladenović, N., Salhi, S.: Solving the planar p-median problem by variable neighborhood and concentric searches. (in English), J. Glob. Optim. (2014)
29. Francis, R.L., Lowe, T.J., Rayco, M.B., Tamir, A.: Aggregation error for location models: survey and analysis. Ann. Oper. Res. 167, 171–208 (2009).
30. Irawan, C.A., Salhi, S., Scaparra, M.P.: An adaptive multiphase approach for large unconditional and conditional p-median problems. Eur. J. Oper. Res. 237, 590–605 (2014)
31. Rabie, H.M., El-Khodary, I.A., Tharwat, A.A.: A particle swarm optimization algorithm for the continuous absolute p-center location problem with Euclidean distance. Int. J. Adv. Comput. Sci. Appl. (IJACSA) 4(12), 101–106 (2013)
32. Mirjalili, S., Mirjalili, S.M., Lewis, A.: Grey wolf optimizer. Adv. Eng. Softw. 69, 46–61 (2014)
33. Ahmed H., Glasgow, J.: Swarm intelligence: concepts, models and applications. In: Technical Report 2012–585 School of Computing, Queen's University, Kingston, Ontario, Canada 2012, vol. K7L3N6

34. Tharwat, A., Ella Hassanien, A., Elnaghi, B.E.: A BA-based algorithm for parameter optimization of support vector machine. Pattern Recognit. Lett. **93**, 13–22 (2017)
35. Sumathi, S., Paneerselvam, S.: Computational Intelligence Paradigms: Theory & Applications using MATLAB. Taylor and Francis Group LLC (2010)
36. Elsevier. (2019, 16-feb-2019). Most Cited Advances in Engineering Software Articles. Available: https://www.journals.elsevier.com/advances-in-engineering-software/most-cited-articles
37. Hassanien A.E., Emary, E.: Swarm Intelligence: Principles, Advances, and Applications. CRC Press (2018)
38. Martinez, W.L., Martinez, A.R., Solka, J.: Exploratory data analysis with MATLAB. Chapman and Hall/CRC (2017).
39. Kennedy, J., Eberhart, R.: Particle swarm optimization. In: Proceedings of IEEE International Conference on Neutral Networks, pp. 1942–1948 (1995).
40. Yang, X.-S.: Nature-Inspired Optimization Algorithms. Elsevier (2014)
41. Shelokar, P.S., Siarry, P., Jayaraman, V.K., Kulkarni, B.D.: Particle swarm and ant colony algorithms hybridized for improved continuous optimization. Appl. Math. Comput. **188**, 129–142 (2007)
42. Reinelt, G.: TSLIB—a traveling salesman library. ORSA J. Comput **3**, 376–384 (1991)
43. Brimberg, J., Hansen, P., Mladenovl, N., Taillard, E.D.: Improvements and comparison of heuristics for solving the uncapacitated multisource weber problem. Oper. Res. **48**(3), 444–460 (2000)
44. Brito, J., Martínez, F.J., Moreno, J.A.: Particle swarm optimization for the continuous p-median problem. In: 6th WSEAS International Conference on Computational Intelligence, Man-Machine Systems and Cybernetics, CIMMACS, pp. 14–16 (2007).

Gene Ontology Analysis of Gene Expression Data Using Hybridized PSO Triclustering

N. Narmadha and R. Rathipriya

Abstract The hybridized PSO Triclustering Model is the combination of Binary Particle Swarm Optimization and Simulated Annealing algorithm to extract highly correlated tricluster from the given 3D Gene Expression Dataset. The proposed hybrid Triclustering algorithms namely HPSO- TriC model generally produce higher quality results than standard meta-heuristic triclustering algorithms. Some of the issues in classical meta-heuristic triclustering models can be overcome in the HPSO-TriC model.

Keywords PSO · Gene ontology · Triclustering

1 Introduction

Each of the standard meta-heuristic algorithms such as Genetic Algorithm (GA), Ant Colony Optimization (ACO), and Particle Swarm Optimization (PSO) has its own pros and cons. In this chapter, PSO was taken into account and studied the performance of the hybrized version of PSO for Triclustering of 3D gene expression data. Generally, PSO aims at faster convergence but, it may lose an optimal solution because of premature convergent of global best (gbest) or stagnation of particles. To overcome some of these limitations in standard PSO based Triclustering models and to increase the convergence speed, a novel hybrid algorithms has been proposed in this chapter for Triclustering of 3D gene expression data.

N. Narmadha (✉) · R. Rathipriya
Department of Computer Science, Periyar University, Salem, India
e-mail: mahanarmadha@gmail.com

R. Rathipriya
e-mail: rathipriyar@gmail.com

© The Author(s), under exclusive license to Springer Nature Switzerland AG 2021
A. E. Hassanien et al. (eds.), *Machine Learning and Big Data Analytics*
Paradigms: Analysis, Applications and Challenges, Studies in Big Data 77,
https://doi.org/10.1007/978-3-030-59338-4_22

437

Hybridization is defined as the combination of two or more different things, aimed at achieving a particular objective or goal. The proposed hybrid triclustering algorithm called HPSO-TriC identifies higher quality tricluster than standard PSO triclustering algorithms by overcoming some of the issues in classical PSO triclustering models

This chapter is organized as follows. Section 2 describes about the literature review needed for the research work. Section 3 details explain the Binary PSO with the limitation. Novel algorithm, namely HPSO-TriC is detailed in Sect. 4. Section 5 provided the experimental analysis and discussed the obtained results. The summary of the proposed model is presented in Sect. 6.

2   Literature Review

This section provides an overview of the related works needed for this chapter. Simulated Annealing based biclustering is used to extract highly correlated user groups from the given preprocessed web usage data [1]. SA tried to discover more significant biclusters [2]. Biclustering Genetic based Simulated Annealing (Genetic SA) is used to predict the missing items in the gene expression data [3]. A hybrid PSO-SA BIClustering (PSO-SA-BIC) that combines features of binary PSO with Simulated Annealing to extract biclusters of gene expression data and ACV used to identify shifting pattern and scaling pattern biclusters [4].

Binary Particle Swarm Optimization is used to retrieve the global optimal bicluster from the web usage data [5]. Hybrid PSO triclustering of 3D gene expression data is used to find the effectively coherent pattern with a high volume of a tricluster [6]. Particle Swarm Optimization is easy to identify the user specified cluster [7].

K means with PSO is used to find the optimal point. The resultant optimal point is taken as a initial cluster for K means to find the final cluster [8]. Many algorithms like K-means, FCM and hierarchical techniques are used for the clustering of gene expression data. But PSO based K-means give good performance for the clustering of gene expression data [9]. Particle Swarm Optimization (PSO) is used to identify the global optimal solution [10].

This literature study clearly shows that Binary PSO, SA, PSO, or Hybrid PSO based approaches mainly used for clustering and biclustering of 3D gene expression data. Moreover, there are many studies they used Hybrid PSO over the evolutionary algorithm such as GA. Therefore, Hybridized PSO Triclustering (HPSO-TriC) Model is proposed in this chapter for 3D gene expression data to extract large volume tricluster with high coherent quality.

3 Binary Particle Swarm Optimization (BPSO)

3.1 Hybrid PSO (BPSO + SA) Triclustering Model

There are two phases in Hybrid PSO. They are:

1. **Binary Particle Swarm Optimization (BPSO)**

Algorithm 1 mentions the step by step procedure for the particle using Binary PSO. The output of this algorithm is treated as optimized tricluster.

Algorithm 1: Binary PSO Triclustering Algorithm

Input   : Swarm of size 's'
 Output : Optimized Triclusters
 Step 1. Initialize the position x_i and velocity v_i of each particle
         Repeat
 Step 2. For each particle
         Evaluate the fitness of each particle
         Update pbest using equation (3)
         Update gbest using equation (4)
          End (for)
 Step 3. For each particle
         Update velocity using equation (1)
         Update position using equation (2)
         Until (stopping condition)
         End (for)

2. **Simulated Annealing (SA)**

Algorithm 2 depicts the Standard Simulated Annealing Algorithm it aims to identify global optimal correlated tricluster [6].

```
Algorithm 2: Function SA( pbest, gbest)
```

If *pbest* position is not changed over a period of time

   i.   Find a new *pbest* position using temperature

   ii. accept the new position as p*best* position with probability

      exp(-($\Delta E/T$) even though current position is worse

      reduce T

   elseif *gbest* position is not changed over a period of time

   i.   Find a new *gbest* position using temperature

   ii. accept the new position as *gbest* position with probability

      exp(-(E/T) even though current position is worse

  e. reduce T

  f. Update *pbest* and *gbest*

3.2 *Particle Swarm Optimization*

Particle Swarm Optimization is an algorithm developed by Kennedy and Eberhart that simulates the social behaviours of bird flocking. In the basic PSO technique, each particle represents a candidate solution to the optimization problem. The best previous position for each particle (the position giving the best fitness value) is called particle best or pbest and the best position among all the particles in its neighbourhood is called global best or gbest. Each particle tries to modify its position using the following information:

- the current positions $x_i(t)$
- the current velocities $v_i(t)$
- the distance between the current position and pbest
- the distance between the current position and the gbest.

The rate of position change for each particle is called velocity of the particle. The velocity and position update are defined in Eq. 1

$$v_{id}(t+1) = w \times v_{id}(t) + c_1 \times r_1(pbest_{id} - x_{id}(t))c_2 \times r_2(gbest_{id} - x_{id}(t))$$

$$(1)$$

In Binary PSO, a particle flies in a search space restricted to zero and one. Therefore, the speed of the particle must be constrained to the interval [0, 1]. A logistic sigmoid transformation function $S(v_i(t + 1))$ is shown in Eq. (2) can be used to limit the speed of the particle.

$$S_{sig}(V_i(t + 1)) = \frac{1}{1 + e^{(V_i(t+1))}} \tag{2}$$

The new position of the particle is obtained using Eq. (3) shown below:

$$x_i = \begin{cases} 1, & \text{if } r_3 < S_{sig}(v_i(t + 1)) \\ 0 & \text{otherwise} \end{cases}, \tag{3}$$

where r_3 is a uniform random number in the range [0, 1], w is the inertia coefficient between [0, 1]; c_1, c_2 are the cognitive parameters; r_1, r_2 are random values in interval [0,1]; $V_i = [v_{i1}, v_{i2}, ..., v_{id}]$ is the velocity vector in which $v_{id}^{(t)}$ is the velocity of the ith particle in the dth dimension at iteration 't'. The personal best of each particle 'i' '$pbest_i$' and the global best '$gbest$' are updated using Eqs. (4) and (5).

$$pbest_i(t + 1) = \begin{cases} pbest_i(t) \ if \ f(x_i(t + 1) \geq f(pbest) \\ x_i(t + 1) \ if \ f(x_i(t + 1) < f(pbest) \end{cases} \tag{4}$$

$$gbest(t + 1) = \min(f(pbest_1, pbest_2, ..., pbest_k) \tag{5}$$

The aim of the triclustering problem is to maximize the Mean Correlation Value (MCV) of the tricluster. The velocities of the particles depend on the inertia coefficient 'w'. Increasing 'w' focus on global search and decreasing 'w' leads to local search. Hence, tuning of the parameter 'w' is an issue and it is problem dependent. The Binary PSO algorithm provides pbest solutions and gbest solution. The step by step procedure of triclustering algorithm using Binary PSO is provided Algorithm 1.

The following fitness function is used to extract the high volume tricluster subject to MCV threshold μ is shown in Eq. 6.

$$\max \ f(T) = |G'| * |S'| * |T'| \text{ subjected to } g\left(G' \cdot S', T'\right) \leq \mu \tag{6}$$

where $g(T) = (1 - \rho(B))$, $|G'|$, $|S'|$ and $|T'|$ are the number of genes, samples and timepionts in the tricluster respectively and μ is the correlation threshold which is defined in Eq. 7.

$$\text{MCV threshold } \mu = \frac{\sum_{i=1}^{N} MCV(T(i))}{N} \tag{7}$$

Otherwise, several experimentations are conducted on the appropriate dataset to choose specific constant value for threshold μ.

3.3 Limitations of PSO

Particle Swarm Optimization algorithms start with a group of randomly generated populations. It has a fitness function to evaluate the population. A very common problem in PSO is that some particles become stagnant over a few iterations which affect the global solution (i.e. gbest). In order to overcome the stagnation of particles, a hybrid algorithm namely HPSO-TriC that combine the features of Binary PSO with SA has been proposed in this chapter. By hybridizing Binary PSO and SA, it is possible to get a better tricluster solution than the solutions obtained from standard BPSO. Binary PSO yields faster convergence when compared to SA, because of the balance between exploration and exploitation in the search space. Moreover, the inertia weight in Binary PSO helps to balance between the global and local search abilities. The large inertia weight facilitates global search while the small inertia weight facilitates local search.

4 Hybrid PSO-TriC

In this section, the hybridized version of BPSO is proposed to overcome the above stated limitations. Hybridized Binary PSO combines BPSO and Simulated Annealing (SA) to identify global optimal correlated tricluster, which combines the advantages of both BPSO (that has strong global-search ability) and SA (that has strong local search ability). This hybrid approach makes full use of the exploration capability of both PSO and SA and offsets the weaknesses of each. Consequently, through the application of SA to Binary PSO, the proposed algorithm is capable of escaping from a local optimum.

4.1 Simulated Annealing (SA): An Overview

Simulated Annealing is a well-established stochastic technique originally developed to model the natural process of crystallization and later adopted to solve optimization problems [4]. SA is a variant of the local neighbourhood search. Traditional local search (e.g. steepest descent for minimization) always moves in a direction of improvement whereas SA allows non-improving moves to avoid getting stuck at a local optimum.

It has the ability to allow the probabilistic acceptance of changes which lead to worse solutions i.e. reversals in fitness. The probability of accepting a reversal is inversely proportional to the size of the reversal with the acceptance of smaller reversals being more probable. This probability also decreases as the search continues or as the system cools allowing eventual convergence on a solution. It is defined by Boltzman's Eq. 8:

$$P(\Delta E)\alpha e^{\frac{-\Delta E}{T}} \tag{8}$$

where ΔE is the difference in energy (fitness) between the old and new states and T is the temperature of the system.

In the virtual environment, the temperature of the system is lowered after a certain predefined number of accepted changes, successes, or total changes, attempts, depending on which is reached first. The rate at which temperature decreases depends on the cooling schedule. In the natural process, the system cools logarithmically, however, this is so time consuming that many simplified cooling schedules have been introduced for practical problem solving, the following simple cooling model is popular:

$$T(k) = \frac{T(k-1)}{(1 + \alpha_{cool})} \tag{9}$$

where $T(k)$ is the current temperature, $T(k-1)$ is the previous temperature, and α_{cool} indicates the cooling rate is shown in Eq. 9.

Each step of the SA algorithm replaces the current solution by a random nearby solution, chosen with a probability that depends on the difference between the corresponding function values and on a global parameter T called the temperature that is gradually decreased during the process.

The main steps of the Standard Simulated Annealing Algorithm are described in algorithm 3.

Algorithm 3:Standard Simulated Annealing (SA) Algorithm

Step 1. Initialize a very high "temperature" and particles.

Step 2. Perturb the placement through a defined move.

Step 3. Calculate the fitness of particles.

Step 4. Depending on the change in score, accept or reject the move. The probability of acceptance depending on the current "temperature" T.

Step 5. Update the temperature value by lowering the temperature. Go back to Step 2.

The proposed hybrid Triclustering model incorporates the SA in BPSO with two different strategies. They are:

i.   Only when the individual particles stagnate in their *pbest* position over a period of time, the number of iterations taken for stagnation checking is 5.
ii. Only when the *gbest* particle stagnates, the number of iterations taken for stagnation checking is 5.

Algorithm 4 shows the proposed Hybrid PSO-Tric model.

Algorithm 4: Hybrid PSO-Tric
Input: Swarm S, size of swarm n.
Output: pbest solution, gbest solution
Initialize $v_i = 0$, pbest$_i$ = particles P_i, gbest = best particle in S
Repeat
For each particle
Step 1. Evaluate the fitness of each particle
Step 2. Update pbest using equation (3)
Step 3. Update gbest using equation (4)
End for
For each particle
Step 4. Update velocity using equation (1)
Step 5. Update position using equation (2)
Step 6. Call SA() module
End for
Until maximum iteration reached.

5   Results and Discussion

This section shows the graphical representation for the Biological Significant like biological process, molecular function, and cellular component. Table 1 describes the parameters used in this chapter. Table 2 shows the 3D gene expression data set from Yeast Cell Cycle data.

Table 1 Description of parameters for BPSO versus HPSO

Parameters	Descriptions
No. of population (np)	24
Mean correlation of the population (MCV)	0–1
Inertia weight (w)	0.7–0.9
Learning factor (c1, c2)	2
Range of velocity [min, max] (v)	−20, 20
Population size (pop size)	24

Table 2 3D microarray dataset description

Dataset	Genes (G)	Sample (S)	Timepoint (T)
CDC15 experiment	8832	9	24
Elutriation	7744	9	14
Pheromone	7744	9	18

Table 3 Comparison of mean volume for BPSO versus HPSO

Dataset	Mean volume for BPSO	Mean volume for HPSO
CDC15	2160	860,319
Elutriation	2079	1260
Pheromone	1456	1248

Table 3 shows the comparison of mean volume for BPSO versus HPSO. From the results, it is evident that HPSO shows a high volume when compared to BPSO. These triclusters contain highly relevant genes, samples over a set of time points of a 3D microarray data. Figure 1 shows the mean volume for BPSO versus HPSO.

From the results, it is evident that HPSO-TriC extracts highly correlated triclusters called optimal tricluster are shown in Table 4. Figure 2 shows the graphical representation of mean MCV for BPSO versus HPSO it is clearly proved that HPSO extract high correlated tricluster.

Table 5 shows the performance of optimal tricluster for HPSO, which contains three different datasets with the number of genes, number of samples, number of time points, volume and optimal correlated tricluster.

Table 6 shows the GST Coverage of Optimal tricluster for HPSO. The table clearly shows the GST coverage here sample coverage is 100% for three different datasets. But the gene coverage (100, 82.98, and 98.03) and time point converge is differs (95.83, 71.42 and 72.22). The graphical representation of GST coverage is shown in Fig. 3.

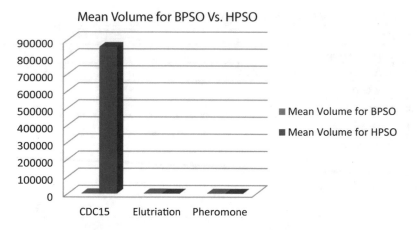

Fig. 1 Graphical representation of mean volume BPSO versus HPSO

Table 4 Comparison of mean MCV BPSO versus HPSO

Dataset	Mean MCV for BPSO	Mean MCV for HPSO
CDC15	0.9864	0.9864
Elutriation	0.9832	0.9700
Pheromone	0.9797	0.9800

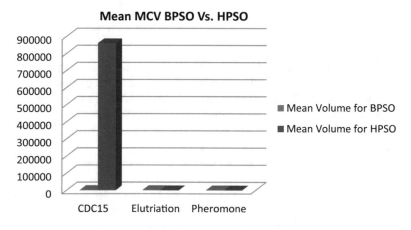

Fig. 2 Graphical representation of mean MCV for BPSO versus HPSO

Table 5 Performance of optimal tricluster for HPSO

Dataset	No. of genes	No. of samples	No. of timepoints	Volume	Optimal correlated tricluster
CDC15	24	9	18	860,319	0.9684
Elutriation	14	9	10	1260	0.9604
Pheromone	16	8	13	1248	0.9784

Table 6 GST coverage of optimal tricluster for HPSO

Dataset	Gene coverage %	Sample coverage %	Timepoint converge %
CDC15	100	100	95.83
Elutriation	82.98	100	71.42
Pheromone	98.03	100	72.22

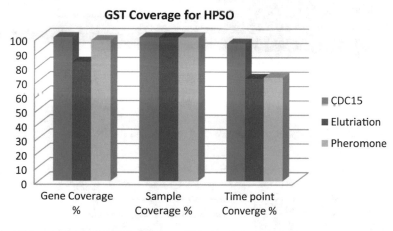

Fig. 3 GST coverage of optimal tricluster for HPSO

Table 7 shows the mean volume and mean MCV for three different datasets. The graphical representation of mean Volume and mean MCV is shown in Figs. 4 and 5. Similarly, Table 8 shows the mean Volume with worst case and best case, mean MCV with worst case and best case for three different dataset are clearly described.

It is proved that based on the result, HPSO gives high correlated tricluster with high volume is clearly shown in graphical representation Fig. 6 shows most of the genes correlated for CDC15 Dataset.

Table 7 Characteristics of initial population using HPSO

Dataset	Mean volume	Mean MCV
CDC15	860,319	0.9864
Elutriation	1260	0.9604
Pheromone	1248	0.9784

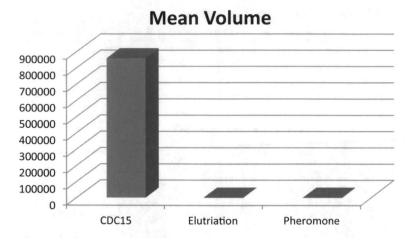

Fig. 4 Graphical representation of mean volume for HPSO Optimal tricluster

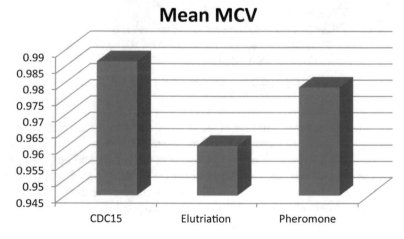

Fig. 5 Graphical representation of mean MCV for HPSO optimal tricluster

Table 8 Characteristics of initial population using HPSO

Dataset	Mean volume		Mean MCV	
	Worst	Best	Worst	Best
CDC15	540	860,319	0.4690	0.9864
Elutriation	297	1260	0.9315	0.9604
Pheromone	480	1248	0.8333	0.9784

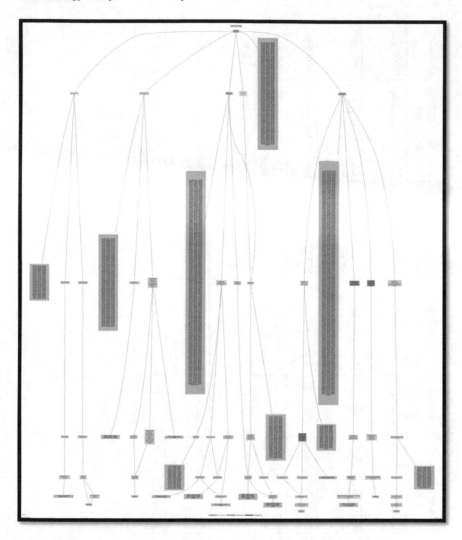

Fig. 6 Graphical representation of HPSO for CDC15 dataset

Tables 9, 10, 11 shows the Biological Significant of CDC15 dataset from the yeast cell cycle for Biological Process, Molecular Function and Cellular Component. This tabulation contains values of Gene Ontology (GO) term in the first column, Cluster Frequency in the second column, Genome frequency in the third column and corrected *P*-Value fourth column.

Similarly, Fig. 7 shows the graphical representation of HPSO for the Elutriation dataset.

Table 9 Biological significant of biological process for CDC15 dataset from yeast cell cycle

Gene ontology term	Cluster frequency with Percentage	Genome frequency with Percentage	Corrected P-value	Gene Ontology term	Cluster frequency with Percentage	Genome frequency with Percentage	Corrected P-value
Cellular component organization	1120 of 3333 genes, 33.6%	2087 of 7166 genes, 29.1%	1.57E−11	Regulation of macromolecule metabolic process	604 of 3333 genes, 18.1%	1122 of 7166 genes, 15.7%	0.00018
Biological regulation	1099 of 3333 genes, 33.0%	2055 of 7166 genes, 28.7%	1.42E−10	Regulation of primary metabolic process	582 of 3333 genes, 17.5%	1083 of 7166 genes, 15.1%	0.00048
Response to stimulus	705 of 3333 genes, 21.2%	1268 of 7166 genes, 17.7%	1.97E−09	Organic substance transport	513 of 3333 genes, 15.4%	945 of 7166 genes, 13.2%	0.00059
Regulation of biological process	914 of 3333 genes, 27.4%	1690 of 7166 genes, 23.6%	2.11E−09	Regulation of metabolic process	635 of 3333 genes, 19.1%	1192 of 7166 genes, 16.6%	0.00063
Cellular component organization or biogenesis	1277 of 3333 genes, 38.3%	2447 of 7166 genes, 34.1%	8.61E−09	Chromosome organization	346 of 3333 genes, 10.4%	614 of 7166 genes, 8.6%	0.00071
Regulation of cellular process	850 of 3333 genes, 25.5%	1579 of 7166 genes, 22.0%	8.81E−08	Regulation of cellular metabolic process	590 of 3333 genes, 17.7%	1105 of 7166 genes, 15.4%	0.0013
Organelle organization	804 of 3333 genes, 24.1%	1496 of 7166 genes, 20.9%	6.35E−07	Cellular process	2499 of 3333 genes, 75.0%	5173 of 7166 genes, 72.2%	0.0017

(continued)

Table 9 (continued)

Gene ontology term	Cluster frequency with Percentage	Genome frequency with Percentage	Corrected P-value	Gene Ontology term	Cluster frequency with Percentage	Genome frequency with Percentage	Corrected P-value
Response to chemical	332 of 3333 genes, 10.0%	567 of 7166 genes, 7.9%	4.88E−06	Response to drug	90 of 3333 genes, 2.7%	134 of 7166 genes, 1.9%	0.00319
Cellular response to stimulus	603 of 3333 genes, 18.1%	1103 of 7166 genes, 15.4%	7.82E−06	Carboxylic acid metabolic process	253 of 3333 genes, 7.6%	439 of 7166 genes, 6.1%	0.00326
Catabolic process	518 of 3333 genes, 15.5%	935 of 7166 genes, 13.0%	1.14E−05	Cellular response to chemical stimulus	225 of 3333 genes, 6.8%	386 of 7166 genes, 5.4%	0.00425
Cellular catabolic process	481 of 3333 genes, 14.4%	864 of 7166 genes, 12.1%	1.93E−05	Macromolecule localization	524 of 3333 genes, 15.7%	979 of 7166 genes, 13.7%	0.00468
Localization	832 of 3333 genes, 25.0%	1573 of 7166 genes, 22.0%	1.98E−05	Negative regulation of biological process	405 of 3333 genes, 12.2%	741 of 7166 genes, 10.3%	0.00582
Cell communication	282 of 3333 genes, 8.5%	477 of 7166 genes, 6.7%	2.58E−05	Signaling	216 of 3333 genes, 6.5%	370 of 7166 genes, 5.2%	0.00601
Establishment of localization	736 of 3333 genes, 22.1%	1382 of 7166 genes, 19.3%	4.73E−05	Oxoacid metabolic process	261 of 3333 genes, 7.8%	457 of 7166 genes, 6.4%	0.00602
Transport	712 of 3333 genes, 21.4%	1336 of 7166 genes, 18.6%	7.70E−05	Organic acid metabolic process	262 of 3333 genes, 7.9%	459 of 7166 genes, 6.4%	0.0061
Organic substance catabolic process	422 of 3333 genes, 12.7%	754 of 7166 genes, 10.5%	8.37E−05	Nitrogen compound transport	453 of 3333 genes, 13.6%	838 of 7166 genes, 11.7%	0.00678
Regulation of nitrogen compound metabolic process	569 of 3333 genes, 17.1%	1049 of 7,166 genes, 14.6%	0.00012	Signal transduction	213 of 3333 genes, 6.4%	365 of 7166 genes, 5.1%	0.00739

Table 10 Biological significant of molecular function for CDC15 dataset from yeast cell cycle

Gene ontology term	Cluster frequency with Percentage	Genome frequency with Percentage	Corrected Pvalue
Catalytic activity	1258 of 3333 genes, 37.7%	2434 of 7166 genes, 34.0%	2.05E−07
Ion binding	851 of 3333 genes, 25.5%	1603 of 7166 genes, 22.4%	1.44E−06
Nucleotide binding	478 of 3333 genes, 14.3%	879 of 7166 genes, 12.3%	0.00041
Nucleoside phosphate binding	478 of 3333 genes, 14.3%	879 of 7166 genes, 12.3%	0.00041
Small molecule binding	511 of 3333 genes, 15.3%	948 of 7166 genes, 13.2%	0.00067
Anion binding	513 of 3333 genes, 15.4%	966 of 7166 genes, 13.5%	0.00677

Tables 12, 13, 14 shows the Biological Significant of Elutriation dataset from the yeast cell cycle for Biological Process, Molecular Function and Cellular Component. This tabulation contains values of Gene Ontology (GO) term in the first column, Cluster Frequency in the second column, Genome frequency in the third column and corrected PValue fourth column. Figure 8 shows the graphical representation of HPSO for pheromone dataset.

Similarly, Tables 15, 16, 17 shows the Biological Significant of Phermone dataset from the yeast cell cycle for Biological Process, Molecular Function and Cellular Component.

6   Conclusion

The comparison of results of standard Binary PSO and hybrid PSO-TriC triclustering models have proven that the hybrid algorithm is effective in identifying global optimal solutions. Hence, hybrid algorithms that combine the features of both SA and Binary PSO algorithms are preferred. Thus, hybrid algorithms are able to identify tricluster with a high correlation degree among genes over samples and time points.

Owing to the complexity of these optimization problems, particularly those of large dataset sizes encountered in most practical settings, hybrid PSO-Tric often perform well. The global optimal tricluster extracted using the proposed model has more biologically significant than the standard Binary PSO.

Table 11 Biological significant of cellular component for CDC15 dataset from yeast cell cycle

Gene ontology term	Cluster frequency with Percentage	Genome frequency with Percentage	Corrected Pvalue	Gene ontology term	Cluster frequency with Percentage	Genome frequency with Percentage	Corrected Pvalue
Organelle	2409 of 3333 genes, 72.3%	4824 of 7166 genes, 67.3%	2.77E−14	Membrane part	847 of 3333 genes, 25.4%	1629 of 7166 genes, 22.7%	0.0002
Intracellular organelle	2408 of 3333 genes, 72.2%	4822 of 7166 genes, 67.3%	2.90E−14	Organelle part	1570 of 3333 genes, 47.1%	3151 of 7166 genes, 44.0%	0.00027
Membrane-bounded organelle	2255 of 3333 genes, 67.7%	4526 of 7166 genes, 63.2%	7.79E−11	Intracellular organelle part	1565 of 3333 genes, 47.0%	3142 of 7166 genes, 43.8%	0.00033
Intracellular membrane-bounded organelle	2225 of 3333 genes, 66.8%	4467 of 7166 genes, 62.3%	2.53E−10	Cell periphery	450 of 3333 genes, 13.5%	824 of 7166 genes, 11.5%	0.00035
Cell	2831 of 3333 genes, 84.9%	5834 of 7166 genes, 81.4%	3.08E−10	Protein-containing complex	1144 of 3333 genes, 34.3%	2256 of 7166 genes, 31.5%	0.00061
Cell part	2828 of 3333 genes, 84.8%	5829 of 7166 genes, 81.3%	4.50E−10	Cytoplasm	2287 of 3333 genes, 68.6%	4729 of 7166 genes, 66.0%	0.00524
Membrane	1077 of 3333 genes, 32.3%	2070 of 7166 genes, 28.9%	1.12E−06	Catalytic complex	405 of 3333 genes, 12.2%	751 of 7166 genes, 10.5%	0.0079
Intracellular	2734 of 3333 genes, 82.0%	5661 of 7166 genes, 79.0%	1.77E−06	Integral component of membrane	648 of 3333 genes, 19.4%	1247 of 7166 genes, 17.4%	0.00988

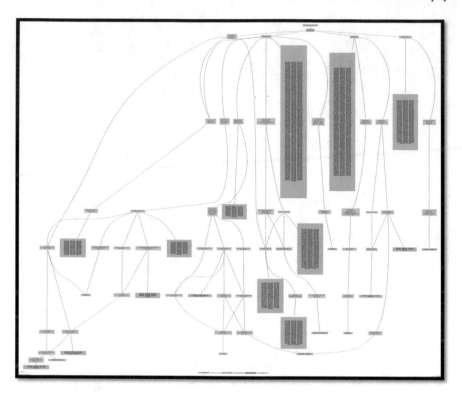

Fig. 7 Graphical representation of HPSO for elutriation dataset

Table 12 Biological significant of biological process for elutriation dataset from yeast cell cycle

Gene ontology term	Cluster frequency with percentage	Genome frequency with percentage	Corrected P value	Gene ontology term	Cluster frequency with percentage	Genome frequency with percentage	Corrected P value
Biological regulation	1091 of 3308 genes, 33.0%	2055 of 7166 genes, 28.7%	1.96E−10	Reproductive process	278 of 3308 genes, 8.4%	486 of 7166 genes, 6.8%	0.001
Regulation of biological process	903 of 3308 genes, 27.3%	1690 of 7166 genes, 23.6%	1.58E−08	Positive regulation of biological process	433 of 3308 genes, 13.1%	794 of 7166 genes, 11.1%	0.00116
Regulation of cellular process	845 of 3308 genes, 25.5%	1579 of 7166 genes, 22.0%	7.22E−08	Cell cycle	429 of 3308 genes, 13.0%	786 of 7166 genes, 11.0%	0.00117
Cellular component organization	1079 of 3308 genes, 32.6%	2087 of 7166 genes, 29.1%	3.53E−06	Regulation of RNA metabolic process	377 of 3308 genes, 11.4%	683 of 7166 genes, 9.5%	0.00143
Regulation of macromolecule metabolic process	602 of 3308 genes, 18.2%	1122 of 7166 genes, 15.7%	9.36E−05	Regulation of nucleic acid-templated transcription	357 of 3308 genes, 10.8%	644 of 7166 genes, 9.0%	0.0017
Regulation of metabolic process	634 of 3308 genes, 19.2%	1192 of 7166 genes, 16.6%	0.00021	Regulation of RNA biosynthetic process	357 of 3308 genes, 10.8%	644 of 7166 genes, 9.0%	0.0017
Regulation of nucleobase-containing compound metabolic process	417 of 3308 genes, 12.6%	756 of 7166 genes, 10.5%	0.00035	Regulation of transcription, DNA-templated	354 of 3308 genes, 10.7%	639 of 7166 genes, 8.9%	0.00208
Cellular response to chemical stimulus	228 of 3,308 genes, 6.9%	386 of 7,166 genes, 5.4%	0.00041	Localization	810 of 3308 genes, 24.5%	1573 of 7166 genes, 22.0%	0.00329
Reproduction	288 of 3308 genes, 8.7%	502 of 7166 genes, 7.0%	0.00041	Regulation of macromolecule biosynthetic process	443 of 3308 genes, 13.4%	821 of 7166 genes, 11.5%	0.00421

(continued)

Table 12 (continued)

Gene ontology term	Cluster frequency with percentage	Genome frequency with percentage	Corrected Pvalue	Gene ontology term	Cluster frequency with percentage	Genome frequency with percentage	Corrected Pvalue
Regulation of primary metabolic process	578 of 3308 genes, 17.5%	1083 of 7166 genes, 15.1%	0.00052	Regulation of cellular macromolecule biosynthetic process	437 of 3308 genes, 13.2%	809 of 7166 genes, 11.3%	0.00429
Regulation of nitrogen compound metabolic process	561 of 3308 genes, 17.0%	1049 of 7166 genes, 14.6%	0.00058	Positive regulation of cellular process	421 of 3308 genes, 12.7%	777 of 7166 genes, 10.8%	0.0045
Regulation of gene expression	471 of 3308 genes, 14.2%	867 of 7166 genes, 12.1%	0.0006	Regulation of biosynthetic process	458 of 3308 genes, 13.8%	852 of 7166 genes, 11.9%	0.00475
Regulation of cellular metabolic process	587 of 3308 genes, 17.7%	1105 of 7166 genes, 15.4%	0.00098	Regulation of cellular biosynthetic process	456 of 3308 genes, 13.8%	850 of 7166 genes, 11.9%	0.00681

Table 13 Biological significant of molecular function for elutriation dataset from yeast cell cycle

Gene ontology term	Cluster frequency with percentage	Genome frequency with percentage	Corrected Pvalue
Catalytic activity	1250 of 3308 genes, 37.8%	2434 of 7166 genes, 34.0%	1.72E−07
Ion binding	836 of 3308 genes, 25.3%	1603 of 7166 genes, 22.4%	3.26E−05
Coenzyme binding	94 of 3308 genes, 2.8%	144 of 7166 genes, 2.0%	0.0027

Table 14 Biological significant of cellular component for elutriation dataset from yeast cell cycle

Gene ontology term	Cluster frequency with percentage	Genome frequency with percentage	Corrected Pvalue
Intracellular organelle	2359 of 3308 genes, 71.3%	4822 of 7166 genes, 67.3%	7.89E−09
Membrane-bounded organelle	2226 of 3308 genes, 67.3%	4526 of 7166 genes, 63.2%	8.34E−09
Organelle	2359 of 3308 genes, 71.3%	4824 of 7166 genes, 67.3%	1.07E−08
Intracellular membrane-bounded organelle	2190 of 3308 genes, 66.2%	4467 of 7166 genes, 62.3%	1.76E−07
Endomembrane system	574 of 3308 genes, 17.4%	1091 of 7166 genes, 15.2%	0.00169
Nucleus	1221 of 3308 genes, 36.9%	2458 of 7166 genes, 34.3%	0.00751

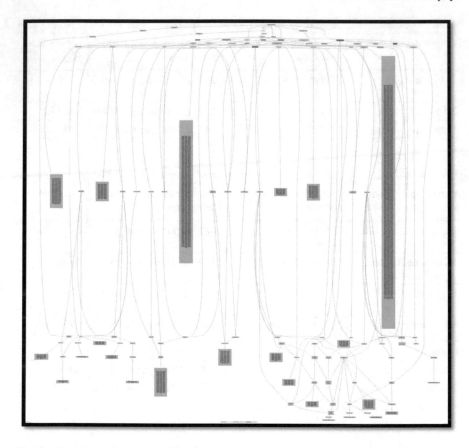

Fig. 8 Graphical representation of optimal tricluster for HPSO for pheromone dataset

Table 15 Biological significant of biological process for pheromone dataset from yeast cell cycle

Gene ontology term	Cluster frequency with percentage	Genome frequency with percentage	Corrected Pvalue	Gene Ontology term	Cluster frequency with percentage	Genome frequency with percentage	Corrected Pvalue
Cellular component organization	1122 of 3291 genes, 34.1%	2087 of 7166 genes, 29.1%	3.41E−14	Negative regulation of biosynthetic process	218 of 3291 genes, 6.6%	370 of 7166 genes, 5.2%	0.00063
Biological regulation	1101 of 3291 genes, 33.5%	2055 of 7166 genes, 28.7%	4.03E−13	Negative regulation of cellular biosynthetic process	218 of 3291 genes, 6.6%	370 of 7166 genes, 5.2%	0.00063
Organelle organization	817 of 3291 genes, 24.8%	1496 of 7166 genes, 20.9%	8.37E−11	Response to stress	434 of 3291 genes, 13.2%	797 of 7166 genes, 11.1%	0.00067
Regulation of biological process	911 of 3291 genes, 27.7%	1690 of 7166 genes, 23.6%	1.19E−10	Regulation of biosynthetic process	461 of 3291 genes, 14.0%	852 of 7166 genes, 11.9%	0.00074
Regulation of cellular process	853 of 3291 genes, 25.9%	1579 of 7166 genes, 22.0%	6.27E−10	Negative regulation of cellular macromolecule biosynthetic process	204 of 3291 genes, 6.2%	344 of 7166 genes, 4.8%	0.00082
Cellular component organization or biogenesis	1260 of 3291 genes, 38.3%	2447 of 7166 genes, 34.1%	2.14E−08	cell cycle	428 of 3291 genes, 13.0%	786 of 7166 genes, 11.0%	0.00084
Negative regulation of cellular process	365 of 3291 genes, 11.1%	634 of 7166 genes, 8.8%	1.75E−06	negative regulation of nitrogen compound metabolic process	251 of 3291 genes, 7.6%	435 of 7166 genes, 6.1%	0.00088
Cellular protein modification process	477 of 3291 genes, 14.5%	857 of 7166 genes, 12.0%	2.62E−06	regulation of macromolecule biosynthetic process	445 of 3291 genes, 13.5%	821 of 7166 genes, 11.5%	0.00096

(continued)

Table 15 (continued)

Gene ontology term	Cluster frequency with percentage	Genome frequency with percentage	Corrected P value	Gene Ontology term	Cluster frequency with percentage	Genome frequency with percentage	Corrected P value
Protein modification process	477 of 3291 genes, 14.5%	857 of 7166 genes, 12.0%	2.62E−06	regulation of cellular macromolecule biosynthetic process	439 of 3291 genes, 13.3%	809 of 7166 genes, 11.3%	0.00098
Cellular response to stimulus	599 of 3291 genes, 18.2%	1103 of 7166 genes, 15.4%	2.93E-06	organic cyclic compound biosynthetic process	601 of 3291 genes, 18.3%	1140 of 7166 genes, 15.9%	0.00113
Response to stimulus	679 of 3291 genes, 20.6%	1268 of 7166 genes, 17.7%	4.38E−06	Regulation of gene expression	467 of 3291 genes, 14.2%	867 of 7166 genes, 12.1%	0.00127
Chromosome organization	352 of 3291 genes, 10.7%	614 of 7166 genes, 8.6%	7.38E−06	Regulation of nucleic acid-templated transcription	356 of 3291 genes, 10.8%	644 of 7166 genes, 9.0%	0.00137
Negative regulation of biological process	416 of 3291 genes, 12.6%	741 of 7166 genes, 10.3%	9.09E−06	Regulation of RNA biosynthetic process	356 of 3291 genes, 10.8%	644 of 7166 genes, 9.0%	0.00137
Cellular process	2483 of 3291 genes, 75.4%	5173 of 7166 genes, 72.2%	2.64E−05	Regulation of transcription, DNA-templated	353 of 3291 genes, 10.7%	639 of 7166 genes, 8.9%	0.00169
Regulation of macromolecule metabolic process	600 of 3291 genes, 18.2%	1122 of 7166 genes, 15.7%	7.38E−05	Negative regulation of cellular metabolic process	264 of 3291 genes, 8.0%	463 of 7166 genes, 6.5%	0.00173
Regulation of cellular metabolic process	591 of 3291 genes, 18.0%	1105 of 7166 genes, 15.4%	9.46E−05	Cellular response to stress	405 of 3291 genes, 12.3%	745 of 7166 genes, 10.4%	0.00238

(continued)

Table 15 (continued)

Gene ontology term	Cluster frequency with percentage	Genome frequency with percentage	Corrected Pvalue
Nucleic acid-templated transcription	407 of 3291 genes, 12.4%	734 of 7166 genes, 10.2%	0.0001
Regulation of metabolic process	633 of 3291 genes, 19.2%	1192 of 7166 genes, 16.6%	0.00011
Transcription, DNA-templated	405 of 3291 genes, 12.3%	731 of 7166 genes, 10.2%	0.00013
Regulation of primary metabolic process	579 of 3291 genes, 17.6%	1083 of 7166 genes, 15.1%	0.00014
RNA biosynthetic process	408 of 3291 genes, 12.4%	738 of 7166 genes, 10.3%	0.00016
Regulation of cellular component organization	262 of 3291 genes, 8.0%	451 of 7166 genes, 6.3%	0.0002
Regulation of nitrogen compound metabolic process	561 of 3291 genes, 17.0%	1049 of 7166 genes, 14.6%	0.00024
Negative regulation of macromolecule biosynthetic process	208 of 3291 genes, 6.3%	348 of 7166 genes, 4.9%	0.00026

Gene Ontology term	Cluster frequency with percentage	Genome frequency with percentage	Corrected Pvalue
Aromatic compound biosynthetic process	565 of 3291 genes, 17.?%	1081 of 7166 genes, 15.1%	0.00337
Positive regulation of catalytic activity	127 of 3291 genes, 3.9%	203 of 7166 genes, 2.8%	0.00354
Heterocycle biosynthetic process	575 of 3291 genes, 17.?%	1094 of 7166 genes, 15.3%	0.00375
Regulation of RNA metabolic process	373 of 3291 genes, 11.?%	683 of 7166 genes, 9.5%	0.0038
Positive regulation of biological process	428 of 3291 genes, 13.0%	794 of 7166 genes, 11.1%	0.00384
Negative regulation of RNA biosynthetic process	174 of 3291 genes, 5.?%	292 of 7166 genes, 4.1%	0.00423
Negative regulation of nucleic acid-templated transcription	174 of 3291 genes, 5.?%	292 of 7166 genes, 4.1%	0.00423
Transcription by RNA polymerase II	301 of 3291 genes, 9.?%	540 of 7166 genes, 7.5%	0.00449

(continued)

Table 15 (continued)

Gene ontology term	Cluster frequency with percentage	Genome frequency with percentage	Corrected P-value	Gene Ontology term	Cluster frequency with percentage	Genome frequency with percentage	Corrected P-value
Negative regulation of nucleobase-containing compound metabolic process	201 of 3291 genes, 6.1%	335 of 7166 genes, 4.7%	0.00029	Chromatin organization	202 of 3291 genes, 6.1%	346 of 7166 genes, 4.8%	0.00455
Regulation of nucleobase-containing compound metabolic process	415 of 3291 genes, 12.6%	756 of 7166 genes, 10.5%	0.00038	Phosphate-containing compound metabolic process	447 of 3291 genes, 13.6%	837 of 7166 genes, 11.7%	0.00846
Regulation of cellular biosynthetic process	461 of 3291 genes, 14.0%	850 of 7166 genes, 11.9%	0.0005	Organic cyclic compound metabolic process	1091 of 3291 genes, 33.2%	2182 of 7166 genes, 30.4%	0.00953

Table 16 Biological significant of molecular function for pheromone dataset from yeast cell cycle

Gene ontology term	Cluster frequency with percentage	Genome frequency with percentage	Corrected Pvalue
Catalytic activity	1263 of 3291 genes, 38.4%	2434 of 7166 genes, 34.0%	2.65E−10
Ion binding	852 of 3291 genes, 25.9%	1603 of 7166 genes, 22.4%	3.30E−08
Protein binding	509 of 3291 genes, 15.5%	929 of 7166 genes, 13.0%	4.68E−06
Enzyme activator activity	90 of 3291 genes, 2.7%	136 of 7166 genes, 1.9%	0.00143
Metal ion binding	443 of 3291 genes, 13.5%	830 of 7166 genes, 11.6%	0.00334
Cation binding	447 of 3291 genes, 13.6%	841 of 7166 genes, 11.7%	0.00541
Sequence-specific DNA binding	157 of 3291 genes, 4.8%	266 of 7166 genes, 3.7%	0.00987

Table 17 Biological significant of cellular component for pheromone dataset from yeast cell cycle

Gene ontology term	Cluster frequency with percentage	Genome frequency with percentage	Corrected Pvalue	Gene ontology term	Cluster frequency with percentage	Genome frequency with Percentage	Corrected Pvalue
Intracellular organelle	2431 of 3291 genes, 73.9%	4822 of 7166 genes, 67.3%	2.27E−25	Chromosomal part	272 of 3291 genes, 8.3%	468 of 7166 genes, 6.5%	2.43E−05
Organelle	2431 of 3291 genes, 73.9%	4824 of 7166 genes, 67.3%	3.71E−25	Nuclear chromosome	213 of 3291 genes, 6.5%	360 of 7166 genes, 5.0%	0.00012
Membrane-bounded organelle	2282 of 3291 genes, 69.3%	4526 of 7166 genes, 63.2%	5.89E−21	Site of polarized growth	168 of 3291 genes, 5.1%	277 of 7166 genes, 3.9%	0.0003
Intracellular membrane-bounded organelle	2253 of 3291 genes, 68.5%	4467 of 7166 genes, 62.3%	2.41E−20	Nucleoplasm	170 of 3291 genes, 5.2%	282 of 7166 genes, 3.9%	0.00045
Cell	2796 of 3291 genes, 85.0%	5834 of 7166 genes, 81.4%	4.28E−10	Nuclear chromosome part	198 of 3291 genes, 6.0%	336 of 7166 genes, 4.7%	0.00053
Cell part	2793 of 3291 genes, 84.9%	5829 of 7166 genes, 81.3%	6.33E−10	Nucleoplasm part	149 of 3291 genes, 4.5%	243 of 7166 genes, 3.4%	0.00055
Nucleus	1268 of 3291 genes, 38.5%	2458 of 7166 genes, 34.3%	1.84E−09	Non-membrane-bounded organelle	759 of 3291 genes, 23.1%	1478 of 7166 genes, 20.6%	0.00125
Intracellular	2715 of 3291 genes, 82.5%	5661 of 7166 genes, 79.0%	8.23E−09	Intracellular non-membrane-bounded organelle	759 of 3291 genes, 23.1%	1478 of 7166 genes, 20.6%	0.00125
Intracellular part	2715 of 3291 genes, 82.5%	5661 of 7166 genes, 79.0%	8.23E−09	Protein-containing complex	1126 of 3291 genes, 34.2%	2256 of 7166 genes, 31.5%	0.00207
Organelle part	1575 of 3291 genes, 47.9%	3151 of 7166 genes, 44.0%	4.77E−07	Cellular bud neck	124 of 3291 genes, 3.8%	200 of 7166 genes, 2.8%	0.00214

(continued)

Table 17 (continued)

Gene ontology term	Cluster frequency with percentage	Genome frequency with percentage	Corrected Pvalue	Gene ontology term	Cluster frequency with percentage	Genome frequency with Percentage	Corrected Pvalue
Intracellular organelle part	1570 of 3291 genes, 47.7%	3142 of 7166 genes, 43.8%	6.10E−07	Transferase complex	229 of 3291 genes, 7.0%	402 of 7166 genes, 5.6%	0.00261
Chromosome	292 of 3291 genes, 8.9%	499 of 7166 genes, 7.0%	2.74E−06	Endomembrane system	569 of 3291 genes, 17.3%	1091 of 7166 genes, 15.2%	0.0036
Cellular bud	162 of 3291 genes, 4.9%	255 of 7166 genes, 3.6%	5.35E−06	Membrane	1035 of 3291 genes, 31.4%	2070 of 7166 genes, 28.9%	0.00479

References

1. Rathipriya, R., Thangavel, K.: Extraction of web usage profiles using simulated annealing based biclustering approach. J. Appl. Inf. Sci. **2**(1), 21–29 (2014)
2. Kenneth Bryan, P.: Application of simulated annealing to the biclustering of gene expression data. IEEE Trans. Inf. Technol. Biomed. **10**, 519–525 (2006)
3. Ramkumar, M., Nanthakumar, G.: Biclustering gene expression data using genetic simulated annealing algorithm. Int. J. Eng. Adv. Technol. (IJEAT) **8**(6), 3717–3720 (2019)
4. Thangavel, K., Bagyamani, J., Rathipriya, R.: Novel hybrid PSO-SA model for biclustering of expression data. In: International Conference on Communication Technology and System Design, vol. 30, pp. 1048–1055. Elsevier (2011)
5. Rathipriya, R., Thangavel, K., Bagyamani, J.: Binary particle swarm optimization based biclustering of web usage data. Int. J. Comput. Appl. **25**(2), 43–49 (2011)
6. Narmadha, N., Rathipriya, R.: Gene ontology analysis of 3D microarray gene expression data using hybrid PSO optimization. Int. J. Innov. Technol. Explor. Eng. (IJITEE) **8**(11), 3890–3896 (2019)
7. Blas, N.G., Tolic, O.L.: Clustering using particle swarm optimization. Int. J. Inf. Theor. Appl. **23**(1), 24–33 (2016)
8. Chouhan, R., Purohit, A.: An approach for document clustering using PSO and K-means algorithm. In: International Conference on Inventive Systems and Control (ICISC), pp. 1380–1384, June 2018
9. LopamudraDey, A.M.: Microarray gene expression data clustering using PSO based K-means algorithm. Int. J. Comput. Sci. Appl. **1**(1), 232–236 (2014)
10. Das, S., Idicula, S.M.: Greedy search-binary PSO hybrid for biclustering gene expression data. Int. J. Comput. Appl. **2**(3), 1–5 (2010)

Modeling, Simulation, Security with Big Data

Experimental Studies of Variations Reduction in Chemometric Model Transfer for FT-NIR Miniaturized Sensors

Mohamed Hossam, Amr Wassal, Mostafa Medhat, and M. Watheq El-Kharashi

Abstract Recent technology trends to miniaturize spectrometers have opened the doors for mass production of spectrometers and for new applications that were not possible before and where the spectrometer can possibly be used as a ubiquitous spectral sensor. However, with the miniaturization from large reliable bench-top to chip-size miniaturized spectrometers and with the associated mass production, new issues have to be addressed such as spectrometers unit-to-unit variations, variations due to changing the measurement setup and variations due to changing the measurement medium. The unit-to-unit variations of the sensors usually result from changing mode of operation, aging, and production tolerances. The aim of this work is to study the issues emerging from the use of miniaturized Fourier Transform Near-Infrared (FT-NIR) spectral sensors and evaluate the influence of these issues on the multivariate classification model used in many applications. In this work, we also introduce a technique to transfer a classification model from a reference calibration sensor to other target sensors to help reducing the effect of the variations and to alleviate the degradation that occurs in the classification results. To validate the effectiveness of the model transfer technique, we developed a Gaussian Process Classification (GPC) model and Soft Independent Modeling Class Analogy (SIMCA) model both using spectral data measured from ultra-high temperature (UHT) pasteurized milk with different levels of fat content. The models aim to classify milk samples according to the percentage of their fat content. Three different experiments were conducted on the models to mimic each type of variations and to test how far they affect the mod-

M. Hossam · M. W. El-Kharashi
Computer and Systems Engineering Department, Ain Shams University, Cairo 11517, Egypt
e-mail: mohamed.hossam@eng.asu.edu.eg; mohamed.hossam@si-ware.com

M. W. El-Kharashi
e-mail: watheq.elkharashi@eng.asu.edu.eg

A. Wassal (✉)
Computer Engineering Department, Cairo University, Cairo 12613, Egypt
e-mail: wassal@eng.cu.edu.eg

M. Hossam · A. Wassal · M. Medhat
Si-Ware Systems, Heliopolis, Cairo 11361, Egypt
e-mail: mostafa.medhat@si-ware.com

© The Author(s), under exclusive license to Springer Nature Switzerland AG 2021 469
A. E. Hassanien et al. (eds.), *Machine Learning and Big Data Analytics Paradigms: Analysis, Applications and Challenges*, Studies in Big Data 77,
https://doi.org/10.1007/978-3-030-59338-4_23

els' accuracy once the transfer technique is applied. Initially, we achieved perfect discrimination between milk classes with 100% classification accuracy. The largest retardation in accuracy appeared while changing the measuring medium reaching 45.4% in one of the cases. However, the proposed calibration transfer technique showed a significant enhancement in most of the cases and standardized the accuracy of all retarded cases to get the accuracy back to over 90%.

Keywords FT-NIR · Partial least squares · Gaussian process classification · Soft independent modeling class analogies · Milk · Model transfer · Unit-to-unit variation

1 Introduction

Spectroscopy is the discipline that focuses on studying the interactions between materials and electromagnetic radiation, and according to these interactions a lot of characteristics and properties of materials are obtained. These interactions are represented in spectra where the condition of each wavelength can be determined, and for a certain chemical compound its spectrum is considered as a unique signature. The motivation for spectroscopy in general came from that it offers a robust fast technique to get a lot of information on the material under test by just interpreting its spectrum. Similar goals are usually achieved using sensitive slow chemical methods and in most cases these methods destroy the sample itself to extract the required information. Thus, it normal to see the adoption of spectroscopic methods in factories and big firms to ensure the quality of their production lines and also its existence in security checks to search for vicious materials. The spectra of materials are measured using instruments called spectrometers and in some cases they are called spectrophotometers if they measure only the visible range of the spectrum (380–750 nm).

In this study, we focus on using near infrared (NIR) spectra to obtain information about the substance we have under test. The main advantage of using NIR is it contains of most of the organic materials' traces and features. In addition to that, NIR spectra are very easy to measure without the need of sample preparation such as preparing impractically extremely thin samples like those required to measure mid infrared (MIR) spectra. However, due to the short wavelength of NIR range, this range is crowded with features and traces of different materials which leads to the overlapping between features and ends up to be very difficult to interpret NIR spectrum without the help of pre-trained models to infer the spectra. These models are also known as chemometric models.

The main goal of this study is to build reliable chemometric models that be used in a real application and develop techniques that can enable us to depend on cheap miniaturized spectrometers without compromising on the accuracy of the models. The miniaturized spectrometers which also known as spectral sensors are the future of the spectroscopy field as they will enable spectrometers to be integrated in complex versatile systems which will open the doors for completely new scenarios of using spectrometers. However, moving from an expensive large bench top instrument to mass produced miniaturized sensors will have a negative impact on the measurements accuracy and will create new challenges on building reliable models. Hence, we present a feasibly study on using calibration transfer technique to overcome variations that may emerge due to the usage of spectral sensors.

2 Theoretical Background

In this section, we will review the main concepts required for building a multivariate model and cover mathematical background of most techniques and algorithms used in this case study. The classification flow starts with data acquisition stage or spectra acquisition in our case followed by preprocessing operations to treat the data before modeling, then the core of this flow is dimension reduction stage and classifier training. After getting the first potential model, a validation stage follows it to approve the model quality or to repeat the modeling again starting from the preprocessing stage with changing some of the modeling parameters. Finally by getting the best performing model according to the validation stage, a final testing stage is applied on the model to judge on the model generality. The classification flow is summarized graphically in Fig. 1.

Now, we got a classification model to use it we have to follow a similar prediction flow which is summarized in Fig.2. The prediction flow is a simplified version of the classification (calibration) flow, it starts with data acquisition then use the same

Classification

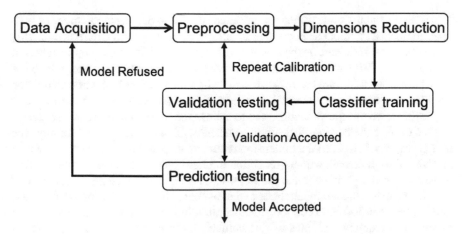

Fig. 1 The calibration flow to build a multivariate classification model

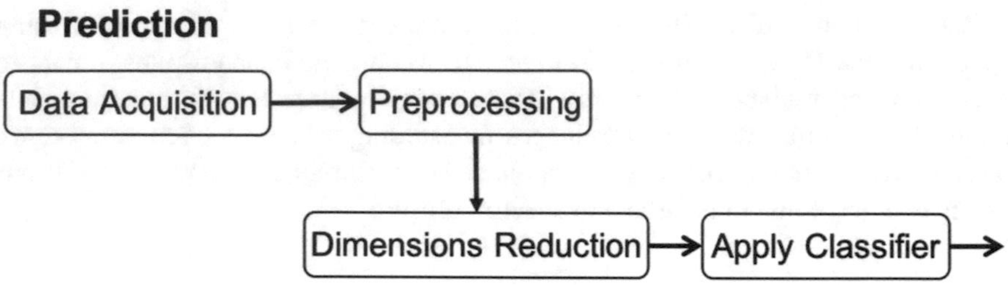

Prediction

Fig. 2 The prediction flow to use the built classification model

preprocessing configured in the calibration. After that, it followed by the dimension reduction and apply the classifier using the same parameters tuned in the calibration.

2.1 Preprocessing

Preprocessing are those techniques and algorithms that used for treating the data before modeling as mentioned previously. The purpose of these treatments can differ to various targets, like alleviating the light scattering effect on the collected spectra, reducing the noise in spectra, or removing the baseline shifts. There are dozens of techniques at can achieve one or multiple of these targets, covering these techniques is beyond the scope of this work and good coverage is available in [1–5].

2.2 Dimension Reduction

The term dimension reduction is usually used to refer to decreasing number of variables in a dataset without losing important information. In other words, it is removing variables of low influence on the calculating the response value and keep variables of high influence. Dimension reduction is a crucial step in the modeling flow which has to be performed with care because mis-choice of number of variables after reduction can lead to overfitting or underfitting models. To evade the problem of overfitting or underfitting, we use the validation step to choose the optimal number of variables.

One of the bad decisions that might be taken while building a model for any type of applications, is to drop the dimension reduction step and use the whole data as it is. Using the whole data without reduction implies considering every wavelength or wavenumber as a variable in the model, where number of wavelengths in a normal spectrum can be counted in hundreds and in our case (the case of using NeoSpectra) they are between 300 and 1000 wavelengths. Building a model with this huge number of variables requires multiples of this number in the calibration dataset, which is considered waste of time and money, and above of all that the model will be vulnerable to overfitting.

2.2.1 Principle Component Analysis

Principle Component Analysis (PCA) is a very popular technique in dimension reduction and considered by data scientists an unsupervised technique since it performs the reduction apart of the response matrix $\mathbf{Y}$. PCA depends only on the dataset $\mathbf{X}$ to reduce it to its principal components. In fact, the number of principal components in any dataset is equal to the number of its original variables, whereas not all principal components are equal in importance. Importance of a principal component is determined according to the amount of variations in this component. Components with low variance are discarded and the others with high variance are kept, and by this way dimensionality can be reduced without losing a lot of variance, i.e., information.

We can understand the idea of PCA intuitively by conceiving PCA as a data projection over the direction of the high variance component. In Fig. 3a, a small dataset is composed of two variables x1 and x2. In this imaginary example data needs to be reduce to a single variable, option 1 is to project all data points in x1 direction to get what shown in Fig. 3b, which is not bad option as it preserves variance 0.2327. Option 2 is to project all data points on the direction of x2 to get points as shown in Fig. 3c with variance equals 0.1975. The best option as it preserves most of data variance is to project data points on the direction of the first principle component to get what shown Fig. 3d which contains the largest variance 0.4023.

Now the compelling question should be how to calculate the principle components. Let us assume $\mathbf{w}_1$ is a weight vector where $\mathbf{t}_1 = \mathbf{X}\mathbf{w}_1$ and $\mathbf{t}_1$ is a vector contains the scores of the first principle component. Scores vector refer to the new value of each point in the dataset after projection. According to the definition of the first principle component, scores vector should contain the maximum variance, thus $\mathbf{w}_1$ should be chosen to maximize variance of $\mathbf{t}_1$ while assuming that $\mathbf{X}$ is mean centered as shown in Eq. 1.

$$\mathbf{w}_1 = \arg\ \max_{\|\mathbf{w}_1\|=1} \left\{ \sum_i (\mathbf{t}_1)_{(i)}^2 \right\} = \arg\ \max_{\|\mathbf{w}_1\|=1} \left\{ \sum_i (\mathbf{X}_{(i)}\mathbf{w}_1)^2 \right\} \tag{1}$$

We can rewrite the equation to be

$$\mathbf{w}_1 = \arg\ \max_{\|\mathbf{w}_1\|=1} \left\{ \|\mathbf{X}\mathbf{w}_1\|^2 \right\} = \arg\ \max_{\|\mathbf{w}_1\|=1} \left\{ \mathbf{w}_1'\mathbf{X}'\mathbf{X}\mathbf{w}_1 \right\} \tag{2}$$

Since $\|\mathbf{w}_1\|$ is conditioned to be 1, i.e., unit vector we can achieve the same result using

$$\mathbf{w}_1 = \max \left\{ \frac{\mathbf{w}_1'\mathbf{X}'\mathbf{X}\mathbf{w}_1}{\mathbf{w}_1'\mathbf{w}_1} \right\} \tag{3}$$

We can calculate $\mathbf{w}_1$ by considering Eq. 3 as a normal optimization problem, or solve it analytically which is the easier approach from the implementation point of

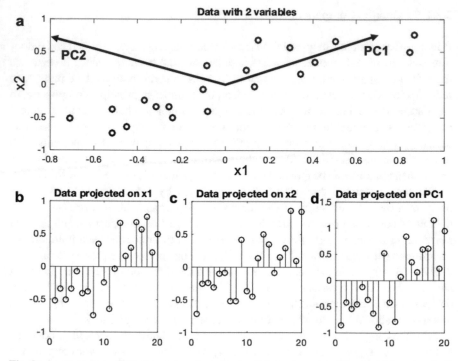

Fig. 3 An example on PCA. In **a** data points of a small dataset consist of two variables x1 and x2. In **b** data points after projecting them on x1 direction. In **c** data points after projecting them on x2 direction In d, data points after projecting them on PC1 direction.

view. Equation 3 is in the form of Rayleigh quotient[6], which states that for any symmetric matrix $\mathbf{M}$ and vector $\mathbf{w}$ where

$$R(\mathbf{M}, \mathbf{w}) = \frac{\mathbf{w'Mw}}{\mathbf{w'w}} \tag{4}$$

$R(\mathbf{M}, \mathbf{w})$ ranges from minimum eigen value ($\lambda_{\min}$) of $\mathbf{M}$ to its maximum eigen value ($\lambda_{\max}$). Let $\mathbf{v}_{\max}$ be the eigen vector of maximum eigen value then it will be the solution of $\max\{R(\mathbf{M}, \mathbf{w})\} = R(\mathbf{M}, \mathbf{v}_{\max}) = \lambda_{\max}$.

Back to Eq. 3, $\mathbf{X'X}$ is always a symmetric matrix which also the can be considered the covariance of $\mathbf{X}$ if $\mathbf{X}$ was mean centered, therefore to maximize the variance of $\mathbf{t}_1$, $\mathbf{w}_1$ must be equal to the $\mathbf{v}_{\max}$ of $\mathbf{X'X}$ and the variance will be equal to $\lambda_{\max}$. In other words, the direction of first principle component is the direction of the maximum eigen vector (the eigen vector of the maximum eigen value) of the covariance matrix of $\mathbf{X}$. Similarly, the weight vector of the second principle component ($\mathbf{w}_2$) is the second maximum eigen vector of the covariance matrix, and so on until computing all principle component using all eigen vectors.

2.2.2 Partial Least Squares

Partial Least Squares (PLS) is by far the most commonly used dimension reduction technique in the multivariate calibration of the spectroscopy applications. Unlike PCA, PLS uses the response matrix $\mathbf{Y}$ as an input to achieve the best dimensionality reduction. PLS also considered to be a supervised technique since it uses $\mathbf{Y}$ in its procedure. As we know from the previous section, PCA main target is reducing the data dimensionality while preserving maximize variance of the data in its components. Similarly, PLS uses a close approach but instead of maximizing the variance of its components, it seeks maximizing the covariance between the PLS components and the response. PLS components are also known as latent variables, and they are chosen according to their covariance with the response variable(s), similar to principle components, latent variables of low covariance are discarded and others with high covariance are used.

PLS models usually show better results than PCA models. This fact can be easily predicted as PLS leverages the covariance between its latent variables and response variable(s), which leads to choose variables that are more relevant to the response. While in the other hand, PCA only maximizes the variance in the principle components, and doesn't consider the meaning of this variance or if it is related to the response or not. However, this doesn't mean that PLS can replace PCA in all applications, since PCA shines in the clustering applications and the applications that don't have response matrix associated with its data.

The common algorithm to compute PLS latent variables for a dataset is called nonlinear iterative partial least squares (NIPALS) algorithm [7–10], and it has two variants PLS1 which is the simpler as it assumes the response matrix to be a single column matrix, i.e., a vector while the other variant is PLS2 which is more general as it consider response matrix to have as many columns as desired.

Initially, PLS defines $\mathbf{X}$ and $\mathbf{Y}$ to be

$$\mathbf{X} = \mathbf{TP'} + \mathbf{E} \tag{5}$$

$$\mathbf{Y} = \mathbf{UQ'} + \mathbf{F} \tag{6}$$

where $\mathbf{X}$ is the observer matrix or the spectral data matrix of size $n \times N$, where n is the number of measured spectra and N is the number of variables (wavelengths). $\mathbf{Y}$ is the response matrix of size $n \times M$, where M is the number of the responses variables. $\mathbf{T}$ and $\mathbf{U}$ are the scores matrices of the observer and the response respectively, and both have the size equals to $n \times p$, where p is the chosen number of latent variables. $\mathbf{P}$ and $\mathbf{Q}$ are the loading matrices for the observer and the response respectively, and $\mathbf{P}$ size is $N \times p$ while $\mathbf{Q}$ size is $M \times p$. Finally, $\mathbf{E}$ is a $n \times N$ observer residual matrix and $\mathbf{F}$ is a $n \times M$ response residual matrix.

$\mathbf{t}_1$ and $\mathbf{u}_1$ denote the first score vectors and the first columns of the matrices $\mathbf{T}$ and $\mathbf{U}$ respectively. To calculate the $\mathbf{t}_1$ and $\mathbf{u}_1$, we will define weight vectors $\mathbf{w}_1$ and $\mathbf{c}_1$, where

$$\mathbf{t}_1 = \mathbf{Xw}_1 \tag{7}$$

$$\mathbf{u}_1 = \mathbf{Yc}_1 \tag{8}$$

$\mathbf{t}_1$ and $\mathbf{u}_1$ should have the maximum covariance as proposed with PLS approach, therefore Eq. 9 is used.

$$\arg \, max_{\|\mathbf{w}_1\|=\|\mathbf{c}_1\|=1}\left\{\left(cov(\mathbf{t}_1, \mathbf{u}_1)\right)^2\right\} = \left\{\left(cov(\mathbf{Xw}_1, \mathbf{Yc}_1)\right)^2\right\} \tag{9}$$

By ensuring that both $\mathbf{X}$ and $\mathbf{Y}$ are mean centered, i.e., they have zero mean.

$$cov(\mathbf{t}_1, \mathbf{u}_1) = \frac{\mathbf{t}_1'\mathbf{u}_1}{n} = \frac{\mathbf{w}_1'\mathbf{X}'\mathbf{Yc}_1}{n} \tag{10}$$

NIPALS proposed an iterative algorithm to compute $\mathbf{t}$ and $\mathbf{u}$

1. $\mathbf{u}_1$ is initiated with any random values.
2. $\mathbf{w}_1 = \frac{\mathbf{X}'\mathbf{u}_1}{\mathbf{u}_1'\mathbf{u}_1}$, this will ensure that $\mathbf{w}_1$ will achieve maximum covariance while being a unit vector.
3. $\mathbf{t}_1 = \mathbf{Xw}_1$
4. $\mathbf{c}_1 = \frac{\mathbf{Y}'\mathbf{t}_1}{\mathbf{t}_1'\mathbf{t}_1}$, this also will ensure that $\mathbf{c}_1$ will achieve maximum covariance while being a unit vector.
5. $\mathbf{u}_1 = \mathbf{Yc}_1$
6. Repeat from 2 to 5 until reaching convergence as $\mathbf{t}_1$ and $\mathbf{u}_1$ don't change anymore.

Now, we are able to compute the first loading vectors in terms of $\mathbf{t}_1$ and $\mathbf{u}_1$. By returning to equations Eq. 5 and 6 and ignoring the residual matrices, loading vectors can be deduced by

$$\mathbf{p}_1 = \frac{\mathbf{X}'\mathbf{t}_1}{\mathbf{t}_1'\mathbf{t}_1} \tag{11}$$

$$\mathbf{q}_1 = \frac{\mathbf{Y}'\mathbf{u}_1}{\mathbf{u}_1'\mathbf{u}_1} \tag{12}$$

To proceed in calculating the next scores and loading vectors, the matrices $\mathbf{X}$ and $\mathbf{Y}$ should be deflated by removing the deduced components as shown in Eqs. 13 and 14.

$$\mathbf{X}_1 = \mathbf{X} - \mathbf{t}_1\mathbf{p}_1' = \mathbf{X} - \mathbf{t}_1\mathbf{t}_1'\mathbf{X}/\mathbf{t}_1'\mathbf{t}_1 \tag{13}$$

$$\mathbf{Y}_1 = \mathbf{Y} - \mathbf{u}_1\mathbf{q}_1' = \mathbf{Y} - \mathbf{u}_1\mathbf{u}_1'\mathbf{Y}/\mathbf{u}_1'\mathbf{u}_1 \tag{14}$$

After deflating the matrices $\mathbf{X}$ and $\mathbf{Y}$ from their first component, we can repeat the same procedure with the deflated matrices $\mathbf{X}_1$ and $\mathbf{Y}_1$ to obtain the second scores and loading vectors. Thereafter, we will deflate $\mathbf{X}_1$ and $\mathbf{Y}_1$ to get $\mathbf{X}_2$ and $\mathbf{Y}_2$, then obtain the third vectors and so on until we reach the required number of latent variables.

Now, we are ready to use the calculated weight $\mathbf{W}$ and loading $\mathbf{P}$ matrices to extract the scores Matrix $\mathbf{T}_t$ from any new coming spectral data $\mathbf{X}_t$. Steps of calculating $\mathbf{T}_t$ are:

1. Remove mean $\mathbf{X}_{t0} = \mathbf{X}_t - \overline{\mathbf{X}}$.
2. $\mathbf{t}_{t1} = \mathbf{X}_{t0}\mathbf{w}_1$.
3. $\mathbf{X}_{t1} = \mathbf{X}_{t0} - \mathbf{t}_{t1}\mathbf{p}'_1$.
4. Repeat 2 and 3 p times (p is the number of chosen latent variables).

2.3　Cross-Validation

We have mentioned the importance of the validation stage in tuning the model parameters until reaching the desired performance. However, dedicating part of the dataset for validation is difficult thing due to the scarce of the reference data and the difficulty of measuring spectral data and associate reference with it. And here it comes the necessity of cross-validation technique, which enable us to use the same calibration data without dedicating a separate part for validation. Therefore, by using the cross-validation technique dataset will be divided into two parts only, calibration part and testing part. The calibration part of the dataset will be used two perform both calibration and cross-validation, while the testing part will be used to preform the final testing of the model.

The idea of the cross-validation is to validate the model in an iterative basis. In the first iteration calibration data is divided into parts, calibration part and validation part. The model is trained on the calibration part and validated with validation part. After that, calibration and validation parts are combined again to form one dataset, and then in the next iteration new calibration and validation parts are produced. In each iteration new validation part is deducted from the data until we have all parts of the dataset are presented once in the validation as shown in Fig. 4. The size of the validation part determine the type of the cross-validation, it will be known as leave one sample out cross-validation if one sample is used in the validation part, while using more than one sample in validation part it will make it K-fold cross-validation.

There are several metrics can be measured to evaluate the model. In regression models, there are some metrics used to measure the prediction error, like root mean square of the error (RMSE) or the standard deviation of the error (SE), some other metrics measure how the model fitting the data, like measuring the coefficient of correlation (R) or coefficient of determination (R^2). While in the classification models, metrics measured number of mis-classified samples or measure the distance of the samples from the cluster center. In case of using cross-validation errors or any other metrics are collected from all the validation iterations. A common convention is used in the metrics naming is to add last character to identify the stage at which this metric is measured, e.g. RMSECV for RMSE measured in cross-validation while for RMSE measured in the testing stage its name will be RMSET.

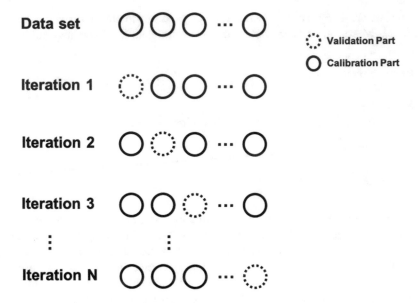

Fig. 4 The procedure of selecting validation part in each iteration of cross-validation

2.4 Soft Independent Modeling Class Analogies

Soft independent modeling class analogies [5, 11] (SIMCA) is a classification technique based on PCA. Basically, the idea of SIMCA consists of blending two metrics to reach a decision on whether an unknown point belongs to a certain class or not. The SIMCA metrics are:

- The Euclidean distance between the point under test and the center of the class.
- The residual projection error of the point on the class hyperplane.

We can elaborate the SIMCA concept by using a simple example. Assume we have spectra to classify between two classes Class A and Class B, and the spectra comprise only three wavelengths. Initially, we will perform PCA on spectra of Class A alone, and the number of principle components are usually chosen to be small and we can use cross validation to determine the optimal number of components. For our example, we will choose 2 principle components to keep things simple. We will repeat the same thing on Class B and choose only one principle component. For any new unknown spectrum subjected to this model, we will project the new spectrum on Class A plane and measure the projection error of the spectrum to be our first metric then after projection we will measure the Euclidean between the projected point and Class A center to be the second metric. The same procedure will be repeated with Class B and measure the two metrics to be compared with those we got with Class A and then decide which class the new spectrum belongs to. The SIMCA example

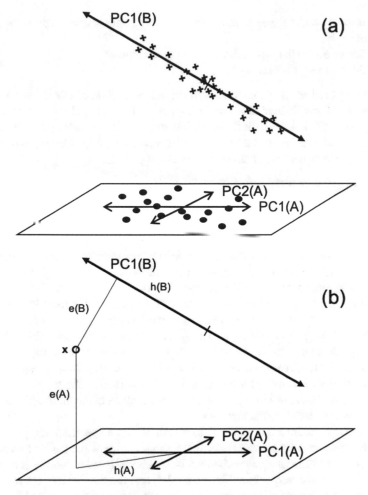

Fig. 5 An illustration of SIMCA example. In **a** the projection of the 3D spectra on the Class A plane and Class B line. In **b** the projection of a new spectrum on Class A and Class B

is illustrated graphically in Fig. 5, where the 2D principle component plane of Class A and 1D line of Class B are represented in Fig. 5a, and the projection of the new spectrum is shown in Fig. 5b.

3 Milk Model

In this case study, we will address a common issue faces the models when they are being generalized to accommodate conditions weren't present during their calibration. The conditions we are trying to study in this case are:

- Variations appear in the calibration spectrometer due to heating, aging or any other factors.
- Variations due to changes in the measuring conditions.
- Variations among different spectrometers.

We have proposed an algorithm to transfer the old calibration of a model to accommodate new conditions instead of recalibrating the model over again [12]. And to verify the proposed solution we built three models to discriminate between milk samples according to their fat levels, then we tested the model in three different experiments, where each experiment imitates one of the stated conditions.

3.1   Calibration Transfer

The main purpose of the algorithm is to offer a standardization method between the sensor used in the model calibration and any other sensors that will use the same model. The algorithm can also mitigate the effect of changing the conditions at which the model is calibrated and the conditions which the model is used in. Originally, calibration transfer was used to transfer the models among large bench-top spectrometers with large differences in their specifications. In our case, the differences between spectrometers are minimal since they are all from the same brand, and having the same design. However, the sensor might be suffering from components and manufacturing variations and tolerances. More importantly, the transfer technique needs to be simple and fast to cope with the large numbers of the miniaturized sensors needed to be subjected to this algorithm.

To transfer a model from sensor A to sensor B, some spectra readings are taken using both sensors from samples having the same nature as the calibration samples and cover most of the samples variations. Next, dimensional reduction is applied to the data in the same way applied in the model, and the data points are transferred from sensor A to sensor B using the linear transformation given in Eq. 15.

$$\mathbf{X}_b^* = A\mathbf{X}_b + B \tag{15}$$

where $\mathbf{X}_b$ is the original data points of sensor B and $\mathbf{X}_b^*$ is the transferred points, while A and B are the coefficients of the linear equation. The coefficients A and B are calculated using the least squares method as per equation 16.

$$\min\left\{ \sum \left[\mathbf{X}_a - \mathbf{X}_b^*\right]^2 \right\} = min\left\{ \sum \left[\mathbf{X}_a - (A\mathbf{X}_b + B)\right]^2 \right\} \tag{16}$$

where $\mathbf{X}_a$ is the data points of sensor A, and the error between $\mathbf{X}_a$ and $\mathbf{X}_b^*$ should be minimized by tuning the valuesof A and B. The coefficients A and B are the output of this step, and for any new sensor to be transferred to sensor A these steps should be repeated to get new coefficients associated with the new sensor. Afterwards, for every new reading taken with sensor B, equation 15 should be applied before using the model calibrated on sensor A.

3.2 Models Setup and Intrumentation

In this section, we will list the number and the nature of the samples in the model, the measuring configuration and the spectrometer configurations, as shown in Table 1.

Samples used to perform the experiments and to build the model were bought from local grocery store and consisted of 11 different ultra-high-temperature (UHT) treated milk packs. The 11 packs were distributed as follows:

- 5 samples of full fat milk
- 4 samples of skimmed milk
- 2 samples of half-and-half milk.

The samples were measured in diffuse reflection configuration through a quartz cuvette with a 10 mm path length. And during the model building and the experiments, we have used 3 DVKs, named D20, D39 and D48.

3.2.1 NeoSpectra

NeoSpectra is the commercial name of the Fourier Transform Near Infrared (FT-NIR) spectrometer from the Egyptian company, Si-ware Systems. NeoSpectra has several variants of spectrometers, while in this case study we have tested and used only NeoSpectra micro development kit (DVK). DVK is a FT-NIR spectrometer based on Micro Electromechanical Systems (MEMS) technology to miniaturize the traditional Michelson interferometer into chip-size component [13, 14]. It has also an application specific integrated circuits (ASICs) to control the MEMS and for data processing, and contains an integrated light source to be used in diffuse reflection measuring configuration as shown in Fig. 6, to end up that the whole instrument is

Table 1 The setup of the milk classification models

Parameter	Description
Samples number	11 UHT milk samples
Spectrometer	3 NeoSpectra micro development kits (DVK)
Measuring configuration	Diffuse Reflection
Measurements number	5 measurements from each sample
Scan time	10 s
Common wavenumber	Enabled, with 1024 points
Analysis tool	MATLAB R2018a

Fig. 6 NeoSpectra micro development kit

Table 2 Summarized specifications of NeoSpectra DVK

Parameter	Conditions	Value	Units
Wavelength range	–	1350–500	nm
Resolution	At $\lambda = 1550$ nm	16	nm
		66.6	cm^{-1}
Typical SNR	2 s scan time, at $\lambda = 2350$ nm	> 2000 : 1	–
Wavelength accuracy	At $\lambda = 1400$ nm	±1.5	nm
Wavelength repeatability	At $\lambda = 1400$ nm	±0.15	nm

integrated into a small spectral sensor. The specifications of the DVK are summarize in Table 2 to include only data relevant to spectroscopy.

3.3 Model Details and Results

The Classification model was developed with the spectral data measured by D48 where its data is plotted in Fig. 7a. Spectra were treated with a second order detrending to remove baseline trends.

3.3.1 Gaussian Process Classification Model

This model is a combination between PLS and Gaussian Process Classification (GPC) [15], where PLS was applied to reduce the dimensions of the dataset to two latent variables. Afterwards, we trained two GPC models using the PLS latent variables. The first classifier is a one-dimensional GPC which uses the first latent variable only as shown in Fig. 7b. While, the second uses the two latent variables together in a two-dimensional GPC model as shown in Fig. 7c. The two models were validated using cross validation and both of them achieved 100% classification accuracy. These two models will be used in the following experiments to test their performance under different conditions.

3.3.2 Soft Independent Modeling Class Analogies Model

The SIMCA model was trained on the first two principle components of the milk classes. The PCA was carried out on each class separately as shown in Fig. 8. The model was validated using cross validation and it achieved 100% classification accuracy, similar to the GPC models.

3.4 Experiment 1

The purpose of this experiment is to study the effect of changing the signal-to-noise ratio (SNR) within a single spectrometer sensor. The effect of changing SNR can be simulated by changing the scan time duration for the sample measurement. All samples were measured using sensor D48 with scan time of 2 s and scan time of 5 s through a quartz cuvette of a 10 mm path length. These two test sets were subjected to the models developed using spectra collected with 10 s scan time on the same sensor D48. From Fig. 9a and b, it is clear that changing the SNR has minimal effect on the GPC classification accuracy, however, it may still be beneficial to test the calibration transfer procedure on this data to obtain the results shown in Figs. 9c and d. The same results were achieved by the SIMCA model to score 100% accuracy before and after the calibration transfer.

3.5 Experiment 2

In this experiment, we were keen on testing the effect of the variations between different sensors. We used the three sensors D20, D39 and D48 to measure all the samples using a scan time duration of 2 s and through the regular cuvette we used in the prior experiment. From Fig. 10a, it is clear that testing with the 1D GPC model D39 achieves a performance poorer than that of D20 while D48 shows the

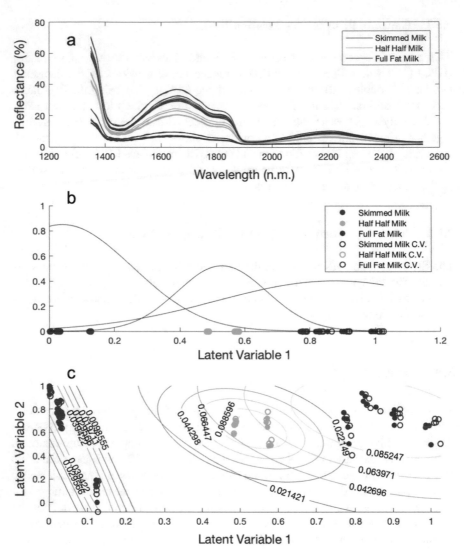

Fig. 7 The cross validation results of the milk model. In **a**, raw spectra of the calibration. In **b**, the scores of LV1, where solid dots represent calibration data, hollow dots represent cross validation data and black lines represent the probability of each class across the space. In **c**, the scores of both LV1 and LV2, and the contour lines represent the probability of each class

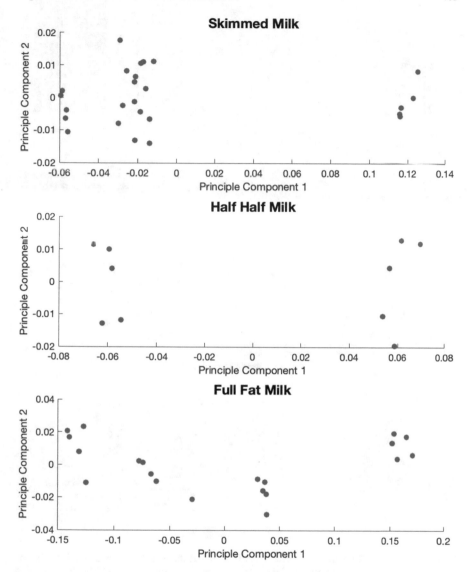

Fig. 8 The scores of the first two principle components for each milk class

best performance since it is the calibration sensor. On the other hand, testing with
the 2D GPC model shows a performance degradation for both D20 and D39, as
shown in Fig. 10b, due to the large spread appear in LV2. Based on these results,
calibration transfer is a must and it showed considerable improvements as illustrated
in Figs. 10c, d and Table 3. The results of the SIMCA model was quite similar to the
results of 1D GPC as shown in Table 4.

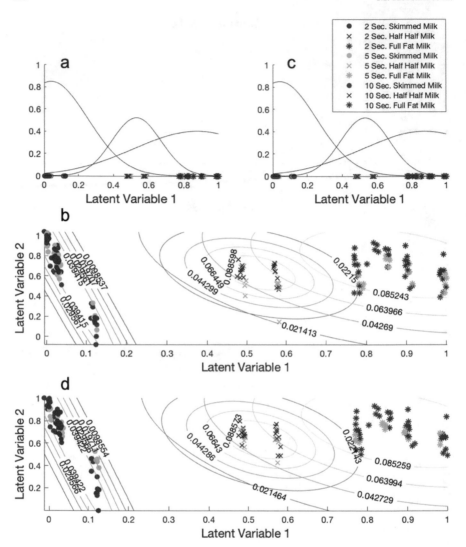

Fig. 9 The testing results of the 1st milk experiment where the dataset comprises different values of SNR. LV1 of the test set collected by changing scan time are plotted in **a** while **c** is the same LV1 after applying calibration transfer. **b** is the 2D GPC model, while **d** is the model after applying calibration transfer

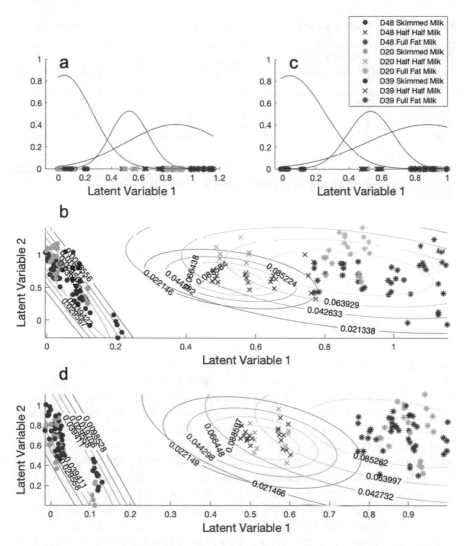

Fig. 10 The testing results of the 2nd milk experiment where the dataset comprises measurements from different DVKs. LV1 of the test set collected by the 3 DVKs are plotted in **a**, while **c** is the same LV1 after applying calibration transfer. **b** is the 2D GPC model, while **d** is the model after applying calibration transfer

Table 3 The classification accuracy of the GPC models for the test data measured using the three DVKs

Sensor	1D GPC model		2D GPC model	
	Before transfer (%)	After transfer (%)	Before transfer (%)	After transfer (%)
D48	100	100	100	100
20	100	100	87.27	98.18
D39	90.9	100	83.64	96.36

Table 4 The classification accuracy of the SIMCA model for the test data measured using the three DVKs.

Sensor	Before transfer (%)	After transfer (%)
D48	100	100
D20	100	100
D39	90.9	100

3.6   Experiment 3

This experiment focuses mainly on studying the variations that may occur because of changing the container's material. Since the model is developed using milk spectra measured through a quartz cuvette, the test data will be measured through a different container which is a glass beaker in this experiment. Similar to prior experiments, the same three sensors were used and their scan time was set to 2 s. From Fig.11a and b, it is clear that changing the measuring medium or container has a great impact on the GPC models classification accuracy even for sensor D48. After applying the calibration transfer procedure, the spread of the data decreases noticeably and the three classes become more confined as shown in Fig. 11c and d. Both models benefit from transfer but the 1D GPC model achieves a better classification accuracy as shown in Table 5. SIMCA model showed similar behavior by being vulnerable to this type of variations and return back to the acceptable region after applying calibration transfer as shown in Table 6.

4   Conclusion

In this study, we introduced the calibration transfer technique and studied the feasibility of using it with chemometric models to mitigate any negative impact that may occur due to using spectral sensors. One of the common issues of the spectral sensors, and we tested its impact on our models in experiment 1, is the degradation of the SNR of the sensors which appears to have minimal effect on the models. On

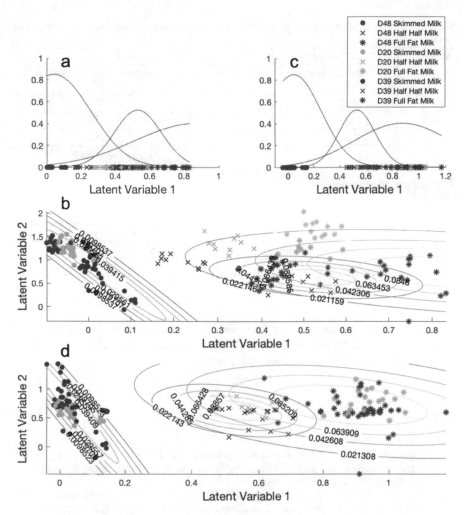

Fig. 11 The testing results of the 3rd milk experiment where the spectra were measured through glass beaker. LV1 of the test set measured through glass beaker by the 3 DVKs are plotted in **a**, while **c** is the same LV1 after applying calibration transfer. **b** is the 2D GPC model, while **d** is the model after applying calibration transfer

the other hand, the usage of mass produced sensors usually leads to unit-to-unit variations between different sensors, and according to experiment 2 these variations can be minimized to almost return to the original accuracy of the models. Finally, we tested the effect of changing the measuring medium in experiment 3, and it appeared that the models were vulnerable this type of variations and this showed the most gain of calibration transfer as it restored the accuracy from less than 50% in some cases to above 90%. The three experiments were conducted and tested using three

Table 5 The classification accuracy of the GPC models for the test data measured through a glass beaker

Sensor	1D GPC model		2D GPC model	
	Before transfer (%)	After transfer (%)	Before transfer (%)	After transfer (%)
D48	45.45	100	50.9	89.1
D20	90.9	92.27	96.36	90.9
D39	47.27	96.36	81.81	90.9

Table 6 The classification accuracy of the SIMCA model for the test data measured through a glass beaker

Sensor	Before transfer (%)	After transfer (%)
D48	58.18	98.1818
D20	63.64	100
D39	85.45	92.73

different models: one-dimensional GPC model, two-dimensional GPC model and SMICA model, the overall performance of the one-dimensional GPC and SIMCA were better than the two-dimensional GPC model.

References

1. Martens, H., Nielsen, J.P., Engelsen, S.B.: Light scattering and light absorbance separated by extended multiplicative signal correction. Application to near-infrared transmission analysis of powder mixtures. Anal. Chem. **75**(3), 394–404 (2003)
2. Rinnan, Å., Van Den Berg, F., Engelsen, S.B.: Review of the most common pre-processing techniques for near-infrared spectra. TrAC Trends Anal. Chem. **28**(10), 1201–1222 (2009)
3. Wold, S., Antti, H., Lindgren, F., Öhman, J.: Orthogonal signal correction of near-infrared spectra. Chemometr. Int. Lab. Syst. **44**(1–2), 175–185 (1998)
4. Westerhuis, J.A., de Jong, S., Smilde, A.K.: Direct orthogonal signal correction. Chemometr. Intell. Lab. Syst. **56**(1), 13–25 (2001)
5. Lavine, B.: A user-friendly guide to multivariate calibration and classification, Tomas Naes, Tomas Isakson, Tom Fearn and Tony Davies, NIR Publications, Chichester, 2002, ISBN 0-9528666-2-5,£ 45.00. J. Chem. J. Chemometr. Soc. **17**(10), 571–572 (2003)
6. Simoncini, V., Eldén, L.: Inexact rayleigh quotient-type methods for eigenvalue computations. BIT Numer. Mathematics **42**(1), 159–182 (2002)
7. Manne, R.: Analysis of two partial-least-squares algorithms for multivariate calibration. Chemometr. Intell. Lab. Syst. **2**(1–3), 187–197 (1987)
8. Helland, I.S.: On the structure of partial least squares regression. Commun. Stat.-Simul. Comput. **17**(2), 581–607 (1988)
9. De Jong, S.: SIMPLS: an alternative approach to partial least squares regression. Chemom. Intell. lab. syst. **18**(3), 251–263 (1993) (Vancouver)

10. Rosipal, R., Krämer, N.: Overview and recent advances in partial least squares. International Statistical and Optimization Perspectives Workshop. Subspace, Latent Structure and Feature Selection, pp. 34–51. Springer, Berlin, Heidelberg (2005)
11. Wold, S.: Pattern recognition by means of disjoint principal components models. Pattern Recognit. **8**(3), 127–139 (1976)
12. Hossam, M., Wassal, A., El-Kharashi, M.W.: Reduction of variations using chemometric model transfer: a case study using FT-NIR miniaturized sensors. In: International Conference on Advanced Machine Learning Technologies and Applications, pp. 272–280. Springer, Cham (2019)
13. Sabry, Y.M., Khalil, D.A.M., Medhat, M., Haddara, H., Saadany, B., Hassan, K.: Si ware systems. Integrated spectral unit. U.S. Patent Application 10/060,791 (2018)
14. Khalil, D.A., Saadany, B.A.: Si ware systems. Interferometer with variable optical path length reference mirror using overlapping depth scan signals. U.S. Patent 8,792,105 (2014)
15. Hensman, J., Matthews, A., Ghahramani, Z.: Scalable variational Gaussian process classification (2015)

Smart Environments Concepts, Applications, and Challenges

Doaa Mohey El-Din, Aboul Ella Hassanein, and Ehab E. Hassanien

Abstract This chapter presents the clear definition of a smart environment, its advantages, previous motivations in various applications, and open research challenges. These challenges are classified into two parts, artificial intelligence, and internet-of-things. They are powerful for researchers and students to select a research topic. This chapter also removes the blurred idea of a smart environment that is limited to the vision of a smart environment definition just interprets with the internet-of-things. It presents the importance of recently used smart environments such as smart homes, smart farming, smart city, smart education, or smart factory. The recent statistics refer to the predication of used smart devices for constructing the smart environments that reach to nine double of a population around a world by 2025. This chapter presents a proposed criterion of building a good smart environment for any domain with respect to two dimensions data and security.

Keywords Smart environment · Internet-of-Things · Artificial intelligence · IoT devices · Sensors

1 Introduction

A smart environment (SA) refers to the simulation management system for the real environment such as smart city or smart parking [1, 2]. It improves decision making

D. M. El-Din (✉) · E. E. Hassanien
Information Systems Department, Faculty of Computers and Artificial Intelligence, Scientific Research Group in Egypt, Cairo University, Cairo, Egypt
e-mail: d.mohey@alumni.fci-cu.edu.eg
URL: http://www.egyptscience.net/

E. E. Hassanien
e-mail: E.Ezat@fci-cu.edu.eg

A. E. Hassanein
Faculty of Computers and Artificial Intelligance, Scientific Research Group in Egypt, Cairo University, Cairo, Egypt
e-mail: aboitcairo@gmail.com

© The Author(s), under exclusive license to Springer Nature Switzerland AG 2021   493
A. E. Hassanien et al. (eds.), *Machine Learning and Big Data Analytics
Paradigms: Analysis, Applications and Challenges*, Studies in Big Data 77,
https://doi.org/10.1007/978-3-030-59338-4_24

remotely and concurrently [3, 4]. There is a blurred defining of a smart environment that refers to the limit on the internet-of-things [5, 6]. However, the real meaning of a smart environment is illustrated in the combination of artificial intelligence and internet-of-things [7, 8]. It means Interconnected several sensors and Internet-of-things devices via the internet. This connection causes of huge data that are required interpreting and processing. The smart key of any smart environment is data [9]. The good interpreting and processing data lead to making good decisions simultaneously.

Internet-of-things (IoT) is a technology supports a new idea for tracking objects, sensing devices, and monitoring things of each environment [10]. IoT allows specific sensors for a specific environment to communicate with other devices such as smartphones via Bluetooth or Wi-Fi to transmit enormous amounts of data to the network. It allows users to have a better meaning of the environment's objects cases, conditions, and problems. It faces several obstacles and challenges in network security, reliability, and consistency.

Artificial intelligence (AI) refers to the intelligence of a machine that can understand and interpret the input data by several algorithms or techniques to support making decisions [11–13]. It is known as a cognitive technology, generates a technology that enables machines to work intelligently for simulating the real environments. However, the powerful of usage of AI, it has many challenges in several decision levels that makes getting the data is hard. Although recent researches try to solve some AI challenges, it is faced with open research challenges until now.

This chapter presents a definition of smart environment technology and the importance of using it. It discusses the relationship between Smart environment, artificial intelligence, and internet-of-things. It shows many smart environment's application in various domains. It also introduces its benefits and challenges to the usage of a smart environment and how to reach a good criterion to construct a new smart environment.

The rest of this chapter is organized as follows. Section 2 examines the smart environment definition and its main architecture and structure. Section 3 presents the importance of the smart environment. Section 4 discusses the benefits of smart environments. Section 5 introduces a comparison between the real smart environment's applications. Section 6 presents the smart environments' challenges. Section 7 introduces a discussion and generated criteria to create a smart application. Finally, the conclusion and future work of this work in Sect. 8.

2 Smart Environment

Recently, there is a confusing definition of a smart environment that is limited to the internet-of-things. It causes a blurred understanding of the smart environment and how to construct it. So, there is a need to discuss the real meaning of a smart environment and how to construct it.

Smart Environment (SA) is defined by that any environment based on the interconnection of IoT sensors to support the interpretation processing of big data extracted

Fig. 1 Smart environment architecture

from multiple IoT sources [3–5]. These data may be the same type or variant data types. These data are captured from Different media depending on the target environment. They are required to make processing Data classification, clustering, fusion, and outliers. Another formal definition of a smart environment is an intelligent agent that perceives the state of the resident and the physical surroundings using sensors and acts on the environment using controllers in such a way that the specified performance measure is optimized [7]. It is an automated management environment that is based on the continuous communication between sensors connects via the internet [8].

2.1 Smart Environment Architecture

The essential architecture for constructing any smart environment consists of five levels, sensors devices, Internet-of-things connections, cloud networks, and extracting big data from the sensory devices as shown in Fig. 1.

The Smart environment gets a benefit from artificial intelligence to interpret and fuse the extracted data to improve analytics for improving the making decisions.

2.2 Smart Environment Structure

The smart environment relies on the specific context, each context has a specific number of sensors S_1 to S_n, sensor's types, features, and conditions. to understanding, processing and extracting the data as shown in Fig. 2. These sensors may be cameras, built-in sensors on devices such as on smart-phones, wearable sensors, or IoT devices for specific domain such as LADAR. That causes of extracting big data with variant types D_1 to D_n and different targets. These data may be images, videos, text or signals. Each data type has several algorithms or techniques This connection requires to be reliable continuously. There are some types of network connection that depends on the target, the number of users and security level. Distributed, centralized, semi-centralized, or Blockchain are types of cloud network. That also requires getting the good and big servers to guarantee the consistency and integrity of these huge data.

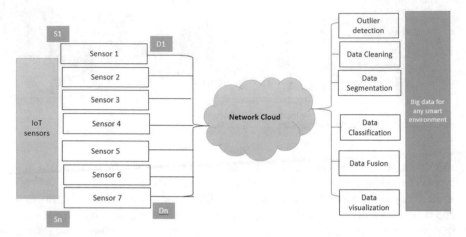

Fig. 2 A detailed smart environment structure

These data require some processing and cleaning for fusing in the same structure for reaching the target. Many challenges have faced fusing and interpreting sensory data.

3 The Importance of Smart Environment

The importance of the smart environment is declared in Making decisions, monitoring controlling, saving things, and dealing with outliers. Recently, the new trend of the industry goes forward to implement a real smart environment. According to Statista, the current usage of 2020, the expected communicated sensors through the internet for smart domains reach 30 billion sensors [39]. According to Statista, this statistic increases to 75 sensors in 2025 [40]. So, smart environment becomes important for research to try solving the problems of the real smart environment implementation (Fig. 3).

The result of evolution usage of sensors is concluded in that the statistics of a number of used IoT sensors are bigger than the number of Population world.

$$\#No. \, of \, used \, IoT \, Devices > \#No. \, of \, Poplution \tag{1}$$

So, the usage of IoT is increasing in the industry to help the automated management control and put values for the extracted big data.

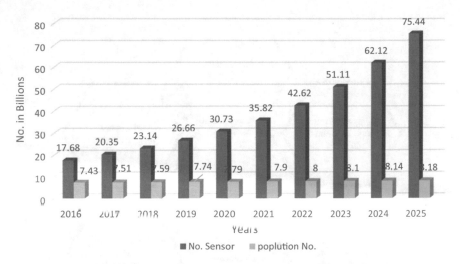

Fig. 3 The statistics between number of Populations around the world and number of used IoT sensors

4 Smart Environment Benefits

Smart technology for simulating the real environment has main six benefits for users and the real-world, like the following, Effective Data-Driven, Improved Making Decisions, Trust communities, Reduce Negative impact for real environments, Communicate many smart environments with each other, and Open economic development chances (Fig. 4).

The effective data-driven is very powerful for tracking data or objects and previous status for any object. Data-driven [12, 13]. These data are valuable in the industry that can use in analytics, marketing, or sales. It also improves the decision making though following the real status for each object. That is very effective in the market [14]. There are several companies try to get or buy this information to improve the marketing section and people's targets. Trust communities are provided based on the confidence of the network of Internet-of-things [15]. the reliable connection between objects reaches trust users, data, communications, and decisions concurrently. İt reduces a negative impact on real environments. The real data provides real decisions, and making a good decision that can save lives, things simulations [16]. So, that reduces the negative effects on real environments such as save lives in fire events in monitoring forests or fainting patient cases in smart health. The connect communication network provides new researches and industry to try communicating several smart environments with each other's to see the full vision in various dimensions [17]. For example, smart parking and smart vehicles connection support management system to see the full vision of streets and around available parking areas and the peak time of traffic congestion. The future trend of research and industry to simulate a full Smart city which includes smart vehicles, smart homes, smart

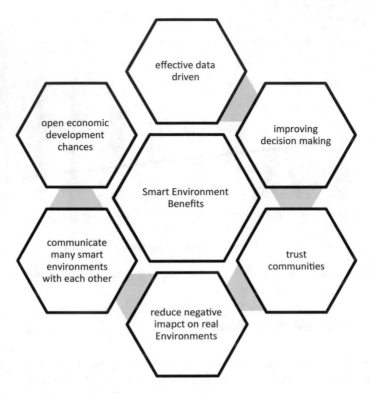

Fig. 4 Smart environment benefits

health, smart education, smart parking, and smart factory. The investment in large scale implementation such as smart transportation, smart city, or smart factory is increasing continuously that reaches millions of dollars to construct infrastructure and tasks [18]. So, that opens new economic development chances and jobs opportunities for improving decision making and making several automated systems. That also requires several research to support the integration, efficiency, and consistency between data analytics, sensors, and decision making in smart environments.

5 Smart Environment Applications

This section presents a comparison between the real smart environment applications and their various objectives, characteristics and conditions that use for making decisions as Table 1. It also presents a comparison between the advantages and current limitation for each smart environment as Tables 2 and 3.

From previous comparisons, finding the motivations try to build various smart environments that are based on fusion multiple sensors on objects or devices for real

Table 1 A comparison study between several researches in smart grid environment

Chapter No.	Domain	Characteristics	Objectives	Conditions	Number of users (SIoT)	Advantages	Disadvantages
[19]	Smart Grid	Includes several characteristics: Communication networks, Cybersecurity, Distributed energy resources, Distribution grid management, Electric transportation, Energy storage, Wide-area monitoring, and Advanced metering infrastructure (AMI)	Improve usage of electricity and low reliability for the national institute of standards and technology	Interoperability challenge for fusing several multiple sensors and develop the smart meters	Few as managers	Improve accuracy and reliability	Less security a fusion problem
[20]	Smart Grid	Real-time readings of electric data as every 15 min, this leads to about 77 billion	Enhance the performance and reduce the cost	Working consistence on the real-time monitoring to be more reliable	Few as managers or grid reader men	Better performance	Unreliability due to the lack of efficient monitoring, fault diagnostic, and automation techniques Inflexibility network

(continued)

Table 1 (continued)

Chapter No.	Domain	Characteristics	Objectives	Conditions	Number of users (SIoT)	Advantages	Disadvantages
[21]	Smart Grid	Quantitative and qualitative dimensions	Reduce usage power control and improve management system	Low power usage condition	Few as managers or grid reader men	Improve quality control	Improving performance
[22]	Smart whether smart meter	Improving the efficiency of education in regard to smart meters	High efficiency	Fusion multiple sensors data	Few	Improves 92% of electricity meters reduced CO_2 emissions and cost savings for electricity customers and utility companies	Requires integrating with variant smart environments to show the impact on other environments. Lack of privacy security of this environment

Table 2 A comparison study between several researches in smart agriculture Environment

Chapter No.	Domain	Characteristics	Objectives	Conditions	Number of users (SioT)	Advantages	Disadvantages
[23]	Smart Agriculture	Maximizing crop production and provide guidance to researchers and engineers	Improving crop yield	Requires high security, good storage, robust infrastructure	Few	Improving accuracy	Improving performance time
[24]	Smart Agriculture	Improves agriculture based on real sensors	Improves farming management system	the roles of agriculture to providing farming	Few	High accuracy	Lack of real datasets
[25]	Smart Agriculture	Get data from several sensors motion detector, light sensor, humidity sensor, temperature sensor, room heater, cooling fan	Targets improve agriculture in real-time	Remote sensing and control irrigation system	Few	Improve the yield of the crops	High cost and deployment of sensor under the soil which causes attenuation of radio frequency (RF) signals
[26]	Smart Farming	Smart farming based on managing business processes and managing the stakeholders	Provide predictive insights in farming operations, drive real-time operational decisions, and redesign business processes	Getting real-time data from multiple Sensors	Many that is between several companies	Improves decision making	Increased uptake of Big Data applications and Big Data governance

(continued)

Table 2 (continued)

Chapter No.	Domain	Characteristics	Objectives	Conditions	Number of users (SioT)	Advantages	Disadvantages
[27]	Smart Weather	Cost-effective, solar-powered automated weather station	Enhance food production in their rural communities	Sensors records in varient weather	Few	Reduced the cost of obtaining accurate, localized scientific weather information	
[28]	Smart Water	Proactive asset maintenance and realtime optimization, smart water systems	Dynamic and rapidly evolving in real-time	Measures the water consumption	Few	Enable varying levels of autonomous decision-making capabilities	Infrastructure security capability limitations

Table 3 A comparison study between several researches in smart Traffic Environment

Chapter No.	Domain	Characteristics	Objectives	Conditions	Number of users (SioT)	Advantages	Disadvantages
[29]	Smart Vehicle	Traffic prediction vehicle prediction model in real-time	Optimize management systems	Extracts predictive Traffic crowd in metropolitan in real-time	Many	Improve predication vehicle prediction	Security and integrity system. And the hardness of the interpretation big data concurrently
[30]	Smart Vehicle	Traffic congestion prediction for detecting specific paths	Using Neural Networks in prediction (Logistic Regression) model	Using hybrid techniques for improving the accuracy	Many	Reaches to 99% accuracy	Integration with various smart environments
[31]	Smart Traffic	It works to predict the traffic congestion that depends on big GPS database in Tunisia	It uses the neural networks model to recognize the speed rate on the roads	Tries to take care of several parameters to be more dynamic system and improve accuracy results.	Many	Reaches to 94% using neural networks approach with respect to 17 hidden layers	Requires enhancing accuracy and performance time
[32]	Smart driving	Interpret behavior of drivers by tracking mobiles	Automated driving	Can recognize driver profile	Few	Automated driving cars based on car sensors Improve performance	Combine car, driver's biologic, psychological, and environmental data

(continued)

Table 3 (continued)

Chapter No.	Domain	Characteristics	Objectives	Conditions	Number of users (SioT)	Advantages	Disadvantages
[33]	Smart parking	It execution based on utilizing the sensor circuitry and cloud server	Providing the reservation parking process from mobile application simultaneously	Requires several sensors and the hardness of ensuring the same people park in the reservation places	Many	Enhancing the parking control	Integration with smart traffic that will enhance the parking management system
[34]	Smart traffic lights	Traffic density using IR sensors and accomplishes dynamic timing slot with different time slots	Monitoring traffic congestion and deal with lights automatically	Deals with traffic jams and congestion. The hardness of remote monitoring	Many users	Reduce grid electricity and realize green operation	Improves performance
[35]	Smart traffic light system	It includes two parts GSM system (Global System for Mobile Communications) is connected to Arduino UNO	Deals with traffic crowd	Reducing waiting time in traffic congestion	Many	Reduces traffic congestion in four lanes based on three lights GSM has better results than UNO Using infrared in emergency cases	Improves accuracy

environments. More than one user that requires interpreting and ordering management for each user. Each user has screen with several conditions, each screen has multiple data targets, conditions, and factors. For example, smart health refers to the observed patients hold determined disease (such as diabetes) that needs observing from physicians or nurses as in Tables 4, 5 and 6.

6 Smart Environment Challenges

The challenges of smart environment are classified into two dimensions data and security. Each dimension has several challenges and problems that requires solving to create a smart environment (Fig. 5).

The Data dimension has five challenges level as the following.

6.1 Big Data Analytics

Data Analytics is defined by stratifying an algorithmic process for extracting ideas. The extracted data from IoT sensors and devices have several characteristics as volume, value, variety, and velocity [56, 57]. The volume of data expresses huge data for each device in each minute or second with respect to the conditions of context domains. The variety refers to the problem of various data type formats. The velocity means the speed changes the data by minutes or seconds. The value means in each record, the data value may be changed at any time. So, they need to create several solutions to improve accuracy and performance in any smart environment.

6.2 Preprocessing Data

The big sensory data with various formats that require to clean with missing data, error data, or outliers. So, many researchers refer to outlier detection and data cleaning processes as the current open researches [58, 59]. Data cleaning is considered the correction process for data and the validation of it. İt makes purifying and tuning for the missing data, conflicting data or validating the data quality. Outlier detection refers to find fault readings from sensors that require classifying if it is just an error or a sudden action requires solving [60].

Table 4 A comparison study between several researches in Smart Health Environments

Chapter no.	Domain	Characteristics	Objectives	Conditions	Number of users (SioT)	Advantages	Disadvantages
[36]	Smart Health	Monitoring patients remotely	Visualize patient cases graphically	Interpertation big data	Few	Visualize patient cases graphyically	Noisy data and redundant features
[37]	Smart Health	It is based on creating surgical prediction multi-model for patients	Visualize and remote control in medical surgeries	Secure network and integrity information	Few	95% accuracy results	Lack of information (requires to expert people)
[38]	Smart Health	Support monitoring healthcare	Interperting disease information from online tweets	Manage diseases remotely and simultaneously	Few	improve accuracy 9%	Improving reliability and integrity
[39]	Smart Medical	Imporves the hospital recommendations	Hospital recommendation based on surveies	Based on 19 recommendation	Few	Enhance patient monitoring	Requires enhancing the accuracy
[40]	Smart Health	Monitoring patients	IoT-based information system	It employs pedestrian dead reckoning, thresholding, and decision trees	Few	Imporves patient monitoring	Requires improving accuracy
[41]	Smart Health	Monitoring patients	IoT-based information system	It employs pedestrian dead reckoning, thresholding, and decision trees	Few	Recognize 4parts: falls, lying, standing, sitting and walking activities	Hardness of fusion with various data types

Table 5 A Comparison study between several researches in various Smart Applications Environments in Smart City Environment

Chapter No.	Domain	Characteristics	Objectives	Conditions	Number of users (SioT)	Advantages	Disadvantages
[42]	Smart City	Management system	Masdar city-Abu Dhabi	Management remotely	Many	Improves management system	Requires optimization system
[43]	Smart City	It supports saving energy and management control	Amsterdam city	Management remotely online	Many	It requires to train the same in several application and cities. Requires more security system	Requires improving accuracy in real-time
[44]	Smart City	Provide more efficient services to citizens	Monitor and optimize existing infrastructure, to increase collaboration among different economic actors,	Smart city management system	Many	Improve urban performance	Lack of integrity
[45]	Smart city	Management system in real-time	Intellectual ability	Reduces Co2	Many	Raise innovation based on knowledgeable and creative human capital.	Lack of real data
[46]	Smart Tourism	A conceptual model of Tourism management system	Management system	It includes multiple layers	Many	Competitive and comparative advantages	Lack of real data

(continued)

Table 5 (continued)

Chapter No.	Domain	Characteristics	Objectives	Conditions	Number of users (SioT)	Advantages	Disadvantages
[47]	Smart Education	A smart education framework It uses mobile application system	Three sub-systems: electronic bookshelves, virtual white space, AND social network with an integrated innovation database	Electronic book-based library,	Many	Adaptive system	Hardness of integration data
[48]	Smart Pollution	It takes care of the air pollution It has two gas sensors namely MQ135 and MQ7, as well as DHT11 which is a dedicated temperature-humidity sensor	advantage of temperature and humidity readings	It monitors the levels of CO, CO2, smoke, alcohol, NH3, temperature and humidity	Few	Monitor pollution management system	Hardness of fusion from multiple sensors
[49]	Smart Military	Tracking military based on wearable t-shirts	Improve management remotely	Tracking soldiers and military objects	Many	Monitoring the sky-running race	Inefficiency
[50]	Smart Military	Wearable military management system based on solar energy	Management army remotely	Measure oxygenation level once a minute	Many	Monitoring army through electronic connection Low power and flexible energy	Hardness of integration data

(continued)

Table 5 (continued)

Chapter No.	Domain	Characteristics	Objectives	Conditions	Number of users (SioT)	Advantages	Disadvantages
[51]	Smart Factory	Creating prototype for smart factory	High defective	Improves management system	Many	Improve adoption system for smart factory and reduces defects problem	Requires high investigation for adding value for factory's objects

Table 6 A comparison study between several researches in Smart Home Environment

Chapter no.	Domain	Characteristics	Objectives	Conditions	Number of users (StoT)	Advantages	Disadvantages
[52]	Smart Home	It works in real-time stream, remotely management, and safety system	It works in real-time stream, remotely management, and safety system	Guaranteeing reliability	Few	Automated system control	Fusion big data with various data types
[53]	Smart Home	It is based on the specific application for managing the home. It does not connect the internet continually	It is based on the specific application for managing the home. It does not connect the internet continually	It provides home automation and safety alarms system	Few	Big data analytics	Requires improving controlling multiple users
[54]	Smart Home	Automated extracted sensors for multiple dataset	Managing dataset	Manage automated devices for smart home	Few	Big data analytics concurrently	Requires improving controlling multiple users
[55]	Smart Home	Describing Frugal Labs IoT Platform (FLIP) for construct smart home	Improve living standard, security and safety	Controlling lighting, home appliances, computers, security camera	Few	Monitoring and controlling Smart Home environment for high flexible	Needs to high security

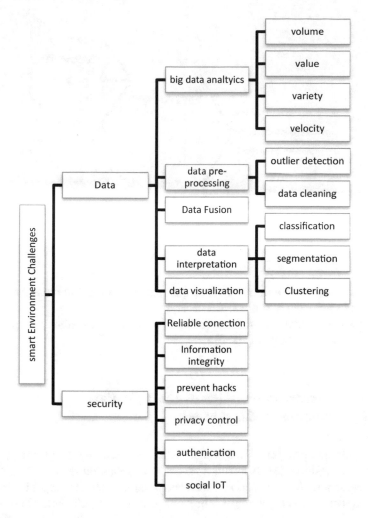

Fig. 5 The smart environment challenges with respect two dimensions data and security

6.3   Data Fusion

Data Fusion has a vital role in understanding the data and combines the parts of meanings to reach the target through several inputs. İt is very important for many applications in various domains [61]. Multi-sensor data fusion refers to the fusing observations from multiple sensors and IoT devices to provide a powerful and perfect description of each sensor [62]. The main problem has faced the research and industry on how to fuse many sensor's data with variant data types. There are two main types of fusion, the same data types, and variant data types. Till now, the main obstacle of

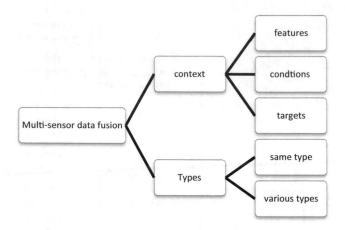

Fig. 6 Multi-sensor data fusion types

data fusion, there is no unification technique that can suitable to more than context whether the same data type or variant data type.

In other words, multi-sensor data fusion is considered a specific domain. It also differs from multiple targets (Fig. 6).

6.4 Data Interpretation (Such as Classification Segmentation, or Clustering)

The interpretation process of data is significant for improving the meanings of data and accuracy results [63]. İt includes several operations on data such as classification, segmentation, or clustering. Data classification means the organizing data in retrieving, sorting or categorizing it to be more powerful and efficient. İt improves the confidability, integrity, and availability. Data segmentation is the operation of getting the data and select the suitable parameters for making processing on it such as image segmentation to determine the tumor. Data clustering is a function of collecting the data in several groups of objects that are known as a cluster.

6.5 Data Visualization

Data Visualization refers to draw a graphical vision for data and its relationships. it is very powerful for visualizing the big data and their relationships that are extracted from multiple sensors [64]. Till now, there are several motivations to enhance the automatic tools for data visualization. That includes several problems such as context

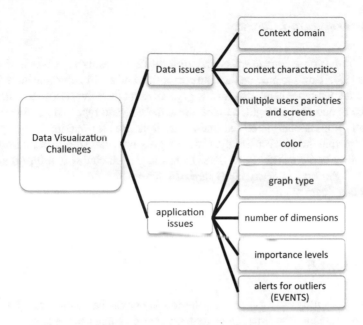

Fig. 7 Data visualization challenges

domain, each context has several characteristics and several users for the same application with various applications. Other problems must be taken into the consideration for constructing applications such as the type of graph, the color of the graph, the explanation details levels (high, medium, or low), the alert types when happening the data outliers (Fig. 7).

The Security dimension faces many challenges that are classified into six types like the following,

6.6 Reliable Connection Continually

Any smart application requires communicating with IoT devices and sensors and being connecting the internet continuously. So, that is very important to be a reliable and robust connection to be online all the time. İt also requires checking on the availability and charging of each IoT device and sensor.

6.7   Information Integrity

It means the data protection in any connection between sensors devices in the internet-of-things [65]. It aims to prevent the data corruption and avoid failures in the extracted sensory data. On other hand, that means guaranteeing the safety data and reach from the source to another one right, correct, the same size and type and holding the trust data. Encryption is considered the main technique uses for making integrity of the data that prevent the data sensitivity. Data integrity provides solutions for improving the reliability and consistency of data. Data integrity includes two types, physical and logical. Physical integrity takes care of fetching and storing right data. Logical Integrity takes care of curing the faults or errors in logical meaning for each specific context.

6.8   Prevent Hacks

The network security requires ensuring the consistency of based on no hacks included [66] whether stealing information or Impersonate personal accounts.

6.9   Privacy Control and Confidentiality

Protecting the Privacy for any smart environment system requires scanning logs for determining any attacks or secret peoples that are offensive on the network. İt also requires guaranteeing the privacy and security of users. İt prevents cyberattacks types. The cyberattack refers to electronic attacks in systems and networks [67]. Several attack types consist of physical cyber-attacks, network attacks, software attacks, software attacks, and encryption attacks. These attacks can impersonate and steal information through some encryption.

6.10   The Authentication Privacy

It aims to protect the personal information for users for example their profiles and password [68]. It takes care of data quality and users verification. There are three types of authentication that are saving profiles through verification applications of questions, passwords quality, image passwords, or CAPTHCA technique. All motivations in authentication target prevent users' profiles from attacks.

6.11 Social Internet-of-Things (SIoT)

Social internet-of-things refers to the number of users that use the same smart environment [69, 70]. It is a new research trend for research due to search for solutions for conflicting decisions from various users or different decisions orders at the same time. Each user has confidential issues and control panels that may be different from others. For example, many users in smart home, that may give a decision to open the light or TV, and another user in the same home decides to off TV. How can order the decisions. Another example another in smart home, if user decides to open the conductor and other user decides to close it. How can solve the conflicting and ordering of decisions from several users with respect to the authorizes and priorities.

The combination of data analytics and Internet-of-things has a big value for improving management control with high accuracy results. The essential significant in any smart environment concludes in the data and how to get it and interpret it.

7 The Proposed Criteria for Construct Robust and Reliable Smart Environment

Previous researches refer to important dimension that should take in the consideration when constructing any smart environments, such as shown in a comparison between thirty chapters in Tables 1, 2. That drives to generate general criteria for constructing any smart environment. These criteria include five steps as shown in Fig. 8.

The combination of data analytics and Internet-of-things has a big value for improving management control with high accuracy results. The essential significant in any smart environment concludes in the data and how to get it and interpret it. The detailed steps of the proposed criteria are illustrated as the following, as Fig. 8. Five steps grantee constructing any smart environment with good quality in various domains, checking the WIFI connection availability all 24 h every second, ensuring the clients' profiles authorized through studying their behaviors, interpreting the extracted big data detection, determining noisy data or anomalies, and information integrity.

Fig. 8 The generated criteria for building any smart environment

8 Conclusion and Future Work

This chapter introduces new research dimensions that require several motivations until now to serve the industry and research. İt removes a blurred ide about the smart environment and a definition discusses it. A smart environment refers to the combination of artificial intelligence and internet-of-things. The main value of a smart environment appears in data. The smart environment challenges are classified into two dimensions data and network security that includes eleven research challenges to reach a reliable and consistent smart environment. No smart environment can neglect these phases and their challenges. However, each research targets one or more challenges to build a smart environment. So, this chapter presents the consistency criteria to reach a good smart environment.

References

1. Bessis, N., Dobre, C.: Big Data and Internet of Things: A Roadmap for Smart Environments. Studies in Componential İntelligence, vol. 546. Springer (2014)
2. Alberti, A.M., et al.: Platforms for smart environments and future ınternet design: a survey. IEEE Access **4**, 1–33 (2016)
3. Cook, D., Das, S.K.: Smart Environments: Technology, Protocols, and Applications. Wiley Series on Parallel and Distributed Computing (2005)
4. Alberti, A.M., et al.: Platforms for smart environments and future internet design: a survey. IEEE Access **7**, 165748–165778 (2019)
5. Camarinha-Matos, L.M., Afsarmanesh, H.: Collaborative systems for smart environments: trends and challenges. In: PRO-VE 2014: Collaborative Systems for Smart Networked Environments, pp. 3–15 (2014)
6. Shahrestani, S.: İnternet-of-Things and Smart Environments. Assistive Technologies for Disability, Dementia, and Aging. Springer (2017)
7. Atziori, L., Iera, A., Morabito, G.: The Internet of Things: a survey. Comput. Netw. **54**(15), 2787–2805 (2010)
8. Bhayani, M., Patel, M., Bhatt, C.: Internet of things (IoT): ın a way of smart world. In: Proceedings of the International Congress on Information and Communication Technology, Advances in İntelligent Systems and computing Book Series, vol. 438, pp. 343–350 (2016)
9. Ahmed, E., et al.: Internet-of-things-based smart environments: state of the art, taxonomy, and open research challenges. IEEE Wirel. Commun. **23**(5), 10–16 (2016)
10. Chin, J., Callaghan, V., Allouch, S.B.: The Internet-of-Things: reflections on the past, present and future from a user-centered and smart environment perspective. J. Ambient Intell. Smart Environ. **11**, 45–69 (2019)
11. Naik, P.: Importance of artificial intelligence with their wider application and technologies in present trends. Int. J. Sci. Res. Comput. Sci. Eng. Inf. Technol. (IJSRCSEIT) **1**(3), 1–9 (2016)
12. Kibria, M.G., et al.: Big data analytics and artificial intelligence in next-generation wireless networks. arXiv:1711.10089v3 [cs.IT] (2018)
13. Vimal Jerald, A., Rabara, S.A., Bai, T.D.P.: Internet of Things (IoT) based smart environment integrating various business applications. Int. J. Comput. Appl. **128**(8), 0975–8887 (2015)
14. Ephzibah, E.P., Dharinya, S.S., Remya, L.: Decision making models through AI for Internet of Things. In: Internet of Things for Industry 4.0, pp. 57–72 (2019)
15. Rebort: IoT for Smart Living Environments, Alliance for Internet of Things Innovation (AIOTI) (2019)

16. Sotala, K.: Advantages of artificial intelligences, uploads, and digital minds. Int. J. Mach. Conscious. **4**(1), 275–291 (2012)
17. Chowdhury, M., Sadek, A.W.: Advantages and limitations of artificial intelligence. In: Artificial Intelligence Applications of Critical Transportation Issues: Why Artificial Intelligence? The National Academies Press (2012)
18. Hemlata, P.G.: Big Data analytics, research. J. Comput. Inf. Technol. Sci. **4**(2), 1–4 (2016)
19. Ghasempour, A.: Internet of Things in smart grid: architecture. Appl. Serv. Key Technol. Chall. Inventions **4**(22), 1–12 (2019)
20. Jaradat, M., Jarrah, M., Bousselham, A., Jararweh, Y., Al-Ayyoub, M.: The internet of energy: smart sensor networks and big data management for smart grid. In: The International Workshop on Networking Algorithms and Technologies for IoT (NAT-IoT 2015), vol. 56, pp. 592–597 (2015)
21. Li, Y., et al.: Smart choice for the smart grid: narrowband: internet of things (NB-IoT). IEEE Internet Things J. 2327–4662 (2017)
22. Wang, Q., Lewandowski, S.: Examining whether smart meters have been used smartly: a case study of residential electricity customers in Vermont. Int. J. Sustainable Green Energy **6**(5), 76–83 (2017)
23. Ayaz, M.: Internet-of-Things (IoT) based smart agriculture: towards making the fields talk. IEEE Access, 1–34 (2019)
24. N.Suma, et al.: IOT based smart agriculture monitoring system. Int. J. Recent Innov. Trends Comput. Commun. **5**(2), 177–181 (2017)
25. Gondchawar, N., Kawitkar, R.S.: IoT based smart agriculture. Int. J. Adv. Res. Comput. Commun. Eng. **5**(6), 838–842 (2016)
26. Wolfert, S., et al.: Big data in smart farming – a review. Agric. Syst. **153**, 69–80 (2017)
27. Adoghe, A.U., et al.: Smart weather station for rural agriculture using meteorological sensors and solar energy. In: Conference: 2017 IAENG WCE International Conference of Electrical and Electronics Engineering, vol.2017 (2017)
28. Rasekh, A., et al.: Smart water networks and cyber security. J. Water Resour. Plann. Manage **142** (7), 1–3 (2016)
29. Stojov, V., Koteli, N., Lameski, P., Zdravevski, E.: Application of machine learning and time-series analysis for air pollution prediction. In: 15th International Conference on Informatics and Information Technologies (CiiT) (2018)
30. Devi, S., Neetha, T.: Machine learning based traffic congestion prediction in IoT based Smart City. Int. Res. J. Eng. Technol. (IRJET) **4**(5), 3442–3445 (2017)
31. Elleuch, W., Wali, A., Alimi, A.M.: Towards and efficient traffic congestion prediction method based on neural networks and big GPS data. IIUM Eng. J. **20**(1), 108–118 (2019)
32. Karaduman, M., Eren, H.: Smart driving in smart city. In: 5th International Istanbul Smart Grid and Cities Congress and Fair (ICSG) (2017)
33. Shruthi, M., Shreya, A., Sujay, M., Chaitanya, K.J.: IoT based smart car parking system. IJSART. **5**(1), 270–272 (2019)
34. Gazal, B., et al.: Smart traffic light control system. In: Third International Conference on Electrical, Electronics, Computer Engineering and their Applications (EECEA) (2016)
35. Thaar Kareem, T.A., Jabbar, M.K.: Design and implementation smart traffic light. Iraqi J. Comput. Inform. **44**(2), 1–5 (2018)
36. Galletta, A., Carnevale, L., Bramanti, A., Fazio, M.: An innovative methodology for Big Data Visualization for telemedicine. IEEE Trans. Ind. Inform. **15**, 490–497 (2018)
37. Radovic, M., Ghalwash, M., Filipovic, N., Obradovic, Z.: Minimum redundancy maximum relevance feature selection approach for temporal gene expression data. BMC Bioinform. **18**(9), 1–14 (2017)
38. Kuang, S., Davison, B.D.: Learning word embeddings with chi-square weights for healthcare tweet classification. Appl. Sci. **7**, 846 (2017)
39. Khoie, M.R., Sattari Tabrizi, T., Khorasani, E.S., Rahimi, S., Marhamati, N.: A hospital recommendation system based on patient satisfaction survey. Appl. Sci. **7**, 966 (2017)

40. Dziak, D., Jachimczyk, B., Kulesza, W.J.: IoT-based information system for healthcare application: design methodology approach. Appl. Sci. **7**, 596 (2017)
41. Sundaravdivel, P., et al.: Everything you wanted to know about smart health care: evaluating the different technologies and components of the internet of things for better health. IEEE Consum. Electron. Mag. **7**(1), 18–28 (2018)
42. Governament of Abu Dhabi: The Abu Dhabi economic vision 2030. Abu Dhabi: Abu Dhabi Council for Economic Development and others (2008)
43. Lee, J.H., Hancock, M.: Toward a framework for smart cities: a comparison of Seoul. Research Chapter. Yonsei University and Stanford University, San Francisco (2012)
44. Marsal-Llacuna, M.L., Colomer-Llinàs, J., Meléndez-Frigola, J.: Lessons in urban monitoring taken from sustainable and livable cities to better address the Smart Cities initiative, Technological Forecasting and Social Change (2014)
45. Zygiaris, S.: Smart city reference model: assisting planners to conceptualize the building of smart city innovation ecosystems. J. Knowl. Econ. **4**(2), 217–231 (2013)
46. Koo, C., et al.: Conceptualization of smart tourism destination competitiveness. Asia Pac. J. Inf. Syst. **26**(4), 367–384 (2016)
47. Salah, A.M., Lela, M., Al-Zubaidy, S.: Smart education environment system. GESJ: Comput. Sci. Telecommun. **4**(44), 21–26 (2014)
48. Abba, S., Beauty, P.: Smart framework for environmental pollution monitoring and control system using IoT-based technology. Sens. Transducers **229**(1), 93 (2019)
49. Perego, P., Moltani, A., Andreoni, G.: Sport monitoring with Smart Wearable System (2012)
50. Jokic, P., Magno, M.: Powering smart wearable systems with flexible solar energy harvesting. In: IEEE International Symposium on Circuits and Systems (ISCAS) (2017)
51. Lee, H.K., Kim, T.: Prototype of IoT enabled smart factory. ICIC Int. **7**(4), 955–960 (2016)
52. Shouran, Z., Ashari, A., Priyambodo, T.: Internet of Things (IoT) of smart home: privacy and security. Int. J. Comput. Appl. **182**(39), 0975–8887 (2019)
53. Kodali, R.K., Jain, V., Bose, S., Boppana, L.: IoT based smart security and home automation system. In: International Conference on Computing, Communication and Automation (ICCCA) (2016)
54. Malche, T., Maheshwary, P.: Internet of Things (IoT) for building smart home system. In: International Conference on I-SMAC (IoT in Social, Mobile, Analytics and Cloud) (I-SMAC) (2017)
55. Hsu, Y.-L., et al.: Design and implementation of a smart home system using multi-sensor data fusion technology. Sensors **17**, 1–21 (2017)
56. Rihai, Y.: Big data and big data analytics: concepts. Types Technol. Int. J. Res. Eng. **5**(9), 524–528 (2018)
57. Ajah, I.A., Nweke, H.F.: Big data and business analytics: trends, platforms, success factors and applications. Big Data Cogn. Comput. **3**, 32 (2019)
58. Alsharif, M.H., Kelechi, A.H., Yahya, K., Chaudhry, S.A.: Machine learning algorithms for smart data analysis in internet of things environment: taxonomies and research trends. Symmetry **12**(1), 88 (2020)
59. Zimek, A., Schubert, E., Kriegel, H.P.: A survey on unsupervised outlier detection in high-dimensional numerical data. Stat. Anal. Data Min. **5**(5), 363–387 (2012)
60. Wang, J., et al.: A survey on data cleaning methods in cyberspace. In: IEEE Second International Conference on Data Science in Cyberspace (DSC) (2017)
61. El Faouzi, N.-E., Klein, L.A.: Data fusion for ITS: techniques and research needs. Transp. Res. Procedia **15**, 495–512 (2016)
62. Andò, B., Baglio, S.: A multisensor data-fusion approach for ADL and fall classification. IEEE Trans. Instrum. Measur. **65**, 1960–1967 (2016)
63. Khalifa, N.E.M., et al.: Deep Iris: deep learning for gender classification through iris patterns. Acta Inform. Med. **27**(2), 96–102 (2019)
64. Zheng, J.G.: Data visualization for business intelligence. In Global Business Intelligence, Chap. 6. Taylor & Francis (2017)

65. El-Din, D.M., Hassanien, A.E. and Hassanien, E.E.: Information integrity for multi-sensors data fusion in smart mobility. In: Toward Social Internet of Things (SIoT): Enabling Technologies, Architectures and Applications (2020)
66. Deogirikar, J., Vidhate, A.: Security attacks in IoT: a survey. In: International conference on I-SMAC (IoT in Social, Mobile, Analytics and Cloud) (2017)
67. Bendovschi, A.: Cyber-attacks – trends, patterns and security countermeasures. Procedia Econ. Finan. **28**, 24–31 (2015)
68. Oladimeji Biodun, S., Chukwudebe, G., Agbakwuru, A.O., Efe, O.C.: Comparative study of multi-factor authentication systems. İnt. J. Adv. Res. Sci. Eng. Technol. **6**(4), 8785–8791(2019)
69. Dutta, D., et al.: Social İnternet of Things (SIoT): transforming smart object to social object. In: Conference NCMAC (2015)
70. Atzori, L., et al.: The social Internet of Things (SIoT) – when social networks meet the Internet of Things: concept, architecture and network characterization. Comput. Netw. **56**(16), 3594–3608 (2012)

Synonym Multi-keyword Search Over Encrypted Data Using Hierarchical Bloom Filters Index

Azza A. Ali and Shereen Saleh

Abstract Search on encrypted data refers to the capability to identify and retrieve a set of objects from an encrypted collection that suit the query without decrypting the data. Users probably search not only exact or fuzzy keyword due to their lack of data content. Therefore, they might be search the same meaning of stored word but different in structure. So, this chapter presents synonym multi-keyword search over encrypted data with secure index represented by hierarchical bloom filters structure. The hierarchical index structure improves the search process, and it can be efficiently maintained and constructed. Extensive analysis acquired through controlled experiments and observations on selected data show that the proposed scheme is efficient, accurate and secure.

Keywords Searchable encryption · Synonym keyword · Bloom filter · Hierarchical bloom filter · Locality sensitive hashing

1 Introduction

Cloud storage service has become more popular as providing a lot of benefits over traditional available storage solutions. At first users save their data on un-trusted third party servers without any encryption, as a result this information is become available to everyone and susceptible to physical and harmful attacks. Users and companies who use storage to store and search data which used in various aspects of their businesses are becoming worried about loss of their data integrity and privacy through unauthorized access threats. The data may include several million records; at least some of these companies wish to keep some records private such as the

A. A. Ali (✉)
Department of Computer Science, College of Science and Humanities, Imam Abdulrahman Bin Faisal University, 31961, Jubail, Kingdom of Saudi Arabia
e-mail: aaaali@iau.edu.sa

S. Saleh
Math and Computer Science, College of Science, Minufiya University, Shibin El Kom, Egypt
e-mail: shereensaleh357@gmail.com

© The Author(s), under exclusive license to Springer Nature Switzerland AG 2021
A. E. Hassanien et al. (eds.), *Machine Learning and Big Data Analytics Paradigms: Analysis, Applications and Challenges*, Studies in Big Data 77, https://doi.org/10.1007/978-3-030-59338-4_25

customer private information. This information may be of value to others who may have a malicious intent. If the adversary be able to obtain such private information, he will create problems for the company, its customers, or both. The main objective of outsourced storage and stored data in the server is to provide confidentiality and to keep privacy of the data. Moreover, firms and big companies tend to keep away from using the outsourced storage service for fear of becoming their data known.

One common method used to protect sensitive information and to comply with privacy regulations or policies is encryption, where the user needs to cipher his information before uploading it on the server. However, use of encrypted data raises other issues, how the user submit a search request to the server to retrieve particular data without disclosure of any information or decrypting all of data. The idea is when a user wants to submit a query; he has to download all the encrypted data. It can be considered as a solution that maintains the privacy of the user, but practically this solution is not convenient due to the huge amount of data that should be downloaded, the client side is limited in computational resources and the bandwidth between the client and the server is limited. Searchable encryption is a recent concept that performs searches over the data which encrypted and stored in un-trusted server without any leak of information. The main idea is to be able to perform an encrypted search query without having to download all the encrypted data. Indeed, searchable encryption is consisting of two phases:

- Kept encrypted data on the un-trusted third party (Store process).
- Submit secure encrypted request to search and retrieve the desired information (Search process).

Confirmation the idea of searching and retrieving over encrypted data will be clarified in the following example:

> Jack has a set of documents and he stores them in an encrypted form into an un-trusted third party server. Jack wants to retrieve documents that contained the word "urgent" among his encrypted documents, he sends an encrypted request to the server where the word that he wants to search about is hidden, in such way the servers doesn't know any information about the content of this request. The server has to send to Jack all encrypted documents containing the word "urgent".

> The previous example deals with a kind of searchable encryption called symmetric encryption. There are other approaches that will be illustrated it in the next section.

Many approaches support search over encrypted data [1–8], and they are based on storing additional information (i.e. metadata) which by itself does not seep out any information about the data, but enables the predicates to be evaluated over the encrypted representation.

With huge increasing in data size, especially with Cloud data, searching mechanisms becomes easier by using bloom filters structure. Recently most papers used bloom filter data structure in encrypted data search [9–13], and it leads to solve most problems of searching over Cloud data.

The proposed scheme (Synonym Multi-Key-SHBlooFi) uses Hierarchical index structure for bloom filters HBlooFi in searching over encrypted data over cloud. The proposed Synonym Multi-Key-SHBlooFi scheme enhances the search process of Wang scheme [9]. Results show that Hierarchical Bloom Filter Index introduces efficient and a scalable solution for searching through a huge number of bloom filters compared with Wang scheme [9].

The rest of the chapter is organized as follows. Section 2 represents the related work such as Wang scheme [9], and hierarchical index structure for bloom filters [14]. Section 3 represents the proposed Synonym Multi keyword search scheme over encrypted data (Synonym Multi-Key-SHBlooFi). Evaluation of the proposed Synonym Multi-Key-SHBlooFi scheme is represented in Sect. 4. Finally, Sect. 5 is the conclusion of this chapter.

2 Related Work

All approaches need to store the data in a server, and there are users whose own the data and perform the encryption process on it then outsource the data to the server, after that perform the search process.

There are two search approaches over encrypted data:

1. Symmetric searchable encryption
   Also called private-key cryptography is one of the oldest and most secure encryption methods. The term "private key" comes from the fact that the key used to encrypt and decrypt data must remain secure because anyone with access to it can read the coded messages. The main attribute of this model of private search is that the sender encrypts a message into cipher-text using a key, and the receiver uses the same key to decrypt it [15].
   People can use this encryption method as either a "stream" cipher or a "block" cipher, depending on the amount of data being encrypted or decrypted at a time. A stream cipher encrypts data one character at a time as it is sent or received, while a block cipher processes fixed blocks of data. The symmetric scheme was introduced by Goldreich and R. ostrovsky [16] and supposes that the user encrypts his data with a secret key, kept it in the un-trusted server, can retrieve his encrypted data then decrypts it with the same key. Several searchable symmetric schemes were done ([1, 2, 16–18]). Not only complexity was improved but also search features such as conjunctive search [15] were added in recent works. Common symmetric encryption algorithms include Data Encryption Standard (DES), Advanced Encryption Standard (AES), and International Data Encryption Algorithm (IDEA) [19]. Everybody who is supposed to be able to read the information must have the key. The problem with this sort of code is that the key has to be given to them over a secure line.

2. Asymmetric searchable encryption

    Data encrypted with the recipient's public key can only be decrypted with the corresponding private key. Data can therefore be transferred without the risk of unauthorized access to the data [20]. This use can also provide assurance of the identity of the sender or recipient of the communication. This is done using a process called digital signing. A message signed with the private key of the sender can be verified by the recipient using the corresponding public key. Certificates for signing communications can also be issued by trusted third parties (such as Certificate Authorities) who can provide further assurance that the owner of a particular key pair is who they say they are [21].

    Currently, there are several applications that need more than a symmetric searchable encryption. It is possible to imagine a user who wants to search in a public data to retrieve encrypted document he didn't store herself. This situation would be impossible to solve unless he has a common secret with the person who encrypted these data. This scheme introduced by Boneh et al. [3], by analogy to cryptographic scheme, allows a number of users who have a public key to store data in the un-trusted server but only the person who had the secret key can perform a search to test of the occurrence of a word. This scheme is exposed to a number of developments either in complexity [22] or search features [23, 24] that allow searching for a set of keyword at once.

    Many schemes have been proposed in research that enable keyword searching on encrypted data in cloud computing. These schemes build their index files based on the actual keywords extracted or variations of keywords that were generated based on the predefined edit distance value. These indexes can only support search with keywords that are identical to the actual keywords or keywords that have very similar structures. Aiming at tolerance of both minor typos and format inconsistencies in the user search input, fuzzy keyword search over encrypted cloud data has been proposed. The users may predefine the edit distance value to a larger number to increase the range of the search result, but at the same time, it will decrease the search quality because more unrelated keywords will be returned in the search results, also, it degrades the search performance because the size of the index will increase.

    As an attempt to enhance query predicates, conjunctive keyword search over encrypted data have also been proposed. Conjunctive keyword search returns "all-or-nothing", which means it only returns those documents in which all the keywords specified by the search query appear; disjunctive keyword search returns undifferentiated results, which means it returns every document that contains a subset of the specific keywords, even only one keyword of interest.

    Other schemes support semantic search (keywords that share similar meanings with the original keywords but have a very different word structure).

    The first construction of searching over encrypted data was proposed by Song et al. in 2000 [1]. The scheme didn't contain an index, it only encrypts each word in the file independently, and thus, the search operation went through the entire file. This method is simple and fast, but it uses a sequential searching over data, and it isn't suitable of large amount of data size. The scheme is too slow in searching for a large number of documents.

In 2002, Goh [2] developed Per-file searchable index schemes. He used a bloom filter to construct the indexes for the data files, which reduce the cost of searching corresponding to number of files.

As a complementary approach, the first searchable encryption system using the public key system is proposed, by Boneh et al. [3], in which server contains encrypted files and keywords. User creates keyword trapdoor T_W using its private key to search w. The server checks T_W with existing encrypted keywords and sends encrypted file that match it. In Boneh scheme, the trapdoor may be memorized and then it well reveals knowledge about the keyword [3]. G. Duntao scheme [25] tried to solve the problem of memorized trapdoor. Based on Boneh's scheme, G. Duntao et al. proposed a temporary keyword search scheme over public key encryption. G. Duntao scheme solves the problem of the memorized trapdoor. G. Duntao scheme divides the time into a few time slides, and generates a temporary trapdoor for corresponding time slides. The trapdoor of a keyword in some time t_1 doesn't reveal anything about the trapdoor at time t_2.

In 2010, Li et al. [4] proposed a first technique using fuzzy keyword search scheme over encrypted cloud data. Li scheme used the metric of edit distance to construct a fuzzy keyword search scheme. In 2011, C. Lu et al. [26] proposed a construction of dictionary-based fuzzy set. This dictionary contains each keyword and its corresponding fuzzy keywords by using edit distance. This scheme reduces each of index size, storage, and the overhead computation, where the dictionary contains all fuzzy keywords.

In 2011, T. Balamuralikrishna et al. [27] proposed a brand new Symbol-based Trie-Traverse searching scheme to enhance the search efficiency, where a multi-way tree is constructed for storing the fuzzy keyword set over a finite symbol set. All the trapdoors sharing a common prefix have common nodes. The fuzzy keyword in the trie can be found by depth first search approach.

In 2011, when N. Cao et al. made the first attempt to define and solve multi-keyword ranked query problem(kNN) [28]. In N. Cao et al., data owner first defines a set of keywords and builds a dictionary containing them. The dictionary contains an index vector p built for each keyword. If the file in dataset contains a keyword, the element $p[i]$ is set to 1, otherwise, $p[i]$ is set to 0. To execute multi keyword ranked query q, firstly sent a trapdoor for query keywords set to cloud server provider (CSP). CSP uses inner product between the trapdoors and $p[i]$ to determine the similarity between them. Finally, the first k results with the highest scores are returned to user. But in this scheme, position of each keyword in the dictionary is fixed, so it must rebuilt it when the number of keywords increased.

In 2013, Tseng et al. [29] proposed IPEKS, where cloud storage provider maintains two lists, C-list and N-list. N-list contains files identifier, and C-list contains files identifiers for pervious searching. CSP searches C-list first to obtain file identifier. If the file identifier isn't in C-list, the CSP searches the N-list to obtain file identifier and then includes this information to C-list.

In 2014, Fu et al. [30] proposed a searchable encrypted scheme which supports both multi-keyword ranked search and synonym-based search. Z. Fu scheme uses a vector space model (VSM) to build document index. Searchable index tree is constructed with the document index vectors. So the related documents can be found by traversing the tree.

In 2016, Jin Shi et al. [31] proposed a new cloud data search strategy based on fuzzy multiple keywords, as the Chinese and English data mixed more and more, and the existing fuzzy retrieval scheme supports only single language retrieval, using English and Chinese comparison table convert Chinese keyword to English keyword to support Chinese and English keyword search. Realize the efficient retrieval of fuzzy multi-keyword in the hybrid data improving the efficiency of retrieval.

In 2018, Ge et al. [32] proposed a verifiable exact keyword search scheme and then extend to fuzzy keyword search scheme. In X. Ge et al. scheme built a linked list with three nodes for the same keyword and produce a fuzzy keyword set for it. One index vector for each fuzzy keyword set was generated to decrease the storage space and computation cost.

2.1 Hierarchical Index Structure for Bloom Filters

Bloom filter is a data structure used to check whether a specific element exists in a specific set or not. Every bloom filter is represented by an array of bit values (0's and 1's) with length m. Values of bloom filter array are created by utilizing a set of k hash functions. The blank bloom filter has value zero in all bits. A new element can be inserted into bloom filter by applying k-hash functions on that new element, and then every result of k-hash functions denotes to a specific position in the bloom filter. The bit value of each position in bloom filter becomes *one* as illustrated in Fig. 1. To check whether that element exists in the set of bloom filters or not, k of hash functions of the specific element is computed and then searching for the positions in the bloom filters with the same values resulted from k hash functions. If any used hash function result index position in the bloom filter has value *zero*, the test element does not exist within the set, with probability 1. But in case of all used hash functions result

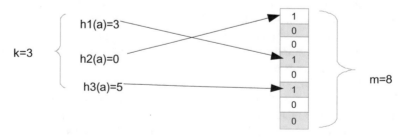

Fig. 1 Using k hash functions with bloom filter

index position in the bloom filter have values one, it means that the test element exist within the bloom filter. It should be taking into consideration that, having one in the specified places in bloom filter does not necessarily mean existence of the element in the set which represented by the bloom filter; it may be a false positive or a true positive.

Dataset index file contains the bloom filter of each file. Bloom filters structure can be utilized to examine existence of specific object in the set (match) or whether the object without doubt not exists within the set (no match). There is a possibility of existence false positives, but false negatives unlikely to occur.

For example, suppose text file F contains set of words as: $F = \{w_1, w_2, \ldots, w_n\}$ and r independent pseudorandom functions $h_i : \{0, 1\}^* \rightarrow \{1, m\}$ where m is the size of bloom filter that represents the text file F. To test whether the word "network" is exists in the bloom filter. It should be calculated h_1 ("network"), h_2 ("network"). If therefore $\forall i \ni [1, r] h_i$ ("network") $= 1$ therefore h_1 "network" $\ni F$ (may be), else if at least one value is equal to zero then "network" $\notin$ F. It can be said that "network" may belong to F because there are false positives since one hash functions gave the same result for different element as explained in Fig. 2.

Wang et al. [9] proposed a privacy-preserving multi-keyword fuzzy search over encrypted data in the Cloud. Wang built an index for each file; this index file contains all words in file and represents it as a bloom filter. The index file supports multiple keyword searches without needing for increasing the index. The search operation in Wang scheme passes all files within the dataset. Thus, the search time of Wang scheme grows linearly with growing of size of the documents set.

Hierarchical Bloom filter Index (HBlooFi) was proposed by Adina [14] as a searching data structure, which can accommodate to tens of thousands of bloom filters with low maintenance cost. This paper suggests a scheme to check existence of specific object in data. Searching over HBlooFi is based on the following idea, first it must be noted that the leaves of the tree are represented the original bloom filters index, and the bloom filters of parent nodes are acquired by applying a bitwise-OR on child nodes. This process keeps going until the root is reached. Query starts from the root bloom filter traverses the tree down the path until reach to the leaf level. To examine that specific object exist in specific node in leaf level, it should satisfy the

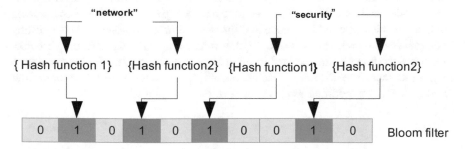

Fig. 2 Example for insert word in bloom filter

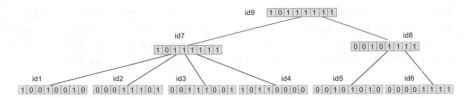

Fig. 3 HBlooFi tree of order 2

condition that all the position's value in bloom filter that match result from k hash functions must be equal one.

Construction of HBlooFi tree is based on the following steps: *firstly*, generating the outer nodes (leaves) of the tree which represent the original Bloom filters index of data. *Secondly*, by using a bitwise-OR on the child nodes, the parent nodes can be obtained within the bloom filters tree. This process keeps going till the root node is reached. The HBlooFi has a feature that every inner nodes bloom filter among the tree represents the union of the sets that formed by the bloom filters among the sub-tree rooted at that node. Figure 3 illustrates an example of HBlooFi of order two. Suppose the threshold $d = 2$(then every inner node has between d and $2d$ child nodes). The leaf nodes of the tree include the actual bloom filters, with identifiers 1, 2, 3, 4, 5, and 6.

Now algorithms of inserting, deleting, and updating Bloom filters will be introduce by HBlooFi structure.

- Insert

In the insertion operation, HBlooFi scheme starts searching for a leaf node within the tree that is "close" to the input bloom filter wanted to be inserted into the tree. Searching for the closest leaf node depends on the distance measured between leaf node bloom filter and the desired inserted bloom filter.

Once the closest node found, if the parent node for the closest node still has no more than $2d$ children's (where d specific threshold), add the new one bloom filter with its child leave nodes, and then the insertion process is completed. On the other hand, if the number of parent child exceeds the specified threshold $2d$; therefore the parent node must be split. Splitting is done and new parent node has been created contains the closest child leaf node and the desired inserted node. A new splitting node (new parent node) has new bloom filter value computed by performing bitwise-OR operation between the desired inserted bloom filter node and the closest bloom filter child node. The desired inserted bloom filter node and the closest bloom filter child node will be modified as child nodes of the new parent node.

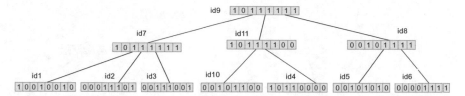

Fig. 4 Bloom filter tree after insert and split

For example, to insert the bloom filter with value "00101100" into the tree of bloom filters in Fig. 3. First search for the nearest node to the new node by using hamming distance as the metric distance. As a result, it is found that distances, measured between the new node and nodes id_7 and id_8, are equal to $four$. Let's assume that we chose id_7 as the nearest node. Therefore the new node is inserted as a child node of id_7 which has more than $four$ child nodes after inserting new one, so id_7 should be split and create new inner node as shown in Fig. 4.

Algorithm 1: insert (new Bloom filter) into tree

1. Input: new Bloom Filter value
 	
//Loop through all nodes in tree to find the child node with the closest value to the new Bloom filter using Hamming distanc.
   2. for i = 1 to tree.size() **do**
   3.   closestNode = tree.get(i)
   4.   currentDistance = closestNode.value.computeDistance(new bloom filter)
 	
// Compute the Hamming distance between this currentNode value and the new Bloom filter value given as Hamming distance = number of positions with different value. Which mean XOR the two values and take the sum or all 1s.
   5.   closestNode.value.xorcardinality (new Bloom filter);
   6. End for
   7. closestNode.children.add(newNode)
   8. newNode.value=new Bloom filter
   9. if !(closestNode. childrens after insertion process are overflow(more than 2*d +1
      children)) where (d=2) represents order of tree **then**
   10. Updata closestNode.value until reach to root node.
   11. Re-compute bitwise OR between closestNode.value and new inserted bloom filter value.
   12. Else
 	
//Split the closestNode by create new node, and move last half children of closestNode to new node.
   13. for i = order + 1 to closestNode.children.size() **do**
   14.   Insert (closestNode.children.get(i)) to newNode created
   15. End for
   16. Re-compute bitwise OR between newNode.value and new inserted node value.
   17. Updata newNode.value until reach to root node.
 	
// Remove the last half of the children for the closestNode
   18.   closestNode.children.clear() (from (order + 1) to closestNode.children.size())

A. A. Ali and S. Saleh

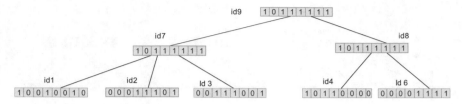

Fig. 5 Bloom filter tree after delete and redistribute

- Delete

The deletion algorithm operates on remove specific node from the bloom filter index tree. If the node, needed to be deleted, is a leaf node, then delete the pointer to this node from its parent node, and check if the parent node is not under flowing (at least 2 children are exist). The bloom filter values for the remaining nodes must be recomputed by using bitwise-OR of their children in the path from the parent node to the root node. If there is an underflow happened (mean there are node contains children under specific threshold after the delete process), then parent node tries to redistribute its child nodes with a sibling and update the sub tree contain the moving node to the root. Or the parent node merges with a sibling, by giving all its entries to the sibling. The bloom filter value in the sibling node is updated, and the delete procedure is called recursively for the parent node, where the delete propagates up to the root.

For Example, assume that id_5 node, with value "00101010" from Fig. 3, is needed to be removed. The final tree result after deletion process of node id_5 and redistribution between id_7 and id_8 is illustrated in Fig. 5.

Algorithm 2: delete (specific node) from tree
//Delete the given node from the index. The deletion moves bottom up.
// remove node from the list of children in its parent
1.   currentNode = node.parent
2.   currentNode.children.remove(node)
// check if underflow at the parent (parent.childrens less than specified order)
3.   **if** (!parent.needMerge()) **do**
// no underflow, update values.
4.   **For** i=1 to parent.child.size() **do**
5.   Parent.value OR parent.child.get(i).value
6.   Re-compute values of the nodes to the root.
7.   **End for**
8.   **Else**
// Merge the childrens between parent node and its sibling; all the childrens from the parent node are given to the sibling, then the sibling node check if need split and updated to be the OR of all children.
9.   **For** i = 0 to parent.children.size() **do**
10.   childToMove = parentNode.children.remove(node.children.size() .get(i)
11.   sibling.children.add(childToMove)
12.   sibling.value.OR(childToMove.value)
13.   childToMove.parent = sibling
14.   **End for**
//Check if sibling is overflow, then create new node and insert the last half of the sibling node children into the new node.
15.
16.   **for** i = order + 1 to siblingNode.children.size(); where order=2 **do**
17.   Insert (siblingNode.children.get(i)) to newNode created
18.   **End for**
19. Re-compute bitwise OR between newNode.value and new inserted node value.
20. Re-compute bitwise OR between siblingNode.value and all its child node values.
21. Updata siblingNode.value and newNode.value until reach to root node.
// Remove the last half of the children for the siblingNode
22. siblingNode.children.clear() (from order + 1 to siblingNode.children.size())

- Update

The input parameters of the update algorithm are the new value of bloom filter and the desired updated leaf node. Once update specific bloom filter value within the tree, all bloom filters values will be updated, using bitwise-OR, within the path from the leaf to the root with the new value.

For Example: to update node id_6 with new value "11001111", values of all the pass nodes, from node id_6 to the root, will be recomputed using bitwise-OR with "11001111" and the final structure of tree is shown in Fig. 6.

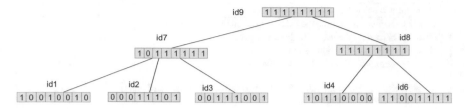

Fig. 6 Bloom filter tree after update

Algorithm 3: update (current node, new Bloom filter) in tree
//Update value of the current node with the new Bloom filter
1. currentNode.value.OR BloomFilter(newValue)
2. Re-compute values to currentNode.parent.
3. currentNode.parent.value.OR currentNode.value
4. Update values of nodes in tree to the root if needed.

2.2 Multi Keyword Search Scheme Over Encrypted Data (Wang Scheme)

In 2014, Wang and et al. presented multi-keyword fuzzy search scheme over encrypted data in the Cloud [9]. Wang scheme builds an index I_D for each file D. The index I_D contains all the keywords in file D. Convert the file to m-bit and then treated as bloom filter [10]. In Wang scheme, read file word by word and convert each word into its bigram vector representation using this bigram as input to specific Locality Sensitive Hashing LSH functions family using the result from this function in insert process within the Bloom filter. Generating query which containing multiple keywords to be searched can be done by following the same steps used to create the index. The search process includes a simple inner product operation done between the index bloom filter values and the query bloom filter values. If the keyword(s) in the query exist in specific file, therefore the bloom filter represent this file contains bit 1 corresponding to the same bit position in query bloom filter which result that the inner product between them can came a high value. The steps of Wang scheme are explained as following steps:

1. *keyGen*: Given a security parameter m, output the secret key $sk(M_1, M_2, S)$, where $M_1, M_2 \varepsilon R_{m \times m}$ are invertible matrices and $S\varepsilon\{0, 1\}m$ is a vector.
2. *BuildIndex(D, SK, L)*: choose L independent LSH functions, build a m-bit bloom filter I_D to represent the index for every file D:

   - Obtain the keywords set from D.
   - For each keyword, using specific LSH functions family to insert it into bloom filter index.

- Encrypt the index I_D using $\textbf{\textit{Index\_Enc}}(SK, I_D)$ and output $\text{Enc}_{sk}(I_D)$.

3. *Index_Enc(SK, I)*: The bloom filter index I is divided into two vectors $\{I', I''\}$ by using the rule: for each element $i_j \varepsilon I$, set $i'_j = i''_j = i_j$ if $s_j \varepsilon S$ is one; else $i'_j = \frac{1}{2}i_j + r$, $i''_j = 1/2i_j - r$, as r is a random number. Then encrypt $\{I', I''\}$ with (M_1, M_2) into $\{M_1^T.I', M_2^T.I''\}$. Output $\text{Enc}_{sk}(I) = \{M_1^T.I', M_2^T.I''\}$ as the secure index.

4. *Trapdoor* : Create bloom filter of length m-bit for the query Q. For every keyword wanted to be searched, insert it into bloom filter using the same LSH functions family used in creating index. Then using $Query - Enc(SK, Q)$ to encrypt query.

5. *Query–Enc(Sk, Q)*: The query Q is divided into two vectors $\{Q', Q''\}$ using the rule: $q'_j = q''_j = q_j$ if $s_j \varepsilon S$ is zero; else $q'_j = 1/2q_j + r'$, $q''_j = 1/2q_j - r'$ where r' is another random number. Then encrypt two vectors $\{Q', Q''\}$ as $\{M_1^{-1}.Q', M_2^{-1}.Q''\}$ which result the trapdoor represented by $\text{Enc}_{sk}(Q) = \{M_1^{-1}.Q', M_2^{-1}.Q''\}$.

6. *Search* $(Enc_{sk}(Q), Enc_{sk}(I_D))$: Perform the inner product operation $\langle \text{Enc}_{sk}(Q), \text{Enc}_{sk}(I_D)\rangle$ as the output of the search process between the query Q and the document D.

Wang scheme is distinguished from the previous search schemes [6, 33] as it enables the update process efficiently, due to the fact that each document is indexed individually. So when trying adding, deleting and modifying file can be done efficiently, compared to the other schemes, including only indexes of the files that will be modified, not affecting other files.

The following figure represents an example that illustrates Wang algorithm, which represent file by bloom filter and encrypt bloom filter using secret key. Then Use inner product to search:

- Between index vector and query vector.
- High result means that keyword is in file.

As illustrated from previous example in Fig. 7:

- Index encryption represents as: ((0.07 0.08 0.77), (1.11 −1.0 −0.16))
- And let query encryption is: ((0.01 0.5 0.8), (−0.2 0.3 −0.7)).

Then inner product between query and index is:

$$[(\text{index encryption}) \text{ inner product (query encryption)}]$$
$$= ((0.07\ 0.08\ 0.77).(0.01\ 0.5\ 0.8)) + ((1.11\ -1.0\ -0.16).(-0.2\ 0.3\ -0.7))$$
$$= 0.2467$$

Follow these steps between keyword search and other files to return result with high inner product between query and index. Which denote that files may be containing the keyword search more than other files.

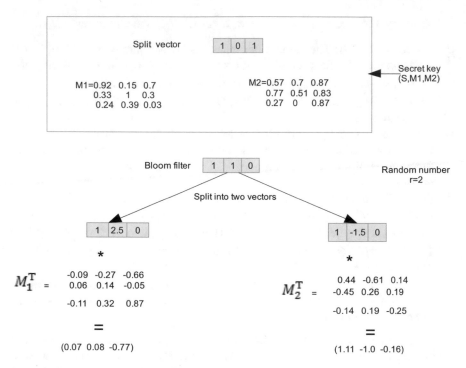

Fig. 7 Example of Wang scheme algorithm

3 The Proposed Synonym Multi Keyword Search Scheme Over Encrypted Data

The proposed Synonym Multi keyword search scheme over encrypted data can be illustrated in the following subsections.

3.1 Constructing the Bloom Filter for Each File in Dataset Collection (Algorithm.4)

First read word by word from file then extract all synonyms of each word and converting each word to its bigram vector and using LSH Euclidean hash function family to insert each word in bloom filter as illustrated in Fig. 8.

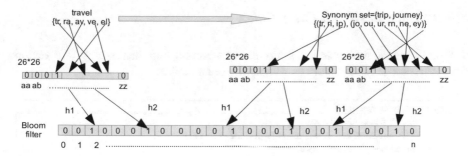

Fig. 8 Construct bloom filter

Algorithm 4: Creating bloom filter
1. **Input:** File (F). **Output:** bloom filter (bf).
2. **For each** word W_i in File (F), find list of all synonym words for W_i .
3. Words$\{W_1, W_2, W_3, \ldots . W_n\}$. Where n is the number of synonym words for W_i .
4. **For** $i = 1\, to\, n$
5. Compute $LSH(W_i)$. Where LSH locality sensitive hash function.
6. Find all positions of bloom filter bf that equal to all the values of $LSH(W_i)$ result.
7. Change position's values in bloom filter bf to be equal one.
8. **End** for
9. **End** for

3.2   Constructing the Bloom Filter Tree (Algorithm.5)

For each document generate a leaf node where store bloom filter, then build the tree
from down to up traversal with all leaf nodes generated where internal nodes created
from bitwise OR of its child's values.

Algorithm 5: Constructing the bloom filter tree
1. **Data:** bloom filter list $bfList$ represent all files in dataset
2. **Sort** bloom filters in $bfList$ according to specific distance (hamming distance).
3. **For** each bloom filter bf of $bfList$ **do**
4. Create leaf node in the tree.
5. Set name and value (bloom filter) for each created leaf node
6. Insert right each bloom filter to leaf node
7. **End** for
Perform bitwise OR bloom filters for all interior nodes, recursively computing them traverse all the tree from down to up.

3.3   Encrypt Bloom Filter Index

Split each bloom filter in leaf node into two vectors, and then encrypt two vectors
with transpose of two matrices. Output the secure index.

Algorithm 6: encrypt tree T
1. **Input**: tree of bloom filters
2.   generate secret key as two matrix M_1, M_2 and split vector S
3. **for** each leaf node in tree **do**
4.   split bloom filter I into two vectors I', I''
5.   **for** each element $i_j \in I$
6.   **If** $s_j = 1$ **(where** $s_j \in S$**)**
7.   set $i'_j = i''_j = ij$ **(where** $i' \in I'$, $i'' \in I''$**)**
8.   **Else**
9. $i'_j = 1/2\, i_j + m,\ i''_j = 1/2\, i_j - m'$ where m is a random number
10.   **End for**
11. $\text{Enc}_{sk}(I) = \{M_1^T.I', M_2^T.I''\}$
End for

3.4   Search

The search algorithm starts from root node traverse down to leaf nodes in tree.
Following steps described in algorithm 4. When algorithm find node is inner node,
first check the condition that the hash result of query has values *one* in that inner node
value. If the condition met then the search function is called recursively for all child
nodes of this inner node until reach to leaf nodes. When reaches to leaf-level which
rooted to specific node, algorithm computes inner product between query bloom filter
value and leaf nodes values and return the identifier of the leaf node that produce
the biggest inner product between its value and the value of the query. Else if the
condition is not met, exit from this sub-tree and search in other inner node value.

Algorithm 7 : Search over encrypted tree
1.   **Input**: Encrypted tree T, query hash result H, two vectors of encrypted query $v1, v2$)
2.   **For** each node in the tree T **do.**
3.   **If** node is interior node N and contain all H.
4.   **for** (i = 0 ; i < $N.nbchildren$; $i + +$) **do**
5.   Multiply $v1, v2$ into leaf node values.
6.   **end for**
7.   set all result of multiplication in hash table HT < result, file   name >
8.   Return file name correspond to max result.
9.   **end if**
10.   **Else if** node is interior node N and not contain all H.
11.   Exit from this inner node
12.   Start search in other inner node in the tree calling *search* algorithm from step3.
13.   **End else if**
14.   **Else**
15.   this node is a leaf
16.   Just call step 5, 7
17.   **End else**
End for

4 Performance Analysis for the Proposed Scheme (Synonym Multi-key-SHBlooFi)

In this section, the overall analysis and experiments of the proposed algorithm will clarify on specific dataset: the Enron Email Dataset [34]. We choose randomly variety number of emails to create dataset. The entire experiment system is implemented by java language on windows 7 with Intel Core i5-2450 M Processor 2.50 GHZ. Using JWNL (Java WordNet Library) which is API used for accessing WordNet dictionary which consider a lexical database for the English language connecting with it and find set of words that have the same meaning for that word. WordNet is containing set of words that are synonyms of each other. Each of these sets is called a synset. Since a word can have more than one meaning, so more than one synset can be contain the same word.

Tables 1 and 2 represent the storage size, construction time and search time regard to the proposed encrypted data searching over BF-tree scheme compared with Wang scheme. Also, the results of Tables 1 and 2 show the memory space, that the index used, for each of the proposed scheme and Wang scheme. The proposed Synonym Multi-Key-SHBlooFi scheme uses little more index storage space compared to Wang scheme. Also, two Tables 1 and 2 demonstrate the searching time with each scheme which is affected by the increasing number of documents uploaded. Two tables illustrate the changes in construction index time once the numbers of files increased. In the proposed Synonym Multi-Key-SHBlooFi scheme, the index construction time is increased compared to the index construction time of Wang scheme.

Table 1 Experimental results obtained from the multi keyword fuzzy search (bloom filter list index) of Wang scheme

Text files size	Size of index in memory			Primitives						Construction time/s	Search time/s
	Objects	Non null references	Null references	Int	Long	Double	Char	Float	Boolean		
1 KB	166	165	328	84	–	32,800		–	–	0.832	0.004
20 KB	270	269	536	136	–	53,600		–	–	1.365	0.007
40 KB	642	641	1280	322	–	128,000		–	–	2.239	0.05
132 KB	1302	1301	2600	652	–	260,000		–	–	4.788	0.061
300 KB	2662	2661	5320	1332	–	532,000		–	–	11.572	0.09
512 KB	5350	5349	10,696	2676	–	1,069,600		–	–	15.592	0.379
1 M	12,674	12,673	25,344	6338	–	2,534,400		–	–	48.377	0.602
2 M	24,054	24,053	48,104	12,028	–	4,810,400		–	–	120.668	1.642
4 M	36,918	36,917	73,832	18,460	–	7,383,200		–	–	347.646	4.712
8 M	74,054	74,053	148,104	37,028	–	14,810,400		–	–	1139.759	43.676

Table 2 Results of the proposed encrypted data searching over the proposed Synonym Multi-Key-SHBlooFi scheme

Text files size	Size of index in memory			Primitives						Construction time/s	Search time/s
	Objects	Non null references	Null references	Int	Long	Double	Char	Float	Boolean		
1 KB	1791	2222	8142	1974	16,400	146,310	589	298	59	1.402	0.00
20 KB	3623	4502	16,482	3988	32,400	296,362	1120	604	121	1.877	0.00
40 KB	7223	8982	32,808	7948	64,400	591,642	2215	1204	241	2.568	0.00
132 KB	14,624	18,192	66,452	16,090	130,400	1,198,848	4543	2437	487	5.909	0.00
300 KB	29,956	37,272	136,161	32,954	266,400	2,456,200	9527	4993	999	13.238	0.00
512 KB	60,228	74,944	273,941	66,252	548,584	4,959,040	19,484	10,039	2009	16.589	0.01
1 M	142,551	177,390	647,033	156,810	1,299,276	1,173,927	64,375	23,758	4751	56.086	0.02
2 M	270,600	336,740	1,227,212	297,663	2,465,732	2,228,466	90,849	45,100	9020	137.026	0.04
4 M	415,304	516,816	1,888,967	456,838	3,784,288	3,420,194	141,125	69,217	13,843	379.429	0.10
8 M	833,100	1,036,740	3,791,409	916,413	7,590,732	6,860,966	286,155	138,850	27,770	1274.042	0.52

Tables 1 and 2 show the results of searching time for the proposed algorithm compared to Wang scheme algorithm. Encrypted data searching over proposed Synonym Multi-Key-SHBlooFi scheme is better than encrypted data searching over Wang scheme. When evaluating search time over the proposed scheme and search over Synonym Multi-Key-SHBlooFi for different number of bloom filters, it seems better than search time in Wang scheme, as there is no need to traverse all bloom filters in tree, only bloom filters that have high probability contained the keyword search have been checked. On the other hand, the search operation in Wang scheme includes passes all the files within the dataset perform inner product function. Thus, the search time of Wang scheme grows linearly with growing of size of the documents set, this can be a result of the search process must re-examine all the files within the dataset before the ultimate result get.

The performance analysis of the proposed scheme is illustrated the remaining of this section. The search cost, the search efficiency, efficiency of index construction, the storage requirements and result accuracy are computed compared with Wang scheme.

4.1 Search Cost

The time consumed for search is given by the invocations number of search process. In the proposed scheme there are three cases:

- **The best case**: in case of the tree does not contain leaf-level, contains only the root node, then only one node will be checked for matching, which can be considered as the best case.
- **Average case**: in case of existing the leaf-level, number of search invocations is equal to $O(d * log_d N)$, where N is the number of nodes, and d is number of checked nodes to find bloom filter that matches a query. In this case, one path will be followed and the nodes of this path will only be checked to find the one that match.
- **Worst case**: the worst case is $O(N)$, it will exist when examining all nodes in the tree to find the matched one.

4.2 Search Efficiency

The search time grows linearly with growing of variety number of the documents with the same size. Figure 9. Shows results of searching time of the proposed scheme compared with Wang scheme, when using different number of documents with the same size. The results from Fig. 9 show that the search time of the proposed scheme is smaller than in Wang scheme for various varieties of documents. This is often as a result of the search process in Wang scheme has to reconsider all the files within

Fig. 9 Searching time for Wang scheme and the proposed scheme

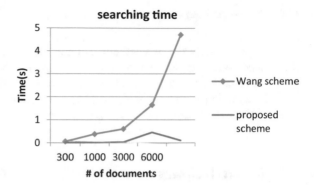

the dataset before the ultimate result get compared to the proposed scheme, as don't need to traverse all bloom filters of files within the same tree.

4.3 Efficiency of Index Construction

Also in Fig. 10, we are able to see the changes in construction index time with the proposed scheme and Wang scheme, once variety number of documents uploaded increases as described in Fig. 10. In the proposed Synonym Multi-Key-SHBlooFi scheme, time is increased compared with time Wang scheme. As the index construction of our proposed scheme includes first creating bloom filters and furthermore building the search tree. Expected it requires more time than the Wang scheme.

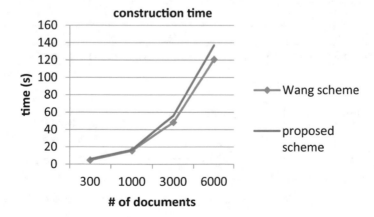

Fig. 10 Construction time for Wang scheme and the proposed scheme

4.4 Storage Requirements

In our experiments, different numbers of files are uploaded as a whole and monitoring the changes within the size of the index file. It is noticed that the index file created by Wang scheme smaller than the index file created by the proposed scheme.

This reduction in index file size proved that Wang scheme is efficient than the proposed scheme in terms of storage requirements.

4.5 Result Accuracy

Measure the search result accuracy using precision and recall performance metrics. Where tp means true positive, fp is false positive and fn is false negative. Where precision is the fraction/percentage of retrieved documents that are relevant to keyword search which equals to $\frac{tp}{tp+fp}$. And recall is the fraction/percentage of relevant docs that were retrieved from search that equals to $\frac{tp}{tp+fn}$. Figure 11 shows the performance metrics of Wang scheme and the proposed scheme according to number of hash functions k. One observation is that precision is very low when k is small. This is because multiple LSH functions are used together to enlarge the gap between query keyword search and words in index. So when the k is small, the gap is not big enough to distinguish the different keywords, and most of the files in the dataset have been returned, which leads to high false positive and low false negative. Another observation is that the recall drops when increasing the k. This is because increasing the k will cause more false negatives.

From the above we can summarize the comparison between the proposed scheme and Wang's scheme as the following:

Wang's scheme	Proposed scheme
Support multi-keyword fuzzy search	Support synonym multi-keyword search

(continued)

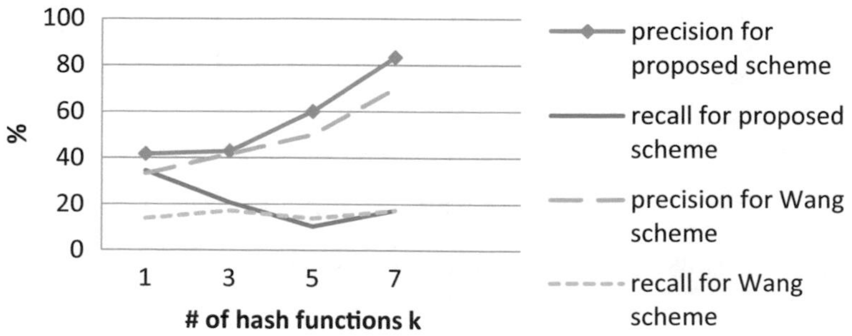

Fig. 11 Precision and recall for Wang scheme and the proposed scheme

(continued)

Wang's scheme	Proposed scheme
Need less storage than proposed scheme	Need little more storage than Wang's scheme
Less efficiency in search time than proposed scheme	Fast and more efficient in search time
Need less construction time than proposed scheme	Need construction time than Wang's scheme
Need complexity search time $O(n)$	Need complexity search time $O(d * log_d N)$

5 Conclusion

Searchable encryption represents a new technology for improving the process of outsourcing encrypted data on storage servers such as Cloud Computing infrastructures. Actually, the Cloud Computing has become a new concept to solve such problems by providing computing services on infrastructures via internet access. One of the main services that provided by Cloud computing is data storage (documents, mails, images, videos …) provided by many companies such as Google, Amazon, and Microsoft… On the one hand, Cloud Computing storage has many advantageous features as the data stored in cloud is available to access every time, everywhere, with reliability and scalability. On the other hand, Cloud computing users don't need to know either the location of the physical server that store the data and perform the search process or the way their data were processed and computed. What matters the most, is the outcome of their search queries.

This chapter introduced a proposed synonym multi-keyword search over encrypted data with secure index scheme for enhancing the searching time over encrypted data.

The proposed scheme was based on combining Wang scheme with secure index represented by hierarchical bloom filters structure. The experiments showed that the proposed scheme improved the search time compared with Wang scheme. Although, the proposed scheme increased the storage needed for the index and the time of the index construction, the proposed scheme achieving the desired goal of fasting search time and also support the synonym multi-keyword search property. In future work, we suggest implement the proposed algorithm in parallel using map-reduce mechanism to reduce construction time.

References

1. Song, D., Wagner, D., Perrig, A.: Practical techniques for searches on encrypted data, In: Proceedings of the IEEE Symposium on Security and Privacy, pp. 44–55 (2000)
2. Goh, E.-J.: Secure indexes. In: Cryptology ePrint Archive on October 7th, pp. 1–18 (2003)

3. Boneh, D., DiCrescenzo, G., Ostrovsky, R., Persiano, G.: Public key encryption with keyword search. In: Proceedings of Euro Crypto'04, pp. 506–522 (2004)

4. Li, J., Wang, Q., Cao, N., Ren, K., Lou, W.: Fuzzy keyword search over encrypted data in cloud computing. In: Proceedings of IEEE INFOCOM, pp. 1–5 (2010)

5. Jin Shi, X., Ping Hu, Sh.: Fuzzy multi-keyword query on encrypted data in the cloud. In: 2016 4th International Conference on Applied Computing and Information Technology/3rd International Conference on Computational Science/Intelligence and Applied Informatics/1st International Conference on Big Data, Cloud Computing, Data Science & Engineering (ACIT-CSII-BCD), pp. 419–425 (2016)

6. Xia Z., Wang, X., Sun, X., Wang, Q.: A secure and dynamic multi-keyword ranked search scheme over encrypted cloud data. IEEE Trans. Parallel Distrib. Syst. pp. 340–352 (2015)

7. Pandiaraja, P., Vijayakumar, P.: Efficient Multi-keyword Search over Encrypted Data in Untrusted Cloud Environment. In: 2017 Second International Conference on Recent Trends and Challenges in Computational Models (ICRTCCM), pp. 251–256 (2017)

8. Orencik, C., Kantarcioglu, M., Savas, E.: A practical and secure multi-keyword search method over encrypted cloud data. In: IEEE sixth International Conference on Cloud Computing(CLOUD), pp. 390–397 (2013)

9. Wang, B., Yu, S., Lou, W., Hou, Y.: Privacy preserving multi-keyword fuzzy search over encrypted data in the cloud. In: IEEE INFOCOM, IEEE Conference on Computer Communications, pp. 2112–2120 (2014)

10. Ali, F., Lu, S.: Searchable encryption with conjunctive field free keyword search scheme. In: 2016 International Conference on Network and Information Systems for Computers (ICNISC), pp. 260–264 (2016)

11. Fu, Z., Wu, X., Guan, Ch., Sun, X., Ren, K.: Toward efficient multi-keyword fuzzy search over encrypted outsourced data with accuracy improvement. IEEE Trans. Inf. Forensics Secur., pp. 2706–2716 (2016)

12. Umer, M., Azim, T., Pervez, Z.: Reducing communication cost of encrypted data search with compressed bloom filters. In: IEEE 16th International Symposium on Network Computing and Applications (NCA), pp. 1–4 (2017)

13. Poon, H., Miri, A.: A low storage phase search scheme based on bloom filters for encrypted cloud services. In: 2015 IEEE 2nd International Conference on Cyber Security and Cloud Computing, pp. 253–259 (2015)

14. Crainiceanu, A.: Bloofi: a hierarchical bloom filter index with applications to distributed data provenance. In: Proceedings of the 2nd International Workshop on Cloud Intelligence Article No. 4 (2013)

15. Sarker, M., Parvez, M.: A cost effective symmetric key cryptographic algorithm for small amount of data. In: 2005 Pakistan Section Multitopic Conference, INMIC (2005)

16. Goldreich O., Ostrovsky, R.: Software protection and simulation on Oblivious RAMs. J. ACM, **43**(3), 431–473 (1996)

17. Curtmola, R., Garay, J., Kamara, S., Ostrovsky, R.: Searchable symmetric encryption: improved definitions and efficient constructions (2006)

18. Chang, Y.C., Mitzenmacher, M.: Privacy preserving keyword searches on remote encrypted data. In: Applied Cryptography and Network Security Conference (ACNS) (2005)

19. Thakur, J., Nagesh. K.: DES, AES and Blowfish: Symmetric key cryptography algorithms simulation based performance analysis. Int. J. Emer. Technol. Adv. Eng. (2011)

20. Nithya, S., Raj, E.: Survey on asymmetric key cryptography algorithms. J. Adv. Comput. Commun. Technol. (2014)

21. Rivest, A., Shamir, A., Adleman, L.: A method for obtaining digital signatures and public-key cryptosystems. Commun. ACM (1978)

22. Abdalla, M., Bellare, M., Catalano, D., Kiltz, E., Kohno, T., Lange, T., Lee, J.M., Neven, G., Paillier, P., Shi, H.: Searchable encryption revisited: consistency properties, relation to anonymous IBE, and extensions. In: Shoup, V. (ed.) CRYPTO 2005. LNCS, vol. 3621, pp. 205–222. Springer, Heidelberg (2005)

23. Conjunctive, subset, and range queries on encrypted data; Dan Boneh, Brent Waters- Theory of Cryptography. Springer (2007)
24. Park, D., Kim, K., Lee, P.: Public key encryption with conjunctive field keyword search. In: Lim, C.H., Yung, M. (eds.) WISA 2004. LNCS, vol. 3325, pp. 73–86. Springer, Heidelberg (2005)
25. Duntao, G., Dawei, H., Haibin, C., Xiaoyuan, Y.: A new public key encryption with temporary keyword search. In: International Conference on Computer, Mechatronics, Control and Electronic Engineering(CMCE), vol. 4, pp. 80–83 (2010)
26. Liu, C., Zhu, L., Li, L., Tan, Y.: Fuzzy keyword search on encrypted cloud storage data with small index. In: IEEE International Conference on Cloud Computing and Intelligence Systems(CCIS), pp. 269–273 (2011)
27. J.Wang, X.Chen, H.Ma, Q.Tang, and J.Li, A verifiable fuzzy keyword search scheme over encrypted data. J. Internet Serv. Inf. Secur. (JISIS) 2(1/2), 49–58 (2011)
28. Cao, N., Wang, C., Li, M., Ren, K., Lou, W.: Privacy-preserving multi-keyword ranked search over encrypted cloud data. In: Proceedings of the IEEE INFOCOM, pp. 829–837 (2011)
29. Tseng F.-K., Chen, R., Paul lin B.: iPEKS: Fast and Secure Cloud Data Retrieval from the Public-Key Encryption with Keyword Search. In: 12th IEEE International Conference on Trust, Security and Privacy in Computing and Communications (TrustCom), pp. 452–458 (2013)
30. Fu, Z., Sun, X., L.N., Zhou, L.: Achieving effective cloud search services: multi-keyword ranked search over encrypted cloud data supporting synonym query, In: IEEE Trans. Consum. Electron.. 164–172 (2014)
31. Shi, X., HU, S.: Fuzzy Multi-keyword query on encrypted data in the cloud. In: 4th International Conference on Applied Computing and Information Technology/3rd International Conference on Computational Science/Intelligence and Applied Informatics/1st International Conference on Big Data, Cloud Computing, Data Science & Engineering (2016)
32. Ge, X., Yu, J., Hu, C., Zhang, H., Hao, R.: Enabling efficient verifiable fuzzy keyword search over encrypted data in cloud computing, IEEE Access 6, 45725–45739 (2018)
33. Jivane, A.: Time efficient privacy-preserving multi-keyword ranked search over encrypted cloud data. In: 2017 IEEE International Conference on Power, Control, Signals and Instrumentation Engineering (ICPCSI), pp. 497–503 (2017)
34. Cohen, W.W.: Enron email dataset. http://www.cs.cmu.edu/~enron/

Assessing the Performance of E-government Services Through Multi-criteria Analysis: The Case of Egypt

Abeer Mosaad Ghareeb, Nagy Ramadan Darwish, and Hesham Hefny

Abstract E-government projects have been mostly supply driven with relatively less information about the performance of e-government services as well as the perception of the citizens regarding them. Less demanding of the available e-government services may be an indication of some difficulties in using or accessing these services. Therefore, the performance of the available e-government services needs to be assessed. This chapter aims to propose a methodology to assess the performance of e-government services. The proposed methodology combines two of the most well-known and extensively used Multi-Criteria Decision Making methods. These methods are PROMETHEE and AHP. The proposed methodology has been applied to assess the performance of e-government services available on Egyptian national portal. The research outcomes inform the policy makers about how e-government services have performed from citizen's perspective and help them to take suitable corrective actions to better the ranks of the underperforming services and fulfill the citizens' needs.

Keywords E-government · PROMETHEE · AHP · Adoption

1 Introduction

The term electronic government (e-government) has captured the attention of researchers and practitioners. There are now several journals and annual conferences

A. M. Ghareeb (✉) · H. Hefny
Department of Computer Science, Faculty of Graduate Studies for Statistical Research, Cairo University, Giza, Egypt
e-mail: abeer_mosad@yahoo.com

H. Hefny
e-mail: hehefny@ieee.org

N. R. Darwish
Department of Information Systems and Technology, Faculty of Graduate Studies for Statistical Research, Cairo University, Giza, Egypt
e-mail: nagyrd@cu.edu.eg

© The Author(s), under exclusive license to Springer Nature Switzerland AG 2021   547
A. E. Hassanien et al. (eds.), *Machine Learning and Big Data Analytics
Paradigms: Analysis, Applications and Challenges*, Studies in Big Data 77,
https://doi.org/10.1007/978-3-030-59338-4_26

devoted to e-government. Yonazi defined e-government as the utilization of ICTs to transform and enhance the relationship of the public sector and its clients through an improved range and quality of service [1]. International organizations such as United Nations and World Bank have also offered definitions for e-government. According to United Nations, E-government is the use and application of Information Technologies in public administration to streamline and integrate workflows and processes, to effectively manage data and information, enhance public service delivery, as well as expand communication channels for engagement and empowerment of people [2]. The adoption has been described as a simple decision of using or not using online services depending on some factors [3]. The other dimensions of adoption include: How frequently services are actually used; Scope of usage; Preference of the online medium over other mediums of transactions with government [3].

The countries around the globe have invested heavily in e-government projects and have experienced substantial progress in provision of online services [4]. However, E-government projects have been mostly supply driven with relatively less information about the performance of the e-government services as well as the perception of the citizens regarding them. However, this paradigm has begun to change. Citizens are placing numerous stresses on governments by less demanding e-government services. Having high demand and use of the available e-government services is crucial to the success of e-government projects. The failure of e-government projects comes with high financial, beneficial and political costs [5].

Citizens' usage of e-government services mainly depends on multiple criteria. Multi-Criteria Decision Making (MCDM) has been widely used for evaluation problems containing multiple criteria. However, the quantitative approach supported by statistical analysis is the most dominant used approach in the area of e-government adoption [6]. Preference Ranking Organization Method for Enrichment Evaluations (PROMETHEE) and Analytical Hierarchy Process (AHP) are widely used MCDM methods. This chapter aims to assess the performance of e-government services using PROMETHEE and AHP. The remainder of this chapter is organized as follows. Section 2 provides an overview of MCDM. Section 3 presents the mathematical background of PROMETHEE. Section 4 provides the proposed methodology. Section 5 presents a case study. Finally, Sect. 6 gives conclusion.

2 Multi-criteria Decision Making

MCDM has grown as a part of operations research concerned with designing computational and mathematical tools for supporting the subjective evaluation of performance criteria by decision-makers [7]. AHP, ANP, DEA, goal programming, TOPSIS, ELECTRE, and PROMETHEE are among numerous MCDM methods. MCDM methods have been broadly classified into two categories: (i) Multiple Attribute Decision Making (MADM) and (ii) Multiple Objective Decision Making (MODM) [8, 9]. MADM is the most well-known branch of decision making [8]. The two main families in MADM methods are those based on the Multi-Attribute Utility

Theory (MAUT) and Outranking methods. MAUT methods allow complete compensation between criteria. AHP is one of the more widely applied MAUT methods. It was proposed by Saaty [10]. It builds on complete aggregation of additive type characteristic of the American school [11]. AHP is based on three principles: (i) Construction of a hierarchy; (ii) Priority setting; and (iii) Synthesis of the priorities [11, 12]. First, AHP breaks down the decision problem into a hierarchy of interrelated decision elements. With AHP, the objectives, criteria and alternatives are arranged in a hierarchy structure. A hierarchy has at least three levels as follows: the overall goal of the problem at the top; multiple criteria in the middle; and alternatives at the bottom. Applications of AHP have been dominant in manufacturing, followed by the environment management and agriculture field, power and energy industry, transportation industry and healthcare [13]. AHP has been also used to propose evaluation methodologies for government websites [14–16].

PROMETHEE is one of the main outranking methods typical of the European school [11]. It is quite simple in conception and application compared to other MCDM methods [15]. PROMETHEE has been widely used for practical MCDM problems in various domains such as finance, transportation, and information technology strategy selection [15]. PROMETHEE has been applied to rank a sample of 13 water development projects in Jordan [17]. An integrated approach combining PROMETHEE and Geometrical Analysis for Interactive Aid methods has been applied for evaluating the performance of 20 national institutions in India [18]. PROMETHEE has some strength in comparison with other MCDM methods [19].

3   PROMETHEE

The mathematical background of PROMETHEE has been described in several references [20–22]. Let us define a multi-criteria problem as follows.

$$Optimizing\{f_1(a), f_2(a), \ldots, f_k(a) | a \in A\} \tag{1}$$

where A is a set of alternatives and f_j, $j = 1, 2, 3, \ldots, k$, is criterion to be maximized or minimized.

PROMETHEE procedure starts by filling the evaluation table. The alternatives are assessed on each criterion where $f_j(a)$ is the performance measure of alternative (a) with respect to jth criterion. Then, a specific preference function $P_j(a, b)$ is associated to each criterion to translate the deviation between the assessments of two alternatives on that criterion into a preference degree as given in Eq. 2.

$$P_j(a, b) = P_j\big(f_j(a) - f_j(b)\big) \tag{2}$$

The preference degree represents the preference of alternative (a) over alternative (b) for criterion f_j. This degree is normalized so that

$$0 \le P_j(a, b) \le 1 \tag{3}$$

There are six known preference functions as follows: Usual; U-shape; V-shape; level; V-shape with indifference area; and Gaussian preference functions [19, 20]. The preference degrees are used to calculate a global preference index $\pi(a, b)$ for each pair of alternatives as given in Eq. 4. It represents the preference of alternative (a) over (b) considering all criteria and taking into account the weights of criteria.

$$\pi(a, b) = \sum_{j=1}^{k} w_j P_j(a, b) \tag{4}$$

where, w_j is the weight of jth criterion and it is a choice of the decision maker and

$$\sum_{j=1}^{k} w_j = 1 \tag{5}$$

Global indices are used to calculate three preference flows for each alternative. Leaving Flow $\varphi^+(a)$ denotes how much alternative (a) dominates the other as given by Eq. 6. Entering flow $\varphi^-(a)$ denotes how much alternative (a) is dominated by other alternatives as given by Eq. 7. Net flow is calculated by subtracting entering flow from leaving flow as given by Eq. 8.

$$\varphi^+(a) = \frac{1}{n-1} \sum_{y \in A} \pi(a, y) \tag{6}$$

$$\varphi^-(a) = \frac{1}{n-1} \sum_{y \in A} \pi(y, a) \tag{7}$$

$$\varphi(a) = \varphi^+(a) - \varphi^-(a) \tag{8}$$

where n is the number of alternatives.

PROMETHEE ranks the alternatives based on the values of the preference flows. Two main PROMETHEE tools can be used. They are PROMETHEE I partial ranking, and PROMETHEE II complete ranking. PROMETHEE I uses the values of leaving and entering flows. Alternative (a) is preferable to alternative (b) if alternative (a) has a greater leaving flow than the leaving flow of alternative (b) and a smaller entering flow than the entering flow of alternative (b). PROMETHEE II uses the net flow to rank the alternatives.

In addition, the uni-criterion net flow can be calculated, as given by Eq. 9, to show the contribution of one criterion to the net flow score of the alternative.

$$\varphi_j(a) = \frac{1}{n-1} \sum_{y \in A} [P_j(a, y) - P_j(y, a)] \tag{9}$$

4   The Proposed Methodology

Many of the previous studies regarding the adoption of e-government services have concerned with identification of the adoption factors. Now, it is quite required to go a further step and introduce methodologies to measure the acceptance rate of the deployed e-government services with respect to specific criteria in order to enhance their adoption in the society. This simply turns the e-government adoption to be a MCDM problem which we are concerned in this chapter. The proposed methodology intends to enhance the acceptance and adoption rate of e-government services and reducing provision-usage gap to capture the advantages of e-government.

The proposed methodology for assessment the performance of e-government services consists of five phases as depicted in Fig. 1. In Phase 1, a set of e-government services are selected for evaluation purpose. In Phase 2, the real needs, expectations and requirements of citizens regarding e-government services are identified and translated into a set of acceptance and adoption criteria. In Phase 3, AHP is used to set up the decision hierarchy. In Phase 4, the PROMETHEEE procedure described in Sect. 3 is implemented. The measurement scale is defined. The criterion can be assessed quantitative or qualitative. Each criterion has to be decided whether it has to minimized or maximized. The preference functions are chosen by decision makers. The evaluation outcomes depend on both choice of preference function and its parameters [23]. PROMETHEE II is used to rank the e-government services.

In Phase 5, the results are analyzed using PROMETHEE rainbow. The basic idea of the PROMETHEE rainbow is to calculate the uni-criterion net flow assigned to each e-government service as given in Eq. 9. For each e-government service, a bar is drawn. The different slices of each bar are colored according to the criteria. Each slice is proportional to the contribution of one criterion to the net flow score of the e-government service.

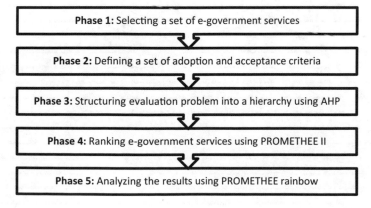

Fig. 1 Proposed methodology to assess the performance of e-government services

5 Case Study

A case study was conducted to demonstrate effectiveness of the proposed method-ology. Egypt provided a perfect context for this study. Egypt started its e-government initiative early, in 2000, with clear vision and top management support. Several successful projects have been undertaken such as Family Card system [24] and University Enrolment project [25]. The country achieved a great deal at the beginning and won some awards such as that Egyptian Government Electronic Tenders Portal has won 2nd place at the United Nations Public Administration Network (UNPAN) 2011 Public Service Award in the category of Preventing and Combating Corruption in the Public Service. [26]. However, Egypt declined in its global e-government development rank over the last years [4, 27]. Figure 2 tracks the status of e-government progress in Egypt from 2003 to 2018. Egypt lags behind other Arab countries [28]. Figure 3 shows Egypt's E-Government Development Index compared with Gulf countries over the period from 2014 to 2018.

E-government implementation in developing countries is generally restricted by existence of a combination of political, legal, technical, economic, and social barriers. Egypt faces some similar generic barriers to those of developing countries in addition to other particular challenges associated to the Egyptian context. Challenges of e-government in Egypt include: Lack of coordination among governmental ministries

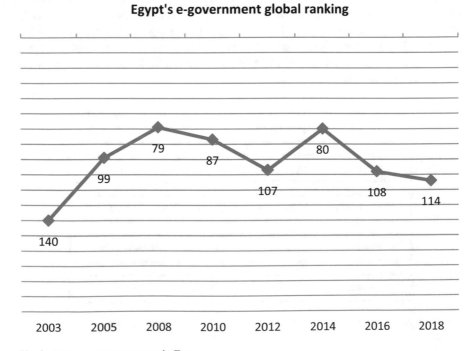

Egypt's e-government global ranking

79 87 99 107 80 108 114 140

2003 2005 2008 2010 2012 2014 2016 2018

Fig. 2 E-government progress in Egypt

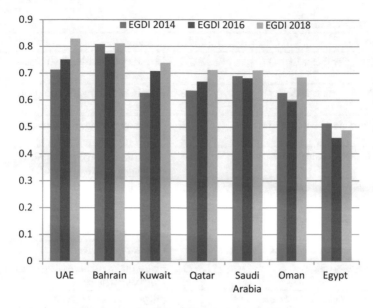

Fig. 3 E-government development index of Egypt and Gulf countries

and organizations. Egyptian government is not corporate. This means that the plan is neither unified nor shared. "The corporate government" is theoretically applied in Egypt but it is not activated [24]; Electronic services are not a priority for citizens. Although e-government projects provide many opportunities, but still there is very low demand for what is supplied by these projects [24]; Poor marketing efforts that are required to change the behavior of the Egyptian citizens [24]; and lack of e-signature mechanism and lack of e-government services quality measurement [28]. Therefore, there is a need to study the e-government development for the Egyptian case to better the position of Egypt and enhance citizens' satisfaction with public services.

5.1 Selecting a Set of E-government Services

The e-government services available on the Egyptian national portal [26] were examined to select a set of services for assessment. Citizen-oriented services were the main focus of this study and the services oriented to other stakeholders such as foreigners and investors were excluded. 44 services were found. 20 of these 44 are either under construction or not found. Running e-government services were further sub-divided into two categories. They were paid and free services. The current study focused on evaluating paid e-government services. Table 1 provides the name, label and provider of each e-government services and Table 2 provides the services' description.

Table 1 E-government services

No	Name	Label	Provider
1	Birth Certificate Extract	BCE	Ministry of Interior-Civil Status Organization
2	Death certificate extract	DCE	Ministry of Interior-Civil Status Organization
3	Marriage document extract	MDE	Ministry of Interior-Civil Status Organization
4	Divorce document extract	DDE	Ministry of Interior-Civil Status Organization
5	Family record extract	FRE	Ministry of Interior-Civil Status Organization
6	Replacement of national ID card	RID	Ministry of Interior-Civil Status Organization
7	Traffic Fines Payment Certificate	TFP	Ministry of Justice
8	Bus reservation	BUS	West middle delta, east delta, and upper Egypt companies
9	Trains tickets reservation	TTR	Egyptian National Railways
10	Equivalence of scientific degree	ESD	Supreme Council of Universities
11	Searching in laws and legalization	SLL	Ministry of Justice
12	Payment of phone bill	PPB	Telecom Egypt
13	Egypt-Air e-ticketing	EAT	Egypt-Air

5.2 Defining a Set of Adoption and Acceptance Criteria

Based on extensive literature review in the area of e-government adoption, nine important acceptance and adoption criteria were defined as given in Table 3. These criteria were cost (CO), payment method (PM), time (TI), responsiveness (RE), usability (US), accessibility (AC), information quality (IQ), marketing campaign (MC) and mandate (MA).

5.3 Structuring the Evaluation Problem into a Hierarchy Using AHP

The decision problem was structured into a 3-level hierarchy as depicted in Fig. 4. Level 1 represented the goal of improving acceptance rate of e-government services. Level 2 contained the nine identified adoption criteria. Level 3 contained 13 paid e-government services.

Table 2 Description of selected e-government services

Service	Description
BCE	Requesting a birth certificate extract through national government portal, delivering it at the address and payment upon receipt instead of going to the Civil Status Office
DCE	Requesting a death certificate extract through national government portal, delivering it at the address and payment upon receipt instead of going to the Civil Status Office
MDE	Requesting a marriage document extract through the national government portal, delivering it at the address and payment upon receipt instead of going to the Civil Status Office
DDE	Requesting a divorce document extract through the national government portal, delivering it at the address and payment upon receipt instead of going to the Civil Status Office
FRE	This service is especially useful for students applying to be enrolled in military and police colleges where a family record extract is a prerequisite. 150 thousand students apply manually at the Civil Status Office within a window of 3 months causing high traffic and long queues. Using the online service to acquire a first time family record extract allows the citizen to apply through the internet anytime anywhere and follow up on the application status
RID	Requesting a replacement of national ID card through national government portal, delivering it at the address and payment upon receipt instead of going to the Civil Status Office
TFP	Requesting a certificate of payment of traffic fines through national government portal, delivering it at the address and payment upon receipt instead of going to the Traffic Department. The service covers the governorates of Cairo and Giza only and takes 72 h of the application time
BUS	Booking tickets for bus lines and payment via credit cards. Then booking information is send via email. The traveller prints the booking information and goes to the bus station on time
TTR	Reserving tickets of Egyptian national railway for first and second classes in air conditional trains. Citizens can pay tickets fees electronically using Visa Master cards
ESD	Equivalence of degrees awarded by educational institutions and universities which don't subject to the law of organization of universities No. (49) for the year 1972 and comparing them with corresponding scientific degrees awarded by universities subject to this law
SLL	Accessing to the latest legislation and provisions, legal bulletins, references and jurisprudence studies, and contract forms. Each participant is allocated a number of points commensurate with the value of the subscription. The subscription is not associated with the number of browsing hours. The points shall be deducted when reviewing legislation, judgment or jurisprudential study
PPB	Allowing the subscribers to pay the value of phone calls electronically through visa and master cards
EAT	Booking airline tickets directly from the Egypt Air web site instead of going to the tourist offices. Electronic payment is available via credit cards. Once the payment is done, ticket(s) will be issued immediately and send to the traveller by email

Table 3 Adoption and acceptance criteria

Criterion	Description
CO	• Ability of e-government services in cost saving • Cost of e-government services should not exceeds the cost of traditional ones
PM	• Payment systems should be convenient with citizens' economic and social conditions, educational level, and their lifestyle
TI	• Ability of e-government services in time saving • Displaying citizen charter online which provides the minimum number of days that a particular public organization takes to deliver the service
RE	• Ability to make inquiries online • Replying to citizens' inquiries or complaints through various e-government delivery channels like emails • Providing on-going technical support to citizens
US	• Supporting a complete set of navigational aids such as main menu, site map, home link, and click-ability identification • Adherence of uniformity in terminology, color, style, labeling, abstraction, and positioning of elements • Easy of reading by taking into account some features like font type and font size • Providing service description, hints, guidelines and examples • Explaining the steps of using the service
AC	• Available through the national portal anytime from anywhere • Concise and easy to remember URL • Easing of locating intended service after accessing intended application • Download speed • Compatibility with a variety of browsers
IQ	• Accurate, consistent, and up to date
MC	• Promotion of services through various channels such as TV talk shows, newspaper ads, and word of mouth • Informing the citizens about the web addresses, the laws and legislations of electronic dealing • Informing the citizens about relative advantages of e-services such as cost and time saving, avoiding long queues, synchronizing and updating records among different governmental organizations, and tracking the status of the conducted service
MA	• Services, that don't involve personal interaction for identity authentication and authorization or that don't require physical inspection, should be totally complete online

5.4 Ranking E-government Services Using PROMETHEE II

All the nine criteria were evaluated using a qualitative scale. The cost and time criteria were minimized. The other criteria were maximized. For mandate criterion, a simple yes/no scale was used. A 7-point scale, ranging from strongly agree (SA) to strongly disagree (SDA), was selected for cost and time criteria. A 7-point scale, ranging from strongly disagree to strongly agree, was used for the rest of the criteria (see Table 4). PROMETHEE method provides a technique to handle the missing or not available values (N/A) at the pairwise comparison level [29]. It set both preference degrees

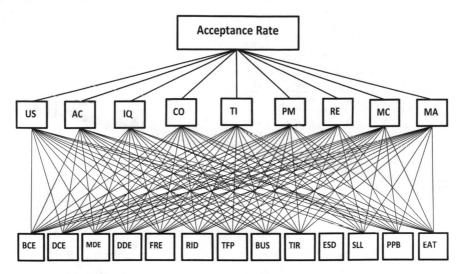

Fig. 4 A 3-level hierarchy structure for assessment problem

Table 4 Measurement scale

Value	Level
1	Strongly disagree (SDA)
2	Disagree (DA)
3	Little disagree (LDA)
4	No comment (NC)
5	Little agree (LA)
6	Agree (A)
7	Strongly agree (SA)

equal to zero as given in Eq. 10.

$$P_j(a, b) = P_j(b, a) = 0 \tag{10}$$

By the time of data collection (which began on 1 January and finished on 31 May, 2018) there has been no change in the design and content of the examined e-government services. Table 5 gives the evaluation data of e-government services.

For all criteria, except mandate, the V-shape preference function was employed. V shape function depends on a preference parameter. This parameter is the lowest value of the difference between two evaluations, above which there is a strict preference of one of the corresponding alternatives. As long as the difference is lower than this threshold, the preference increases linearly with the difference [23]. The values of measurement scale ranged from 1 to 7. So, when the preference threshold is set to 6.00, a small difference, as much as larger one, is accounted. Usual preference function was a good choice for mandate criterion, which included a small number

Table 5 Evaluation table

	US	AC	IQ	CO	TI	PM	RE	MC	MA
BCE	LA	A	NC	SDA	NC	SA	A	DS	NO
DCE	LA	A	NC	DS	NC	SA	A	DS	NO
MDE	LA	A	NC	DS	NC	SA	A	DS	NO
DDE	LA	A	NC	DS	NC	SA	A	DS	NO
FRE	A	A	NC	LA	A	SA	A	DS	NO
RID	A	A	NC	A	A	SA	A	DS	NO
TFPC	NC	DA	A	A	NC	SA	NC	DS	NO
BUS	LDA	LDA	LDA	A	A	DS	NC	DS	NO
TTR	A	A	A	A	A	DS	A	DS	NO
ESD	A	A	A	A	A	A	NC	NC	YES
SLL	LDA	A	NC	N/A	A	NC	NC	DS	NO
PPB	A	A	SA	A	A	DS	SA	NC	NO
EAT	LA	A	A	A	A	DS	LDA	D	NO

of evaluation levels. All weights were set to be equal. The multi-criteria decision aid
software Visual PROMETHEE 1.4 was used to compute preference flows and rank
the e-government services. Ranks and preference flows are shown in Table 6. ESD
had the higher net flow score. It dominated the others. PPB, RID, and FRE were very
close to every other. They had good positive scores. TTR was the closest to zero. It
was more average positive item. DCE, MDE, and DDE had the same negative score.

Table 6 Ranks and preference flows

Rank	Service	$\varphi(a)$	$\varphi^+(a)$	$\varphi^-(a)$
1	ESD	0.2361	0.2747	0.0386
2	PPB	0.1157	0.1883	0.0725
3	RID	0.0957	0.1296	0.0340
4	FRE	0.0772	0.1219	0.0448
5	TTR	0.0355	0.1173	0.0818
6	DCE	−0.0386	0.0756	0.1142
6	MDE	−0.0386	0. 0756	0.1142
6	DDE	−0.0386	0. 0756	0.1142
7	EAT	−0.0448	0.0880	0.1327
8	BCE	−0.0571	0.0741	0.1312
9	TFP	−0.0648	0.0957	0.1605
10	SLL	−0.0926	0.0417	0.1343
11	BUS	−0.1852	0.0463	0.2315

EAT, BCE, TFP, and SLL had also very close but negative scores. BUS service was dominated by all the others.

5.5 Analyzing Results Using PROMETHEE Rainbow

Based on the values of the uni-criterion net flow (see Table 7), PROMETHEE rainbow was constructed (see Fig. 5). To highlight good and bad features, a specific color was assigned to each criterion as follows: Blue for cost; lime for payment method; fuchsia for time; red for responsiveness; yellow for usability; aqua for accessibility; green for information quality; purple for marketing campaign; and grey color for mandate criterion.

PROMETHEE rainbow demonstrated that ESD revealed very little weakness compared to other e-government services. All criteria, except responsiveness, contributed positively to its net flow score. Mandate is the most important feature of this service. Payment method is the most criterion of the positive effect on the net flow score of RID, FRE, DCE, MDE, DDE, BCE, and TFP services. On the other side, payment method is the most criterion of the negative effect on the net flow score of PPB, TTR, EAT, and BUS services. Cost may be the most criterion of the negative effect on the citizens' acceptance and adoption to Civil Status Office online services. Accessibility is the most influential weakness of TFP service. BUS service showed little strengths represented in a limited positive affect of cost and time criteria. However, the rest of the criteria were drawn in the form of large downward slices which means that large negative contribution to its net flow score.

6 Conclusion

A methodology to assess the performance of e-government services through multi-criteria analysis has been proposed. The proposed methodology contains five phases: (1) Selecting a set of e-government services; (2) Defining a set of adoption and acceptance criteria; (3) Structuring evaluation problem into a hierarchy using AHP; (4) Ranking e-government services using PROMETHEE II; and (5) Analyzing the results. The proposed methodology has been applied to assess the performance of citizen-oriented e-government services available on Egyptian National Portal. Each e-government service has been analysed against a pre-defined set of acceptance criteria covering application and service aspects.

Most of the examined e-government services have inadequacy regarding some criteria. Although some e-government services tend to be performing quite acceptably, there is a room for improvement. Mandate seems to be a very influential criterion. The vast majority of the examined services employ traditional and electronic ways and this is in favour of the traditional one. Payment method can be considered the most criterion of the positive or negative impact on e-government services. If

Table 7 Uni-criterion net flow values

	US	AC	IQ	CO	TI	PM	RE	MC	MA
ESD	0.1806	0.0972	0.2222	0.2500	0.1389	0.1528	−0.2222	0.3056	1.0000
PPB	0.1806	0.0972	0.4028	0.2500	0.1389	−0.5694	0.3194	0.3056	−0.0833
RID	0.1806	0.0972	−0.1389	0.2500	0.1389	0.3333	0.1389	−0.0556	−0.0833
FRE	0.1806	0.0972	−0.1389	0.0833	0.1389	0.3333	0.1389	−0.0556	−0.0833
TTR	0.1806	0.0972	0.2222	0.2500	0.1389	−0.5694	0.1389	−0.0556	−0.0833
DCE	0.0000	0.0972	−0.1389	−0.4167	−0.2222	0.3333	0.1389	−0.0556	−0.0833
MDE	0.0000	0.0972	−0.1389	−0.4167	−0.2222	0.3333	0.1389	−0.0556	−0.0833
DDE	0.0000	0.0972	−0.1389	−0.4167	−0.2222	0.3333	0.1389	−0.0556	−0.0833
EAT	0.0000	0.0972	0.2222	0.2500	0.1389	−0.5694	−0.4028	−0.0556	−0.0833
BCE	0.0000	0.0972	−0.1389	−0.5833	−0.2222	0.3333	0.1389	−0.0556	−0.0833
TFP	−0.1806	−0.625	0.2222	0.2500	−0.2222	0.3333	−0.2222	−0.0556	−0.0833
SLL	−0.3611	0.0972	−0.1389	0.0000	0.1389	−0.2083	−0.2222	−0.0556	−0.0833
BUS	−0.3611	−0.4444	−0.1389	0.2500	0.1389	−0.5694	−0.2222	−0.0556	−0.0833

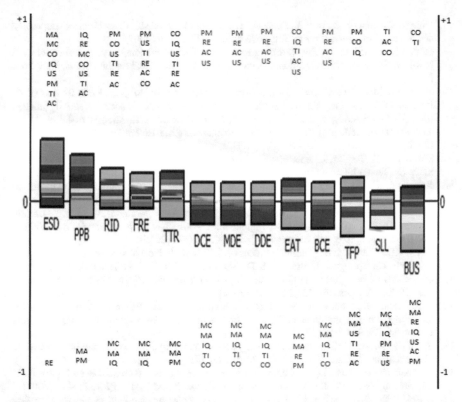

Fig. 5 PROMETHEE rainbow

the cost of an online service exceeds the cost of the traditional one, the extent to which the citizen accepts this increase is affected by the amount of increase and the frequency of the demand of the service. Live chat is an effective way to communicate with citizens while using the services. However, this feature doesn't take enough attention.

Given the scarcity in literature in performance measurement of e-government services, it is hoped that the proposed methodology shared in this chapter fills the gap left in the literature and provide support for improving citizens' adoption of e-government services.

References

1. Yonazi, J.J.: Enhancing adoption of e-government initiatives in Tanzania. Ph.D. Thesis, University of Groningen, SOM Research School (2010)
2. United Nations: E-government for the Future We Want. United Nations, Department of Economic and Social Affairs, New York (2014)

3. Kumar, V., Mukerji, B., Butt, I., Persaud, A.: Factors for successful E-government adoption: a conceptual framework. Elctron. J. E-government **5**(1), 63–76 (2007)
4. United Nations: Gearing E-government to Support Transformation Towards Sustainable and Resilient Societies. United Nations, Department of Economic and Social Affairs, New York (2018)
5. Heeks, R.: Most E-government for Development Projects Fail: How Can Risks Be Reduced?. Institute of Development Policy and Management, University of Manchester, IDPM) (2003)
6. Irani, Z., Weerakkody, V., Kamal, M., Hindi, N.M., Osaman, I.H., Anouze, A.-L., et al.: An analysis of methodologies utilized in e-government research. J. Enterp. Inf. Manag. **25**, 298–313 (2012)
7. Zavadskas, E., Turskis, Z., Kildiene, S.: State of art surveys of overviews on MCDM/MADM methods. Technol. Econ. Dev. Econ. **20**, 165–179 (2014)
8. Kahraman, C.: Fuzzy Multi-criteria Decision Making-Theory and Applications with Recent Developments, 1st edn. Springer, US (2008)
9. Mardani, A., Jusoh, A., Nor, K.M., Khalifah, Z., Zakwan, N., Valipou, A.: Multiple criteria decision making techniques and their applications—a review of the literature from 2000 to 2014. Econ. Res. Ekonomska Istrazivanja **28**(1), 516–571 (2015)
10. Saaty, T.L.: The Analytic Hierarchy Process. McGraw-Hill, New York (1980)
11. Macharis, C., Springael, J., Brucker, K.D., Verbeke, A.: PROMETHEE and AHP: the design of operational synergies in multicriteria analysis. strengthening PROMETHEE with the ideas of AHP. Eur. J. Oper. Res. **153**, 307–317 (2004)
12. Turcksin, L., Bernardini, A., Macharis, C.: A combined AHP-PROMETHEE approach for selecting the most appropriate policy scenario to stimulate a clean vehicle fleet. Procedia Soc. Behav. Sci. **20**, 954–965 (2011)
13. Sipahi, S., Timor, M.: The analytic hierarchy process and analytic network process: an overview of applications. Manag. Decis. **48**(5), 775–808 (2010)
14. Buyukozkan, G., Ruan, D.: Evaluation government websites based on a fuzzy multiple criteria decision approach. Int. J. Uncertain., Fuzziness Knowl.-Based Syst. **15**(3), 321–343 (2007)
15. Bilsel, R.U., Buyukozkan, G., Ruan, D.: A fuzzy preference ranking model for a quality evaluation of hospital websites. Int. J. Intell. Syst. **21**, 1181–1197 (2006)
16. Fu, H.-P., Ho, Y.-C., Chen, R.C., Chang, T.-H., Chien, P.-H.: Factors affecting the adoption of electronic marketplaces: a fuzzy AHP analysis. Int. J. Oper. Prod. Manag. **26**(12), 1301–1324 (2006)
17. Al-Kloub, B., Al-Shemmeri, T., Pearman, A.D., Brans, J.P., Mareschal, B.: Application of multicriteria analysis to rank and evaluation of water development projects (the case of Jordan). In: Multi-objective Programming and Goal Programming. Springer, Berlin, Heidelberg (1996)
18. Ranjan, R., Chakraborty, S.: Performance evaluation of India technical institutions using PROMETHEE-GAIA approach. Inf. Educ. **14**(1), 103–125 (2015)
19. Behzadian, M., Kazemzadeh, R., Albadvi, A., Aghdasi, M.: PROMETHEE: a comprehensive literature review methodologies and applications. Eur. J. Oper. Res. **200**(1), 198–215 (2010)
20. Brans, J.P., Vincke, P.H.: A preference ranking organization method the PROMETHEE method for multiple criteria decision making). Manage. Sci. **31**(6), 647–656 (1985)
21. Brans, J., Vincke, P., Mareschal, B.: How to select and how to rank projects: the PROMETHEE method. Eur. J. Oper. Res. **24**, 228–238 (1986)
22. Brans, J.P., Mareschal, B.: PROMETHEE methods. In: Multiple Criteria Decision Analysis, pp. 164–195. Springer (2005)
23. Podvezko, V., Podviezko, A.: Dependence of multi-criteria evaluation result on choice of preference functions and their parameters. Technol. Econ. Dev. Econ. **16**(1), 143–158 (2010)
24. Abdelkader, A.A.: A Manifest of barriers to successful E-government: cases from the Egyptian programme. Int. J. Bus. Soc. Sci. **6**(1), (2015)
25. El-Baradei, L., Shamma, H.M., Saada, N.: Examining the marketing of e-government services in Egypt. Int. J. Bus. Pub. Manag. **2**(2), 12–22 (2012)
26. Egyptian Government Portal: www.egypt.gov.eg, last accessed 2018/5/31

27. United Nations: E-government in Support of Sustainable Development. United Nations Department of Economic and Social Affairs, New York (2016)
28. Gebba, T.R., Zakaria, M.R.: E-government in Egypt: an analysis of practices and challenges. Int. J. Bus. Res. Dev. **4**(2), 11–25 (2015)
29. Fernandez, N.G.: The management of missing values in PROMETHEE method. Master Thesis, Universitat Politecnic De Catalunya, Barcelona, Spain (2013)

IoTIwC: IoT Industrial Wireless Controller

Tassnim Awad, Walaa Mohamed, and Mohammad M. Abdellatif

Abstract Industrial controller systems are crucially essential to the cutting edge power systems industries. Industrial controllers link the integrated technologies of a computer, communication devices, and electric devices. The communication systems act as a physical intermediary layer for transferring, controlling, and acquirement of data within the system from distant locations. This chapter discusses the Supervisory Control And Data Acquisition (SCADA) systems and proposes a similar system that is an IoT based industrial wireless controller. The proposed system can control multiple devices through the network without the need to be physically near the devices. Because it uses simple and cheap devices, the system is low cost and easy to install. Additionally, the system is modular because extra microcontrollers can be easily added to the system to control more devices should the need arise.

Keywords IoT · Industrial controller · IIoT · SCADA

1 Introduction

A SCADA (Supervisory Control And Data Acquisition) is an automation control system that is used in industries such as energy, oil, gas, water, power, and many more. Normally, the system has a centralized location that monitors and controls entire sites, ranging from an industrial plant to a complex of plants across a whole country. SCADA could be defined as the process of controlling and supervising data collection and processing. This word is used to describe real time systems where data is gathered, processed, and maintained in real time. Some examples may include monitoring over a power plant or an irrigation system. To further illustrate the example above and to show how the system works, a home station could be fitted with that system to monitor different substations or remote stations. If an error occurs in one of the stations, the home station could analyze the data and make sure that the error may not be critical and fixable. The system uses a client server architecture

T. Awad · W. Mohamed · M. M. Abdellatif (✉)
Electrical Engineering Department, The British University in Egypt, Cairo, Egypt
e-mail: Mohammad.Abdellatif@bue.edu.eg

© The Author(s), under exclusive license to Springer Nature Switzerland AG 2021   565
A. E. Hassanien et al. (eds.), *Machine Learning and Big Data Analytics*
Paradigms: Analysis, Applications and Challenges, Studies in Big Data 77,
https://doi.org/10.1007/978-3-030-59338-4_27

and has several elements, most notably, input–output sever, SCADA server, Human machine interfaces, and a control server.

The first generation of SCADA systems was designed with no networking infrastructure as it wasn't quite developed by that time. Then came the Next generation of SCADA system where the concept of networking was introduced. SCADA systems utilized networking for load balancing for efficient resource use and for increasing the dependability of the system. The third generation of SCADA system followed the same design principles as the previous generation but it became a more open system architecture compared to the self-tailored versions done by the companies. There are a-lot of transmission technologies used in SCADA systems, for example, coaxial cables which are used by TVs providers to send data, twisted metal pair as in telephone lines, fiber optic cables which is a relatively expensive option but allows real time communication and disaster recovery because of the high speeds it could transmit data through, it could also utilize satellite communications. All these communication methods require a solid infrastructure and to deploy them in a rural area would cost a fortune. So, we need to relay on wireless communication to reduce the Installation cost.

As the current SCADA systems use the same frequency as those of the TV channels. There was a conflict with major TV companies as it may negatively affect their income from advertisements, It uses the channels between 54 and 862 MHz for maximum coverage which also happens to be the TV channels, The bandwidth is about 6–8 MHz per channel and It identifies the free channels and send on those only to prevent any conflict with any used TV brands. Another method that is used is utilizing the already existing infrastructure of the cellular networks as one of the building blocks for wireless SCADA systems, with base stations connecting to the main network and sending data through it. The base stations can utilize components in the SCADA system to identify the free bands to send data upon so it won't interfere with any Calls or bands which are quite busy at the time. This would allow the SCADA system to cover about 100 km of area and it could provide the same level of service relative to DSL. However, this will add extra cost to the system because of the subscription charges of the cellular service.

This chapter proposes an IoT based SCADA system which is made with low cost devices and that can be controlled wirelessly through Wi-Fi from any device in the network. The proposed system is modular which simplifies expanding the system to control more devices without the need of reconfiguring the system as a whole.

The rest of the chapter is organized as follows. Section 2 gives a background and literature review on SCADA systems. Section 3 describes the system architecture. Section 4 presents ideas for future work. And finally, Sect. 5 concludes the chapter.

2 Background and Literature Review

In this section, we describe some of the earlier work done on web-based remote monitoring as well as some of the state of the art available on the same topic.

2.1 Web-Based Remote Monitoring

One of the early work done on the subject was that of Bertocco et al. [1], where the authors described client–server architecture for remote monitoring of instrumentation over the Internet. The proposed solution allowed multi-user, and multi-instrument sessions by means of a queuing and instrument locking capability. A queue mechanism has been added to the remote environment along with the possibility for each client to query the actual server load. The communication between the server and clients can be obtained either at instrument level or by means of encoded requests in order to reduce the network-imposed overhead.

Tso et al [2] presented a study that indicates that a while a number of frameworks related to global systems have been described in contemporary publications, the detailed structure and formulation of the central-monitoring mechanism of such a partnership system has not received as much attention as it deserves. The proposed framework of a service network is characterized by its coordinating as well as monitoring capabilities. The main feature of the presented system is its rule-based reasoning capability to convert a job request from clients into basic tasks which are to be carried out by a group of virtual agents equipped with various defined capabilities.

Tommila et al. [3] discussed new ways of implementing existing functions and defined that new functionality, e.g. management of hierarchical structures and exception handling should be included in the basic control platform and engineering tools. The current 'flat' collection of application modules like loops and sequences had to be organized in a more hierarchical fashion based on process structure. Each process system is seen as an intelligent resource capable of performing different processing tasks. The interaction mechanisms between different automation activities are defined on the basis of object-oriented analysis and design and emerging international standards. A standardized distribution middleware takes care of the needs specific to the control domain. Above that, a higher level working environment for the other system components of the control platform is needed.

Yang et al. [4] reported a study on Networked Control System (NCS) historical review, recent revolution and research issues on NCS. Fast development and major use of the Internet, a global information platform has been created for control engineers allowing them to do the following:

- Monitoring the condition of machinery via the Internet.
- Remotely control machine.
  They also addressed many new challenges to control system designers. These challenges are summarized as follows:
- Overcoming Web-related traffic delay, i.e. dealing with Internet latency and data loss.
- Web-related safety and security, i.e. ensuring the safety and security of remote control and stopping any malicious attacks and misoperation.
- Collaborate with skilled operators situated in geographically diverse location.

Kalaitzakis et al. [5] developed a SCADA based remote monitoring system for renewable energy systems. It is based on client/server architecture and it does not require a physical connection, e.g. through network, serial communication port or 14 standard interface such as the IEEE-488 of the monitored system with data collection server.

Kimura et al. [6] reported remote monitoring system as one component of manufacturing support system. The proposed remote monitoring system can support single-night unmanned night time operations for diversified manufacturing from the operator's home as the remote site. According to the results, the remote monitoring system performed quite well for providing backup of manufacturing systems during unmanned nighttime operation.

Crowley et al. [7] experimentally explored the implementation of a wireless sensor network with Global system for Mobile (GSM) based communication for real-time temperature logging of seafood production. Subsequently, the network was developed and applied to the monitoring of whelk catches from harvest to delivery at the processing plant. The GSM communication was shown to have performed very well, especially in circumstances where problems with poor network coverage were expected to be encountered.

De la Rosa et al. [8] addressed the challenges and trends in the development of web-based distributed Power Quality (PQ) measurement and analysis using smart sensors. Registered users can configure the sensors, adjust sensitivity levels, and specify deployment location and email notification addresses. The developed website also provides a number of ways to view data from single or aggregated monitors. The authors addressed low cost Internet power monitor, which is cost-effective at the single user level. In addition, the reliance on standard web browsers eliminated the need for significant investment in software and hardware infrastructure that is typically required for other measurement systems.

Ong et al. [9] demonstrated existing SCADA with Java-based application in power systems. The authors also addressed the design issue in Graphical User interface (GUI). The proposed Web-based access tool can not only be used for SCADA Systems via intranet, or internet, but also can be readily used for information exchange among market operators via Internet.

Sung et al. [10] designed a test bed for an Internet-based Computer Aided Design (CAD) and Computer Aided Manufacturing (CAM) system. It was specifically designed to be a networked, automated system with a seamless communication flow from a client-side designer to a server side machining service. This includes a Web-based design tool in which Design-for-Manufacturing information and machining rules constrain the designer to parts that can be manufatured, a geometric representation called SIF-DSG for unambiguous communication between client-side designer and server-side process planner in an automated process planning system with several sub modules that convert an incoming design to a set of tool-paths for execution on a 3-axis Computer Numeric Control (CNC) milling machine.

Altun et al. [11] presented a study on Internet based process control via Internet. The study is to show that any process can be managed remotely with ease. Need for remote managing could appear in health-critical or dangerous conditions, being far away from job, etc. It could be extremely useful for managers to check, administer, or just for taking information as if using visual phone.

The scope of Internet based process control has been clearly specified by Yang et al. [12]. Internet-based control is only an extra control level added to the existing process control hierarchy. The objective is to enhance rather than to replace computer-based process control systems. Six essential design issues have been fully investigated which form the method for design of such Internet-based process control systems. The design issues include requirement specification, architecture selection, web-based user interface design and control over the Internet with time delay, concurrent user access, and safety checking.

Su et al. [13] presented a two WAN model on distribution management system. An integrated DMS consists of networked hardware and software capable of monitoring and controlling the operations of substations and feeders. Building a communication model allows one to determine if leased line capacity or system hardware speeds will cause a bottleneck in the system. The model contains sufficient details about the traffic load and their performance characteristics. WAN modeling is aimed to verify whether hardware design could accommodate the communications load and to avoid overpaying for network equipment. Simulation results indicate that, to cover feeder automation functions, a WAN with distributed processing capability would provide better SCADA performance than an extension of the old centralized system.

2.2   State of the Art

There are many implementation of the classic SCADA system. For example, the SCADA framework that the Lexington-Fayette Urban County Government (LFUCG) [14] depended on to run a system of 80 pump stations and two wastewater treatment plants has worked dependably for almost 20 years. In any case, time was incurring significant damage as new parts were progressively hard to discover and a great part of the framework was out of date. A large number of the current SCADA PCs and HMI programming running the plant's checking framework were old and in the need of substitution. Following a necessary appraisal, LFUCG chose to overhaul the electrical and SCADA frameworks at the Town Branch and West Hickman Creek wastewater treatment plants and supplant every one of the 80 remote terminal units (RTU) at each pump station.

Since LFUCG had a few unique offices engaged with the venture (IT, administration, tasks from each plant and pump stations gathering, and so on.) CDM Smith led workshops with all LFUCG partners and nearby gear producers to learn and decide the proper SCADA framework design and programming required for the undertaking. P&ID illustrations were produced starting with no outside help to demonstrate the current treatment plant process, joined with instrumentation and control frameworks which help convey control works amongst process and SCADA architects, and additionally the general temporary worker.

CDM Smith assessed the current telemetry framework; gave suggestions to supplanting gear; and assessed a few pump stations, the repeater and antenna wire area, and the general surveying programming/equipment. Because of this assessment, LFUCG continued with a full plan for the substitution of existing equipment at 80 pump station destinations. As a major aspect of the outline, our group built up a methodology so all destinations stayed in task amid the entire switchover process.

Changes incorporated the utilization of cutting edge SCADA gear at both wastewater treatment plants for reliability, institutionalization, and long term viability; SCADA control usefulness for West Hickman (the current SCADA framework checked plant activity just); and update the current restrictive RTUs with a Rockwell RTU-based framework and the radio correspondence hardware.

The system mentioned above is shown in Fig. 1. It is bulky and requires special installation expertise to function as intended. The new SCADA framework configuration gives numerous advantages, including disentangled tasks, equipment/programming institutionalization over the whole association, new control plans, capacity to effectively oblige future development, promptly accessible help and extra parts, and equipment/programming adaptability through open design items. Moreover, the new frameworks perm it remote control and progressed mechanized methodologies at LFUCG's WWTPs and pump stations like composed pump station task, and enhanced staff efficiency. Different advantages incorporate protection of staff time and vitality, enhanced unwavering quality, improved basic leadership, and tempest readiness and recuperation bolster planning LFUCG for the following two many years of administration.

Advancement of SCADA frameworks, lately, shows the nearness of a few unmistakable pattern s. Progressively heterogeneous structure applies to their development, both as far as equipment assets, and as far as correspondence systems utilized as a part of them. The regularly reason structure in vast SCADA frameworks circulated in extensive topographical locales is PC—PLC approach with the incorporated WEB server. In littler frameworks, especially in inquire about research centers and labs for remove adapting frequently connected PC—DAQ board or installed control board approach.

The transmission of communication through the Internet permits worldwide access and remote observing of the framework. This has just turned into a standard element in present day SCADA frameworks. Electronic SCADA framework utilizes the Internet to exchange information between the RTUs and the MTU as well as between the administrators' workstations and the MTU. The association over the Internet requires the utilization of extra assets to shield the framework from unapproved access and programmer assaults.

Fig. 1 Smart motor control centers (MCC) [13]

Remote interchanges is quickly developing fragment of the correspondences business, with the possibility to give fast and top notch data More and all the more regularly in SCADA frameworks a remote correspondence innovations are utilized for short range (Wi-Fi, Bluetooth, ZigBee), and for long range (Private Radio Networks—PRN, Satellite, 3G, 4G) information transmission. Remote SCADA replaces or stretches out the Fieldbus to the Internet. It is required in those applications when wire line correspondences to the remote site are restrictively costly or it is excessively tedious, making it impossible to build wire line interchanges. It can lessen the cost of introducing the framework. It is likewise simple to grow.

New patterns in educating and learning techniques, in which mixed learning is a standout amongst the most encouraging, can profit by remote labs as significant academic additional items. Examinations led in a genuine research facility are without a doubt the basic learning background. In any case, remote lab offices enable the understudies to get to the research center framework at nonworking hours. From the perspective of the instructing foundation that offered administrations, this satisfies the understudies in particular.

As it were: mechanization or programmed control of the utilization of various control frameworks for gear, for example, apparatus, plant procedures, boilers and heaters for warm treatment, combination of phone systems, administration and alteration of boats, air ship and different applications, and vehicles with a negligible human mediation which permits a completely robotized process.

Mechanization is accomplished by different means, including, pneumatic, mechanical, water driven, electrical, and electronic and PC gear, for the most part also. Confounded frameworks like present day industrial facilities, planes, and ships, which are frequently joined with every one of these procedures the upsides of mechanization are work sparing, practical power, sparing material expenses and enhancing quality, exactness and accuracy.

2.3 Industrial Internet of Things (IIoT)

With the increasing popularity of the Internet of Things, the idea came to incorporate SCADA systems with the internet of things. This gave birth to the so called Industrial Internet of Things (IIoT) [14].

The industrial internet of things refers to interconnected sensors, instruments, and other devices networked together with computers' industrial applications, including manufacturing and energy management. This connectivity allows for data collection, exchange, and analysis, potentially facilitating improvements in productivity and efficiency as well as other economic benefits. The IIoT is an evolution of a distributed control system (DCS) that allows for a higher degree of automation by using cloud computing to refine and optimize the process controls.

IIoT has all of SCADA capacities. The connection between the whole frameworks through the system, in which all gadgets of the framework can gather/trade information with each other. Obviously, this information can be broken down and prepared when SCADA is working.

IIoT is a much cheaper replacement due to the cost reduction in both the price of the equipment and the installation. Additionally, since most of the communication will be done wirelessly, there is another cost reduction from the lack of wired connections.

Moreover, the system can easily incorporate sensors as well as actuator to monitor and control a large spectrum of devices. And there are always new sensors and actuators that are being built to handle the need of the user.

However, as the connection is wireless, security may be an issue as these wireless devices can be hacked remotely if they are not secured well enough.

The IIoT is enabled by technologies such as cybersecurity, cloud computing, edge computing, mobile technologies, machine-to-machine, 3D printing, advanced robotics, big data, internet of things, RFID technology, and cognitive computing. Five of the most important ones are described below:

- **Cyber-physical systems (CPS)**: the basic technology platform for IoT and IIoT and therefore the main enabler to connect physical machines that were previously disconnected. CPS integrates the dynamics of the physical process with those of software and communication, providing abstractions and modeling, design, and analysis techniques for integrated the whole [15].
- **Cloud computing**: With cloud computing IT services can be delivered in which resources are retrieved from the Internet as opposed to direct connection to a server. Files can be kept on cloud-based storage systems rather than on local storage devices [16].
- **Edge computing**: A distributed computing paradigm which brings computer data storage closer to the location where it is needed [17]. In contrast to cloud computing, edge computing refers to decentralized data processing at the edge of the network. The industrial internet requires more of an edge-plus-cloud architecture rather than one based on purely centralized cloud; in order to transform productivity, products and services in the industrial world.
- **Big data analytics**: Big data analytics is the process of examining large and varied data sets, or big data.
- **Artificial intelligence and machine learning**: Artificial intelligence (AI) is a field within computer science in which intelligent machines are created that work and react like humans. Machine learning is a core part of AI, allows the software to become more accurate with predicting outcomes without explicitly being programmed.

IIoT systems are often conceived as a layered modular architecture of digital technology. The device layer refers to the physical components: CPS, sensors or machines. The network layer consists of physical network buses, cloud computing and communication protocols that aggregate and transport the data to the service layer, which consists of applications that manipulate and combine data into information that can be displayed on the driver dashboard. The top-most stratum of the stack is the content layer or the user interface.

As a rule, when you say "SCADA" consider the generation procedure itself or administration depicted certain guidelines. Basically, the SCADA framework is generally the primary framework through the framework (s) progressively. An exemplary mill SCADA connecting model show is (Controller, Sensor, and Actuator)—(OPC Server)—(SCADA applications)—(once in a while chose parameters over the Internet). Internet of things so far the framework for everything that Internet gets to have. It is perilous to give access to creation line gear specifically finished the Internet.

Web proceeds SCADA (or SCADA Web), where you can carry your framework into the cloud and get in touch with each other. Internet of things represents the Internet of things when it is utilized for modem purposes, it would call it IIoT (Internet Industrial of Things). For the most part, individuals contrasted with SCADA IIoT as opposed to the Internet of things.

IIoT has all SCADA capacities. As it were, the SCADA IIoT with extra highlights, that is. The connection between the whole frameworks through the system, in which all gadgets of the framework can gather/trade information with each other. Obviously, this information can be broke down and prepared when SCADA is working.

2.4   IoT Versus IIoT

Pros of Internet of Things

I.   Cost savings
     The electronic gadgets impart effectively, spare and spare expenses and vitality; this is the reason it is valuable for individuals in day to day schedules. By empowering information and correspondence between electronic gadgets and their interpretation in a coveted way, the IP gadget makes our frameworks more effective.
II. Monitor
     The second most merit of the online group is the follow up. The correct measure of consumable or air quality in your home can likewise furnish you with more data that has not yet been gathered. For instance, in the event that you realize that you have a low measure of drain shading or printer, sooner rather than later you will spare another outing to the store. What's more, checking the item stream can likewise enhance security.

Cons of Internet of Things

I.   Safety
     Envision if a famous programmer changed your income. Or on the other hand, if the store consequently sends an equal item you don't care for or an item that has officially lapsed. Thus, security is at last in the hands of shoppers to confirm the whole mechanization.
II. The prospects for low employment
     With online computerized substance and day by day exercises, there will be less interest for HR and less instructed representatives, which can make an issue of work in the public eye.

Pros of IIoT

I. Accuracy
    Generation and execution information is accessi ble progressively and is trans-
    mitted carefully. There is no compelling reason to enter or translate the bomb
    and the chief, paying little mind to whether it is a Pumper or a Senior Office.
II. Reliability
    Remote observing encourages administrators to wait for tools. Thus, the creation
    levels are more solid. Tank levels are all the more even. The wear on the hardware
    is limited.

Cons of IIoT

I. Price.
    Remote observing encourages administrators to wait for gear. Thus, the creation
    levels are more solid. Tank levels are all the more even. The wear on the hardware
    is limited.
II. Network Performance
    On the off chance that you pick remote observing, you should realize that the
    information is accessible. Guarantee that you have a framework that delivers
    issues identified with crest utilization, scope organization, dormancy, unwa-
    vering quality, and security. A fragmented or separate system keeps away
    from an assortment of activity loads coming about because of non-associated
    applications.

The institutional Internet group is as of now a major piece of regular daily existence, and a large number of us don't know it. As the innovation step by step to
progress and advance, the utilization of the online group is likewise utilized for some
essential associations. It is our errand to choose the amount of our everyday life is
prepared to control the innovation. At the point when this is done legitimately, it
naturally adjusts to our necessities and advantages society in general.

SCADA remote checking has numerous persuading points of interest. By comprehension and tending to challenges, administrators can effectively actualize this
innovation and utilize their prizes.

3 System Architecture

In this section, we discuss the proposed IIoT system that we have given the name
IoT industrial Wireless Controller. A block diagram of the proposed IIoT system is
shown in Fig. 2.

The system is compromised of a PC, multiple microcontrollers, relays, and
multiple electric devices to be controlled.

The PC is used to as a terminal that allows the user to control the electrical devices
through a local web page hosted on each of the microcontroller. The PC can connect

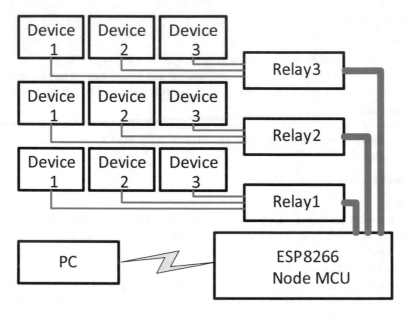

Fig. 2 Block diagram of the SCADA system

to multiple microcontrollers wirelessly using the local WiFi network. Moreover, each of the microcontrollers can connect to multiple devices through relays.

The microcontroller chosen for this work is the ESP8266 Node MCU [18] which is shown in Fig. 3.

NodeMCU is a low cost open source IoT platform. It includes a firmware which runs on the ESP8266 Wi-Fi SoC from Espressif Systems, and hardware which is based on the ESP-12 module. The term "NodeMCU" by default refers to the firmware

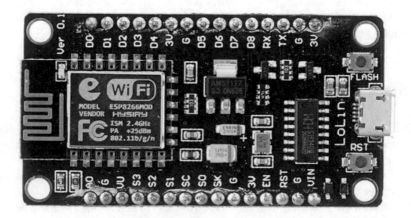

Fig. 3 ESP8266 NodeMCU microcontroller

rather than the development kits. The firmware uses the Lua scripting language. It is based on the eLua project, and built on the Espressif Non-OS SDK for ESP8266. It is used in many open source projects, such as lua-cjson and SPIFFS. The ESP8266 became popular in IoT diy projects as it is simple and easy to use.

In this work, the NodeMCU receive commands from the user\s terminal through its Wi-Fi module. Based on these commands, it controls the on/off operation of multiple devices. These devices are connected to the microcontroller through a relay module.

Figure 4 shows the relay module used in this work. Each relay can be connected to up to 4 devices, and each microcontroller can be connected to up to 4 relays. The aim is to place one microcontroller in a room and have it controlling up to 16 devices per room. Should there be more devices to be controlled, additional microcontroller can be easily added.

The user can control the devices from a web page that is hosted on the micro-controller. This page can be accessed from any PC within the same Wi-Fi network. The user will then have the option to switch on or off any device connected to that microcontroller.

Fig. 4 Relay module

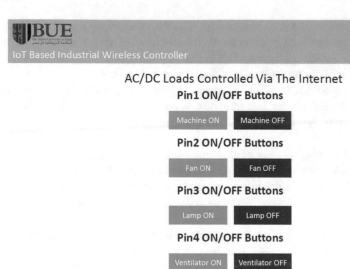

Fig. 5 Example webpage

Once the microcontroller joins the WiFi network, it acquires an IP address. All the user have to do next is to connect to that microcontroller from a browser using that IP address. The webpage will be displayed and allows the user to control all the devices connected to that microcontroller. An example of a basic webpage is shown in Fig. 5.

The figure shows how you can turn on or off four devices that are connected to the microcontroller through a relay. The buttons allows for the control of a device connected to one of the pins of the microcontroller. Pressing the button will send a signal to the connected device to either turn it on or off.

Figures 6 and 7 show an example of the complete setup in real life. The figure show a microcontroller connected to the PC and four relays. Each relay is connected to one device.

The example shows a lamp and a fan connected to two different relays through two different microcontrollers. Once the respective button is pressed on the respective webpage of the respective microcontroller, the device is turned on.

Ideally the microcontroller should be powered by a battery or an independent power source. In the figure, for the sake of simplicity, it is powered by the USB cable from the PC. However the controlling link is still the WiFi link.

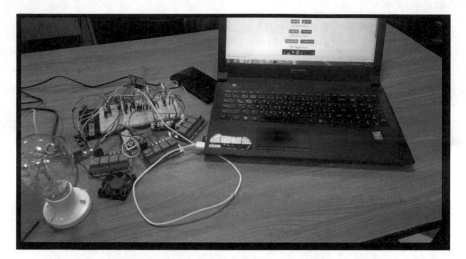

Fig. 6 Complete setup OFF

Fig. 7 Complete setup ON

4 Future Work

Since this is in its early stages, the prototype is disorganized and there are many wires. This will all be enhanced in the next stage. The device will be packaged in a way that makes it easier to install and operate.

Additionally, the website is hosted locally on each of the microcontrollers. This means that it can only be controlled from the local network. This is good because it

Table 1 Layered modular architecture IIoT

Content layer	User interface devices (e.g. screens, tablets, smart glasses)
Service layer	Applications, software to analyze data and transform it into information
Network layer	Communications protocols, wifi, cloud computing
Device layer	Hardware: CPS, machines, sensors

increases the security of the system. However, it will mean that the person has to be physically located on the premises to be able to control the devices.

Ideally, the website should be accessed from anywhere in the world. Which means that the security of the system will have to be increased and an authentication method will have to be added.

5   Conclusion

This chapter proposed an IoT based industrial wireless controller system. The proposed system was shown to be able to control multiple devices with the very low overhead and infrastructure. It is easy to install with very low cost as it uses basic devices and microcontrollers. Additionally, it can be controlled through the internet which increases the range of control with no boundaries. Moreover, the system is modular, which means that whenever the need arises to control more devices, another module can be added to the existing system without the need to change the whole system.

References

1. Bertocco, M., et al.: A client-server architecture for distributed measurement systems. IEEE Trans. Instrum. Measur. **47**(5), 1143–1148 (1998)
2. Tso, S.K., et al.: A framework for developing an agent-based collaborative service-support system in a manufacturing information network. Eng. Appl. Artif. Intell. **12**(1), 43–57 (1999)
3. Tommila, T., Ventä, O., Koskinen, K.: Next generation industrial automation–needs and opportunities. Autom.Technol. Rev. **2001**, 34–41 (2001)
4. Yang, T.C.: Networked control system: a brief survey. IEE Proc.-Control Theory Appl. **153**(4), 403–412 (2006)
5. Kalaitzakis, K., Koutroulis, E., Vlachos, V.: Development of a data acquisition system for remote monitoring of renewable energy systems. Measurement **34**(2), 75–83 (2003)
6. Kimura, T., Kanda, Y.: Development of a remote monitoring system for a manufacturing support system for small and medium-sized enterprises. Comput. Ind. **56**(1), 3–12 (2005)
7. Crowley, K., et al.: Web-based real-time temperature monitoring of shellfish catches using a wireless sensor network. Sens. Actuators A: Phys. **122**(2), 222–230 (2005).
8. De La Rosa, J.J.G., et al.: A web-based distributed measurement system for electrical power quality assessment. Measurement **43**(6), 771–780 (2010).
9. Ong, Y.S., Gooi, H.B., Lee, S.F.: Java-based applications for accessing power system data via intranet, extranet and internet. Int. J. Electr. Power Energy Syst. **23**(4), 273–284 (2001)

10. Sung, H.A., et al.: "CyberCut: an Internet-based CAD/CAD system." Transactions of the ASME. J. Comput. Inf. Sci. Eng. **1**, 52–59 (2001)
11. Altun, Z.G., et al.: Process control via internet. J. Integr. Des. Process Sci. **5**(2), 111–122 (2001)
12. Yang, ShuangHua, X. Chen, and J. L. Alty. "Design issues and implementation of internet-based process control systems." Control engineering practice 11.6 (2003): 709–720.
13. Su, C-L., C-N. Lu, and M-C. Lin. "Wide area network performance study of a distribution management system." International Journal of Electrical Power & Energy Systems 22.1 (2000): 9–14.
14. https://www.cdmsmith.com/en/Client-Solutions/Projects/Lexington-Fayette-SCADA
15. Boyes, H., Hallaq, B., Cunningham, J., Watson, T.: The industrial internet of things (IIoT): an analysis framework. Comput. Ind. **101**, 1–12 (2018). https://doi.org/10.1016/j.compind.2018.04.015.ISSN0166-3615
16. Staff, Investopedia (2011–01–18). "Cloud Computing". Investopedia.
17. Hamilton, E.: What is Edge Computing: The Network Edge Explained. cloudwards.net.
18. https://www.nodemcu.com/index_en.html

Mohammad M. Abdellatif is a full-time Lecturer at the British University in Egypt. Dr. Mohammad holds a B.Sc. '04 in Electronics and Communications Engineering from Assiut University, Assiut, Egypt, an M.Sc. '06 in Telecommunication Engineering from King Fahd University of Petroleum and Minerals, Dhahran, Saudi Arabia, and a Ph.D. '15 in Telecommunications engineering from the electrical and computer engineering department of the University of Porto, Portugal. Mohammad has received a four-year scholarship from the Foundation of science and technology of Portugal (FCT) for his Ph.D., as well as a one year CMU Porto joint research project award (SELF-PVP), His research interests are in the fields of IoT, WSN, and wireless communications.

Applying Software Defined Network Concepts for Securing the Client Data Signals Over the Optical Transport Network of Egypt

Kamel H. Rahouma and Ayman A. Elsayed

Abstract The physical layer of the Optical Transport Network (OTN) is the weakest layer in the network, as anyone can access the optical cables from any unauthorized location of the network and stat his attack by using any type of the vulnerabilities. The paper discusses the security threats and the practical challenges in the Egyptian optical network and presents a new technique to protect the client's data on the physical layer. A new security layer is added to the OTN frames in case of any intrusion detection in the optical layer. The design of the proposed security layer is done by using a structure of XOR, a Linear Feedback Shift Register (LFSR), and Random Number Generator (RNG) in a non-synchronous model. We propose the security model for different rates in the OTN and wavelength division multiplexing (WDM) system. The proposed model is implemented on the basis of protecting the important client signals only over the optical layers by passing these signals into extra layer called security layer, and before forming the final frame of the OTN system, this done by adding a new card in the Network Element (NE) to perform this job and by using the software defined network (SDN) concept of the centralized controller for all the network to find the intrusions in the optical layers. The encryption techniques of the client signals over the OTN are done between the source and the destination stations only and the signals are encrypted in the entire routes between both sides. The centralized controller of the SDN is used to manage the cryptographic model by distributing the encryption and decryptions keys to the source and the destination stations of the client signals. At the same time it is used to automatic detection of any intrusions in the OTN sections by continues tracing of the variations in the optical to signal network ratio (OSNR) in the OTN, these variations are proportionally related to the risks of the optical hacking and may be new intrusion is started. The results show that using the centralized controller of the SDN in the proposed model of the OTN encryption schemes is providing a high security against any wiretapping attack

K. H. Rahouma (✉)
Electrical Engineering Department, Faculty of Engineering, Minia University, Minia, Egypt
e-mail: kamel_rahouma@yahoo.com

A. A. Elsayed
Telecom Egypt, Cairo, Egypt
e-mail: aymanelmelm@gmail.com

© The Author(s), under exclusive license to Springer Nature Switzerland AG 2021
A. E. Hassanien et al. (eds.), *Machine Learning and Big Data Analytics Paradigms: Analysis, Applications and Challenges*, Studies in Big Data 77,
https://doi.org/10.1007/978-3-030-59338-4_28

at the same time the processes of detecting the intrusions in the optical layer over all the network become easier than before, and we can found that If any unauthorized attacker has the ability to access the fiber cables from any unmonitored location, the centralized controller of the SDN in the OTN will detect the variations in the OSNR in of the intruded section of the network and will automatically enable the check phase and according to the results of the check phase it will activate the cryptographic techniques for the selected client signals which passing through this intruded section, and the attacker will find encrypted data signals only and will need many years to find one the right key to perform the decryption process.

Keywords Linear feedback shift registers (LFSR) · Random number generators · XOR · Optical transport network (OTN) · Wiretapping. software defined network (SDN) · Centralized controller · Optical signal to noise ratio (OSNR)

1 Introduction

The transportation of the client data signals across the global communication networks is considered a critical issue to most of the network operators in how they can keep the confidentiality of the contents of these data through its complete journey in a global network, also in how they can avail the network resources to carry and maintain the performance of a hug amount of data bits from the source to the destination.

The Optical Transport Network (OTN) Technology was designed to transfer multiple types of clients' signals with different data rates through different wavelengths and over the same fiber optic cable by using Dense Wave Division Multiplexing (DWDM) technology. The process of transporting the client data signals from the source to the destination in the core transmission network includes many steps to map these signals into the generic OTN frames which depend on the data types and client data rates, where the client data is received at the client port in the Network element of the optical transmission network, the transponder card encapsulates these signals in the payload container and at the same time it adds the required overheads to perform the optical payload unit (OPU), the next step it multiplexes many of low payload rates to higher optical data units (ODU) rates, and the final step in the mapping process is to perform the optical transport unit (OTU) frames of the optical network.

The security of the client data while it is travelling journey across the OTN of the core Transmission network depends only on the standard encryption algorithms of the client signals at the application layers. These signals travel through different layers in the optical transmission networks, and the operators of these networks rely mostly on the security algorithms of the client signals at the applications layers only.

However; the problem with the majority of the optical transmission networks is that any attacker can access the physical layer from the optical fiber cables and split the optical signal by wiretapping it, after that he will be able to keep a live copy of the optical signal, and by trying different types of the reverse engineering in the de-mapping processes on this live copy according to the standard structure of the OTN frames, the attacker will reach to the original layer of the client signals. On the other hand, by using different types of the decryption algorithms in the service layer, the probability that the attacker can reach to original contents of the client data will be very high, and he will be able to break the encryption technique in the optical network from the first try by 100% percent success [1].

The necessity to secure the important client data signals while it was traveling through the optical transmission network remains crucial especially on the optical cables which are passing through long routes, and are vulnerable to be split and wiretapped by the interested attackers from any undetermined place [2].

Another important attack which targets the optical core networks is the jamming attacks; this attack is done by accessing the optical fiber links from any place in the network and additional harmful data signals is inserted inside the contents OTN frames. The aim the jamming attacks are to make service degradations and the results are misleading or modifying of the original data contents of the client data signals [3].

Most of the previous studies conducted many solutions to the security problem of the physical layer in the optical network, one of these studies proposed to make security profiles for every user in the optical network and creates upper category from these users which called gold users, where the gold user will have $1 + 2$ protection links against any security attacks in the physical layer of the optical network, as an example for every user link there is 2 other extra links, and the user has the ability to reroutes his traffic on these extra links in case of the original link is not safe, this solution is very costly and wastes the resources of the optical network especially in case of many gold users [1].

Other studies discussed how to use the XOR and optical LFSR to secure the optical data signals despite the difficulties on how they may face the distribution of the encryption keys between the different NE's in the optical network [4, 5].

More studies discuss new algorithms for the optical data encryption that uses quantum noise inherent in laser light and modulations algorithms that use two different cycles of the M-ary phase-shift-keyed (PSK) signal, a technique which allows the receiver who owns the short secret key to transform M-ary signals to standard modulated signals. The attacker, who does not own the secret-key, will be forced to try different M-ary measurements for several times, which is very difficult [2, 6].

Despite there were a many numbers of the studies proposed many years ago on the optical security, the physical layer of optical networks are still the weakest loop in communication networks from security perspective overview.

In this paper, we study a new proposed model for securing the client's data over the optical network, and this will be done by adding new security layer through extra separate module in the NE's, which will be used to make the required functions of the encryption algorithms for the selected clients' signals as per customer choices.

By adding this security layer for the different client frames in the OTN structure, the wavelengths over optical cables will be more secured from any unauthorized access and wiretaps hacking. On the other hand, implementing security algorithms in a separate stage in the OTN system for certain client signals only as an option will reduce the complexity of transmission systems, especially in the optical networks with huge capacity.

All the encryption algorithms in the proposed security layer will be built by using XOR gates and LFSR, and by considering that the client signal 10/100 Gb/s is equal to the same capacity of the wavelengths in the DWDM systems.

The proposed security layer will be enabled according to the results of the intrusion detection phase in the optical network. The detection of any intrusions in the network will be done by using centralized security controller (CSC) for all the optical network elements, the main job of the CSC is to monitor the variations of the optical to signal ratio (OSNR) of the target links in the optical network. In case of any up-normal changes in the OSNR values in one link of the optical network, it will be processed according to the risk of new attack in this link will be stated, and according to the investigations of the network operator about these changes the CSC will enable the security layer for the client data signals which travel through the intruded link or it will make reroute to the services to other links until the network operator finishes the investigations about root cause of these changes.

The paper is structured as following: Sect. 2 presents the current structure of the OTN frames in the optical transmission network Sect. 3 presents the proposed security model in the OTN system Sect. 4 presents proposed model analysis, Sect. 5 presents the numerical analysis and the results of the proposed model, and finally Sect. 6 presents the conclusion and the future work of the paper.

2   The Structure of the OTN Frames in the Optical Network

There are different mapping stages of the client signals inside the NE's of the transmission network consist of:

Stage1: The transponder card which receives the client signals with rates 10/100 Gb/s transforms it to the electrical domain by using Small Form-Factor Pluggable (SFP) module with the suitable laser frequency for the client signal. The next step in this stage is the mapping of the client signals to be part of the structure of the OTN frames, this is done by putting the client data in containers with fixed size according to the client data capacity. These containers are used to form the optical payload units (OPU$_k$) (where k represents the capacity of the OTN frame with k = 2 represents a frame with data rates equal to 10 Gb/s and k = 4, OTN frame with data rates equal

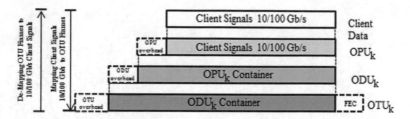

Fig. 1 The construction of the OTN Frame

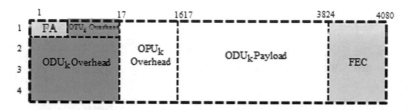

Fig. 2 The structure of the OTU_k frame

to 100 Gb/s) by adding its overheads, the next step is to form the optical data unit ODU_k. Finally, on adding the final overheads and alignments words with forward errors corrections (FEC), the optical transport unit (OTU_k) will be formed (see Fig. 1), and by converting the OTU_k to the optical domain it will ready it be multiplexed as a wavelength in the WDM transmission system [7]. The final frame of the OTU_k in the OTN system is shaped in 4 different rows of 4080 bytes and repeated every certain period (12.191 μs for OTU_2 frame and 1.167 μs for OTU_4 frame) [8]. This frame consists of bytes that stand for frame alignment word (FA), different layered overheads bytes (OH), payload data and finally bytes that represent forward error correction algorithms. After that, this frame is converted to standard optical signals with standard wavelengths (see Fig. 1) (Fig. 2) [9].

Stage2: Giving that the network consists of two network elements with two directions in the optical network, the 2nd step in the optical transmission system processes is the multiplexing of many standard optical signals to form one beam of laser which consists of many wavelengths and different standard frequencies in the c band.

Stage3: The last stage in the transmission network inside the NE's is the amplification process that uses any type of the optical amplifiers such Erbium Doped Fiber Amplifiers (EDFAs) with the suitable gains according to different factors on the network such as the travelling distance between the two network elements and the attenuations on the available fiber cable.

The transmission model in this paper consists of two network elements connected by a fiber cable with 40 wavelengths (W1 to W40) in the DWDM system and client signals with different bit rates (see Fig. 3).

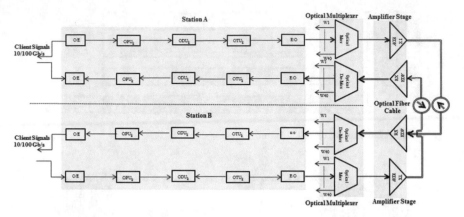

Fig. 3 The transmission model of two Network elements

The problem in the current model from the security perspective can be explained as follows:

Assume that Alice, in location A wants to send a message to Bob, in another location B Where A&B are connected through the pervious optical transmission network. Then, the attacker Eve in location C can split the laser signals of the fiber cable from any point of the transmission network into two paths by using optical splitters where the first path of the laser signals will be returned back to the original route of the network to transfer original message of Alice to Bob and the 2nd path will be used by Eve's equipment where Eve can have a live copy of the OTU frames without affecting the original routes of the network and in case Eve's hacking system made a little effort in the de-mapping process according to the standards of the OTN structures, Eve will be able to break the optical network structure and will know the contents of the messages between Alice and Bob while they are not realizing that, all their communications messages are known by third person who is Eve.

As shown in the previous optical transmission network, there aren't any security algorithms that are implemented to prevent the interested attacker such Eve from understanding the contents of the optical signals even though he can access the fiber cable and keep a live copy of the client signal.

3 The Proposed System Model

The proposed model consists of the intrusion detection in the physical layer of the optical links and the security model which performs the cryptographic algorithms to the client data signals.

3.1 Intrusion Detection in the Physical Layer of the Optical Network

In the recently years the software defined network (SDN) technologies attracted many researchers to apply its concepts on many fields in the network, the most advantages of the SDN concept is it transforms the control parameters, the protocols and the functions of any network to be programmable and this done by separating the control plane from the data plane. Extending the principles of the SDN to the optical transport network can provide new framework in the applications, coordination and orchestrations of the higher order optical layer [10].

In our security model the principles of the SDN in the intrusions detection and requested response in the physical layers of the optical network, and this done by continues monitoring the variations in the optical signal to noise ratio (OSNR) in the different optical links by using proposed software defined security (SDS) for all the physical layers of the optical transport network. In the SDS model the control plane is separated from the data plane by using centralized Security Controller (CSC) to perform this function in all the optical sections. The CSC measures automatically the values of the OSNR for the different optical links by monitoring the performance of these links from change management data base (CMDB) of the operation support system (OSS) server and from the different network management systems (NMS's) of the optical transport network. The OSNR values for every links can be calculated according to the following equations [11]:

$$OSNR \ (DB) \ = \ P_{out} - P_{ASE} \tag{1}$$

where: P_{out} is the output power of the amplifier, P_{ASE} is the total amplified spontaneous emission (ASE) noise power.

$$
\begin{aligned}
P_{ASE} &= NF + G + N_{in} \\
&= 10 \log 2n_{sp} + 10 \log(G - 1) + 10 \log hvB_o
\end{aligned} \tag{2}
$$

where: NF is the external noise index, G is the amplifier gain, and N_{in} is the input noise power $N_{in} = 10 \log hvB_o$ which means input noise equal the power of photon.

$$G = P_{out} - P_{in} \tag{3}$$

where: P_{out} is the output power, and P_{in} is the input power of the amplifier.

Figure 4 shows the proposed model for the SDS model which performs the function of monitoring the variations of the OSNR in the optical network, and this done by using centralized security controller to perform this function only as following sequence:

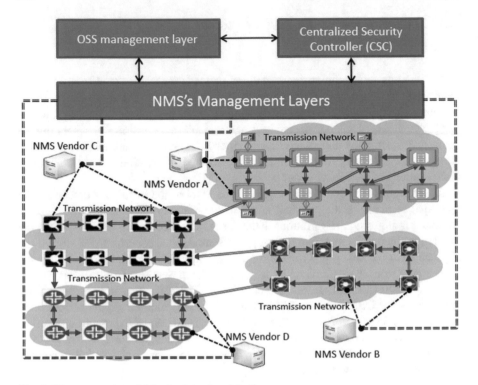

Fig. 4 The proposed model for the intrusion detection

- The operation support system (OSS) of the transmission network collects the performances of the optical links from the Network management layers (NMS's)
- The centralized security controller (CSC) communicates with the change management database (CMDB) of the OSS server and checks any variations of the performances of the mentioned optical links
- In case of any changes in the performance the CSC will start and calculate the OSNR of this link after that it will compare the results with the threshold values of this link
- If the values of the OSNR are less than the threshold values for this link the CSC will send notifications to the network operator about these changes
- The network operator will decide if these changes are determined by defined root cause or not determined
- If the root cause is unknown of these changes the CSC will enable the security layer for the selected traffic which passing through the defected link.

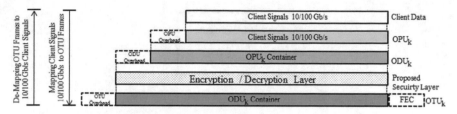

Fig. 5 The proposed security layer in the OTN frame structure

3.2 The Proposed Security Layer

After the detection of the intrusion in the optical link and according to the previous proposed model the CSC will enable the proposed security model to secure the client signals over the OTN system at the source and destination NE's only in the DWDM transmission network. The security techniques will be against the wiretapping of the optical signals and is done by adding new security layer in the mapping process of the client signal and before forming the final frame of the OTN system. According to the proposed model the security layer is implemented after forming the ODU_k bytes of the frame and before the final OTU_k stage (see Fig. 5).

To achieve this security layer, there are two types of the Pseudo Random Generators (RNG) that could be used to generate the keys of the encryption/decryption processes. The 1st one is synchronizes RNG's while the 2nd is a non-synchronies RNG's, provided that both types of the RNG's systems are working with the same linear feedback shift register (LFSR) and by using XOR operations to perform the encryption/decryption processes, the only difference between them is that in the first type of the RNG the source and destination stations are synchronized with the same clock and it generates the same keys in both stations as result of using the same clock for synchronization, While in the 2nd type of the RNG the source and destination stations is working separately in every station with different keys generation.

3.2.1 Generating the Encryption Key

In our model the security over the entire optical network is managed by the centralized security controller (CSC) which is part of the software defined security (SDS) of the optical network (see Fig. 6). As soon as the CSC discovers any intrusions in optical links it will transmit the trigger pluses and the initial symmetric key to the source and destinations NE's of the selected client data signals. The transmission of the initial keys to the NE's is performed by using secure transmission lines, and the generation of the initial symmetric keys by the CSC may include Cryptographic Secure Pseudo-Random Number Generator (CSRNG). The source and the destination NE's will initiate its security model by using the initial symmetric key which was generated and transmitted before by the CSC, and the proposed security model starts to generate its

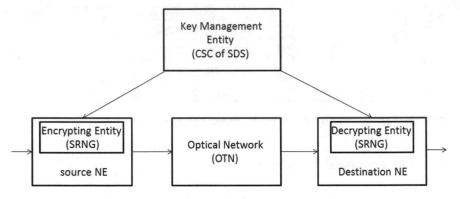

Fig. 6 The key management entity in the optical network

encryptions- decryptions keys according to the combinations between the data plain bits and the initial key [12].

For generating the initial key a chaotic map has been used with the following equation [B]:

$$f(x_i) = \begin{cases} x_i/\gamma & x_i \in [0, \gamma] \\ (1 - x_i)/(1 - \gamma) & x_i \in (\gamma, 1] \end{cases} \tag{4}$$

where: γ is the control value, x_0 is the initial state and their value included in the interval (0, 1).

One important characteristic of the selected chaotic map is it has no periodic widows and maintains the chaoticity in the whole parameters space, as any selected key will meet the required chaoticity according to the space the cryptographic system parameters. One weak point of the chaotic map is its digital applications depends on finite word length, to overcome this disadvantage point of chaotic map in our model we use the Pseudo Random Number Generator (PRNG) as it disturbs the chaotic orbits and the randomness of this system can be improved [13].

3.2.2 The Proposed Security Model Implementation

The proposed security layer is performed according to the traditional techniques of the Encryption—Decryption processes as the following concepts [14, 15]:

- Consider the length plaintext of the data as m, the output sequence of the security layer as y, and by using secret key for the encryption operations as k
- the processes of securing the plaintext m at the source station is done according to $y_i = m_i \oplus k_i$ where i is the bit number in the plaintext frame and $\oplus$ is the XOR operation

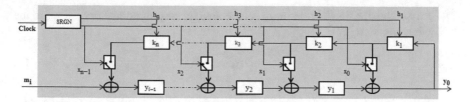

Fig. 7 The proposed model of the security layer

- To retrieve the original plaintext at the destination station, the decryption process will be implemented by using the same secret key according to $m_i = y_i \oplus k_i$.

The transmission over the optical networks takes place by transmitting the bits within periodic frames with different frame rates (in the synchronize digital hierarchy system (SDH) is 125 μs, and the OTN is varying from 12.191 μs for OTU$_2$frame to 1.167 μs for OTU$_4$frame) which means that for every 1 s there are 8000 frames in the SDH system, 82,027 frames for OTU$_2$ and 598,802 frames for OTU$_4$ will be transmitted from source to destination [7–9, 14–18].

The proposed model for the new security layer of the OTN frame is to use dual Synchronize Pseudo Random Number Generator (SRNG) to generate the same secret key and the same polynomial of the LFSR with the trigger with every trigger clock pulse from the CSC, and by considering that both SRNG's as synchronized with the same distributed clock pluses on the optical transmission network from the CSC, both source and destination stations will have the same generator polynomials P_n and the same seeds of initializing key words k_n which was received before from the CSC to the LFSR's [16].

By assuming that the SRNG generates random polynomial with degree n and the CSC transmits random initialized key word for the LFSR with length n bits with every trigger clock pulse. The proposed model for the encryption system is shown in the following figure (see Fig. 7).

The proposed security model consists of:

- Dual synchronized random number generator (SRNG) which generates random polynomial $P_n = f(x_n)$ with degree n, and used to enable or disable the switches in the proposed model to implement different possibilities of the random number generator and
- CSC generates random key k_n with stream of bits h_1 to h_n as well as to initialize the LFSR with every reset cycle of the encryption period.
- Flip-flops to keep the encryption bits with every bit shift encryptions.
- XOR to implement the encryptions/decryptions algorithms.
- Plaintext which is the data of the ODU$_k$ and equal to $M = N * m$ where N is the number of the reset cycles and m is the length of the part of the plain text which will be encrypted/decrypted by the same initialized key and the generated polynomial.
- Encrypted data y_i.

As assumed, the SRNG randomly selects one prime polynomial according to the following equation [17]:

$$P \leq \psi(n-1)/n \tag{5}$$

where $\psi(\cdot)$ is Euler function, n is the degree of the polynomial which is limited by the range of $n_{min} \leq n \leq n_{max}$, and n_{min} is equal to the key length which is used to encrypt data message m.

Due to the high speed of the OTN system the length of the message of the ODU_k becomes very long and encrypting this length with the same encryption key will be very difficult. Therefore, the original message M which is equal to ODU_k in every frame with length L_M will be divided to many equal data parts with length $L_m = L_M/N$, where N is the number of the reset cycles to encrypt/decrypt the messages with length l_m as per the following condition [17]:

$$n_{min} \geq \left(\frac{L_M}{N} - 1\right) \tag{6}$$

The uncertainty of the generated polynomial from the SRNG and the initialized secret key from the CSC are measured by the entropy of the probability of true polynomial may be detected as the following equation [18]:

$$H_1 = \sum_{i=0}^{n} P_I \left(\log 1/P_I\right) \tag{7}$$

where P_i is the probability to guess one polynomial of degree n out of 2^{n-1} polynomials, and the entropy of the probability to guess the initialized secret key of the LFSR as per the following equation:

$$H_2 = \sum_{i=0}^{n} k_i \left(\log 1/k_i\right) \tag{8}$$

where k_i is the probability of detecting one secret keyword of length l_k out of 2^n keys. If we considered the generations of the polynomials and the initialized keys are independent, then the joint entropy of the polynomial and the secret keys of the whole system will be $H = H_1 + H_2$. Afterwards, the generated polynomial can be selected randomly with every clock pulse from 2^{n-1} polynomials, which can be stored in buffers with the required size. The implementation of the proposed model in the OTN transmission system in our case is done between 2 NE's source and destination (see Fig. 8).

where the client data signals are mapped into the OTN frames according to its bit rates, the ODU_k data is encrypted according to the proposed security model of Fig. 5

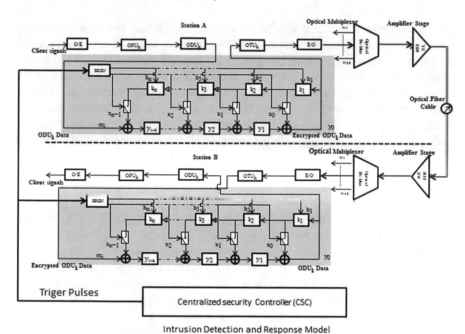

Fig. 8 The Implementation of the security layer in the OTN system

and the encrypted data (Encrypted ODU_k) is returned back to fill the frame of the OTU_k in the OTN system, after that it will be multiplexed optically with other wavelengths, and the amplifier stage will be done over the optical fiber cables to reach the destination station. In the receive side, the same processes will be done but in reverse actions. After the preamplifier stage in the receive direction, the de-multiplexing stage will isolate the OTU_k and forward it to the de-mapping operations until it reaches the encrypted ODU_k, then, the decryption processes will start searching for the original ODU_k by using the same model of the encryption system, as shown in Fig. 6 whereas the dual SRNG polynomial with the same secret key in both sides of source and destination stations is used.

4 The Analysis of the Proposed Security Model

There are many security mechanisms used in the optical transport network, one of these mechanisms is securing Only the Perimeter at the client side, in this mechanism the client data are protected by intrusion provision system (IPS), customer edge switches (CES) or firewalls at the client equipment only. While the most vulnerable points in the network is the physical layer of the optical transport network where this mechanism leaves the internal optical links open to any security threat.

The other one of the security mechanisms in the optical network is the distributed and Uncoordinated Security mechanisms, this technique uses different and independent security algorithms which are applied for the different sections in the optical network, the mechanism increases the complexity and the need resources of the security management system, and reduces the optical network performance [19].

In our model we used the concept of software defined network security (SDS) for applying the security mechanism in the optical network by implementing centralized security controller (CSC) over the entire optical transport network, the CSC not only used for monitor the security breaches and makes the required security decisions but also it optimizes resources which utilized for the security algorithms in the optical network. By using the software defined security concept in the optical network it centralized the security policy management, the coordination and eliminated the dependence on the vendors' security mechanisms. Having only one centralized controller in the control plane in the SDS of our model is considered weakness point from security perspective, where implementing only one CSC for all the optical network will increase the risks of the attacks on the links between this controller and the switches of the SDS of the optical network, and this may isolate the CSC from acting its function in monitoring and provides the required security decision to the different NE's in the optical network [F]. to overcome this weakness point in our proposed model its suggested to use 2 or more CSC in the SDS of the optical network, the first CSC is set as active working controller in the SDS network and the other one is as idle protection controller, the 2 CSC's in the network should be synchronized and should exchange its data between each other and this done by using a certain types of the data replications between the 2 servers of the CSC's.

With every variation in the OSNR values the CSC will check these changes with the network operators and in case of the root cause of these changes are unknown the CSC will send trigger pulses and the initialized keys to activate the security layer in the source and destination stations and it will define the clock rates of the proposed security model. With every clock pulse, the dual synchronized Random number generator (RNG) will generate two independent outputs. The 1st output is the generation of one Random polynomial with degree n out of 2^{n-1} polynomials to enable or disable the switches of the LFSR security model. The 2nd output is the generation of one initialized secret key which was received before from the CSC with length $l_k = n$ out of 2^n keys to initialize and rest the key of the LFSR model with every frame in the OTN transmission system.

Afterwards, the reset cycle starts with the trigger of every clock in both sides of source and destinations and the system makes the encryptions with the same polynomial function and the produced keys from the combinations of the initialized key and the shifted data bits of the original data message with length $l_m = l_n$.

In case the attacker has the ability to manage and access the fiber cable by splitting the optical laser signals and take a live copy of the OTN frames to start the de-mapping process, he will only get the encrypted ODU_k instead of the original ODU_k which included in the OTN frame (see Fig. 9).

The attacker will need to be aware of 4 variables to understand the encrypted ODU_k, and to be able to retrieve the original ODU_k of the OTN frames.

Fig. 9 The Structure of the OTN 10/100 Gb/s frame with encrypted ODU

- The 1st variable is to know the algorithms of the dual SRNG with its synchronization clock.
- The 2nd variable is to know the secret keyword which used in the reset cycle with every trigger of the clock pulse which has probability $p_k = 1/2^n$ and entropy equal $\sum_{i=0}^{n} k_i (\log 1/k_i)$ as listed in Eq. (4) to initialize the original message.
- The 3rd variable is to know the polynomial degree, with the probability $p_P = 1/2^{n-1}$ and entropy $\sum_{i=0}^{n} P_i (\log 1/P_i) \sum_{i=0}^{n-1} P_i (\log 1/P_i)$ as listed in Eq. (5)
- The 4th variable is to know the generation and clock rate of the dual SRNG.

The time required to break the proposed system and to guess the right key to decrypt the data can be estimated by T^E where [17]:

$$T^E \geq N \cdot F \cdot \tau \cdot 2^{n-1} \tag{9}$$

where: τ is the time required for the attacker system to guess the right key for the decryption process, 2^{n-1} is the number of the right polynomials functions with degree n and the right number of initialized secret key with length $l_k = n$ with probability to find right polynomial function with degree n and right initialized secret key with length l_k is equal to $1/2^{n-1}$, F is the number of the initialized cycles in 1 s, and finally N is the number of the secret keys which is used in the same encryption cycle.

The weakness points in our proposed model as following:

- The transmission of the initial key from the CSC of the control plane to the NE's needs a secure transmission channels which may not be available and cost more resources and this can be solved by make more researches in using the reserved bytes for the future use in the OTN frames to carry the initial keys to the targeted NE'S.
- Using only one centralized controller for all the security system in the optical transport network is very risky as this controller may be damaged or may be hacked by any interested attackers, the solution of this problem is to use one CSC as active controller and using other one to be protection for the first one. Other solution to this problem is using cloud principles by sharing the functions of the CSC between the many controllers in the optical network.

5   The Numerical Analysis of the Proposed Model

In order to be able to test the usefulness of the proposed model in protecting the client signals over the OTN system, we will calculate the time required to break the security model for one hypothetical case which is the clock rate is equal to the same rate of the frames repetition in the OTN system and estimated by the same rate of the frames in the SDH system as equal to 125 μs, 12.191 μs for OTU_2 and 1.167 μs for OTU_4 frame. Then the OTN system will transmit 82,027 frames every 1 s for 10 Gb/s, and 598,802 frames every 1 s for 100 Gb/s as a result, the dual SRNG will generate 82,027 polynomials and initialized secret kays every 1 s in case of 10 Gb/s and 598,802 polynomials and initialized secret keys every 1 s in case 100 Gb/s as well. Table 1 gives the results from Eq. (5) 31 of the primitive polynomials of the SRNG for the polynomial degree with range $10 \leq n \leq 31$ [8].

The generated polynomials increased in huge rate with every little step increase in the n variable. From Table 1 we can find that for the 10 Gb/s bit rate to achieve 82,027 polynomials and initialized secret keys in 1 s, n is estimated to be equal to or greater than 21 ($n \geq 21$), and for 100 Gb/s bit rate to achieve 598,802 polynomials and initialized secret keys in 1 s, n is estimated to be equal to or greater than 29 ($n \geq 29$). To measure the time required to break the proposed security system for $l_m = l_M /N$, we will consider n = 21 for 10 Gb/s and n = 29 for 100 Gb/s [8].

For 10 Gb/s client signals and n = 21, the length of the ODU_2 equal to 10.037 Gb/s then

$$N = 10.037 \text{ Gbits/s}/n = 4779 \times 10^5 \qquad (10)$$

For 100 Gb/s client signals and n = 29 the length of the ODU_4 is equal to 104.7944 Gb/s, then

Table 1   The number of the generated primitive polynomial for $10 \leq n \leq 31$

N	2^n	$\psi(n-1)$	P	N	2^n	$\psi(n-1)$	P
10	1024	600	60	21	2,097,152	1,778,112	84,672
11	2048	1936	176	22	4,194,304	2,640,704	120,032
12	4096	1728	144	23	8,388,608	8,210,080	356,960
13	8192	8190	630	24	16,777,216	6,635,520	276,480
14	16,384	10,584	756	25	33,554,432	32,400,000	1,296,000
15	32,768	27,000	1800	26	67,108,864	44,717,400	1,719,900
16	65,536	32,768	2048	27	134,000,000	113,000,000	4,202,496
17	131,072	131,070	7710	28	268,000,000	133,000,000	4,741,632
18	262,144	139,968	7776	29	537,000,000	534,000,000	18,407,808
19	524,288	524,286	27,594	30	1,070,000,000	535,000,000	17,820,000
20	1,048,576	480,000	24,000	31	2,150,000,000	2,150,000,000	69,273,666

$$N = 104.7944 \text{ Gbits/s} \big/ n = 3742 \times 10^6 \qquad (11)$$

From Eq. (10) the time required to break the security system and to guess the correct key for the encryption process of the 10 Gbit/s client signals in the proposed model by considering n = 21, proposed $\tau = 10^{-12}$ s, and F = 82027 frames/s will be:

$$
\begin{aligned}
T^E &\geq F \cdot N \cdot \tau \cdot 2^{n-1} \\
&- 82027 \times 4779 \times 10^5 \times 10^{-12} \times 2^{20} \\
&= 475.75 \text{ days} = 1.3 \text{ years}
\end{aligned}
\qquad (12)
$$

From Eq. (10) the time required to break the security system and to guess the correct key for the encryption process of the 100 Gbit/s client signals in the proposed model by considering n = 29, proposed $\tau = 10^{-12}$ s, and F = 598802 frames/s will be:

$$
\begin{aligned}
T^E &\geq F \cdot N \cdot \tau \cdot 2^{n-1} \\
&= 598802 \times 3742 \times 10^6 \times 10^{-12} \times 2^{28} \\
&= 6961665 \text{ days} = 19073 \text{ years}
\end{aligned}
\qquad (13)
$$

From Eqs. (12) and (13) we found that to guess one of the keys that will be used in the decryption process of the ODU_k in the proposed security model in the OTN frames it will take 1.3 years for 10 Gb/s client signals and 19,073 years for 100 Gb/s which is not possible in the practical life to do that, and this indicates the our proposed system makes it very difficult to break the structure of the OTN frames by any interested hacker in the normal conditions and understanding the contents of the client data in the OTN frames will be impossible from any hacker.

6 The Future Work

The future work of this paper is to use the machine learning technology with the software defined network to perform the optical cryptographic system, and to make the optical transport network smarter. Using the 2 technologies in the same network will permit to apply the clouding principles on the optical network and will provide the ability to transform the traditional optical network to be part of a huge optical cloud network. Also this will drive the security system in the optical transport network to the automation. This will be done by using neural network to monitor and trace any changes in the values of the Optical Signal to Noise Ratio (OSNR) over all sections of the optical network and by using the centralize control plane of the SDN the traffic will be rerouted to other safe routes according to the available resources in the optical cloud network this will be used to enables automatic optical cryptographic system in the sections of the optical network which may has any strange change in its OSNR value which is likely to be wiretapping attack. By using machine learning technology and the SDN in the optical cryptographic system the detection of the attack and the response will be done automatically in the affected sections only and this technique will provide robust network security system in the optical network, and at the same time will reduce the complexity and the cost of using cryptographic system over all optical wavelengths without usefulness in the sections of the network which have no attacks.

7 The Conclusion

In this paper, we addressed, for the first time, one of the practical challenges that face the security of the optical physical layer associated with the OTN transmission network and this was done by adding new layer for the security system in the stages of mapping client signals into the OTN frame. For the first time we used the principles of the software defined network (SDN) by separating the control plane from data plane of the security system in optical network, and this used to detect any intrusions in the physical layers of the optical transport network by automatically monitoring the changes in the optical signal to noise ratio. The results were very good where in case of any attacker has the ability to access the fiber cables from any unauthorized place the control plane of the security system will detect the variations of the OSNR in intruded optical link after that the CSC of the control plane in the SDS will activate the security layer for the selected services only at the source and destination NE's, by enabling the proposed security layer the attacker will only able to take copy the encrypted data signals without understanding what is the contents of this data, and this is as a result of using our encryption algorithm which was implemented before on this data, the attacker will need many years to find out one right key to perform the decryption processes. By using the SDN concepts in the proposed security model the system become partially automated in detecting any intrusions, the proposed model is based on the idea of protecting the important client signals only on the optical physical layers according to the requests of the customers. This system can be achieved by passing the selected client signals only through extra layer, called the security layer and before forming the final frame of the OTN system. This system can be executed on many sections in the transmission network of Egypt to protect the important client signals for certain customers, also the system can be used in the military services over the public transmission network, and finally the system will be very useful in the near future especially with the needs to transfer a huge amount of the clients data through the optical network as the result of the tremendous progress in using the internet of things (IOT) technology and machine to machine communication.

References

1. Skorin-Kapov, N., et al.: Physical-layer security in evolving optical networks. IEEE Commun. Mag. **54**(8), 110–117 (2016)
2. Fok, M.P., et al.: Optical layer security in fiber-optic networks. IEEE Trans. Inf. Forensics Secur. **6**(3), 725–736 (2011)
3. Furdek, M., et al.: An overview of security challenges in communication networks. In: 2016 8th International Workshop on Resilient Networks Design and Modeling (RNDM). IEEE (2016)
4. Han, M., Kim, Y.: Unpredictable 16 bits LFSR-based true random number generator. In: 2017 International SoC Design Conference (ISOCC). IEEE (2017)
5. Liu, X.-B., et al.: A study on reconstruction of linear scrambler using dual words of channel encoder. IEEE Trans. Inf. Forensics Secur. **8**(3), 542–552 (2013)
6. Dimitriadou, E., Zoiros, K.E.: All-optical XOR gate using single quantum-dot SOA and optical filter. J. Lightwave Technol. **31**(23), 3813–3021 (2013)
7. Mobilen, E., Bernardo, R., Monte, L.R.: 100 Gbit/s optical transport network 40 nm test chip design and prototyping. In: 2017 SBMO/IEEE MTT-S International Microwave and Optoelectronics Conference (IMOC). IEEE (2017)
8. Barlow, G.: A G. 709 Optical Transport Network Tutorial. Innocor Ltd. Capturado em: https://www.innocor.com/pdf_files/g709_tutorial.pdf (2003)
9. Loprieno, G., Losio, G.: Timeslot encryption in an optical transport network. U.S. Patent No. 8,942,379. 27 Jan. 2015.
10. Ji, Y., et al.: Prospects and research issues in multi-dimensional all optical networks. Sci. China Inf. Sci. **59**(10), 101301 (2016)
11. Chen, X., et al.: An OSNR calculating method based on network topology for optical network. In: 2017 16th International Conference on Optical Communications and Networks (ICOCN). IEEE (2017)
12. Lohr, J.: Adaptive traffic encryption for optical networks. U.S. Patent Application No. 10/091,171
13. Pérez-Resa, A., et al.: Using a chaotic cipher to encrypt Ethernet traffic. In: 2018 IEEE International Symposium on Circuits and Systems (ISCAS). IEEE (2018)
14. Liu, X.-B., et al.: A study on reconstruction of linear scrambler using dual words of channel encoder. IEEE Trans. Inf. Forensics Secur. **8**(3), 542–552 (2013)
15. Liu, X.-B., et al.: Investigation on scrambler reconstruction with minimum a priori knowledge. In: 2011 IEEE Global Telecommunications Conference (GLOBECOM 2011). IEEE (2011)
16. Kumar, M.C., Praveen Kumar, Y.G., Kurian, M.Z.: Design and implementation of logical scrambler architecture for OTN protocol. Proc. IJARCET **3**(4), 1260–1262)2014)
17. Engelmann, A., Jukan, A.: Computationally secure optical transmission systems with optical encryption at line rate. arXiv preprint arXiv:1610.01315 (2016).
18. Carter, T.: An Introduction to Information Theory and Entropy. Complex Systems Summer School, Santa Fe (2007)
19. Liyanage, M, et al.: Opportunities and challenges of software-defined mobile networks in network security. IEEE Secur. Privacy **14**(4), 34–44 (2016)

Watermarking 3D Printing Data Based on Coyote Optimization Algorithm

Mourad R. Mouhamed, Mona M. Soliman, Ashraf Darwish, and Aboul Ella Hassanien

Abstract The main objective of this work is developing 3D printing Data Protection Using Watermarking approach that considers watermarking problem as an optimization problem. 3D objects watermarking inhabits a challenging obstacle. The existence of many 3D objects representations act one reason for this challenge. The 3D models watermarking research state is furthermore in its opening as opposed to published work in video and image watermarking. This work propose a 3D watermarking approach by utilizing Coyote Optimization Algorithm (COA) in optimizing statistical watermarking embedding for 3D mesh model. Coyote optimization algorithm (COA) consider a recent fast and stable meta heuristic algorithm. This proposed approach aims to introduce an intelligent layer on the watermarking process. The approach starts by selecting the best vertices that will carry the watermark bits using k-means clustering method. Followed by watermark embedding step using COA in finding the best local statistical measure modification value. Finally we extract the embedded watermark without any need of the original model. The proposed approach is validated using different visual fidelity and robustness measures. The experimental results of the proposed approach will be compared with other state of the art approaches to prove its superiority in embedding and extraction of watermark bits sequence with respect to both robustness and imperceptibility.

Keywords 3D mesh model · Watermark · Point of interest (pois) · Clustering · k-means · Coyote Optimization Algorithm (COA)

M. R. Mouhamed (✉) · A. Darwish
Faculty of Science, Helwan University, Cairo, Egypt
e-mail: mouradraafat@yahoo.com

A. Darwish
e-mail: shraf.darwish.eg@ieee.org

M. M. Soliman · A. E. Hassanien
Faculty of Computers and Information, Cairo University, Cairo, Egypt
e-mail: mona.solyman@fci-cu.edu.eg

A. E. Hassanien
e-mail: aboitcairo@gmail.com

© The Author(s), under exclusive license to Springer Nature Switzerland AG 2021   603
A. E. Hassanien et al. (eds.), *Machine Learning and Big Data Analytics*
Paradigms: Analysis, Applications and Challenges, Studies in Big Data 77,
https://doi.org/10.1007/978-3-030-59338-4_29

1 Introduction

Due to the recent revolution of technology and artificial intelligence, many application were appeared most of them uses all types of data like sound, image, text and recently 3D objects. The fresh trend of 3D imagery and communication seen in many fields, such as medicine, architecture and entertainment. 3D object models represented as a 3D points cloud, parametric surfaces, and mesh model. The most supported format is the last one. In this model type, the surface is formed of a polygons set joined to form the shape of model [1]. The mesh is composed primarily of the polygons vertices's coordinates and the vertices connectivity within each polygon. But other topological information also can be obtained like the polygons adjacency or connectivity. The most form that usually used is the polygonal mesh due to its flexibility and simplicity.

Watermarking offers a system for the privacy of copyright or the declaration of ownership of digital content by inserting information in data. The basic idea is to embed a piece of copyright-related information (i.e. the watermark) into the functional part of a mesh model[Watermarking 3D Triangular Mesh Models Using Intelligent Vertex Selection]. The embedded watermark should be robust against various attacks on the watermarked model and also be imperceptible to human eyes. Since 3-D mesh watermarking techniques were introduced, there have been several attempts to improve the performance in terms of impracticality and robustness. Robustness is achieved when the watermark can be retrieved even after the watermarked model has been processed or attacked intentionally specific algorithms.

The optimal watermarking design for an assigned application regularly solves the trade-off between the main competing measures (e.g. robustness and fidelity). Thus, the problem of watermarking can be expressed as a problem od optimization. In mathematics, optimization means the choice of the best part from any set of possible options. without losing the generality, it indicates getting the best possible values of some objective function in the given domain, including various domain types and different types variety objective functions. Bio-inspired intelligence techniques such as genetic algorithm, differential evolution, particle swarm, neural networks, etc. have been employed to solve the watermark problem optimally.

In the last decade, a huge number of nature-inspired metaheuristics had been introduced [1]. Coyote Optimization Algorithm [2] is a population-based algorithm inspired on the Canis latrans species classified as both evolutionary heuristic swarm intelligence, where it is inspired the coyote's behavior [2]. The Coyote optimizations algorithm holds the coyote's social organization and its adaptation to the environment. That contributes to a various structure of algorithm opposed to the metaheuristics from literature. Also, it presents a new mechanism for balancing exploitation and exploration during the process of optimization.

This paper focuses on introducing a robust and blind mesh watermarking approach. The proposed approach use k-means clustering in selecting best vertex positions to be watermark carriers. k-means clustering algorithm aims to detect set of points based on prominent area calculation. Points in such area are considered as model signature.

The watermark bits stream are only embedded on these specified set of vertices. Such set of vertices are known as Points of Interest (POIs). Watermark bits sequence are inserted by modifying the statistical distribution of the radial component for theses POIs only. For these statistical modification we will use Coyote Optimization Algorithm (COA) to get the optimal value of the controlling parameter λ. The proposed approach focuses on introducing a new robust 3D mesh watermarking authentication approaches by ensuring a minimal distortion of the mesh model at the same time ensuring a high robustness of extracted watermark.

The reminder of this paper is organized as follow: Sect. 2 explore the related work proposed by researchers for 3D model watermarking. Section 3 introduces in more details the proposed approach with complete explanation of COA. Experimental results and comparison analysis is introduced in Sect. 4. Finally, Sect. 5 contains the conclusion and the perspective of future research.

2 Related Work

Many techniques that make the transmitted 3D data during the internet more secure one of these techniques is 3D watermarking. More than one review paper had published in this topic [1], The researchers broadly divided the watermark techniques into three generations depends on the domain of the data in spatial or in transform domain. With first generation the researchers are using the data in spatial domain. Ohbuchi et al. [3] was the first technique proposed for the first generation. He chose the ratio between the height of the triangle and its opposite edge length to create a watermarking method that is intrinsically invariant to similarity transformations. The weakness of the techniques of this generation that it is not robust enough. To increase robustness, Yu et al. [4] and Cho et al. [5], instead of inserting the watermark into a single vertex, embed each watermark bit into a group of vertices. Bors [6] uses a neighborhood localized measure to select the vertices that give small embedding distortion and watermark these vertices by local perturbations. Jing et al. [7] proposed a non-blind method using the average of normal to establish a new coordinates system. The signature is inserted into the selected vertices according to area of the two adjacent rings and in function of the curvature of the facets.In the same context, Zhan et al. [8] proposed a blind watermarking algorithm based on vertex curvature The watermark is embedded into vertex bins by modulating the mean fluctuation values of the bins. Recently Jinj et al. [9] describes a novel blind watermarking approach for models of 3D point cloud using feature points to find the embedded watermark. The points with greater curvature mean are judged to be feature points to carry watermarking information; non-feature points are used to build a new coordinate system in which the model of the 3D point cloud is divided into bins including different distances. The 3D watermarking techniques at the second generation works on the data in the transform domain. These techniques are more robust than the other of spatial domain but they are giving less imperceptibility.Frequency analysis based algorithms can achieve excellent results on both watermark robustness and imperceptibility. By utilizing the analysis of spectral by Karni et al. [10], Ohbuchi et al. [11] presented a watermarking algorithm, that embedded the bitstream of the watermark into the low frequencies of Karni et al.'s decomposition. This approach is non-blind, thus the original model is needed during the extraction process of the watermark. Also, Praun et al. [12] presented a robust and non-blind watermarking approach utilizing an edge collapse based multiresolution decomposition. Sighting at robustness against pose deformation or mesh modification, Yang et al. [13] presented another approach algorithm based on Laplacian coordinates, where a bitstream of the watermark is hidden via altering the Laplacian vectors lengths histogram. Xiaoqing et al. [14] proposed another blind and robust 3D polygonal mesh watermark algorithm. They added two types of watermark inside the polygonal mesh model. First type via inserting the watermark bitstream into DCT (Discrete Cosine Transform) frequency domain in some feature patches which are performed utilizing the segmentation of watershed. The second type is based on redundancy information of the 3D polygonal mesh model, like vertex order and vertex coordinates. This approach achieved weak resistance to re-meshing and simplification attack. On another hand, Wang et al. [15] proposed a watermarking technique guarantee a high capacity of the watermark bit stream in which the information of watermark is embedding at various levels of resolution of a semi-regular mesh wavelet decomposition. Hamidi et al. [16] introduced a watermarking approach, that the bitstream of the watermark was embedded by modifying the wavelet coefficients wavelet decomposition.

The last generation became to solve the trade-off between the power of the previous two generation such imperceptibility and the robustness. A new layer of intelligent had added to previous generation to solve this issue. Seoud et al. [17] introduced a robust watermarking method based on a spherical wavelet transformation. They applied the watermarking to 3D compressed model using a multilayered feed-forward neural network (MLFF). Hu et al. [18] proposed a similar histogram-based method for watermarking 3D polygonal meshes by using quadratic programming. Recently, in [19] Soliman introduced a technique that utilizes the Genetic Algorithm (GA) and spherical coordinates to optimize the watermarking method of the 3D polygonal mesh. In her presented technique, the distances from the center of the polygonal mesh model and vertices are adjusted according to the code of watermark using optimization of the Genetic algorithm (GA) in the system of spherical Coordinates. This method is more resistant to Gaussian noise, compared with Cho's method. Luo and Bors [20] used the Quadratic Selective vertex placement scheme in order to find the best location of each vertex after modifying the statistics of the distances.

3   Basic Knowledge

This section presented the basic knowledge and allgorithm that used in the proposed work the first section introduce the COA followed by the brief definition about POIs detection.

3.1 Coyote Optimization Algorithm (COA)

Leandro and Juliano proposed the Coyote Optimization Algorithm (COA), which is a metaheuristic optimization algorithm inspired on the canis latrines kinds. It offers a new structure of algorithm and mechanism for making a balance between exploitation and exploration process. (COA) is categorized as both evolutionary heuristic and swarm intelligence, it has inspired on the behavior of coyotes.

The indifference with the Grey Wolf Optimizer (GWO), which is inspired on the Canis lupus species, the COA has a distinct structure of its algorithm setup and it does not focus on the social hierarchy and dominance rules of these animals, even though the alpha is employed as the leader of a pack (as explained forward). Further, the COA focus on the social structure and experiences exchange by the coyotes instead of only hunting preys as it happens in the GWO [21].

Coyotes are a species from Canis Latrans, most of whom in North America. COA is a population-based algorithm based on evolutionary heuristic and swarms intelligence, inspired by adaptation of coyotes to the environment and social conditions. COA provides a balance between exploration and exploitation in the process of the optimization problem. In COA, coyotes form a difference pack and each pack leader is called alpha. In COA, N population classified into Nc and Np. Np is the packs' number while Nc is the coyote number in the pack. Firstly, the coyote number is identical inside each pack. Consequently, the total population is calculated via multiplying the coyote number in the pack (Nc) and the pack number (Np). With COA, during every optimization problems solving, each coyote expressed as a solution and the coyotes social status, that consists of several decision variables like social situation, temperature, snow depth, gender, the hardness of snowpack is the cost of the objective function. The cth coyote social situation for the pth pack in the tth time would be represented as;

$$SOC_{c,t}^{p,t} = \mathbf{x} = (x_1, x_2, \ldots, x_D) \tag{1}$$

The coyotes adaptation to conditions of environment expressed as the fitness function cost and it specified by $fit_c^{p,t} \in R$

Then the coyote's population is started. Every coyote has arbitrary social situations at the beginning because it stands as a stochastic algorithm. Therefore , arbitrary values for the of the pth pack c^{th} coyote for the jth dimension would be denoted as;

$$SOC_{c,t}^{p,t} = lb_j - r_j * (ub_j - lb_j) \tag{2}$$

where, ub_j is the upper bounds and lb_j is lower bounds for the jth variable of decision respectively. r_j is a arbitrary number adopting range of uniform probability of [0, 1]. The coyote's adaptation to the social situations of the environment as follows;

$$fit_c^{p,t} = f(SOC_c^{p,t}) \tag{3}$$

The coyotes join another pack or leave their packs. The go out coyotes from pack happen among probability P_e, which will be determined as;

$$P_e = 0.005 * N_c^2 \tag{4}$$

Every pack consists of 14 coyotes at maximum where Pe cannot be larger than 1. Therefore, interaction and cultural diversity are given among the coyotes. Alpha will be presented for p^{th} pack in the tth time;

$$alpha^{p,t} = \{SOC_c^{p,t} | arg_{c=1,2,3,...,N_c} \min f(SOC_c^{p,t})\} \tag{5}$$

All coyotes information is determined like a social leaning as follows.

$$cult_j^{p,t} = \begin{cases} O_{\frac{N_c+1}{2} j}^{p,t}, & N_c \text{ is odd;} \\ \frac{O_{\frac{N_c+1}{2} j}^{p,t} + O_{\frac{N_c}{2} j}^{p,t}}{2}, & N_c \text{ is even.} \end{cases} \tag{6}$$

wherever, $O^{p,t}$ is showing the ordered social situations belong to coyotes. The social learning is a determination of median social situations for total coyotes. Every coyote's age is shown as $age_j^{p,t} \in N$ and calculates in COA. Depending on arbitrarily picked parents and social situations, the new coyote birth is shown as follows:

$$pup_j^{p,t} = \begin{cases} SOC_{r_1,j}^{p,t}, & rnd_j < P_s \text{ or } j = j_1; \\ SOC_{r_2,j}^{p,t}, & rnd_j \geq P_s + P_a \text{ or } j = j_2; \\ R_j, & \text{otherwise.} \end{cases} \tag{7}$$

where, r_1 and r_2 are arbitrary coyotes belong to P^{th} pack and j_1 and j_2 are random dimension in the problem. P_s and P_a are association and scatter probability respectively which is provides social learning. R_j indicate to an arbitrary value within uniform probability range $[0, 1]$. P_s and P_a will be presented as follow:

$$P_s = \frac{1}{D} \tag{8}$$

where, D is search space dimension.

$$P_a = (1 - P_s)/2 \tag{9}$$

There are two effects in COA first one packet effect (ζ) and an alpha effect (ε) to represent social communications in the packs amongst coyotes. Both of them represent social variety but the second one is from a stochastic coyote tothe pack (cr1) to alpha and the second one is from an arbitrary coyote (cr2) to the pack social learning. The selection of arbitrary coyotes will be done using uniform distribution probability. ε and ζ will be defined as follow;

$$\varepsilon = alpha^{p,t} - soc_{cr1}^{p,t} \tag{10}$$

$$\zeta = cult^{p,t} - soc_{cr2}^{p,t} \tag{11}$$

Therefore, the coyote new social situation with the pack and the alpha importance will be described as follow;

$$new\_soc_c^{p,t} = r_1 \cdot \varepsilon + r_2 \cdot \zeta \tag{12}$$

where, r_1 and r_2 are the alpha coyote and the pack importance weights. Both r_1 and r_2 are arbitrary values within the range of uniform distribution [0, 1].Then, the new social condition is determined;

$$new\_fit_c^{p,t} = f(new\_soc_c^{p,t}) \tag{13}$$

COA indicates whether the new social condition is more useful than earlier and this is displayed as follows:

$$new\_soc_c^{p,t+1} = \begin{cases} new\_soc_c^{p,t}, & new\_fit_c^{p,t+1} < fit_c^{p,t}; \\ soc_c^{p,t}, & \text{elsewhere.} \end{cases} \tag{14}$$

The coyote social situation, which adjusts itself best to the situation of the environment will be used as a solution to the global problem.

3.2   The Points of Interest (POIs)

Due to the great improvement of technology, the representation of the data in three dimensions widely used in many applications like scientific visualization, manufacturing, computer vision, engineering design, virtual reality,architectural walk through, and video gaming. The 3D objects consist of a huge number of components like thousands of vertices and faces,and that make researchers to detect the more interest components to deal with this objects instead of to deal with whole object that for many applications like watermark, and mesh simplification. 3D points of interest (POIs), further mention salient points or points of feature, are appropriate points in visual perception. There is much importance of POIs that make it very helpful in tasks of geometry processing, such as viewpoint selection, segmentation of mesh, shape enhancement, mesh registration, visual attention guidance and shape retrieval. Although the complicated relationship between geometric descriptors and POIs and, there is similar agreement POIs would be distinguished from a geometric view [22]. Many different techniques [23, 24] had implemented and proposed by many directions of research, that can detect the POIs of 3D mesh models.

3.3   3D Printing Overview

3D printing (or additive manufacturing) is the process of creating a 3-dimensional object by depositing layers of material. This enables the manufacturing of complex 3D objects in a cost effective and automated manner [25].

Recently, the 3D printing content industry has grown rapidly due to the development of advanced 3D printing technology, the emergence of low-cost 3D printers and reduced production costs. As happened earlier in the music and video markets, copyright issues inevitably occur with the expansion of the content industry. Therefore, copyright protection of content is very important in 3D printing environments, even though many challenges still remain to be resolved, such as the strength of printed materials and printing accuracy [26].

In Giao et al. [27] proposed a watermarking algorithm for 3D printing models by embedding the data of watermark into the the 3D printing model feature points. Which had determined by the process of 3D slicing along the 3D printing model Z axis. The bitstream of watermark is inserted into a 3D printing model feature points by changing the length of vector of these feature points in XY plane based on the length of reference. The feature point XY coordinates will be then modified according to the modified length of vector that has carried the watermark.

4   The Proposed Approach of 3D Watermarking Using COA

This proposed work aims to insert watermark sequence over set of interest points. Such points are considered as a signature of each mesh model that selected based on prominent area calculation and k-means clustering algorithm. After determining POI ,the proposed approach will modify the statistical distribution of the radial component for theses POIs only. For these statistical modification we will use COA to get the optimal value of the controlling parameter λ. This parameter controlling is required to provide a good balance between two main watermark requirements (e.g. robustness and imperceptibility). The last phase of the proposed approach is the blind extraction of the inserted watermark bit sequence. The extraction algorithm will obey the similar steps of the embedding process to prove the existence of the watermark. Figure 1 shows the details of the proposed approach. More details will be described in the following subsections

4.1   POIs Selection Phase

The main purpose of this phase is providing an intelligent layer in the selection of watermark carrier vertices. The proposed approach depends on the most prominent area calculation. Points in such area are considered as (POIs), this points set has

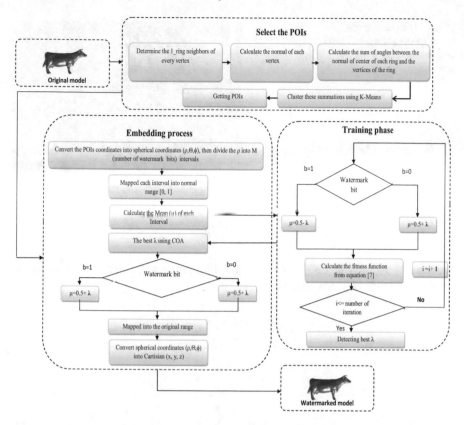

Fig. 1 The general architecture of the proposed approach and its phases

been mentioned as an object signature. Firstly the algorithm determines every point normal and the points normal upon the first ring neighborhood as presented in Fig. 2.

V_1 has been considered as a vertex with normal (n_1),$(V_2:V_7)$ are the vertices of the first ring neighbor of V_1 with values of normals registered as $(n_2:n_7)$, the proposed approach determines the angle θ between n_1 (normal of V_1)and n_2 (normal of V_2). This computation will be iterated for every one ring vertices. K_means algorithm clusters the values of these summations [28]. The cluster with high summations indicate these areas are more prominent, that means the ability to detect points of Interest (POIs) from these areas.

4.2 Embedding Watermark Phase

In [5] the statistical distribution of the radial component of each vertex of the mesh model. The proposed approach will modify the statistical distribution of the radial

Fig. 2 The entire angle
between the normal of center
and the first ring vertices
normals

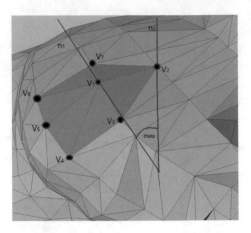

component for only the POIs. In the embedding process the center of mass for the
mesh model will be calculated as (x_c, y_c, z_c):

$$x_c = \frac{1}{N} \sum_{i=1}^{N} x_i \tag{15}$$

$$y_c = \frac{1}{N} \sum_{i=1}^{N} y_i \tag{16}$$

$$z_c = \frac{1}{N} \sum_{i=1}^{N} z_i \tag{17}$$

Where **N** is the Number of all vertices. The model vertices will be transformed
to center of mass (x_c, y_c, z_c), the new coordinates in Cartesian form will be $(x-x_c,$
$y-y_c, z-z_c)$, these new coordinates will be converted into spherical coordinates using
Eq. 6. The radial component of the spherical coordinates for the POIs will be modified
according to Cho [5] method. The main objective from this work to calculate the best
parameter λ that will be used during the modification process, the radial component
will be divided into equal distinct interval according to its magnitude as follow:

$$L_n = \rho_{n,j} | \rho_{min} + \alpha.n < \rho i < \rho_{min} + \alpha.(n+1) \tag{18}$$

where $0 \leq n \leq N - 1, 0 \leq i \leq P - 1$ and $0 \leq j \leq K_n - 1$, K_n is the number of vertex
norms belonging to the nth bin and $\rho_{n,j}$ is the jth vertex norm of the nth bin. The
magnitudes of these $\rho_{n,j}$ norms will be mapped into normalized range [0, 1] using
the following Eq. 19:

$$\hat{\rho_{n,j}} = \frac{\rho_{n,j} - \min_{\rho_{n,j} \in k_n} \{\rho_{n,j}\}}{\max_{\rho_{n,j} \in k_n} \{\rho_{n,j}\} - \min_{\rho_{n,j} \in k_n} \{\rho_{n,j}\}} \tag{19}$$

The mean μ of each bin will be calculated and due to the normalization process, the value of this μ will be around 0.5. now the step of embedding will be begin by modifying this μ via transforming vertex norms by the histogram mapping function [5] as follows:

$$\hat{\mu} = \begin{cases} 0.5 + \lambda, & \text{if } w = 1; \\ 0.5 - \lambda, & \text{if} w = 0. \end{cases} \tag{20}$$

The value λ that will be used in modification process will be calculated using COA. The watermarked mesh model transparency and robustness should be measured in order to formulate a proper fitness function. In the proposed approach a new fitness function described as follow:

$$F = (m/\sum_{i=1}^{m}(NC_i)) + 1/VSNR \tag{21}$$

where m is the number of attacks, NC_i is the normal correlation that measure the robustness of watermark against ith attack, NC can calculated using

$$NC = \frac{\sum \acute{W} \times W}{\sqrt{\sum \acute{W}^2 + \sum W^2}} \tag{22}$$

where $\acute{W}$ depicts the watermark after the happening of attack and W is the initial watermark.

Also VSNR is the measurement of imperceptibility where it is been calculated as follow:

$$VSNR = 20\log\frac{\max|V|}{RMSE(V_w, V)} \tag{23}$$

where

$$RMSE = \frac{1}{N}\sum|V - V_w| \tag{24}$$

$$V = \sqrt{x^2 + y^2 + z^2} \tag{25}$$

The range of this parameter λ is $[0.01, 0.1]$, this range considered where the value of mean at normalized data be around 0.5 so it is not needed to increase more than one., COA find a best value for this parameter which will be within this range that gave minimum value for fitness function that solves the trade-off between imperceptibility and robustness.

After modification of the bin's mean the radial value within each bin are shifted correspond to the change in μ and then converted to the original range again. The whole steps are formulated in algorithm 1 as follow:

During the training phase a different values of parameter λ has been recorded, Table 3 describes the variation of $\lambda's$ value at different size of watermark bits. This

---

Algorithm 1 POIs detection and embedding of watermark algorithm

---
1: **Input:** The vertices v_i of original model in Cartesian form and Watermark bits.
2: **Process:** For every vertex v_i: determine the normal n_i.
3: For every vertex v_i: determine the angles summation between normal of $v'_i s$ and every vertex normal included in the first ring neighbor using the following equation.

$$sum = \sum_{i=2}^{n} arccos \frac{n_i \cdot n_1}{\|n_i\| \|n_1\|} \tag{26}$$

where n_i is the vertices normal of the first ring around n_1 see figure 2.
4: Clustering summations (sum) using $k - means$ to detect the POIs to carry the watermark bits.
5: Transform all the coordinates (x, y, z) to the center of mass of the model (x_c, y_c, z_c).
6: Convert the selected vertices (POIs) into spherical coordinates (r, Θ, ϕ):

$$r = \sqrt{x - x_c^2 + y - y_c^2 + z - z_c^2} \tag{27}$$

$$\Theta = tan^{-1} \frac{y - y_c}{x - x_c} \tag{28}$$

$$\phi = tan^{-1} \frac{z - z_c}{r} \tag{29}$$

7: Sort and divide the radial component (r) to M (length of watermark) equal intervals see Eq 18.
8: For each interval: transform into uniform range [0,1].
9: Calculate the best Parameter λ using COA with fitness function .
10: For each interval L_n: modify their mean corresponding to:
11: **if** $w_i = 0$ **then**
12:     $\mu = 0.5 + \lambda$
13: **else**
14:     **if** $w_i = 0$ **then**
15:         $\mu = 0.5 - \lambda$
16:     **end if**
17: **end if**
18: For Each interval L_i: transform into original range to get the $\acute{r}$.
19: Convert each new vertex spherical coordinate $(\acute{r}, \Theta, \phi)$ to the Cartesian coordinates $(\acute{x}, \acute{y}, \acute{z})$.
20: **Output:** The Cartesian coordinates $(\acute{x}, \acute{y}, \acute{z})$ of the marked mesh model.

---

step show's that no standard value can be considered during the embedding process. Cho's mentioned that when this value decrease that give good imperceptibility with bad robustness (Fig. 3).

4.3 Watermarking Extraction Phase

The proposed 3D mesh watermarking approach is blind which means it doesn't use the original model to extract the embedded watermark, . The extraction algorithm will obey the similar steps of the embedding process to prove the existence of the watermark. Algorithm (2) presents in details the process of watermark extraction.

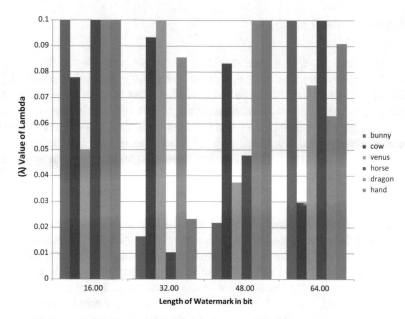

Fig. 3 The value of parameter lambda at different size of watermark

---

Algorithm 2 The watermark Extracting algorithm

1: **Input:** The vertices v_i of Watermarked model in Cartesian form.
2: **Process:**For every vertex v_i: determine the normal n_i.
3: For every vertex v_i: determine the angles summation between normal of $v_i's$ and every vertex normal included in the first ring neighbor using the following equation.

$$sum = \sum_{i=2}^{n} arccos \frac{n_i \cdot n_1}{\|n_i\| \|n_1\|}$$
(30)

where n_i is the verticcs normal of the first ring around n_1 see figure 2.
4: Clustering summations (sum) using $k - means$ to detect the POIs to carry the watermark bits.
5: Transform all the coordinates (x, y, z) to the center of mass of the model (x_c, y_c, z_c).
6: Convert the selected vertices (POIs) into spherical coordinates $(\acute{r}, \Theta, \phi)$:
7: Sort and divide the radial component $\acute{r}$ to M (length of watermark) equal intervals
8: For each interval: transform into uniform range [0,1]
9: For each interval: Calculate the μ.
10: **if** $\mu < 0.5$ **then**
11:     w_i=0
12: **else**
13:    **if** $\mu > 0.5$ **then**
14:       w_i=1
15:    **end if**
16: **end if**
17: **Output:**Watermark bit stream.

---

5 Experimental Results

In this experimental section, we provide a complete analysis for the proposed approach of watermark insertion over set of selection points using an intelligent modification of statistical distribution of the radial component of these points. This experimental result section is classified into two parts that discuss in details the results of the proposed approach in terms the visual fidelity of watermarked model, and the robustness of embedded watermark after different types of attacks. All these experimental results is reported and compared to other well known methods to prove its efficiency in providing a good watermark approach with respect to watermark requirements criteria (e.g. imperceptibility against robustness). We presented a comparison between the proposed approach using COA and GA as two different optimization techniques with two lengths of watermark bits (i.e 32 and 64 bits) on another hand a comparison between the proposed approach using COA and Cho's method at different two lengths of watermark bits (i.e 16 and 48 bits) is introduced. The experiments applied on set of 3D mesh models shown in Table 1, where these models act a benchmark models that used in most of watermark techniques. These models includes Bunny, Cow, Venus, Dragon, Hand and Horse models. The proposed approach had analyzed the embedding capacity at different number of embedded watermark bits, (i.e 16, 32, 48, 64) bits. The fidelity of the proposed approach are calculated for different mesh models at these capacities

5.1 Visual Fidelity

It is sought that the watermark embedded on a particular set of vertices resides invisibly on the mesh, either to protect the aesthetic appearance or to protects the watermark from any potential attacker to be discovered, which might easy to destroy. The degree to which a watermark can be secret is assigned to as the good of the watermark.

To calculate the proposed approach imperceptibility, there are two standard measurements to determine the imperceptibility. First one is Hausdrouff Distance (*HD*)

Table 1 The description of the data set

Model	Number of vertices	Number of faces
Cow	2904	5804
Dragon	50,000	100,000
Bunny	34,835	69,666
Venus	100,759	201,514
Horse	11,2642	225,280
hand	36,619	72,958

Table 2 The HD using the proposed approach compared with Nasima's Method at different size of watermark bits

Model	Method	No. of bits			
		16	32	48	64
Bunny	Proposed	0.55	0.22	0.18	0.13
	NAsima	10.7	5.8	3.9	2.7
Cow	Proposed	0.26	0.1	0.08	0.13
	NAsima	1.5	0.74	0.5	0.41
Horse	Proposed	0.39	0.11	0.14	0.26
	NAsima	9.9	3.6	3.6	2.7

and the second one is vertex signal to noise ratio ($VSNR$). Here, vertex signal to noise ratio ($VSNR$), as formulated in Eqs. (23), (24) and (25), have been used to represent the distortion has been inflicted on the model by the proposed watermarking approach.

The HD is measured between the original mesh model and watermarked one using the following formula

$$HD = \max\{h(T_1), h(T_2)\} \tag{31}$$

where $T_1 = (M, M')$ and $T_1 = (M', M)$, (where M' and M are watermarked mesh and original mesh respectively). $h(T_1) = \max\{\min(d(a, M'))\}$, a in M $h(T_2) = \max\{\min(d(b, M))\}$, b in M'.

For visual fidelity we perform the comparison using both measures HD and $VSNR$ on different mesh models. Table 2 shows HD for the proposed approach against Nasima approach proposed in [19] at different watermark bit capacity (e.g. 16, 32, 48, and 64 bits). Table 3 presents the watermark fidelity comparison between the proposed approach and Cho's method [5] at two watermark bit capacity (e.g. 16 bits, and 48 bits). Table 4 indicates a best imperceptibility with the proposed approach. The proposed approach using k-means and COA had compared with respect to **HD** and **VSNR** against to other meta-heuristic method (e.g. GA).

This comparison shows that the proposed approach give less haussdrouf distance and greater $VSNR$ than Cho's method [5] and nasima method [1]. Also using COA is giving better results than GA as an optimization algorithm at different mesh models that means the proposed approach provide high imperceptibility with good visual fidelity.

5.2 Watermark Robustness Analysis

The experimental results evaluates the robustness of the proposed approach with respect to various types of attacks. These attacks consist of geometric Rotation,

Table 3 The VSNR and HD using the proposed approach compared with Cho Method at different size of watermark bits

Model	16 bits				48 bits			
	Cho		Proposed		Cho		Proposed	
	HD	VSNR	HD	VSNR	HD	VSNR	HD	VSNR
Bunny	1.56	102	0.55	122	1.56	102	0.13	150
Cow	0.69	96	0.28	114	0.18	121	0.08	137
Dragon	1.77	102	0.94	117	1.48	107	0.5	128
Hand	3.06	91	1.48	106	0.99	113	0.53	125

Table 4 The VSNR and HD by Applying COA and GA at different size of watermark bits

Model	32 bits				64 bits			
	GA		COA		GA		COA	
	HD	VSNR	HD	VSNR	HD	VSNR	HD	VSNR
Bunny	0.22	138	0.22	138	0.14	147	0.14	148
Cow	0.14	127	0.1	133	0.21	122	0.13	132
Dragon	0.4	132	0.5	129	0.78	120	0.37	134
Hand	0.47	128	0.45	130	0.52	126	0.47	128

transformation, Translation, Scaling, smoothing and noise attack. Also connectivity attacks such as simplification and subdivision (butterfly type). The robustness is evaluated mathematically as in case of geometric attacks or by applying the attack to the watermarked model, then extract the watermark from the watermarked model. Normal Correlation **NC** between the extracted watermark after attacks and initial watermark will be calculated as a measure of robustness. The normal correlation **NC** has been determined using the Eq. (22).

The **NC** value is between 1 and 0, where if it is approaching 1 that indicating a high correlation between extracted watermark and original one. This means more efficiency of the watermark technique with respect to robustness requirement. The value of **NC** between [0.75 and 1] indicate high relation between the detected watermark after the specific attack and the original one, while the value in range [0.25–0.75] indicate medium relation and weak in between [0 and 0.25].

5.2.1 Noise Attack Results

A random noise added to four watermarked mesh models Bunny, Cow, Dragon and hand with three levels 0.1, 0.3 and 0.5%. Tables 5 and 6 present a comparison between the **NC** values detected using the proposed approach with COA against GA and a comparison between the **NC** values detected using the proposed approach with COA

Table 5 The NC the proposed approach using COA against using GA at different length of watermark bits random noise attack

Model	Type of attack	Ratio (%)	32 bits		64 bits	
			COA	GA	COA	GA
Bunny	Noise	0.1	1	1	0.97	0.94
		0.3	0.94	0.87	0.79	0.71
		0.5	0.94	0.94	0.65	0.65
Cow	Noise	0.1	1	1	0.91	0.94
		0.3	1	1	1	0.69
		0.5	1	1	1	0.56
Dragon	Noise	0.1	1	1	1	1
		0.3	0.94	0.88	0.97	1
		0.5	0.84	0.94	0.93	1
Hand	Noise	0.1	1	1	1	1
		0.3	1	1	1	1
		0.5	1	0.94	1	0.97

Table 6 The NC the proposed approach using COA and *Cho's* method at different length of watermark bitsn for random noise attack

Model	Type of attack	Ratio (%)	16 bits		48 bits	
			COA	Cho	COA	Cho
Bunny	Noise	0.1	1	1	1	1
		0.3	1	1	0.92	1
		0.5	1	1	0.87	0.84
Cow	Noise	0.1	1	1	1	1
		0.3	1	1	0.96	0.92
		0.5	1	1	0.92	0.83
Dragon	Noise	0.1	1	1	1	1
		0.3	0.94	0.88	1	1
		0.5	0.84	0.94	1	1
Hand	Noise	0.1	1	1	1	1
		0.3	1	1	1	1
		0.5	1	0.94	0.96	0.96

against Cho's method respectively. The results show that the proposed method using COA more robust than other methods using GA or Cho's method at most of levels.

Figure 4 shows visually the effect of three levels of Noise attack affected on Bunny model which watermarked using proposed against Cho's method.

a)   b)   c)

d)   e)   f)

Fig. 4 a and d represent the watermarked models using proposed approach against Cho's method respectively, from b and c and e and f are two levels of noise attack 0.5 and 0.7% of bunny which had been watermarked using proposed approach and Cho's method

5.2.2 Smoothing Attack Results

Three levels of smoothing attacks has applied on the previous four models. This type of attack utilized using Gapriel smoothing function using matlab at three levels of iterations 5, 25 and 45 iterations. Tables 7 and 8 present a comparison between the **NC** values detected using the proposed approach with COA against GA and a comparison between the **NC** values detected using the proposed approach with COA against Cho's method respectively. The results show that the proposed method using COA more robust than using GA at many levels, but cho's method was more robust than the proposed on the other hand the value of *NC* still in intermediate value [0.25 ~ 0.75] at most of levels (Fig. 5).

6 Conclusions

In the this paper a blind 3D watermarking approach was proposed based on third generation watermark models. The watermark bit sequence was embedding into mesh model by modifying the distribution of POIs vertex norms. In this work an optimal parameter selection using COA was proposed. COA depends on using a proposed fitness function consisting of two components measuring robustness rate of extracted watermark, and the visual fidelity of mesh model. The embedding water-

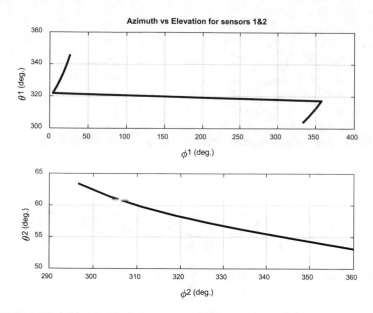

Fig. 6 Emitter azimuth vs elevation wrt sensors 1 and 2

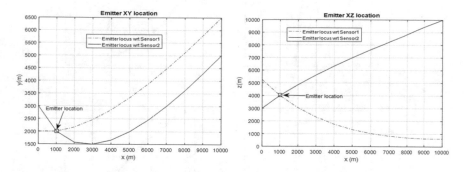

Fig. 7 a Emitter locus with respect to sensor1, **b** Emitter locus with respect to sensor2

The 2D x–y and x–z locations of sensors 1 and 2 are described in Fig. 11 while the 3D sensors and emitter locations are shown in Fig. 12.

5.4 Platform (D): Moving Sensors and a Moving Emitter

d1. Linear Motion

Sensors and emitter are all moving such that: sensor1 movement trajectory is [xs1 = 3500:200:5500, ys1 = 2000:300:5000, zs1 = 4000:200:6000], sensor2 movement

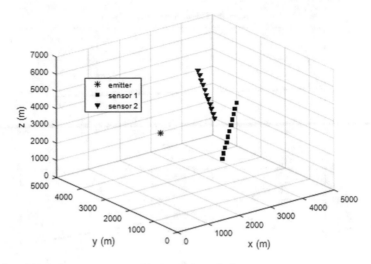

Fig. 8 The 3D moving sensors and stationary emitter platform

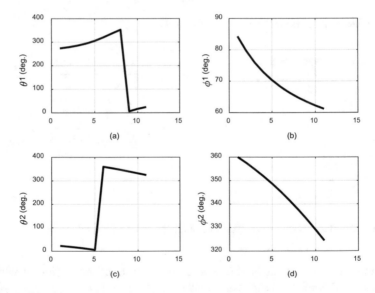

Fig. 9 Azimuth and elevation of the emitter wrt sensors 1 and 2

trajectory is [xs2 = 1000:100:2000, ys2 = 1500:200:3500, zs2 = 2000:100:3000], and the emitter movement trajectory is [xe = 2000:200:4000, ye = 2000:100:3000, ze = 1500:100:2500]. Figure 13 shows the azimuth and elevation angles of emitter wrt sensors 1 and 2. Figure 14 illustrates the sensors angles in space.

Figure 15 shows the estimated x–y and x–z emitter motion track while the 3D linear trajectories of the emitter and sensors are shown in Fig. 16.

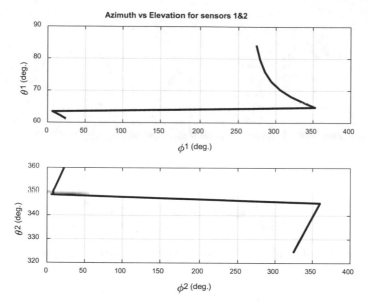

Fig. 10 Emitter azimuth vs elevation wrt sensors 1 and 2

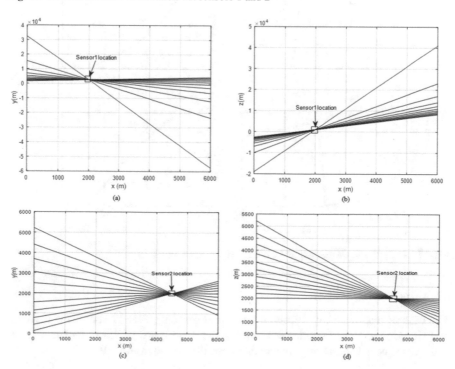

Fig. 11 The moving emitter loci with respect to stationary sensors in **a** xy of sensor 1, **b** xz of sensor 1, **c** xy of sensor 2, **d** xz of sensor2

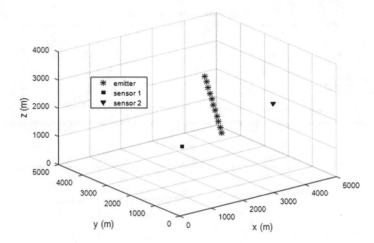

Fig. 12 The 3D locations of the moving emitter with respect to the stationary sensors

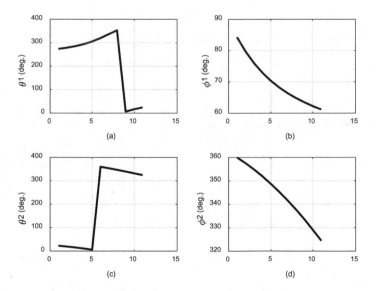

Fig. 13 Azimuth and elevation of the emitter wrt sensors 1 and 2

d2. Maneuvering Motion

The case of maneuvering objects (emitter and sensors), Fig. 17 shows the azimuth and elevation angles of emitter wrt sensors 1 and 2. Figure 18 illustrates the sensors angles in space. The 3D objects locations are illustrated in Fig. 19.

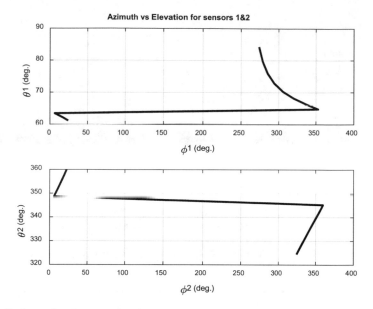

Fig. 14 Emitter azimuth versus elevation wrt sensors 1 and 2

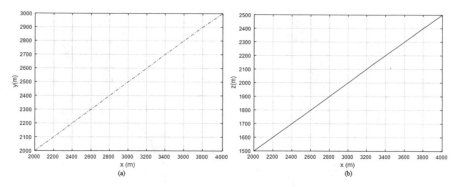

Fig. 15 **a** x-y emitter motion track, **b** x-z emitter motion track

6 Errors in TOA and AOA Measurements

In practice, the emitter collected data, AOA, and TOA used in geolocation process suffers from measurement error.

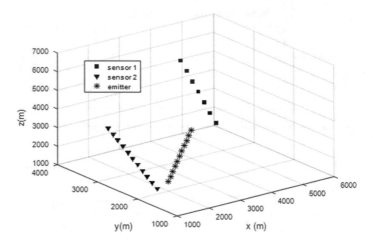

Fig. 16 The 3D locations of a linearly moving emitter and linearly moving sensors

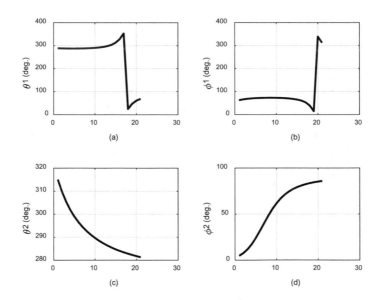

Fig. 17 Azimuth and elevation of the emitter wrt sensors 1 and 2

6.1 Errors in TOA Measurements

From definition of TOA and TDOA shown in Fig. 1, and based on Eq. (8), the time spent by the emitter signal to reach each sensor is given by:

$$\text{TOA}_{1(2)} = \frac{\sqrt{\left(x_e - x_{s1(2)}\right)^2 + \left(y_e - y_{s1(2)}\right)^2 + \left(z_e - z_{s1(2)}\right)^2}}{c} \qquad (15)$$

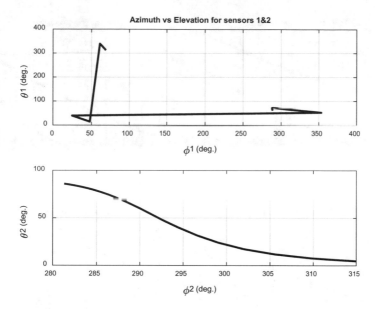

Fig. 18 Emitter azimuth versus elevation wrt sensors 1 and 2

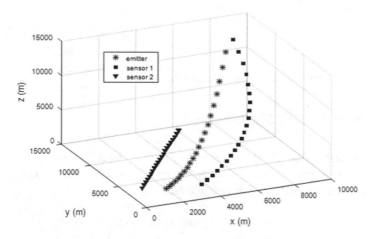

Fig. 19 The 3D locations of a maneuvering emitter and maneuvering sensors

where s1(2) is for sensor1 or sensor2, and c is the speed of light. Assuming that the TOA error is e = 1% of the calculated value in Eq. (15), the measured values of the moving platforms (b-c-d1-d2) are compared with the calculated values as shown in Figs. 20 (a, b, c, and d).

From TOA figures shown above we conclude that the measured values can be considered with small errors.

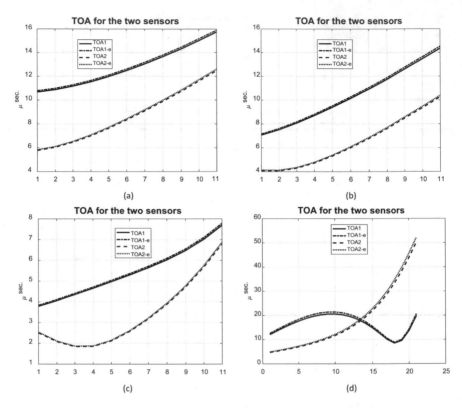

Fig. 20 TOA of sensors 1 and 2 versus the measured values

6.2   Errors in AOA Measurements

AOA described in Eqs. (9, and 10) are the error free values. In real measurements, errors are always present. If e_θ, and $e_\varnothing$ are the measurement errors in azimuth and elevation, the measured angles in practice are,

$$e_\theta = \theta - \theta_0 \text{ and } e_\varnothing = \varnothing - \varnothing_0 \tag{16}$$

where $\theta, \theta_0, \varnothing$, and $\varnothing_0$ are the real and error free values of azimuth and elevation angles. Taking the sine of both sides of Eq. (15) [16]:

$$L_1 \sin(e_\theta) = L_1 \sin(\theta - \theta_0), \text{ and}$$
$$L_2 \sin(e_\varnothing) = L_2 \sin(\varnothing - \varnothing_0) \tag{17}$$

where:

$$L_1 = \sqrt{(x_e - x_s)^2 + (y_e - y_s)^2} \tag{18}$$

$$L_2 = \sqrt{(x_e - x_s)^2 + (y_e - y_s)^2 + (z_e - z_s)^2} \tag{19}$$

Re-writing Eq. (16):

$$L_1 \sin(e_\theta) = L_1[\sin(\theta)\cos(\theta_0) - \cos(\theta)\sin(\theta_0)], \; and$$
$$L_2 \sin(e_\emptyset) = L_2[\sin(\emptyset)\cos(\emptyset_0) - \cos(\emptyset)\sin(\emptyset_0)] \tag{20}$$

From definition of AOA $(\theta_0, \; and \; \emptyset)_0$ as shown in Fig. 2:

$$L_1 \sin(e_\theta) = (x_e - x_s)\sin(\theta) - (y_e - y_s)\cos(\theta), \; and$$
$$L_2 \sin(e_\emptyset) = \frac{(xe - xs)\sin(\emptyset)}{\cos(\emptyset)} - (z_e - z_s)\cos(\emptyset) \tag{21}$$

Assuming that $e_\theta, \; and \; e_\emptyset$ are small measured values, i.e. $\ll 1$, so:

$$\sin(e_\theta) \cong e_\theta \; and \; \sin(e_\emptyset) \cong e_\emptyset$$

Finally, errors in angles measurements are expressed as:

$$e_{\theta 1} = \frac{(x_e - x_{s1})\sin\theta_{s1}}{L_{11}} - \frac{(y_e - y_{s1})\cos\theta_{s1}}{L_{11}} \tag{22}$$

$$e_{\theta 2} = \frac{(x_e - x_{s2})\sin\theta_{s2}}{L_{12}} - \frac{(y_e - y_{s2})\cos\theta_{s2}}{L_{12}} \tag{23}$$

$$e_{\emptyset 1} = \frac{(x_e - x_{s1})\sin\emptyset_{s1}}{L_{21}\cos\theta_{s1}} - \frac{(y_e - y_{s1})\cos\emptyset_{s1}}{L_{21}} \tag{24}$$

$$e_{\emptyset 2} = \frac{(x_e - x_{s2})\sin\emptyset_{s2}}{L_{22}\cos\theta_{s2}} - \frac{(y_e - y_{s2})\cos\emptyset_{s2}}{L_{22}} \tag{25}$$

where:

$e_{\theta 1}, s1, L_{11}, e_{\emptyset 1},$ and L_{21} are for sensor 1,

and,

$e_{\theta 2}, s2, L_{12}, e_{\emptyset 2},$ and L_{22} are for sensor 2.

The simulation results presented in Sect. 5 are now re-calculated considering the measurement errors in AOA. The same platforms and objects coordinates are used to illustrate the effect of errors on the emitter location.

The MoMo algorithm is modified using Eqs. (22–25) to demonstrate the effect of the AOA measurements errors on the emitter location in different platforms. For fixed emitter platforms (a) and (b), the error is found to be less than 1% of the measured angle. Figures 21, 22, and 23 present the location of moving emitter platforms c, d1, and d2without and with angle errors.

From the shown figures, the measurements error is considerably small. Platforms (d1), and (d2) are affected mostly in the x-y location. Objects maneuvering makes the continuous angle measurements suffer from little higher errors. From Fig. 22a, it is seen that at the beginning of emitter maneuver the angle measurement error has higher values. In Fig. 23a, all objects are maneuvering so, at some measuring instant the error sign is changed and the real values are less than the error free values.

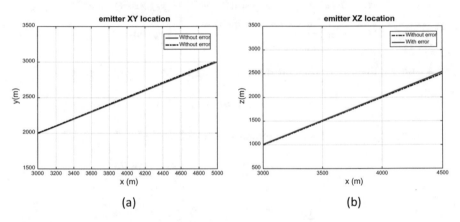

(a)   (b)

Fig. 21 Platform (c) emitter location without and with AOA errors, **a** x-y plane. **b** x-z plane

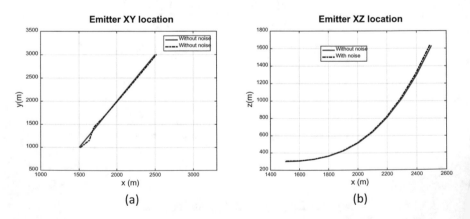

(a)   (b)

Fig. 22 Platform (d1) emitter location without and with AOA errors, **a** x-y plane. **b** x-z plane

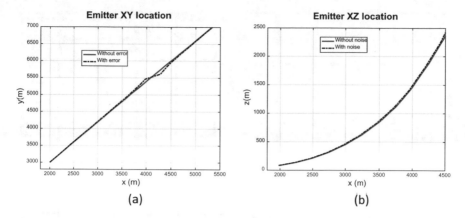

Fig. 23 Platform(d2) emitter location without and with AOA errors, **a** x-y plane. **b** x-z plane

7 Resulting Position Error and Comparative Analysis

7.1 MoMo Resulting Emitter Position Error

The emitter position error in meters, as a measure of the algorithm accuracy, is used and calculated as shown below, where R_e is the position error.

$$(R_e) = \sqrt{(\hat{x}_e - x_e)^2 + (\hat{y}_e - y_e)^2 + (\hat{z}_e - z_e)^2} \tag{26}$$

From Eq. (15):

$$\Delta R_e = \sqrt{\Delta x_e^2 + \Delta y_e^2 + \Delta z_e^2} \tag{27}$$

where $\hat{x}_e, \hat{y}_e$, and $\hat{z}_e$ are the emitter location estimated values, x_e, y_e, and z_e are the emitter position true values, and $\Delta R_e, \Delta x_e, \Delta y_e$, and Δz_e, are the emitter position error, and errors in x, y, and z coordinates. The mean values of the emitter 3D coordinates $\hat{x}_e, \hat{y}_e$, and $\hat{z}_e$ are calculated by the MoMo algorithm. Different values of standard deviations of the MoMo input parameters azimuth θ, elevation $\emptyset$, and time of arrival TOA are tried to evaluate the MoMo performance. For all platform examples stated in section (V) standard deviations are selected to be 0.05, 0.1, 0.15, and 0.2 for all parameters. For each platform example, the emitter position error, in meters, is computed by averaging 1000 simulation iterations. More than 1000 iterations have non-significant enhancements. The emitter position error against standard deviation values for the emitter sensors platform examples a, b, c, d1, and d2 are illustrated in Figs. 24a–and e.

From Fig. 24a, MoMo algorithm achieves precise geolocation estimation for stationary platforms. The accuracy decreases when the platforms start moving but

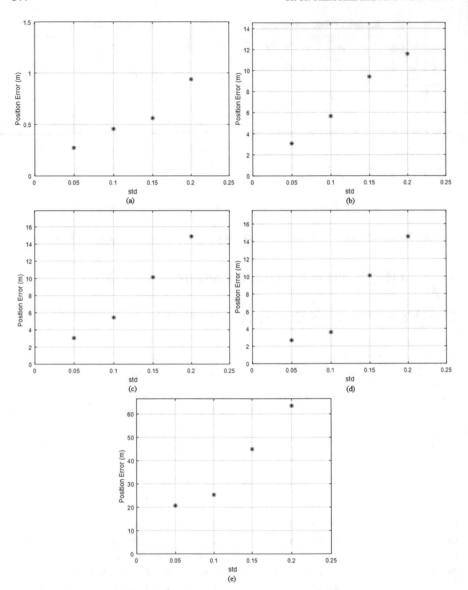

Fig. 24 **a** Stationary platform a, **b** Stationary emitter-moving sensors platform b, **c** Moving emitter-stationary sensors platform c, **d** Maneuvering emitter-moving sensors platform D1, and **e** Maneuvering emitter-maneuvering sensors platform D2

the error values are considered good. For linearly moving sensors and emitter, it is shown from Fig. 24b, c that the output errors are approximately the same. As for nonlinear motion platforms, maneuvering sensors and emitter, the output error is higher due to the complex motion platforms of sensors and emitter. Considering the platform d2 example, the emitter and sensors motion ranges reached about 12,000 m. For such positions, the MoMo output error is accepted. From previously introduced results analysis, it is found that the MoMo algorithm realized remarkable emitter geolocation estimation results.

To enhance the emitter position error of d2 platform, we suggest considering the frequency of arrival FOA of the RF emitted signal with respect to the sensors. That means, redesign the algorithm to utilize the data offered by FOA beside AOA, and TOA to reduce the emitter position resulting error and enhance the geolocation process.

7.2 Comparison with Previous Work

In this subsection, the results of the work done by some researchers, as examples, are compared with that obtained from the MoMo algorithm using the same platform conditions. Huai and Lee [17], gave geolocation platforms as shown in Fig. 25. The location estimation error in distance according to different standard deviations of TDOA and AOA are summarized in Table 2. Standard deviations of AOAs θ, and ∅, are assumed to be equal and have the value of 0.2°. Two stds of TDOA, 10 ns, and 30 ns are considered during estimation process. In our work, considering the same geolocation scheme and sensors-emitter platforms the location estimation error in distance is calculated and stated in Table 2. Gudrun [18] presented 2D sensors-emitter platforms as shown in Fig. 26.

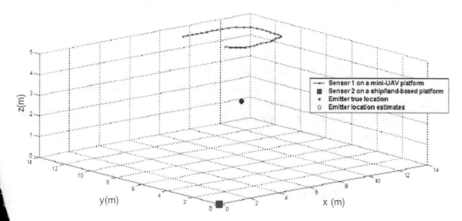

Fig. 25 Huai sensors-emitter platforms

Table. 2 Location estimation error in distance

	Location estimation error in distance (m)	
	Huai and Lee (2004) research results	MoMo algorithm results
Std of TDOA in nsec	10 15	10
	30 25	18

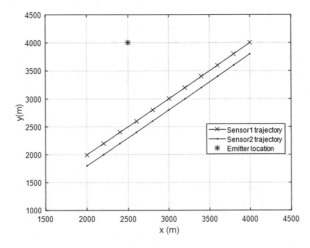

Fig. 26 Gudrun sensors and emitter Platforms

The system accuracy for 10 km distance could reach 10 m. Momentary error for 10 km distance is 500 m. Applying MoMo algorithm for the Gudrun case, the accuracy reaches 6.4 m, and the momentary error is 60 m. Tufan and Tuncer [19], introduced sensors-emitter platforms as shown in Fig. 27. The position error is calculated for different TDOA and AOA standard deviations. The best result is obtained using Cramer-Rao Lower Bound, CRLB, method. The used TDOA standard deviation is 1 ns, The resulting position error is varying from 25 up to 55 m if the AOA standard deviation is changed from 1 up to 5°. If, the AOA standard deviation is fixed to be 5°. The position error varies from 50 up to 83 m if the TDOA standard deviation is changed from 0.5 up to 1.5 ns. The technical report published by ITU-R y [20], introduced 2D measurements of maximum position estimation error of 50 m. Applying MoMo algorithm assuming the same sensors and emitter platforms, the maximum position error is found to be 5 m.

8 Conclusions and Future Work

The MoMo algorithm is found to be dynamic, reliable, and achieved good results Geolocation of RF emitting source is done in both 2D and 3D coordination systems

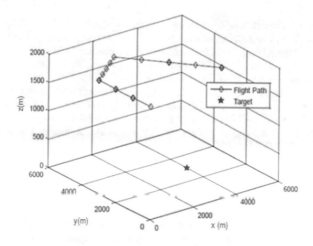

Fig. 27 Tufan and Tuncer sensors and emitter platforms

Different platforms including simple and complex platform motions are considered. The distance position error is reduced compared with other techniques. The research is now opened to study different moving and maneuvering platforms.

References

1. Okello, N.: Emitter geolocation with Multiple UAVs. In: 9th International Conference on Information Fusion, Florence (2006)
2. Scerri, P., Glinton, R., Owens, S., Scerri, D., Sycara, K.: Geolocation of RF Emitters by Many UAVs. Carnegie Mellon University, Pittsburgh (2007)
3. Kaune, R., Musicki,D., Koch, W.: On Passive Emitter Tracking in Sensor Networks, Sensor Fusion and its Applications (Thomas, C. (ed.)) (2010). ISBN: 978-953-307-101-5
4. Fisher, G.W.: Robust geolocation techniques for multiple receiver systems. M.Sc. Thesis, Department of Electrical and Computer Engineering, Graduate Faculty, Baylor University, USA (2011)
5. Bamberger, R.J., Moore, J.G., Goonasekeram, R.P., Scheidt, D.H.: Autonomous geolocation of RF emitters using small, unmanned platforms. Johns Hopkins APL Technical Digest **32**(3), 636–646 (2013)
6. Thoresen, T., Moen, J., Engebråten, S.A., Kristiansen, L.B., Nordmoen, J.H., Olafsen, H.K., Gullbekk, H., Hoelster, I.T., Bakstad, L.H.: Distribuerte COTS UAS for PDOA WiFi geolokalisering med Android smarttelefoner. Technical report, Forsvarets forskningsinstitutt, FFI-rapport 14/00958
7. Eric J. Bailey. (2015). Single Platform Geolocation of Radio Frequency Emitters. MSc. thesis, Air Force Institute of Technology, Ohio, USA
8. Kim, Y.H., Kim, D.-G., Kim, H.-.N.: Two-step estimator for moving-emitter geolocation using time difference of arrival/frequency-difference of arrival measurements. IET Radar Sonar and Navigation **9**(7), 881–887
9. Li, X., Deng, Z.D., Rauchenstein, L.T., Carlson, T.G.: Source localization algorithms and applications using time of arrival and time difference of arrival measurements. Rev. Sci. Instrum. **87**, 041502, 1–12 (2016)

10. Liu, Z., Zhao, Y., Hu, D., Liu, C.: A moving source localization method for distributed passive sensor using TDOA and FDOA measurements. Int. J. Antennas Propag. Article ID 8625039, 12 p (2016). https://doi.org/10.1155/2016/8625039. Accessed 18 Mar 2018
11. Boukerche, A., Oliveira, H.A., Nakamura, E.F., Loureiro, A.A.: Localization systems for wireless sensor networks. IEEE Wirel. Commun. **14**, 6–12 (2007)
12. Alfandi, O., Bochem, A., Bulert, K., Maier, A., Hogrefe, D.: Received signal strength indication for movement detection. In: Eighth International Conference on Mobile Computing and Ubiquitous Networking (ICMU), Hakodate, Japan, pp. 82–83 (2015)
13. Fowler, M.L., Hu, X.: Signal models for TDOA/FDOA estimation. IEEE Trans. Aerosp. Electron. Syst. **44**(4), 1543–1550
14. Muˇsicki, D., Koch, W.: Geolocation using TDOA and FDOA measurements. In: 11th International Conference on Information Fusion, pp. 1987–94, Germany (2008)
15. Lee, B.H., Chan, Y.T., Chan, F., Du, H., Dilkes, F.A.: Doppler frequency geolocation of uncooperative radars. In: MILCOM 2007—IEEE Military Communications Conference, USA (2007)
16. Lee, J., Liu, J.: Passive emitter AOA determination and geolocation using a digital interferometer. In: RTO SET Symposium on Passive and LPI Radio Frequency Sensor, Poland, pp. 23–25 (2001)
17. Du, H.-J., Lee, J.P.Y.: Simulation of multi-platform geolocation using a hybrid TDOA/AOA method. Technical Report, Ministry of Defence R&D Canada, Ottawa (2004)
18. Høye, G.: Analyses of the geolocation accuracy that can be obtained from shipborne sensors by use of time difference of arrival (TDOA), scanphase, and angle of arrival (AOA) measurements. Forsvarets forsknings institutt Norwegian Defence Research Establishment (FFI), Norway (2010)
19. Tufan, B., Engin Tuncer, T.: Combination of emitter localization techniques with angle, frequency and time difference of arrival. In: IEEE 21st Signal Processing and Communications Applications Conference (SIU), Turkey (2013)
20. International Telecommunication Union, Radio Sector: Comparison of time difference-of-arrival and angle-of-arrival methods of signal geolocation. Report ITU-R SM.2211-2, Switzerland (2018)

Printed in the United States
by Baker & Taylor Publisher Services

Table 7 The NC the proposed approach using COA and GA at different length of watermark bits for smoothing attack

Model	Type of attack	No. of iterations	32 bits		64 bits	
			COA	GA	COA	GA
Bunny	Smoothing	5	0.6	0.16	0.25	0.32
		25	0.4	0.14	0.2	0.19
		45	0.25	0.02	0.19	0.4
venus	Smoothing	5	0.94	0.87	0.71	0.81
		25	0.7	0.19	0.22	0.2
		45	0.57	0.18	0.27	0.27
Dragon	Smoothing	5	0.5	0.33	0.4	0.46
		25	0.3	0.29	0.1	0.04
		45	0.15	0.25	0.08	0.11
Hand	Smoothing	5	0.7	0.7	0.53	0.5
		25	0.32	0.09	0.23	0.2
		45	0.07	0.13	0.03	0.04

Table 8 The NC the proposed approach using COA and Cho's method at different length of watermark bits smoothing attack

Model	Type of attack	No. of iterations	16 bits		48 bits	
			COA	Cho	COA	Cho
Bunny	Smoothing	5	0.62	1	0.33	0.83
		25	0.49	0.88	0.33	0.25
		45	0.52	0.75	0.32	0.03
Venus	Smoothing	5	0.88	1	0.88	1
		25	0.77	1	0.22	0.34
		45	0.42	1	0.13	0.14
Dragon	Smoothing	5	0.63	1	0.4	0.46
		25	0.38	1	0.1	0.04
		45	0.49	1	0.08	0.11
Hand	Smoothing	5	1	1	0.71	0.91
		25	1	1	0.16	0.26
		45	0.63	0.87	0.15	0.01

a)   b)   c)

d)   e)   f)

Fig. 5 The originl model a and d, while b and e the watermarked model using proposed and cho's respectively, from c and f smoothing attack at 25 iterations of bunny which had been watermarked using proposed approach and Cho's method respectively

mark sequence is modifying only set of POIs. Such points are calculated using k-means clustering. The experimental results showed that the proposed approach gave good fidelity compared with the most known watermarking method, also it gave a good robustness against some geometrical attack like noise and smoothing attacks.

More improvements can be introduced to the proposed approach by introducing other optimization method. Also the fitness function can be developed to include more watermark requirements like: watermark capacity and including more attacks like connectivity attacks. We can use the proposed approach in ensuring the security for one of recent applications like 3D printing, and 3D medical images.

References

1. Medimegh, N., Belaid, S., Werghi, N.: A survey of the 3D triangular mesh watermarking techniques. Int. J. Multimed. **1**(1) (2015)
2. Pierezan, J., Coelho, L.D.S.: Coyote optimization algorithm: a new metaheuristic for global optimization problems. In: 2018 IEEE Congress on Evolutionary Computation (CEC), pp. 1–8 (2018)
3. Ohbuchi, R., Masuada, H., Aono, M.: Data embedding algorithms for geometrical and non-geometrical targets in three-dimensional polygon models. Comput. Commun. **21**, 1344–1354 (1998)
4. Yu, Z., Ip, H.S., Kwok, L.F.: A robust watermarking scheme for 3D triangular mesh models. Pattern Recogn. **36**(11), 2603–2614 (2003)

5. Cho, J.-W., Prost, R., Jung, H.Y.: An oblivious watermarking for 3-D polygonal meshes using distribution of vertex norms. IEEE Trans. Sig. Proc. **55**(1), 142–155 (2007)
6. Bors, A.G.: Watermarking mesh-based representations of 3-D objects using local moments. IEEE Trans. Image Proc. **15**(3), 687–701 (2006)
7. Jing, L., Yinghui, W., Wenjuan, H., Ye, L.: A new watermarking method of 3D mesh model. Telkomnika Indonesian. J. Electr. Eng. pp. 1610–1617 (2014)
8. Zhan, Y.Z., Li, Y.T., Wang, X.Y., Oian, Y.: A blind watermarking algorithm for 3D mesh models based on vertex curvature. J. Zhejiang Univ-Sci. C (Comput. Electron.) **15**, 351–362 (2014)
9. Jinj, L., Yajie, Y., Douli, M., Yinghui, W., Zhigeng, P.: A watermarking method for 3D models based on feature vertex localization. IEEE Access **6**, 5122–5134 (2018)
10. Zachi, K., Craig, G.: Spectral compression of mesh geometry. In: SIGGRAPH, pp. 279–286 (2002)
11. Ohbuchi, R., Mukaiyama, A., Takahashi, S.: A frequencydomain approach to watermarking 3D shapes. Comput. Graph. Forum **21**(3), 373–382 (2002)
12. Praun, E., Hoppe, H., Finkelstein, A.: Robust mesh watermarking. SIGGRAPH, pp. 49–56 (1999)
13. Yang, Y., Ivrissimtzis, I.: Polygonal mesh watermarking using Laplacian coordinates. Computer Graphics Forum (Proceeding Eurographics/ ACM SIGGRAPH Symposium on Geometry Processing **29**, 1585–1593 (2010)
14. Xiaoqing, F., Wenyu, Z., Yanan, L.: Double watermarks of 3D mesh model based on feature segmentation and redundancy information. Multimedia Tools Appl. **68**, 497–515 (2014)
15. Yinghui, W., Jing, L., Yajie, Y., Douli, M., Ruijiao, L.: 3D model watermarking algorithm robust to geometric attacks. IET Image Process. **11**(10), 822–832 (2017)
16. Hamidi, M., El Haziti, M., Cheri, H., Aboutajdine, D.: A robust blind 3-D mesh watermarking based on wavelet transform for copyright protection. Proc, pp. 1–6. ATSIP, Fez, Morocco (2017)
17. El-seoud, S., Rumman, N., Tajeddin, I., Hatatneh, K., Gutl, C.: Robust digital watermarking for compressed 3D models based on polygonal representation. Int. J. Comput. Appl. **61**(4) (2013)
18. Hu, R., Alface, P., Macq, B.: Constrained optimisation of 3d polygonal mesh watermarking by quadratic programming. In: IEEE International Conference on Acoustics, Speech and Signal Processing, pp. 1501–1504 (2009)
19. Mona, M., Aboul Ella, H., Hoda, M.: A robust 3D mesh watermarking approach using genetic algorithms. In: Proceedings of the 7th IEEE International Conference Intelligent Systems IS2014, vol. 323, September 24–26, Warsaw, Poland, pp. 731–741 (2014)
20. Luo, M., Bors, A.: Shape watermarking based on minimizing the quadric error metric. In: IEEE International Conference on Shape Modeling and Applications, pp. 103–110 (2009)
21. Pierezan, J., Coelho, L.D.S.: Coyote optimization algorithm: a new metaheuristic for global optimization problems. In: Proceeding IEEE Congress on Evolutionary Computation (CEC), pp. 1–8 (2018)
22. Shu, Z., Xin, S., Xu, X., Liu, L., Kavan, L.: Detecting 3D points of interest using multiple features and stacked auto-encoder. IEEE Trans. Vis. Comput. Graph. (2018). https://doi.org/10.1109/TVCG.2018.2848628
23. Sun, J., Ovsjanikov, M., Guibas, L.: A concise and provably informative multi-scale signature based on heat diffusion. In: Eurographics Symposium on Geometry Processing (SGP), pp. 1383–1392 (2009)
24. Castellani, U., Cristani, M., Fantoni, S., Murino, V.: Sparse points matching by combining 3D mesh saliency with statistical descriptors. Comput. Graph. Forum **27**(2), 643–652 (2008)
25. Moore, S., Armstrong, P., Mcdonald, T., Yampolskiy, M.: Ulnerability analysis of desktop 3d printer software. In: Resilience Week (RWS), IEEE, pp. 46–51 (2016)
26. Jong-Uk, H., Do-Gon, K., Sunghee C., Heung-Kyu, L.: 3D print-scan resilient watermarking using a histogram-based circular shift coding structure. In: Proceeding 3rd ACM Workshop Information Hiding Multimedia Security, pp. 115–121 (2015)

27. Giao, N.P., Suk-Huan, L., Ki-Ryong, K.: A 3D printing model watermarking algorithm based on 3D slicing and feature points. Electronics **7**(2), 23 (2018)
28. Mourad, R.M., Mona, S.M., Ashraf, D., Aboul Ella, H.: Detecting interest points detection of 3D mesh model using K means and shape curvature. In: International Conference on Advanced Intelligent Systems and Informatics (2018)

A 3D Geolocation Analysis of an RF Emitter Source with Two RF Sensors Based on Time and Angle of Arrival

Kamel H. Rahouma and Aya S. A. Mostafa

Abstract The three-dimensional geolocation of a radio frequency RF emitting source is commonly determined using two RF sensors. Even today, most re-searchers are working on one of the three emitter-sensors motion platforms: stationary sensors–stationary emitter, moving sensors–stationary emitter, or stationary sensors–moving emitter. The present work is aimed to investigate a fourth scenario of moving RF sensors to find a moving RF emitter in space. A mathematical analysis is to consider the different cases and scenarios. Also, a corresponding algorithm is designed to simulate this analysis. We consider straight line and maneuvering motions of both the emitter and sensors. The present algorithm uses a hybrid situation of angle of arrival (AOA) and time of arrival (TOA) of the emitter RF signal to estimate the 3D moving emitter geolocation. Measurement errors of AOA and TOA are investigated and compared with the calculated values. We test the algorithm for long and short distances and it is found to be dynamic and reliable. The algorithm is tested for different values of AOAs, and TOAs with different standard deviations. Relatively small resulting emitter position error is detected. A MATLAB programming environment is utilized to build up the algorithm, carrying out calculations and presenting the output results and figures. Some of the applications of our analysis and algorithm will be presented.

Keywords 3D geolocation · Moving sensors moving emitter platforms · AOA estimation · TOA estimation · Hybrid AOA and TOA estimation

K. H. Rahouma (✉) · A. S. A. Mostafa
Department of Electrical Engineering, Faculty of Engineering, Minia University, Minia, Egypt
e-mail: kamel_rahouma@yahoo.com

A. S. A. Mostafa
e-mail: ayasami89@yahoo.com

© The Author(s), under exclusive license to Springer Nature Switzerland AG 2021
A. E. Hassanien et al. (eds.), *Machine Learning and Big Data Analytics Paradigms: Analysis, Applications and Challenges*, Studies in Big Data 77,
https://doi.org/10.1007/978-3-030-59338-4_30

1   Introduction

Geolocation is referred to methods used for locating an object or set of objects in the space. In general, space means the most widely used coordinate system, latitude, longitude, and elevation. It means, x, y, and z. In this research and similar, geolocation of an RF emitting source is carried out using RF sensors in 2D or 3D coordination [1–2]. Emitter and sensors movements mostly obey four possible platforms: (a) Stationary sensors–stationary emitter, (b) Moving sensors–stationary emitter, (c) Stationary sensors–moving emitter, and (d) Moving sensors–moving emitter. All of these platforms are presented with 2D or 3D coordinates. The first three platforms are studied in many types of research [3–10], and others. The last platform is rarely discussed. Creating such a dynamic platform is not simple and is considered a problem to be solved. This platform is important for military and space applications. Nowadays, having high speed computers and powerful programing environments make it possible to simulate and study such platform. Depending on the features of the RF signal, there are many techniques used for geolocating the RF emitting source. Received signal strength (RSS), time difference of arrival (TDOA), frequency difference of arrival (FDOA), and angle of arrival (AOA) are the methods used individually or combined with each other to carry out the geolocation process. The present work uses a hybrid application of TDOA and AOA methods. A generalized algorithm, named moving sensors moving emitter (MoMo) is designed to solve such a problem and carry out the following tasks: (1) Simulate the emitter–sensors motion platforms. (2) Calculate AOAs and TOAs for the sensors, and (3) Applying a hybrid AOA/TOA method to locate the emitter in 2D and 3D coordinates. This paper is divided into six sections. Section 1 is an introductory section. Section 2 presents literature preview. Section 3 explains the four signal techniques of geolocating the emitter source. Section 4 presents the algorithm of the proposed geolocation system. Results of the systems are discussed in Sect. 5. Section 6 shows the resulting MoMo position error measurements for different platforms. A comparison with earlier work is discussed in this section. Section 7 depicts the conclusion and at the end a list of utilized references in this paper is given.

2   Literature Review

Most of RF emitter location estimation methods relies on those described in Sect. 1. Researchers used different platforms to study the geolocation problem of an RF emitter. Kaune et al. [3] used a hybrid method of TDOA and FDOA with two scenarios, fixed emitter moving sensor and fixed sensors moving emitter. Fisher [4] investigated multiple moving sensors platforms to locate a stationary emitter using AOA, TDOA, and FDOA methods. Bamberger et al. [5] built a moving sensor emitter platform using unmanned small aircraft system (UAS), but they assumed a horizontal flat platform. This means that no altitude is considered leaving the system acting as a 2D system. Thoresen et al. [6], used UAS system applying the power difference of arrival PDOA assuming a ground fixed RF emitter. Bailey [7], carried out the geolocation process for a single stationary RF emitter using a joint AOA/FOA method. The used platform is a single moving sensor platform. Kim et al. [8] presented an approach for estimating the 2D position of a moving emitter using a combined TDOA and FDOA method utilizing three sensors. The process is carried out for stationary and moving sensors. Li et al. [9] carried out an emitter location estimation using TOA and TDOA method utilizing three sensors for 2D location and 4 sensors for 3D location estimation. The tested platforms are either all fixed, or fixed sensors moving emitter. Lui et al. [10] used an array of 6 moving sensors to find a moving emitting source. They did the estimation process using the combined TDOA and FDOA method.

3 Geolocation Estimation Methods

The Emitter signal received by the sensors is investigated based on the signal strength (RSS), time of arrival (TOA), frequency of arrival (FOA), and angle of arrival (AOA) at each sensor. It is known that, the received signal strength (RSS) method depends on the signal amplitude. Factors due to physical environment like, attenuation, fading, shadowing and multipath reflection of the received emitter signal make geolocation process unreliable [11, 12]. So, RSS method is not commonly used for geolocation application. Thus, RSS method will not discussed here. In the following subsections, we will introduce the other three methods.

3.1 Time Difference of Arrival TDOA

Estimation of the emitter position using TDOA method is based on the time difference of arrival measurement between the two sensor locations [13], see Fig. 1.

Based on Fig. 1, TDOA of the signal between sensor1 and sensor2 is calculated as:

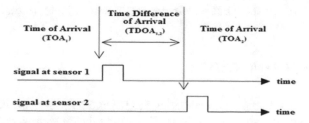

Fig. 1 The concept of TDOA

$$\Delta t_{2,1} = \frac{D_{s2} - D_{s1}}{C} \tag{1}$$

where: D_{s1}, and D_{s2} are the distances between emitter and sensors1 and sensor2 respectively, C is the speed of light, and $\Delta t_{1,2}$ is the time difference of signal arrival between sensor1 and sensor2 respectively, assuming that the signal arrived at sensor1 first. The distance between sensors and emitter in space is calculated as:

$$D_{s1(2)} = \sqrt{\left(x_e - x_{s1(2)}\right)^2 + \left(y_e - y_{s1(2)}\right)^2 + \left(z_e - z_{s1(2)}\right)^2} \tag{2}$$

where e is denoting the emitter, s for sensors, and 1(2) refers to sensor1 or sensor2.

3.2 Frequency Difference of Arrival FDOA

The Doppler frequency shift f_d in Hz is given by [14–15]:

$$f_d = \frac{f_0}{C} V_s \cos(\alpha) \tag{3}$$

where V_s = sensor speed (m/s), C = speed of radio waves (m/s), f_0 = frequency of radio transmission (Hz), and α = angle between the sensor speed vector and a line to the emitter (°). Then, the FDOA is calculated from (3) as follows:

$$FDOA = f_{ds1} - f_{ds2} \tag{4}$$

Equation (4) is rewritten as:

$$FDOA = \frac{f_0}{C}(V_s 1 \cos(\alpha 1) - V_s 2 \cos(\alpha 2)) \tag{5}$$

where $V_{s1(2)}$ = sensor1 or sensor2 speed (m/s), C = speed of radio waves (m/s), f_0 = frequency of radio transmission (Hz), and $\alpha 1(2)$ = angle between sensor1 or sensor2 speed vector and the line from the sensor to the emitter (rad).

3.3 Angle of Arrival AOA

The angle of arrival method is estimating the RF emitter source position based on the direction that the signal came from. Estimating the AOA in space is based on measuring the angle of arrival from the emitter to the sensor of the received signal

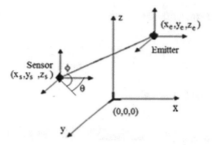

Fig. 2 The RF emitter signal angles of arrival at the RF sensor

in azimuth, θ, and the angle of arrival in elevation ϕ. Thus, neglecting the noise interference, the two angles of arrival are given by [16]:

$$\theta = \arctan\left(\frac{y_e - y_s}{x_e - x_s}\right) \tag{6}$$

$$\emptyset = \arctan\left(\frac{z_e - z_s}{\sqrt{(y_e - y_s)^2 + (x_e - x_s)^2}}\right) \tag{7}$$

Figure 2 shows a possible locations of the emitter and the sensors in space to define the angles of arrival θ, and $\emptyset$.

In the following section, we will discuss the motion of emitter and sensors as point objects. Stationary means zero speed, i.e., fixed object. The trajectory of a moving object is chosen arbitrarily. We select linear motion of sensors and emitter for simplicity, and most of the flying objects like aircrafts and rockets are moving approximately linearly. Emitter and sensors maneuverability motion is also tested.

4 The Proposed Geolocation System

The proposed algorithm aims to determine the location of the emitter source in the space, i.e. determining the emitter position coordinates x_e, y_e, and z_e. The algorithm is designed using equations of calculating AOA, and TOA. To estimate the emitter coordinates, we have to solve three equations for the three variables x_e, y_e, and z_e. Equation (2) is rearranged as follows:

$$\left(TOA_{1(2)} \cdot c\right)^2 = \left(x_e - x_{s1(2)}\right)^2 + \left(y_e - y_{s1(2)}\right)^2 + \left(z_e - z_{s1(2)}\right)^2 \tag{8}$$

Equation (6) can be rewritten as:

$$\tan \theta_{1(2)} = \frac{y_e - y_{s1(2)}}{x_e - x_{s1(2)}} \tag{9}$$

Equation (7) is rearranged to be:

$$(\tan \varnothing_{1(2)})^2 = \frac{\left(z_e - z_{s1(2)}\right)^2}{\left(x_e - x_{s1(2)}\right)^2 + \left(y_e - y_{s1(2)}\right)^2} \tag{10}$$

Sensors coordinates $x_{s1(2)}$, $y_{s1(2)}$, and $z_{s1(2)}$ are assumed to be known. The values of TOA and AOAs $(\theta, \varnothing)$ are continuously measured during the process. Thus, solving the three Eqs. (8, 9, and 10) for the three unknown emitter coordinates x_e, y_e, and z_e we get:

$$x_e = \frac{TOA_{1(2)}.C}{\sqrt{\left(1 + (\tan \theta_{1(2)})^2\right)\left(1 + (\tan \varnothing_{1(2)})^2\right)}} + x_{s1(2)} \tag{11}$$

Then, y_e is calculated using Eq. (9, and 11):

$$y_e = \left(x_e - x_{s1(2)}\right) \tan \theta_{1(2)} + y_{s1(2)} \tag{12}$$

z_e coordinate is calculated using Eqs. (8, 11, and 12):

$$ze = (TOA1(2).C)2 - (xe - xs1(2))2 + (ye - ys1(2))2 + z_{s1(2)} \tag{13}$$

Emitter coordinate z_e is calculated using Eqs. (10, 11, and 12):

$$z_e = \tan \varnothing_{1(2)}\sqrt{\left(x_e - x_{s1(2)}\right)^2 + \left(y_e - y_{s1(2)}\right)^2} \tag{14}$$

The emitter 3D coordinates x_e, y_e, and z_e are now determined. Equations (11, 12, and 13) and/or Eq. (14) are the calculation steps of the proposed algorithm.

5   Simulation Results

The algorithm simulates the four platforms previously mentioned in section (I). The following is the result of applying the algorithm to an arbitrary example of each platform. We will assume that all coordinates are in meters, and all angles are in degrees. A stationary object is located in a fixed point in 2D or 3D while the moving one is represented in 2D or 3D by continuously changed coordinates. For example, 3D coordinates of a stationary emitter are [100,300,200] which means that xe =

Table. 1 Emitter azimuth and elevation	Sensor1	$\theta_1 = 84.2894°$	$\phi_1 = 26.4512°$
	Sensor2	$\theta_2 = 24.4440°$	$\phi_2 = 22.4799°$

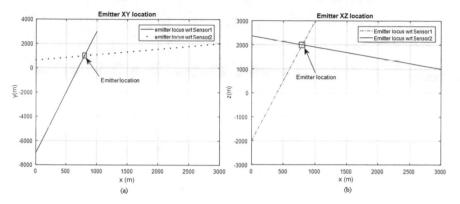

Fig. 3 Fixed sensors and fixed emitter: **a** xy emitter location, **b** emitter xz location

100, ye = 300, and ze = 200. The linear motion of an object, as an example, is xe = 1000:100:2000, ye = 500:200:2500, ze = 100:50:600. It means that the emitter is moving linearly in x coordinate starting from xe = 1000, up to 2000 with a step equals 100. The same will be the motion in the y and z coordinates.

5.1 Platform (a): Stationary Sensors and Stationary Emitter

We suppose that the emitter and sensors are located at: emitter [800; 1000; 2000], sensor1 [1000; 3000; 3000], and sensor2 [3000; 2000; 1000]. Azimuth and elevation angles of the emitter from sensors 1 and 2 point of view are shown in Table 1.

The 2D x-y and x-z locations are shown in Fig. 3 and the 3D locations are shown in Fig. 4.

5.2 Platform (b): Moving Sensors and Stationary Emitter

The stationary emitter coordinates are [xe = 1000, ye = 2000, ze = 4000]. Sensor1 and sensor2 are moving such that sensor1 motion trajectory is [xs1 = 3000:200:5000, ys1 = 2000:200:4000, zs1 = 1500:100:2500], and sensor2 movement trajectory is [xs2 = 2000:200:4000, ys2 = 1000:200:3000, zs2 = 5000:100:6000]. Figure 5 shows the azimuth and elevation angles of emitter wrt sensors 1 and 2. Figure 6 illustrates the sensors angles in space. The 2D x-y and x-z locations are described in Fig. 7 while the 3D locations are shown in Fig. 8.

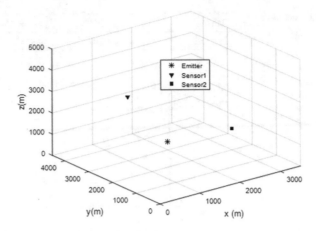

Fig. 4 Fixed sensors and fixed emitter in 3D locations

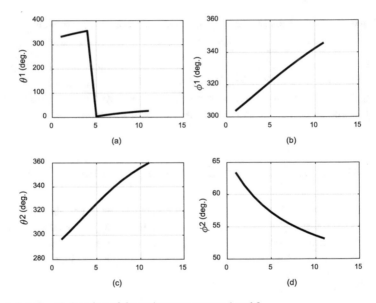

Fig. 5 Azimuth and elevation of the emitter wrt sensors 1 and 2

5.3 Platform (c): Stationary Sensors and a Moving Emitter

Sensors are stationary and emitter is moving such that, sensor1 coordinates are [xs1 = 2000, ys1 = 2500, zs1 = 1000], sensor2 coordinates are [xs2 = 4500, ys2 = 2000, zs2 = 2000]. The emitter movement trajectory is [xe = 2100:100:3100, ye = 1000:200:3000, ze = 2000:100:3000]. Figure 9 shows the azimuth and elevation angles of emitter wrt sensors 1 and 2. Figure 10 illustrates the sensors angles in space.